Materials Science
in Energy Technology

MATERIALS SCIENCE AND TECHNOLOGY

EDITORS

ALLEN M. ALPER
GTE Sylvania Inc.
Precision Materials Group
Chemical & Metallurgical Division
Towanda, Pennsylvania

A. S. NOWICK
Henry Krumb School of Mines
Columbia University
New York, New York

A. S. Nowick and B. S. Berry, ANELASTIC RELAXATION IN CRYSTALLINE SOLIDS, 1972

E. A. Nesbitt and J. H. Wernick, RARE EARTH PERMANENT MAGNETS, 1973

W. E. Wallace, RARE EARTH INTERMETALLICS, 1973

J. C. Phillips, BONDS AND BANDS IN SEMICONDUCTORS, 1973

J. H. Richardson and R. V. Peterson (editors), SYSTEMATIC MATERIALS ANALYSIS, VOLUMES I, II, AND III, 1974; IV, 1978

A. J. Freeman and J. B. Darby, Jr. (editors), THE ACTINIDES: ELECTRONIC STRUCTURE AND RELATED PROPERTIES, VOLUMES I AND II, 1974

A. S. Nowick and J. J. Burton (editors), DIFFUSION IN SOLIDS: RECENT DEVELOPMENTS, 1975

J. W. Matthews (editor), EPITAXIAL GROWTH, PARTS A AND B, 1975

J. M. Blakely (editor), SURFACE PHYSICS OF MATERIALS, VOLUMES I AND II, 1975

G. A. Chadwick and D. A. Smith (editors), GRAIN BOUNDARY STRUCTURE AND PROPERTIES, 1975

John W. Hastie, HIGH TEMPERATURE VAPORS: SCIENCE AND TECHNOLOGY, 1975

John K. Tien and George S. Ansell (editors), ALLOY AND MICROSTRUCTURAL DESIGN, 1976

M. T. Sprackling, THE PLASTIC DEFORMATION OF SIMPLE IONIC CRYSTALS, 1976

James J. Burton and Robert L. Garten (editors), ADVANCED MATERIALS IN CATALYSIS, 1977

Gerald Burns, INTRODUCTION TO GROUP THEORY WITH APPLICATIONS, 1977

L. H. Schwartz and J. B. Cohen, DIFFRACTION FROM MATERIALS, 1977

Zenji Nishiyama, MARTENSITIC TRANSFORMATION, 1978

Paul Hagenmuller and W. van Gool (editors), SOLID ELECTROLYTES: GENERAL PRINCIPLES, CHARACTERIZATION, MATERIALS, APPLICATIONS, 1978

G. G. Libowitz and M. S. Whittingham, MATERIALS SCIENCE IN ENERGY TECHNOLOGY, 1979

In preparation

Otto Buck, John K. Tien, and Harris L. Marcus (editors), ELECTRON AND POSITRON SPECTROSCOPIES IN MATERIALS SCIENCE AND ENGINEERING

Materials Science in Energy Technology

Edited by

G. G. LIBOWITZ
CORPORATE RESEARCH CENTER
ALLIED CHEMICAL CORPORATION
MORRISTOWN, NEW JERSEY

M. S. WHITTINGHAM
CORPORATE RESEARCH LABORATORIES
EXXON RESEARCH AND ENGINEERING COMPANY
LINDEN, NEW JERSEY

ACADEMIC PRESS   New York San Francisco London   1979
A Subsidiary of Harcourt Brace Jovanovich, Publishers

ACADEMIC PRESS, INC.
111 Fifth Avenue, New York, New York 10003

United Kingdom Edition published by
ACADEMIC PRESS, INC. (LONDON) LTD.
24/28 Oval Road, London NW1   7DX

Library of Congress Cataloging in Publication Data

Main entry under title:

Materials science in energy technology.

(Materials science and technology)
Includes bibliographies.

1.   Materials.   2.   Energy industries.
I.   Libowitz, George C.   II.   Whittingham, Michael
Stanley,   Date
TA403.6.M376   620.1'1   78–51235
ISBN   0–12–447550–7

PRINTED IN THE UNITED STATES OF AMERICA

79 80 81 82   9 8 7 6 5 4 3 2 1

Contents

List of Contributors

Numbers in parentheses indicate the pages on which the authors' contributions begin.

M. R. BEASLEY (491), Departments of Applied Physics and Electrical Engineering, Stanford University, Stanford, California 94305

H. BEHRET (381), Battelle-Institut e.V., Frankfurt/Main, Germany

H. BINDER (381), Battelle-Institut e.V., Frankfurt/Main, Germany

H. KENT BOWEN (181), Department of Materials Science and Engineering, Ceramics Division, Massachusetts Institute of Technology, Cambridge, Massachusetts 02139

T. H. GEBALLE (491), Departments of Applied Physics and Materials Science, Stanford University, Stanford, California 94305 and Bell Laboratories, Murray Hill, New Jersey 07974

DIETER M. GRUEN (325), Argonne National Laboratory, Argonne, Illinois 60439

JOHN P. HOWE* (31), Materials, Advanced Energy Systems, General Atomic Company, San Diego, California 92093

G. G. LIBOWITZ (427), Corporate Research Center, Allied Chemical Corporation, Morristown, New Jersey 07960

J. J. LOFERSKI (201), Division of Engineering, Brown University, Providence, Rhode Island 02912

HOWARD L. RECHT (263), Research and Technology, Energy Systems Group, Rockwell International Corporation, Canoga Park, California 91304

G. SANDSTEDE (381), Battelle-Institut e.V., Frankfurt/Main, Germany

* Present Address: Energy Center, University of California at San Diego, La Jolla, California 92093.

MASSOUD T. SIMNAD (31), Advanced Energy Systems, General Atomic Company, San Diego, California 92037

J. H. SINFELT (1), Corporate Research Laboratories, Exxon Research and Engineering Company, Linden, New Jersey 07036

M. S. WHITTINGHAM (455), Corporate Research Laboratories, Exxon Research and Engineering Company, Linden, New Jersey 07036

Preface

There is an ever-increasing demand for more energy in our society and since the "energy crisis" of 1973, the general public has become aware that our present sources of energy are not limitless. Although there are some disagreements as to the nature and extent of the energy shortages, there is little question that there will be significant changes in our energy technologies in the future. Even if the energy sources remain the same, more efficient means of converting them into useful energy forms will be required. It is not generally realized that in most cases the major hindrances to the development of both present and future energy sources are materials limitations. Therefore, the major purpose of this book is twofold; first, to acquaint those involved in energy technology with the fundamental properties of materials relevant to each form of energy; second, to suggest to materials scientists, including physicists, chemists, and metallurgists, where their areas of expertise may be of value in solving energy related problems.

Each chapter reviews the basic properties of those materials that are important to a particular energy application, and it also attempts to identify the areas of research required for the development of new materials for that application. The first six chapters deal with methods of producing energy, and they are arranged in approximate chronological order as to when the methods were (or will be) first utilized. Fossil fuels have been the major energy source for about a century (wood, before about 1880). Nevertheless, there is still much room for improvement in terms of catalyst materials for production of fuel as discussed in the first chapter, particularly in the area of resid and coal liquids conversion. Nuclear fission has become a significant source of energy in the last decade and the wide scope of materials associated with this technology is reviewed in Chapter 2. Magnetohydrodynamic generators hold out the promise of more efficient production of electricity from fossil fuels; although one such generator has been operated in the USSR using natural gas, there are many high-temperature materials problems that must be solved before they become a feasible source of energy using realistic fuels. Ceramic materials for this application are considered in Chapter 3.

Solar energy has been used for hot water and space heating in individual homes, but its large scale use for electricity production, using photovoltaic cells, awaits the development of new materials that will permit the more economic conversion of this source of energy, as discussed in Chapter 4. Geothermal energy has always been available, but its use has been very limited because of severe corrosion problems, as pointed out in Chapter 5. The maximum utilization of this form of energy depends significantly on materials development. The sixth chapter is concerned with controlled thermonuclear fusion, which is one of the most promising future sources of energy, but also the most difficult to develop because of the very severe materials requirements involved.

The last four chapters cover the materials aspects of energy conversion, storage, and transmission. Fuel cells convert fossil and combustible fuels to electrical energy, while hydrogen energy systems convert electrical or solar energy to a combustible fuel. The latter also provide a means for storing and transmitting energy. Although batteries have been used extensively for storing electrical energy on a small scale in the past, advances in solid state science have provided exciting new concepts for more efficient low-cost batteries. Finally, superconducting materials for electrical energy storage and transmission are reviewed (as well as the use of superconducting magnets to confine plasmas in thermonuclear reactors and MHD generators).

It is apparent that not all of the materials aspects of energy production, conversion, and storage have been covered in this book. For example, important materials for the storage of thermal energy such as low melting eutectic salts, or the high-strength materials required for flywheel storage are only briefly mentioned. In addition, other materials that have not been considered come to mind, such as high-temperature alloys for gas turbines or those used in coal conversion and drilling for oil exploration. Undoubtedly, as new energy sources are developed, new materials problems will come to the fore. Nevertheless, we believe that this book gives sufficiently extensive coverage of the relation between materials science and energy technology to be useful for some time to come to scientists and engineers involved or interested in the various aspects of energy development.

Materials Science
in Energy Technology

Chapter 1

Heterogeneous Catalysis in Fossil Fuel Conversion

J. H. SINFELT

CORPORATE RESEARCH LABORATORIES
EXXON RESEARCH AND ENGINEERING COMPANY
LINDEN, NEW JERSEY

I. INTRODUCTION

Heterogeneous catalysis has for many years played a vital role in industrial processes for the manufacture of fuels. This is particularly evident in the petroleum industry, where the demand for high quality gasoline and heating oil has been met very successfully through the application of heterogeneous catalysis. Processes such as catalytic cracking and catalytic reform-

1

ing have had an enormous impact for the past several decades. Over the years these processes have improved markedly, as a result of research directed to new or improved catalyst systems. In addition to these major petroleum fuels processes, and such related processes as hydrocracking and isomerization, there are other important applications of heterogeneous catalysis in the fuels area. These include steam reforming of petroleum naphthas to produce methane-rich heating gas and the Fischer–Tropsch process for synthesis of gasoline from carbon monoxide and hydrogen. The former has been of interest for some time because of limitations in availability of natural gas, while the latter was mainly of interest in Germany during World War II. Currently, other applications of heterogeneous catalysis are actively being considered by various groups concerned with extending fuel capabilities through processes such as coal gasification and liquefaction.

The present article begins with a simple discussion of heterogeneous catalysis, which includes some consideration of the nature of catalytic materials and of catalytic processes in general. The bulk of the article is then concerned with specific catalytic processes involved in the manufacture of fuels. The processes are considered primarily with regard to the nature of the reactions and catalysts involved, but some consideration is also given to important technological features of the processes.

II. GENERAL ASPECTS OF HETEROGENEOUS CATALYSIS

A. Nature of Catalytic Materials

In general, catalysts of practical interest have high surface areas. To achieve high surface areas, it is necessary to prepare catalysts in a very finely divided state or in a highly porous form [1]. In the case of nonporous solids, the surface area is confined to the external surface of the particles. The specific surface area (i.e., the surface area per unit weight of material) is then inversely proportional to the particle size, if the particles are perfect spheres or cubes. With porous solids, however, most of the surface area resides in the walls of the pores and is frequently termed the "internal surface" of the solid. In the case of highly porous solids, the internal surface area is commonly several orders of magnitude higher than the external surface area of a particle or granule.

For the classification of various types of catalysts, it is convenient to divide them into two broad categories, metals and nonmetals. In the first category, the most commonly used metals are those in Group VIII and Group IB of the Periodic Table. In the second category, oxides are the most common catalysts. In either of these two categories, the catalyst is often supported on a carrier to achieve higher dispersion and improved resistance to sintering. For highly dispersed catalysts, the carrier is commonly a high surface area, porous, refractory material such as alumina or silica.

The application of supported metal catalysts in industrial processes is well known. Such catalysts are commonly prepared by impregnation and precipitation methods. In the impregnation method, the carrier is contacted with a solution of a salt of the desired metal. The solute deposits on the carrier, and the resulting material is dried and often calcined at higher temperature. The next step is reduction of the material deposited on the carrier to the metallic form. This is commonly done at elevated temperature in a stream of hydrogen. The final material then consists of an assembly of small metal crystallites dispersed on the surface of the carrier. A typical example would be the preparation of a nickel–silica catalyst, in which an aqueous solution of nickel nitrate would probably be used for the impregnation. The hydrogen reduction step would be conducted at a temperature of about 400°C. Another example would be the preparation of a platinum–alumina catalyst, in which the impregnation would be made with a chloroplatinic acid solution. In this case, the catalyst would usually be calcined in air at about 500–550°C prior to reduction in hydrogen.

In the preparation of supported metals by precipitation methods, a hydroxide or carbonate of a metal may be precipitated from a solution of a metal salt onto a carrier suspended in the solution. Thus, a nickel–silica catalyst could be prepared starting with a nickel nitrate solution in which silica is suspended or slurried. Addition of alkali to the solution causes precipitation of nickel hydroxide onto the silica. The material is then dried and reduced in hydrogen at elevated temperature to form the catalyst. In another procedure, termed "coprecipitation," the carrier is precipitated simultaneously with the active component. Thus, in the preparation of a nickel–alumina catalyst, one may start with an aqueous solution of nickel and aluminum nitrates. On addition of alkali, hydroxides of nickel and aluminum are coprecipitated. The precipitate is then filtered, washed repeatedly, and dried. After calcination at elevated temperature (400–450°C) and subsequent exposure to a stream of hydrogen at about 500°C, the material consists of small nickel crystallites dispersed on alumina.

In the case of nonmetallic catalysts, oxides find the widest application. Such catalysts can be prepared by methods involving precipitation or gel formation. For example, an aluminum oxide catalyst which is highly active for the dehydration of alcohols can be prepared by addition of an aqueous solution of ammonia to an aluminum nitrate solution [2]. This treatment yields a finely divided precipitate of aluminum hydroxide, which is dried and then calcined at elevated temperature (400–600°C) to produce the desired aluminum oxide. An alternate procedure for the formation of the aluminum hydroxide precipitate is the hydrolysis of aluminum isopropoxide [3]. Surface areas of the order of 200 m^2/gm are readily obtained for aluminas prepared by these procedures.

Oxides such as alumina, silica, or chromia are often prepared in the form of gels [4–7]. In the formation of a gel from solution, finely dispersed colloidal particles are generated first. However, instead of remaining in a colloidal suspension as freely moving particles or settling out of solution, the particles are joined together in some form of continuous structure throughout the solution volume. Many gels contain an extremely high fraction of liquid (>99 vol%) within the structure. In such cases, the solid component of the gel is probably fibrillar in form [8]. The formation of a gel, as opposed to a crystalloidal type of precipitate, depends on the detailed conditions of preparation. A classical example of gel formation of interest in catalysis is the preparation of chromia gel by the slow addition of ammonium hydroxide solution to a dilute solution of chromic nitrate [6, 7]. The highly dilute gel which forms is dried to remove most of the water, leaving an amorphous, black solid. Care must be exercised in heating the chromia, since too rapid heating may result in the formation of highly crystalline α–chromia, a bright-green solid. If the chromia is heated slowly through the temperature range 275–375°C, the crystallization can be prevented. Chromia gels heated to temperatures of 300–450°C may give catalysts with surface areas as high as 300 m^2/gm. It has been suggested that the fibrillar structure of chromia gel is due to the formation of a condensation polymer of chromium hydroxide [6, 9, 10].

Like metals, nonmetallic catalyst entities are also frequently deposited on supports to achieve a higher degree of dispersion of the active component. Oxides of metals such as chromium, molybdenum, vanadium, manganese, copper, bismuth, iron, cobalt, and nickel, among others, are commonly supported on refractory oxide carriers in catalytic applications. The catalysts may be prepared by impregnation or coprecipitation techniques. For example, a chromia–alumina catalyst can be prepared by impregnating activated alumina with an aqueous solution of either chromic acid, chromium nitrate, ammonium chromate or ammonium dichromate. The material is then dried at a temperature of 110–150°C and subsequently calcined at temperatures of 350–550°C [11]. Alternatively, a coprecipitated chromia–alumina catalyst can be prepared by adding ammonium hydroxide to a solution of chromic and aluminum nitrates. The material is then dried and calcined in the same manner as impregnated chromia–alumina.

In the foregoing paragraphs, an attempt has been made to describe catalytic materials in a general way, with emphasis on their high surface areas and methods of preparation. The high degree of dispersion is the single most important feature distinguishing catalysts as a class of materials. In addition to their importance as catalysts, highly dispersed solids present some very challenging questions with regard to the chemistry and physics of the solid state, especially when the degree of dispersion is so high that the surface totally dominates the bulk.

B. Nature of Catalytic Processes

A catalytic process involves a sequence of reaction steps in which active catalytic centers participating in the steps are continually being regenerated, so that many molecules of product are formed per active center. Such a sequence is termed a closed sequence [12]. In heterogeneous catalysis, the active center is a site on the surface of the catalyst or a surface complex of the site with a reactant molecule. In the case of the overall reaction, A + B → C, the following sequence of steps may be visualized:

$$A + S \rightarrow A\text{--}S$$

$$A\text{--}S + B \rightarrow C + S$$

In this simple sequence the reactant A is adsorbed on an active center S to form the surface complex A–S, which in turn reacts with the reactant B to yield the product C and regenerate S. This sequence illustrates features common to all catalytic processes, namely, the generation of a reactive intermediate, the formation of product from the intermediate, and the regeneration of the active catalyst site. The sequence of steps in a catalytic process is repeated over and over again using the same active centers on the catalyst surface.

For heterogeneous catalysis to occur, chemisorption of at least one reactant is generally required. The reactant molecule is activated in the chemisorption step, and the chemical nature of the surface is an important factor. In general, catalytic activity is related to the strength of adsorption of a reactant on the surface. Maximum catalytic activity results when chemisorption of the reactant is fast but not very strong. If the adsorption bond is too strong, the catalyst will tend to be fully covered by a stable surface compound which does not readily undergo reaction. On the other hand, if chemisorption of the reactant is slow, the reaction may be limited severely by the adsorption step. Optimum catalytic activity corresponds to an intermediate degree of coverage of the surface by the adsorption complex undergoing reaction [13, 14]. In the case of metal catalyzed reactions, this optimum condition is frequently found among the metals of Group VIII, and consequently these metals are particularly important in catalysis.

Perhaps the most intriguing aspect of catalysis is the specificity observed. It has been known for a long time that the chemical nature of a solid surface determines its catalytic properties. If the role of the surface were simply one of concentrating reactants in a physically adsorbed layer, then any high surface area solid should function as a catalyst for a given reaction. That this is not the case is evident from consideration of a particular example, namely, the partial oxidation of ethylene to ethylene oxide. The only known catalysts for this reaction contain silver as the active component [15]. On other surfaces the predominant reaction is complete oxidation to carbon dioxide and water. Another excellent example illustrating the specific influence of the catalyst is the reaction of ethanol over chemically different surfaces. On copper catalysts ethanol undergoes dehydrogenation to yield acetaldehyde, while on alumina it undergoes dehydration to form ethylene [16]. Similarly, on metals formic acid exhibits primarily dehydrogenation to yield carbon dioxide and hydrogen, while on certain oxides the observed reaction is predominantly dehydration to produce carbon monoxide and water [17]. These latter examples illustrate differences between metal and oxide surfaces, and as such are especially pertinent in a discussion of the application of heterogeneous catalysis in fuels processes. In major petroleum processes, for example, metals and oxides—in particular, oxides exhibiting pronounced surface acidity—both play important catalytic roles. On metals hydrocarbons undergo predominantly reactions such as hydrogenation, dehydrogenation, aromatization, and hydrogenolysis, all of which involve hydrogen as a reactant or product. On acidic oxides, by contrast, hydrocarbons undergo predominantly isomerization and cracking reactions of a type generally associated with carbonium ion chemistry.

The discussion here has been intended only to give a very general summary of some of the main features of catalytic phenomena, and to present some generalizations relating catalytic activity to the chemisorption properties of surfaces. Although much has been learned about catalysis during the past couple of decades, the level of understanding has not progressed to the point where predictions based on first principles can be made. In particular, there is no really fundamental understanding of why a given substance is, or is not, a good catalyst for a given reaction. Despite these limitations, major advances in catalytic technology are continually emerging, as evidenced by the many impressive applications of catalysis in industry.

III. MAJOR PETROLEUM FUELS PROCESSES

A. Catalytic Cracking

1. Process Description

Catalytic cracking is a process in which high molecular weight hydrocarbons are converted to low molecular weight hydrocarbons [18–20]. Generally speaking, the feed stock in catalytic cracking is a petroleum fraction which may contain hydrocarbons with boiling points in the range from about 250° to 600°C. In practice, feed stocks may contain hydrocarbons spanning only a part of this boiling range, and are commonly referred to as light gas oils, heavy gas oils, and vacuum gas oils, corresponding to low, intermediate, and high boiling fractions. The major products of catalytic cracking include a gaseous fraction (mostly C_3 and C_4), gasoline (C_5–C_{10}), and distillate fuels ($C_{10}+$). The products contain both saturated and unsaturated hydrocarbons. The gaseous product is a mixture of paraffins and olefins, while the gasoline contains aromatics and cycloparaffins as well. As the molecular weight increases, the content of cyclic hydrocarbons in the product increases. The catalyst employed in the process is commonly an aluminosilicate, the catalytic activity of which is related to its acidic properties.

Catalytic cracking is commonly conducted at temperatures in the range 450–525°C. The cracking reactions are accompanied by the formation of a highly unsaturated hydrocarbon residue on the surface of the catalyst. This residue, commonly designated as coke, decreases the cracking activity. This requires that reaction periods be very short, of the order of minutes, and that provision be made for continual regeneration of the catalyst. Regeneration is accomplished very effectively by burning the coke off the catalyst with an oxygen-containing gas. Careful control of temperature during regeneration is necessary to avoid damaging the catalyst. The early cracking units, which resulted from the work of Eugene Houdry [19], consisted of several fixed bed reactors which could be cycled between reaction, purging, and regeneration periods. In this way one reactor could be regenerated while another was being used for cracking. The first commercial unit of this type went on stream in 1936. While the original Houdry cracking units operated successfully, they were eventually supplanted by moving bed processes [21] and by the fluid bed process [22]. In these latter processes, the cracking and regeneration reactions are conducted in separate vessels and the catalyst is continuously circulated from the one vessel to the other. In contrast to the fixed bed process, these latter processes are truly continuous, and operations are simpler. In the fluid process, the catalyst is used in the form of a very finely divided powder. When a gas is passed through a bed of such a powdered catalyst at a suitable velocity, the catalyst becomes fluidized and behaves very much like a liquid. The flow of catalyst between reactor and regenerator occurs because of gravitational forces. A column of fluidized catalyst has a hydrostatic head associated with it in the same manner as a liquid. Standpipes based on this principle are used to build up pressures on the catalyst to achieve transfer of the catalyst between the reactor and regenerator vessels. This has the advantage of eliminating mechanical devices for circulating the catalyst. These important features of the fluid catalytic cracking process are discussed in detail elsewhere [19]. The fluid process was first commercialized in 1942 and is today the dominant process in the field of catalytic cracking.

2. Reactions in Catalytic Cracking

The conversion of high molecular weight hydrocarbons to lower molecular weight products may be accomplished by thermal cracking in the absence of a catalyst. In the presence of a cracking catalyst, the overall rates of reaction are much higher. However, the function of the cracking catalyst is more than one of nonselectively accelerating the thermal cracking reactions [18, 20, 23]. Major differences in product distribution are observed between catalytic cracking and thermal cracking. For example, in catalytic cracking the formation of methane and C_2 hydrocarbons is much lower than in thermal cracking. The products of catalytic cracking tend to be limited to C_3 and larger molecules. Furthermore, extensive production of branched-chain paraffins is observed in catalytic cracking, along with significant formation of aromatics from saturated hydrocarbons. Finally, the olefins formed in catalytic cracking undergo secondary reactions such as hydrogen transfer and polymerization. Clearly, the nature of the reactions occurring in catalytic cracking is quite different from that in thermal cracking.

The reactions on a cracking catalyst are a consequence of the acidic character of the surface and are well rationalized in terms of a mechanism involving carbonium ion intermediates, whereas thermal cracking reactions involve free radical intermediates [23–25]. A carbonium ion is an electron-deficient hydrocarbon species which may be formed from a paraffin by abstraction of a hydride ion:

$$C_nH_{2n+2} \rightarrow C_nH_{2n+1}{}^+ + H^-$$

Alternatively, a carbonium ion may be formed from an olefin by addition of a proton:

$$C_nH_{2n} + H^+ \rightarrow C_nH_{2n+1}{}^+$$

The concept of carbonium ion intermediates in hydrocarbon reactions was first proposed by Whitmore [26] for reactions in solution, and was later extended by various workers [23–25] to reactions occurring on solid acid surfaces. While it is common to discuss carbonium ions as if they are free ions, it must be remembered that an associated anion is involved. In general, the associated anion is the catalyst. In the case of a cracking catalyst, the species which is considered to be a carbonium ion is really a highly polarized hydrocarbon portion of a hydrocarbon–catalyst complex. The greater the electronegativity of the catalyst portion of the complex, the more the hydrocarbon portion will behave like a free carbonium ion [18].

Reactions involving carbonium ion intermediates can be rationalized on the basis of a few simple rules concerning the stability and reactivity of carbonium ions [20]. The stability of carbonium ions decreases in the order: tertiary > secondary > primary. In the case of primary ions, the stability of C_3 and higher carbon number ions is much greater than that of ethyl and methyl ions, the stability of the latter being the lowest. With regard to reactivity, carbonium ions readily undergo simple reactions involving the shift of a hydrogen or methyl group from a given carbon atom to an adjacent, electron-deficient carbon atom. As a result of such shifts, the point of charge in the ion readily moves from one carbon atom to another. A carbonium ion may extract a hydride ion from a hydrocarbon molecule to form another carbonium ion, the original ion becoming a neutral hydrocarbon molecule:

$$R^+ + R_1H \rightarrow RH + R_1^+$$

By elimination of a proton, a carbonium ion can produce an olefin. Finally, of particular importance in cracking, a carbonium ion can undergo beta scission, in which the carbon–carbon bond in the beta position to the charged carbon atom is ruptured. This yields an olefin and a smaller carbonium ion:

$$C_n-C^+-C\vdots C-C-C \rightarrow C_n-C=C + C^+-C-C$$

The beta scission process is the simplest mode of scission, since it involves rearrangement of electrons only. No shifting of atoms or groups is required during the process. By a principle of least motion, it is therefore the most favorable way for scission to occur.

The application of carbonium ion theory in catalytic cracking is readily illustrated by a specific example, namely, the catalytic cracking of hexadecane [23]. The reaction is initiated by the removal of a hydride ion from the hexadecane molecule to form a hexadecyl carbonium ion. One way this can occur is by hydride transfer from a hexadecane molecule to a carbonium ion at the surface:

$$C_{16}H_{34} + C_nH_{2n+1}^+ \rightarrow C_{16}H_{33}^+ + C_nH_{2n+2}$$

At the surface of the catalyst there will in general be a significant concentration of small carbonium ions which may enter into such a reaction with the hexadecane. The hexadecyl ions present on the surface are predominantly secondary and tertiary ions, due to the low stability of primary ions. By a beta scission process, the hexadecyl carbonium ions are cracked to lower carbon number species, e.g.,

$$C_8H_{17}-C^+H-CH_2\vdots C_6H_{13} \rightarrow C_8H_{17}-CH=CH_2 + C_6H_{13}^+$$

The primary ion formed in this reaction rearranges to a secondary ion which then undergoes further cracking via beta scission to yield still lower carbon number species. The process is repeated until a carbonium ion is formed which can no longer undergo beta scission to form species limited to C_3 or higher carbon number fragments. This carbonium ion can then deprotonate to an olefin. Alternatively, it can abstract a hydride ion from a fresh hexadecane molecule to generate a new hexadecyl carbonium ion. The sequence of reaction steps just described is then repeated.

On the whole, the carbonium ion theory has been very successful in accounting for the essential features of catalytic cracking reactions, even to the extent of calculating product distributions [23]. However, some questions of mechanistic detail remain. Among these, there is the question of the nature of the initiating step in the cracking of saturated hydrocarbons. One view holds that a carbonium ion is initially formed at the surface by interaction of the catalyst with an unsaturated hydrocarbon molecule present in the system. This carbonium ion then extracts a hydride ion from a saturated hydrocarbon to initiate the cracking reaction. Subsequent steps in the cracking sequence then occur in the manner just described for hexadecane cracking.

3. Cracking Catalysts

As indicated already, the catalysts commonly employed in catalytic cracking are aluminosilicates. The first catalysts of this type were naturally occurring bentonite clays, which were acid leached to remove part of the alumina and impurities such as CaO, MgO, and Fe_2O_3 [18]. The leached material was then washed, dried, and activated by calcination at elevated temperature. Because the natural clay catalysts had thermal stability problems and were sensitive to deactivation by sulfur, synthetic silica–alumina catalysts were developed. The synthetic catalysts are commonly prepared by coprecipitation or cogelation of hydrous silica and alumina from mixed solutions of sodium silicate and aluminum sulfate [18]. The sodium ion must then be removed by ion exchange, e.g., by treatment with a dilute solution of ammonium sulfate or chloride, after which the material is washed thoroughly, dried, and finally calcined at a temperature of 600–700°C. These synthetic catalysts are amorphous materials in the sense that they do not give good diffraction patterns characteristic of well crystallized solids. During the past decade, however, crystalline forms of aluminosilicates have emerged as outstanding cracking catalysts [27]. These materials are often referred to as molecular sieves or zeolites, and they are really porous crystals. The face of a single crystal of a zeolite has a port or hole in it, which leads into a cavity extending through the entire crystal. The holes are uniform and are a characteristic feature of the crystal structure [28].

The surface of an aluminosilicate cracking catalyst possesses acidic properties, as is readily demonstrated by its affinity for basic molecules such as ammonia, pyridine, and quinoline. The extent of adsorption of such molecules on aluminosilicates can, in fact, be employed as a measure of the surface acidity [29, 30]. Studies have shown a good correlation between cracking activity and acidity determined in this way [30]. The detailed nature of the acidity of aluminosilicates has been the subject of a great deal of discussion. In considering the origin of the acidity of silica–alumina cracking catalysts, it is important to realize at the outset that such catalysts do not behave simply as mechanical mixtures of silica and alumina. This is evident from the fact that silica alone is virtually inactive for cracking and that alumina alone is much less effective than silica–alumina. The catalyst actually behaves like a complex of the two substances. Typical silica–alumina cracking catalysts contain silica in excess, the amount of alumina being 10–25 wt%. In such materials the structure may be viewed as consisting of tetrahedrally coordinated silicon and aluminum atoms linked through the sharing of oxygen atoms at the corners of SiO_4 and AlO_4 tetrahedra. For aluminum-containing tetrahedra bonded at all four corners to silicon atoms in tetrahedral coordination, there is an excess of one unit of negative charge. This arises because a trivalent aluminum atom has been substituted for tetravalent silicon in a silica structure. Consequently, there must be present a compensating positive charge to provide electroneutrality. A proton coordinated to the structure satisfies this requirement, and is strongly acidic in its behavior. The structure of a silica–alumina catalyst dried at low temperature (100°C) is illustrated in Fig. 1 [31], which depicts a tetrahedrally coordinated

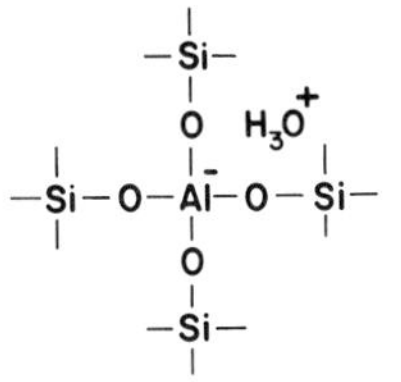

FIG. 1.   Structure of a silica–alumina catalyst dried at low temperature (100°C), exhibiting Bronsted acid sites [31].

aluminum atom in a matrix of SiO_4 tetrahedra with a compensating positive charge (H_3O^+). The latter species is a proton donor and hence functions as a Bronsted acid site in the catalyst. Removal of water from the structure by heating at elevated temperature (say, 500°C) leads to a new structure depicted in Fig. 2, in which the aluminum atom is coordinated to only three oxygen atoms [31]. The aluminum atom in this state is unsaturated and can serve as an acceptor of a pair of electrons. Hence, it functions as a Lewis acid site.

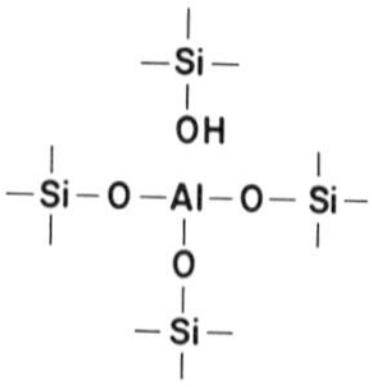

$$
\begin{array}{c}
| \\
-\text{Si}- \\
| \\
\text{OH} \\
| \qquad\qquad | \\
-\text{Si}-\text{O}-\text{Al}-\text{O}-\text{Si}- \\
| \qquad | \qquad | \\
\text{O} \\
| \\
-\text{Si}- \\
|
\end{array}
$$

FIG. 2.   Structure of a silica–alumina catalyst heated at elevated temperature (500°C), exhibiting Lewis acid sites [31].

B. Catalytic Reforming

1. Process Description

Catalytic reforming is a process for the conversion of virgin naphtha fractions of crude oil to high antiknock quality products for inclusion in motor gasoline [32, 33]. The virgin naphthas consist primarily of paraffins and cycloparaffins with boiling points in the approximate range 50–200°C. The major objective of the process is to achieve the selective conversion of these saturated hydrocarbons to aromatic hydrocarbons. The process is conducted at temperatures in the range 425–525°C and pressures of 10–35 atm, using a catalyst such as platinum supported on alumina. A typical catalytic reforming unit consists of a series of fixed bed reactors which are operated adiabatically. The naphtha feedstock is preheated to reaction temperature, and on contact with the catalyst in the first reactor there is a sharp drop in temperature of the order of 75–100°C due to the highly endothermic nature of the reaction. In the first reactor, the major reaction is the dehydrogenation of cycloparaffins to aromatics. The effluent from the first reactor is then reheated prior to entering the second reactor, where further reaction occurs resulting in another temperature drop. This process of reheating followed by reaction is repeated until the hydrocarbon stream has passed through the last reactor. A catalytic reformer will commonly have three or four reactors in series, depending on the requirements of the particular unit. The effluent from the final reactor passes to a separator where the product is split into gas and liquid fractions. The gaseous fraction consists of hydrogen and C_1 through C_4 hydrocarbons, while the liquid product is essentially composed of C_5 through about C_{10} hydrocarbons. A portion of the gas, which generally has a hydrogen concentration of the order of 70–90 mole %, is recycled to the inlet of the first reactor. Typically the stream entering the first reactor contains about 5–10 moles of recycle gas per mole of naphtha feed. Combining the hydrogen-rich recycle gas with the naphtha feed to the first reactor results in a high partial pressure of hydrogen in the system, which is crucial for the maintenance of catalyst activity. The high hydrogen partial pressure serves to retard fouling of the catalyst surface by hydrocarbon residues. Such residues are formed via extensive dehydrogenation of chemisorbed hydrocarbons to highly unsaturated species, which in turn readily undergo condensation or polymerization reactions. The hydrogen serves to inhibit the formation of highly unsaturated species on the surface, and also to remove hydrocarbon residues via hydrogenolysis reactions [34–37]. Since increasing hydrogen pressure also has the effect of decreasing the yield of aromatic hydrocarbons, the choice of hydrogen pressure for a reformer is a matter of balancing product yields against catalyst deactivation rates.

2. Catalytic Reforming Reactions

The major reactions in catalytic reforming are: (a) dehydrogenation of cyclohexanes to aromatics, (b) isomerization of *n*-alkanes to branched alkanes, (c) dehydroisomerization of alkylcyclopentanes to aromatics, (d) dehydrocyclization of alkanes to aromatics, and (e) hydrocracking of alkanes and cycloalkanes to low molecular weight alkanes [32, 33].

The major part of the antiknock improvement obtained in the catalytic reforming of petroleum naphthas is due to the formation of aromatics. Lesser contributions result from the isomerization of straight-chain paraffins to branched paraffins and hydrocracking of high molecular weight components to lower molecular weight components in the gasoline boiling range.

Of all the reactions taking place in catalytic reforming, the dehydrogenation of cyclohexanes occurs by far the most readily. Isomerization reactions also occur readily, but not nearly so fast as the dehydrogenation of cyclohexanes. The limiting reactions in most catalytic reforming operations are hydrocracking and dehydrocyclization, which generally occur at much lower rates.

From the standpoint of thermodynamics, the reactions yielding aromatic hydrocarbons as products are favored by high temperatures and low hydrogen partial pressures. However, catalyst deactivation due to formation of carbonaceous residues on the surface places a practical upper limit on temperature and a lower limit on hydrogen partial pressure in catalytic reforming operations. The equilibria of isomerization reactions are much less temperature sensitive than the equilibria of aromatization reactions, since the heats of reaction of the former are relatively small. Interestingly, the products of isomerization of *n*-alkanes are limited almost exclusively to singly branched isomers in catalytic reforming. Thermodynamic considerations would indicate the presence of appreciable quantities of dimethylbutanes at equilibrium (about 30–35% of the total hexane isomers at 500°C), but such quantities of these isomers are not observed in reforming. While equilibrium tends to be established rather readily in the case of *n*-hexane and the methylpentanes, this does not appear to be true for the dimethylbutanes. This suggests that a strong kinetic barrier resists the rearrangement of singly branched to doubly branched isomers [33], which apparently does not exist in the case of the rearrangement of the normal structure to the singly branched structures.

The dehydrogenation of paraffins to olefins, while it does not take place to a large extent at typical reforming conditions (equilibrium conversion of *n*-hexane to 1-hexene is about 0.3 % at 510°C and 17 atm hydrogen partial pressure), is nevertheless of considerable importance, since olefins appear to be intermediates in some of the reactions [32, 33, 38, 39]. The formation of olefins from paraffins, similar to the formation of aromatics, is favored by the combination of high temperature and low hydrogen partial pressure. The thermodynamics of olefin formation can play an important role in determining the rates of those reactions which proceed via olefin intermediates, since thermodynamics sets an upper limit on the attainable concentration of olefin in the system.

From the standpoint of reaction mechanisms, it is instructive to consider a reaction scheme proposed by Mills *et al.* [38] to describe the reforming of C_6 hydrocarbons (Fig. 3). An important feature of the reaction scheme is the bifunctional nature of a reforming catalyst. Such a catalyst possesses two distinct types of active components. One of these is a metal, commonly platinum, and the other is an acidic component, generally alumina. The former provides hydrogenation–dehydrogenation centers, while the latter provides acidic centers which catalyze hydrocarbon rearrangement reactions involving a change in the carbon skeleton of the molecule. The vertical

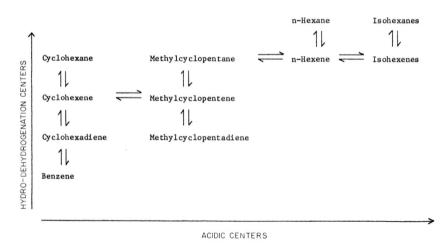

FIG. 3.   Reaction paths in reforming C_6 hydrocarbons [38].

reaction paths in the figure take place on the hydrogenation–dehydrogenation centers of the catalyst and the horizontal reaction paths on the acidic centers. According to the mechanism, the conversion of methylcyclopentane to benzene, for example, first involves dehydrogenation to methylcyclopentenes on hydrogenation–dehydrogenation centers of the catalyst, followed by isomerization to cyclohexene on acidic centers. The cyclohexene then returns to hydrogenation–dehydrogenation centers where it can either be hydrogenated to cyclohexane or further dehydrogenated to benzene, the relative amounts of these products depending on reaction conditions. The mode of transport of olefin intermediates between metal and acidic sites must be considered in this type of reaction scheme. In the case of isomerization of alkanes or dehydroisomerization of alkylcyclopentanes, a reaction sequence involving transport of olefin intermediates from metal to acidic sites via the gas phase gives a good account of much experimental data [32]. In particular, kinetic data on the isomerization of *n*-pentane to isopentane and the dehydroisomerization of methylcyclopentane to benzene on platinum–alumina are well described by rate equations derived from a reaction sequence involving olefin intermediates. At typical reforming conditions, the rate limiting step is the isomerization of the olefin intermediate on the acid sites of the catalyst. The formation of the olefin intermediate on the platinum sites occurs very fast, so that an equilibrium is established between the reactant hydrocarbon and the olefin in the gas phase. The isomerization activity of the catalyst is thus controlled by the acidity of the alumina, which can be increased by the addition of halogens such as chlorine or fluorine to the catalyst.

3. Platinum–Alumina Catalysts

Platinum on alumina catalysts have been used extensively in reforming. In such catalysts the amount of platinum present is commonly in the range of 0.1–1.0 wt%. The catalysts frequently contain comparable percentages of halogens such as chlorine or fluorine. A simple method of preparing such a catalyst involves impregnation of alumina with chloroplatinic acid, followed by calcination in air at temperatures in the range 550–600°C [32]. The aluminas commonly used have surface areas in the range 150–300 m^2/gm. The catalysts are generally used commercially in the form of pellets, extrudates, or spheres about 1/16–1/8 in. in size. Smaller sizes lead to excessive pressure drop in commercial fixed bed reactors.

Measurements of the chemisorption of gases such as hydrogen and carbon monoxide have shown that the platinum in platinum on alumina reforming catalysts is very highly dispersed [40, 41]. The platinum is present in the form of extremely small crystallites or clusters of the order of 10 Å in size.

The alumina support in reforming catalysts has acidic sites on its surface, as can be readily demonstrated by its affinity for basic molecules such as ammonia and pyridine. Evidence exists for the presence of both Lewis and Bronsted acid sites on the surface of alumina. The former refer to sites that can accept a pair of electrons and may be identified with incompletely coordinated aluminum atoms in the surface [42]. The latter refer to protonic acid sites, i.e., proton donor sites [43]. The Bronsted sites on alumina are often visualized as hydrated Lewis acid sites. The relative concentrations of the two types of sites depend on the temperature of dehydration of the alumina. Treatment of alumina with halogens such as chlorine or fluorine enhances the acidity of alumina, in the sense that it increases the activity of alumina for the catalysis of typical acid catalyzed reactions such as the skeletal isomerization of olefins and various cracking reactions of hydrocarbons.

4. Recent Developments in Reforming Catalysts

In recent years some major improvements in reforming catalysts have been made. The new catalysts are probably best described as multicomponent catalysts. An example of such a catalyst is the platinum–rhenium–alumina catalyst first announced by Chevron [44]. The catalyst generally contains an amount of rhenium comparable to the amount of platinum present. A major advantage of the catalyst over platinum–alumina is the much improved activity maintenance of the catalyst in naphtha reforming. This makes it possible to operate for a longer period of time before regeneration of the catalyst is required, i.e., longer reforming cycle lengths can be obtained with platinum–rhenium than with platinum. Also, lower pressure operation is possible for platinum–rhenium for a given cycle length. This has the advantage that lower pressures give higher yields of liquid reformate for inclusion in gasoline. The choice of the best way to take advantage of the improved activity maintenance depends on the particular refinery situation.

From a fundamental point of view, a question of interest regarding the platinum–rhenium catalyst is the physical and chemical nature of the rhenium in the catalyst. Very little information bearing on this question has appeared in the literature. Whether or not a highly dispersed alloy of platinum and rhenium is present on the alumina is a matter of some interest. Bulk platinum and rhenium have different crystal structures and do not exhibit complete miscibility. However, these considerations may not have much significance in the case of highly dispersed metals. Recent studies by Johnson and LeRoy [45] indicate that a platinum–rhenium alloy is not formed, but rather that the rhenium is present as a highly dispersed oxide at typical reforming conditions.

Recently, the term "bimetallic clusters" (or, more generally, "polymetallic clusters") has been proposed by the writer in referring to highly dispersed metallic entities comprising atoms of two or more metals [46–49]. Such clusters may be so small, of the order of 10 Å in size, that virtually every atom in the cluster is a surface atom. In general, the clusters are supported on a high surface area carrier such as silica or alumina, and the total metal concentration may be of the order of 1% or lower. These novel systems introduce much flexibility in the design of catalysts. Virtually any property of a metal catalyst (including activity, selectivity, or stability) may be influenced by combining it with one or more other metals in the form of polymetallic clusters. Systems of interest are not limited to combinations of metals which form bulk alloys, which implies a strong effect of the degree of dispersion on the stability of such systems. The "polymetallic cluster" concept has been investigated extensively by the writer using simple model catalysts and reactions. Bimetallic systems of a Group VIII metal (e.g., ruthenium or osmium) with a Group IB metal such as copper have proved to be very useful model systems [47]. Such studies have demonstrated a strong element of catalytic specificity, i.e., the clusters behave very differently for different types of hydrocarbon reactions.

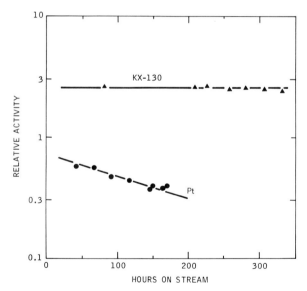

FIG. 4.   Comparison of activities of KX-130 and platinum–alumina catalysts for the reforming of a highly paraffinic naphtha to produce 102.5 research octane number reformate. Reaction temperature and pressure were 500°C and 11 atm., respectively [50]. KX-130 is a catalyst developed at Exxon Research and Engineering Co. as part of a research program on polymetallic cluster catalysts [46].

The concept of polymetallic clusters has introduced some intriguing features related to the solid state chemistry and physics of highly dispersed metals and would appear to be applicable to a variety of combinations of metallic elements. Research based on this concept at the Exxon Corporate Research Laboratories has recently led to a new reforming catalyst [46]. While the composition of the catalyst, designated KX 130, has not yet been disclosed, data on its performance have been released [50]. The catalyst is several-fold more active than conventional platinum catalysts and its activity maintenance is outstanding (Fig. 4). In these respects it is also superior to platinum–rhenium catalysts. The superior activity maintenance of the KX 130 reforming catalyst is related to the very low rate of deposition of carbonaceous residues on the surface of the catalyst relative to that observed on platinum or platinum–rhenium catalysts. The catalyst is thus especially well suited to low pressure reforming operations where fouling of the catalyst by carbonaceous residues is more severe.

C. Hydrocracking

1. Process Description

Hydrocracking is a term which refers essentially to catalytic cracking in the presence of hydrogen [51–55]. As a consequence of introducing hydrogen with the feedstock, and of employing a catalyst with an active hydrogenation component, hydrocracking gives a more saturated product than does catalytic cracking. Hydrocracking, like catalytic cracking, leads to a decrease in molecular weight of the hydrocarbon reactants. The process has attracted interest because of its great flexibility. It is able to handle a wide variety of feedstocks, including heavy naphthas, gas oil fractions, and even petroleum residua. Correspondingly, it has the ability to produce a variety of products, including isobutane, gasoline components, and high quality middle distillate fuels. The fact that the products are hydrogenated means improved storage stability relative to that of unsaturated products. Typically, hydrocracking is a fixed bed process which may be conducted in one or two stages. Catalysts for hydrocracking commonly consist of a strongly acidic component, such as an aluminosilicate, and a hydrogenation component. The latter may be a Group VIII noble metal or an oxide or sulfide of such transition metals as molybdenum, tungsten, cobalt, or nickel.

Hydrocracking is commonly conducted at temperatures in the approximate range of 300–500°C and at pressures of approximately 30–300 atm. The formation of hydrocarbon residues on the catalyst is greatly suppressed by the high hydrogen pressures employed in hydrocracking, and consequently hydrocracking can be conducted for extended periods of time before regeneration of the catalyst is required. Regeneration of the catalyst is accomplished by burning off carbonaceous residues under carefully controlled conditions. Good control of temperature during regeneration is necessary to avoid damaging the catalyst. The usual feed stocks employed in hydrocracking contain significant amounts of sulfur and nitrogen compounds which act as catalyst poisons. Consequently, an important requirement of a hydrocracking catalyst is resistance to poisoning by such compounds. During the hydrocracking process sulfur and nitrogen-containing compounds are converted largely to hydrogen sulfide and ammonia, respectively. As indicated in the previous paragraph, hydrocracking may be conducted in two stages. In the first stage of such an operation, extensive removal of sulfur and nitrogen compounds is accomplished, using a catalyst such as cobalt molybdate on alumina. The second stage then contains a more active hydrocracking catalyst to accomplish the desired hydrocarbon conversion on the purified feed stream.

Hydrocracking processes have commonly been operated to maximize production of motor gasoline components from low quality petroleum distillates such as gas oils. In operations conducted to give essentially complete conversion to products in the gasoline boiling range, approximately half of the total product can be blended directly into premium gasoline while the remaining half may be upgraded by catalytic reforming for inclusion in the gasoline pool. Although such operations are common, they are by no means typical of all hydrocracking operations. Frequently, there is interest in operating hydrocracking units to obtain good yields of jet fuels and high quality middle distillates. The nature of the operation, of course, depends on the particular refinery situation, and may vary markedly from one location to another.

2. Reactions and Catalysts

A discussion of hydrocracking reactions and catalysts involves many of the same considerations applicable to catalytic cracking and catalytic reforming. Typical hydrocracking catalysts are similar to reforming catalysts in their bifunctional nature, i.e., they possess hydrogenation–dehydrogenation sites as well as acidic sites responsible for cracking and isomerization reactions [56, 57]. The acidic sites are similar to those in cracking catalysts, and indeed the acidic component of a hydrocracking catalyst may be identical with the catalyst used in catalytic cracking. In general, the acidic component of a hydrocracking catalyst has much higher cracking activity than the corresponding component of a reforming catalyst. In reforming, the acidity of the catalyst is less pronounced, the objective being to promote isomerization activity with as little cracking activity as possible.

Hydrocracking catalysts comprising a Group VIII noble metal such as platinum or palladium dispersed on an aluminosilicate have attracted considerable attention. Such catalysts are characterized by high hydrogenation–dehydrogenation activity relative to acid activity, although this characteristic is less marked than with typical reforming catalysts owing to the higher acid activity of aluminosilicates relative to alumina. Mechanistically, the catalysts for hydrocracking are commonly viewed to function in the same manner as a reforming catalyst, in that the data can be rationalized readily in terms of a reaction sequence involving an olefin intermediate. The olefin is formed on metal sites, from which it migrates to acid sites. The olefin adsorbed on the acid sites, presumably as a carbonium ion, may rearrange to form an isomer or crack to form smaller hydrocarbon fragments [57]. A hydrocracking catalyst differs from a reforming catalyst, of course, in exhibiting a higher degree of fragmentation relative to isomerization and aromatization reactions. However, it is emphasized that bifunctional hydrocracking catalysts of the type under discussion do exhibit extensive isomerization of saturated hydrocarbons as well as fragmentation. Thus, in the hydrocracking of *n*-hexadecane, extensive amounts of hexadecane isomers are formed in the primary reaction products [57]. In this respect the behavior is very different from that of the aluminosilicate catalysts employed in catalytic cracking, in which case isomers of the reactant hydrocarbon are not formed. Thus, in the catalytic cracking of *n*-hexadecane on silica alumina, hexadecane isomers are not observed [23]. Interestingly, however, the catalytic cracking of an olefin is very different in this respect, as shown by the fact that *n*-hexadecene yields significant amounts of hexadecene isomers [58]. Furthermore, the distribution of products by carbon number from 1 to 15 in the hydrocracking of *n*-hexadecane is very different from that found in the catalytic cracking of the same hydrocarbon, the former exhibiting negligible secondary cracking reactions compared to the latter. However, if one compares the product distribution by carbon number from *n*-hexadecane hydrocracking with that from the catalytic cracking of *n*-hexadecene, there is a definite similarity. In both cases the probability of rupture of carbon–carbon bonds is highest at the center of the molecule, decreasing progressively from the center to the end. These observations are clearly consistent with the view that hydrocracking of paraffins proceeds via olefin intermediates on catalysts possessing a very active hydrogenation–dehydrogenation component. On an acidic catalyst surface the reactions of olefins are very different from those of paraffins, the former exhibiting extensive isomerization and less extensive secondary cracking.

IV. OTHER FUELS PROCESSES

A. Steam Reforming

When hydrocarbons are contacted with steam at elevated temperatures over a suitable catalyst, the overall process occurring is complex. It is commonly dissected into two major parts [59, 60]. One part involves decomposition of the hydrocarbon, which may be represented in an overall way by the equation:

$$C_nH_m + nH_2O \rightarrow nCO + (n + m/2)H_2$$

This part of the process is highly endothermic. The other part involves the reactions:

$$CO + H_2O \rightleftharpoons CO_2 + H_2$$

$$CO + 3H_2 \rightleftharpoons CH_4 + H_2O$$

These reactions are both exothermic. Equilibria are readily established, and hence determine the composition of the product gas. Steam reforming is applied commercially for the production of methane or hydrogen, and the process is commonly conducted with a nickel catalyst [61–64]. The relative amounts of methane and hydrogen formed depend on the operating temperature, lower temperatures favoring the former and higher temperatures the latter.

When methane is the desired product, the process is conducted at a temperature of about 400°C using a petroleum naphtha as the feedstock. Steam reforming for production of methane is of interest when there is a need for a gaseous fuel with a high heating value, which might well apply in situations where the availability of methane-rich natural gas is severely limited. The heating value of methane is approximately 1000 Btu/ft³, which is much higher than that of hydrogen and carbon monoxide. In steam reforming of petroleum naphthas for methane production, it is necessary to employ an excess of steam to suppress formation of carbonaceous residues on the catalyst. A reactor pressure of 30–60 atm is commonly employed in the process. Because of the sensitivity of the nickel catalyst to poisoning by sulfur compounds, it is necessary to use naphtha feed stocks which are virtually free of sulfur.

When steam reforming is conducted to produce hydrogen as the primary product, a higher temperature (about 750–850°C) is employed. In such operations methane itself is commonly employed as the primary reactant. The hydrogen, or carbon monoxide–hydrogen synthesis gas mixture, produced in high temperature steam reforming has uses more valuable than as a gaseous fuel. For example, the hydrogen is vital for the synthesis of ammonia for the production of nitric acid used in the manufacture of fertilizers.

Because of the complexity of the steam reforming process, not much can be said with regard to mechanistic details. In the steam reforming of higher hydrocarbons, the initial decomposition of the hydrocarbon apparently involves multiple rupture of carbon–carbon and carbon–hydrogen bonds to give chemisorbed monocarbon fragments CH_x. Molecules of water are presumably dissociatively chemisorbed prior to reaction with the monocarbon fragments. The following sequence of steps, apart from the initial decomposition of the hydrocarbon, may be visualized [59]:

$$H_2O \rightleftharpoons O(ads) + H_2$$

$$CH_x(ads) + O(ads) \rightleftharpoons CO + (x/2)H_2$$

$$CO + O(ads) \rightleftharpoons CO_2$$

The symbol "(ads)" signifies an adsorbed species. With regard to the initial decomposition of the hydrocarbon reactants leading to the adsorbed monocarbon fragments, one might adopt the view that the surface processes would be similar to those involved in the initial stages of hydrocarbon hydrogenolysis, which have been considered in some detail [48, 65–67]. The differences between steam reforming and hydrogenolysis would then reside in the subsequent reactions of the surface monocarbon fragments. In the presence of water, the monocarbon fragments yield CO, CO_2, and H_2 in addition to CH_4, whereas in the presence of hydrogen the only product is CH_4.

B. Coal Gasification

As is well known, the rapidly growing demand for energy has led to a serious problem with regard to the supply of clean fuels, i.e., fuels whose combustion does not result in extensive pollution of the atmosphere. Reserves of natural gas, the cleanest fossil fuel, appear to be severely limited, at least by comparison with the reserves of coal. However, combustion of coal presents pollution problems because of the presence of sulfur and nitrogen compounds in the material. The extensive interest in coal gasification stems from these considerations. The process involves the chemical transformation of solid coal, by reaction with steam and oxygen, into a gas composed of carbon monoxide, hydrogen, methane, carbon dioxide, and sulfur compounds [68]. Temperatures of 900–1300°C are required to obtain sufficiently high rates of gasification. Operating pressures in the approximate range of 30–70 atm are considered to be practical commercially. Prior to the gasification step, the coal is ground to a powder. The major reactions occurring in gasification are:

$$C + O_2 \rightarrow CO_2$$
$$C + H_2O \rightarrow CO + H_2$$
$$C + 2H_2 \rightarrow CH_4$$

The second reaction between carbon and steam is highly endothermic (31.4 kcal/mole) and is the slowest of the three reactions. The first and third reactions are highly exothermic (94.1 and 17.9 kcal/mole, repectively) and provide the heat for the second reaction. From the standpoint of economy, it is obviously advantageous to obtain as much heat as possible from the third reaction, in preference to the first reaction. To this end, it is desirable to minimize the amount of oxygen required in gasification. Clearly, it would be desirable to have the third reaction proceed to as large an extent as possible during the gasification step. This would eliminate the need to carry out a separate step in which carbon monoxide is reacted with hydrogen to form methane. The improved heat economy would decrease the amount of coal needed for combustion to supply the heat for the carbon–steam reaction yielding carbon monoxide and hydrogen. While the objective of completing the conversion of coal to methane in the gasification reactor in one step is desirable, it has not been achieved thus far. Present technology involves several additional processing steps to accomplish the desired degree of conversion to methane, as discussed in subsequent paragraphs.

Thus far we have considered some general aspects of coal gasification, but have not said anything about the role of catalysis in the process. Generally, the gasification step discussed in the previous paragraph is conducted in the absence of a catalyst, except for the M. W. Kellogg process employing a molten salt medium [68]. The active catalytic component of the molten salt medium is an alkali metal carbonate. However, except for this process, the application of catalysts has been limited to steps subsequent to the gasification step. Typically, the product from the gasifier, after tar and dust are removed from the stream, is contacted with a "shift" catalyst on which the following reaction occurs:

$$CO + H_2O \rightarrow CO_2 + H_2$$

A common shift catalyst is iron oxide, which may contain a promoter such as chromium oxide. The reaction is exothermic (10.5 kcal/mole) and is commonly conducted at a temperature of about 450°C. The objective of this step is to increase the H_2/CO mole ratio to 3/1, which is the required stoichiometry for the subsequent methanation reaction:

$$CO + 3H_2 \rightarrow CH_4 + H_2O$$

Prior to the methanation step, carbon dioxide and various sulfur compounds are removed from the gas leaving the shift converter. The removal of sulfur compounds prior to the methanation step is essential to avoid severe poisoning of the nickel catalysts used. The methanation reaction is a highly exothermic reaction (49.3 kcal/mole) and is conducted at temperatures in the approximate range 250–400°C.

As seen from the previous paragraphs, the conversion of coal to methane rich gas involves a number of steps, including gasification, water gas shift catalysis, gas purification, and methanation. This technology has existed for a long time [69, 70]. Attempts to improve the technology have been concentrated on the application of catalysis in the coal gasification step [71]. As already mentioned, alkali metal carbonates such as K_2CO_3 are catalysts for the gasification reaction. The use of such catalysts makes it possible to lower gasification temperatures by several hundred degrees. This has the potential advantage of improving the thermodynamics of methane formation and decreasing the heat requirements for the gasification step, if high enough rates of methane formation can be attained. Research to date has led to increased rates of gasification with catalysts such as the alkali metal carbonates. However, catalysis of methane formation during gasification has generally been unsatisfactory. Attempts to use dual catalyst systems, in which one promotes gasification and the other methanation, have been considered. Here the key problem is to find a methanation catalyst which is not poisoned by impurities present in the gasification reactor. For example, an active methanation catalyst such as nickel is highly susceptible to sulfur poisoning. Nevertheless, the approach of searching for poison resistant dual catalyst systems would seem to be a good one for further research.

C. Fischer–Tropsch Synthesis

The Fischer–Tropsch synthesis refers to the reaction of hydrogen with carbon monoxide to yield product molecules, primarily hydrocarbons, containing more than one carbon atom [72, 73]. It was discovered by Fischer and Tropsch in Germany in 1923 [69], and was employed extensively by the Germans for the production of gasoline during World War II. Products other than hydrocarbons, including alcohols, aldehydes, ketones, acids, and esters, are also formed. The distribution of products depends on the catalyst and process conditions employed. Active catalysts for the synthesis include iron, cobalt, nickel, and ruthenium. The synthesis is conducted at temperatures in the range 150–300°C and at pressures of 1–100 atm. Stoichiometric equations for the formation of paraffinic hydrocarbons may be written as follows:

$$(2n + 1)H_2 + nCO \rightarrow C_nH_{2n+2} + nH_2O$$

$$(n + 1)H_2 + 2nCO \rightarrow C_nH_{2n+2} + nCO_2$$

The paraffins which are formed are mainly n-paraffins. The desired products are those boiling in the gasoline and heating oil range, which contain from five to about 18 carbon atoms per molecule.

The reactions occurring in the Fischer–Tropsch synthesis are highly exothermic. Removal of the heat of reaction presents a difficult engineering problem. Another factor of major importance in the process is the susceptibility of the catalysts to sulfur poisoning, as is the case with methanation catalysts. Removal of sulfur compounds from the carbon monoxide–hydrogen synthesis gas to a level less than about 1 ppm is required. In early economic evaluations of the Fischer–Tropsch synthesis, one of the highly negative factors was the extent of methane formation in the process. However, with the current shortage of natural gas, this may no longer be a debit.

Iron catalysts are commonly employed in studies of the Fischer–Tropsch synthesis . The catalysts are similar to those employed in ammonia synthesis, consisting of unsupported iron granules containing small amounts of an electronic promoter such as K_2O and a structural promoter such as Al_2O_3. The electronic promoter affects the intrinsic catalytic properties of the iron, while the structural promoter affects the surface area by inhibiting sintering of the metal. The conditions commonly employed in Fischer–Tropsch synthesis are favorable for the formation of iron carbides, which themselves are good catalysts for the synthesis. Interestingly, iron nitrides, which are other examples of interstitial iron compounds, are also good Fischer–Tropsch catalysts.

With regard to the mechanism of Fischer–Tropsch synthesis, it has been suggested that hydrogen and carbon monoxide interact on the surface to form a chemisorbed species of the following type [74–77]:

$$H-_*C_*-OH$$

The asterisks represent chemisorption bonds. The elimination of water between adjacently adsorbed species of this type then leads to formation of a carbon–carbon bond:

$$H-_*C_*-OH + H-_*C_*-OH \xrightarrow{-H_2O} H-_*C_*-_*C_*-OH$$

The resulting species is then hydrogenated to form species of the type:

$$H_3C-_*C_*-OH \quad \text{or} \quad H_3C-CH-OH$$
$$*$$

Similar steps involving elimination of water and hydrogenation may then be repeated in building up the carbon chain. The growth of the carbon chain may be terminated by processes such as the following:

$$R-_*C_*-OH \longrightarrow \begin{array}{l} \longrightarrow RCHO \\[1em] \xrightarrow{+2H} RCH_2OH \end{array}$$

$$R-CH-OH + H \xrightarrow{-H_2O} R-CH \xrightarrow{+2H} R-CH_3$$

These processes lead to the production of aldehydes, alcohols, and hydrocarbons.

D. Coal Liquefaction

Concern about the supply of clean-burning fuels has also stimulated interest in coal liquefaction processes. The conversion of coal to liquid fuels involves increasing the hydrogen content of the material. Coal is generally considered to be composed of layers of aromatic, hydroaromatic, or carbocyclic ring structures [78]. The basic structural unit is believed to consist of three or four such rings with substituents, and to have a molecular weight of 300–500. These condensed ring structures are presumably linked by small aliphatic groups, biphenyl type linkages, ether linkages, and sulfide and disulfide bonds. Six-membered rings predominate in the structure, although five-membered rings are also present in coal. In the liquefaction of coal, it is not sufficient to break down the coal structure into the basic condensed ring units. The condensed ring structure itself must be broken to yield a product that is liquid at room temperature.

Examples of coal liquefaction processes are the H-Coal [79] and Synthoil [80] processes, currently being developed by Hydrocarbon Research Inc. and the Bureau of Mines, respectively. In both processes a mixture of coal, oil and hydrogen is contacted with a catalyst such as cobalt molybdate. In the H-Coal process, an oil slurry of finely ground coal in admixture with hydrogen flows through an ebulliating or fluidized bed of particulate catalyst. In the Synthoil process there is a highly turbulent flow of coal, oil, and hydrogen through a fixed bed of pellets of cobalt molybdate catalyst. In both these processes the oil employed serves as a hydrogen donor solvent, i.e., a hydrogen transfer reaction occurs between the solvent and aromatic rings in the coal structure. The hydrogen transfer to the aromatic portions of the coal structure apparently leads to solubilization of "coal molecules" in the donor oil. Subsequent contact of these dissolved species with the surface of the catalyst results in a hydrogenolysis reaction which breaks the condensed ring structure. The catalyst is also very effective for hydrodesulfurization, i.e., for conversion of the sulfur in the coal to H_2S via reaction with hydrogen. Coal liquefaction processes of the type described here operate at a temperature of about 450°C and a pressure of about 50–100 atm. The amount of hydrogen added to coal during liquefaction may vary significantly, depending on the operating conditions employed. If the hydrogen content is increased by only 2–3 wt%, the product of liquefaction is a heavy oil which finds ready application for firing boilers in electric power generating plants. If the hydrogen content is increased more extensively, by 6 wt% or more, distillable light oils and gasoline fractions are produced. Volumetric yields of product of 3–4 barrels of oil per ton of coal are claimed for coal liquefaction processes, the volumetric yield increasing with increased addition of hydrogen to the coal. The heavy oil resulting from mild hydrogenation of coal is preferred for power generator fuel because of lower cost (due to lower hydrogen consumption in its production), and because of the higher energy density, a factor of importance in transportation and storage.

Another method of coal liquefaction which has generated some interest involves hydrocracking of coal by contacting it with $ZnCl_2$, as in the process developed by the Consolidation Coal Company [81]. In this process coal is mixed with $ZnCl_2$ and slurried in a distillate oil. The slurry is then heated to a temperature of about 450–475°C in the presence of hydrogen at a pressure of about 200–300 atm. Under these conditions the $ZnCl_2$ is molten. During hydrocracking, the $ZnCl_2$ becomes contaminated with impurities such as ZnS and $ZnCl_2·4NH_3$ formed by reaction with H_2S and NH_3 evolved from the coal. The molten zinc chloride also contains residual carbon. Regeneration is accomplished by combustion of the impurities in the melt. This is accompanied by vaporization of the zinc chloride, which must be recondensed for further use in hydrocracking. While the $ZnCl_2$ process has been shown to work in the laboratory, it has not been demonstrated in large-scale operations. The process requires amounts of zinc chloride comparable in weight to the amount of coal being processed. Nevertheless, the process appears to hold some promise, and work in this area is continuing. In addition to zinc chloride, Lewis acids such as $AlCl_3$ and $SbCl_3$ have been investigated for the process. Although a number of Lewis acid systems have been found to work in the process, very little appears to have been done to elucidate the mechanistic aspects.

V. OUTLOOK

Heterogeneous catalysis will continue to play a vital role in processes for the manufacture of fuels. During the past three or four decades heterogeneous catalysis has played a dominant role in commercial processes for the production of various types of fuels from petroleum, and will continue to do so in the future. Based on the past record, one can anticipate that significant improvements will continually be made in such basic processes as catalytic cracking and catalytic reforming. In addition, heterogeneous catalysis should play a major role in the development of processes for the manufacture of fuels from sources other than petroleum. The great concern over the fuel supply problem has stimulated much interest in the development of processes for converting coal into gaseous and liquid fuels. This provides a major challenge for the catalytic chemist and chemical engineer concerned with improving the efficiencies of such processes. Overall, heterogeneous catalysis should continue to have a very exciting future in its application to fuels processing.

REFERENCES

[1]   J. H. Sinfelt, *Ann. Rev. Mater. Sci.* **2**, 641 (1972).
[2]   F. G. Ciapetta and C. J. Plank, *in* "Catalysis" (P. H. Emmett, ed.), Vol. 1, p. 327. Van Nostrand-Reinhold, Princeton, New Jersey, 1954.
[3]   H. Pines and W. O. Haag, *J. Am. Chem. Soc.* **82**, 2471 (1960).
[4]   P. B. Elkin, C. G. Shull, and L. C. Roess, *Ind. Eng. Chem.* **37**, 327 (1945).
[5]   C. J. Plank, U.S. Patent 2,499,680 (1950).
[6]   R. L. Burwell, Jr., A. B. Littlewood, M. Cardew, G. Pass, and C. T. H. Stoddart, *J. Am. Chem. Soc.* **82**, 6272 (1960).
[7]   P. W. Selwood, M. Ellis, and C. F. Davis, Jr., *J. Am. Chem. Soc.* **72**, 3549 (1950).
[8]   P. H. Hermans, *in* "Colliod Science" (H. R. Kruyt, ed.), Vol. 2, p. 483. Elsevier, New York, 1949.
[9]   D. Cornet and R. L. Burwell, Jr., *J. Am. Chem. Soc.* **90**, 2489 (1968).
[10] R. L. Burwell, Jr., G. L. Haller, K. C. Taylor, and J. F. Read, *Adv. Catal.* **20**, 1 (1969).
[11] H. Pines and C. T. Chen, *J. Am. Chem. Soc.* **82**, 3562 (1960).
[12] M. Boudart, "Kinetics of Chemical Processes," p. 61. Prentice-Hall, Englewood Cliffs, New Jersey, 1968.
[13] M. Boudart, *Chem. Eng. Progr.* **57**, No. 8, 33 (1961).
[14] J. H. Sinfelt, *AIChE J.* **19**, No. 4, 673 (1973).
[15] R. H. Griffith and J. D. F. Marsh, "Contact Catalysis," 3rd ed., p. 238. Oxford Univ. Press., London and New York, 1957.
[16] S. J. Thomson and G. Webb, "Heterogeneous Catalysis," p. 3. Wiley, New York, 1968.
[17] P. Mars, *in* "The Mechanism of Heterogeneous Catalysis" (J. H. de Boer, ed.), pp. 52–53. Elsevier, Amsterdam, 1960.
[18] R. C. Hansford, *Adv. Catal.* **4**, 1 (1952).
[19] R. V. Shankland, *Adv. Catal.* **6**, 271 (1954).
[20] H. H. Voge, *in* "Catalysis" (P. H. Emmett, ed.), Vol. 6, pp. 407–493. Van Nostrand-Reinhold, Princeton, New Jersey, 1958.
[21] T. P. Simpson, L. P. Evans, C. V. Hornberg, and J. W. Payne, *Proc. Am. Pet. Inst.* **24** (III), 83 (1943).
[22] E. V. Murphree, C. L. Brown, H. G. M. Fischer, E. J. Gohr, and W. J. Sweeney, *Ind. Eng. Chem.* **35**, 768 (1943).
[23] B. S. Greensfelder, H. H. Voge, and G. M. Good, *Ind. Eng. Chem.* **41**, 2573 (1949).
[24] R. C. Hansford, *Ind. Eng. Chem.* **39**, 849 (1947).
[25] C. L. Thomas, *Ind. Eng. Chem.* **41**, 2564 (1949).
[26] F. C. Whitmore, *J. Am. Chem. Soc.* **54**, 3274 (1932).

[27] J. N. Miale, N. Y. Chen, and P. B. Weisz, *J. Catal.* **6**, 278 (1966).

[28] D. W. Breck, *J. Chem. Ed* **41**, No. 12, 678 (1964).

[29] M. W. Tamele, *Disc. Faraday Soc.* **8**, 270 (1950).

[30] T. H. Milliken, G. A. Mills, and A. G. Oblad, *Disc. Faraday Soc.* **8**, 279 (1950).

[31] J. J. Fripiat, A. Léonard, and J. B. Uytterhoeven, *J. Phys. Chem.* **69**, 3274 (1965).

[32] J. H. Sinfelt, *Adv. Chem. Eng.* **5**, 37 (1964).

[33] F. G. Ciapetta, R. M. Dobres, and R. W. Baker, *in* " Catalysis " (P. H. Emmett, ed.), Vol. 6, p. 495. Van Nostrand-Reinhold, Princeton, New Jersey, 1958.

[34] J. H. Sinfelt and J. C. Rohrer, *J. Phys. Chem.* **65**, 978 (1961).

[35] J. C. Rohrer, H. Hurwitz, and J. H. Sinfelt, *J. Phys. Chem.* **65**, 1458 (1961).

[36] J. C. Rohrer and J. H. Sinfelt, *J. Phys. Chem.* **66**, 1193 (1962).

[37] J. H. Sinfelt, *Chem. Eng. Sci.* **23**, 1181 (1968).

[38] G. A. Mills, H. Heinemann, T. H. Milliken, and A. G. Oblad, *Ind. Eng. Chem.* **45**, 134 (1953).

[39] J. H. Sinfelt, H. Hurwitz, and J. C. Rohrer, *J. Phys. Chem.* **64**, 892 (1960).

[40] L. Spenadel and M. Boudart, *J. Phys. Chem.* **64**, 204 (1960).

[41] J. J. Keavney and S. F. Adler, *J. Phys. Chem.* **64**, 208 (1960).

[42] G. N. Lewis, "Valence and the Structure of Atoms and Molecules," p. 141. Chemical Catalog Co., New York, 1923.

[43] W. F. Luder and S. Zuffanti, "The Electronic Theory of Acids and Bases," pp. 1–106. Wiley, New York, 1946.

[44] R. L. Jacobson, H. E. Kluksdahl, C. S. McCoy, and R. W. Davis, *Proc. Am. Pet. Inst. Div. Refining* **49**, 504 (1969).

[45] M. F. L. Johnson and V. M. LeRoy, *J. Catal.* **35**, 434 (1974).

[46] J. H. Sinfelt, *Chem. Eng. News* p. 18 (July 3, 1972).

[47] J. H. Sinfelt, *J. Catal.* **29**, 308 (1973).

[48] J. H. Sinfelt, *Catal. Rev.—Sci. and Eng.* **9**, No. 1, 147 (1974).

[49] J. H. Sinfelt, *Critical Rev. Solid State Sci.* **4**, 311 (1974).

[50] R. R. Cecil, W. S. Kmak, J. H. Sinfelt, and L. W. Chambers, *Proc. Am. Pet. Inst. Div. Refining* **52**, 203 (1972).

[51] R. A. Flinn, O. A. Larson, and H. Beuther, *Ind. Eng. Chem.* **52**, 153 (1960).

[52] R. C. Archibald, B. S. Greensfelder, G. Holzman, and D. H. Rowe, *Ind. Eng. Chem.* **52**, 745 (1960).

[53] J. W. Scott, Jr., U.S. Patent 2,944,005 (1960).

[54] F. G. Ciapetta, U.S. Patent 2,945,806 (1960).

[55] H. L. Coonradt, F. G. Ciapetta, W. E. Garwood, W. K. Leaman, and J. N. Miale, *Ind. Eng. Chem.* **53**, 727 (1961).

[56] O. A. Larson, D. S. MacIver, H. H. Tobin, and R. A. Flinn, *I&EC Process Design Develop.* **1**, No. 4, 300 (1962).

[57] H. L. Coonradt and W. E. Garwood, *I&EC Process Design Develop.* **3**, No. 1, 38 (1964).

[58] G. Egloff, J. C. Morrell, C. L. Thomas, and H. S. Block, *J. Am. Chem. Soc.* **61**, 3571 (1939).

[59] J. R. Rostrup-Nielsen, *J. Catal.* **31**, 173 (1973).

[60] J. H. Sinfelt and W. F. Taylor, U.S. Patent 3,450,514 (1969).

[61] W. F. Taylor and J. H. Sinfelt, U.S. Patent 3,320,182 (1967).

[62] W. F. Taylor and J. H. Sinfelt, U.S. Patent 3,407,149 (1968).

[63] W. F. Taylor and J. H. Sinfelt, U.S. Patent 3,404,100 (1968).

[64] W. F. Taylor and J. H. Sinfelt, U.S. Patent 3,395,104 (1968).

[65] A. Cimino, M. Boudart, and H. S. Taylor, *J. Phys. Chem.* **58**, 796 (1954).

[66] J. H. Sinfelt, *J. Catal.* **27**, 468 (1972).

[67] J. H. Sinfelt, *Adv. Catal.* **23**, 91 (1973).

[68] G. A. Mills, *Environ. Sci. Technol.* **5**, No. 12, 1178 (1971).
[69] S. J. Green, "Industrial Catalysis," pp. 453–487. Macmillan, New York, 1928.
[70] H. Perry, *Sci. Am.* **230**, No. 3, 19 (1974).
[71] W. P. Haynes, S. J. Gasior, and A. J. Forney, *Preprints Div. Fuel Chem. Am. Chem. Soc.* **18**, No. 2, 1 (1973).
[72] H. H. Storch, N. Golumbic, and R. B. Anderson, "The Fischer-Tropsch and Related Syntheses." Wiley, New York, 1951.
[73] G. C. Bond, "Catalysis by Metals," pp. 353–370. Academic Press, New York, 1962.
[74] W. K. Hall, R. J. Kokes, and P. H. Emmett, *J. Am. Chem. Soc.* **82**, 1027 (1960).
[75] W. K. Hall, R. J. Kokes, and P. H. Emmett, *J. Am. Chem. Soc.* **79**, 2983 (1957).
[76] W. K. Hall, R. J. Kokes, and P. H. Emmett, *J. Am. Chem. Soc.* **79**, 2989 (1957).
[77] J. T. Kummer, H. H. Podgurski, W. B. Spencer, and P. H. Emmett, *J. Am. Chem. Soc.* **73**, 564 (1951).
[78] P. H. Given, *Fuel* **39**, 147 (1960).
[79] Hydrocarbon Research, Inc., British Patent 1,289,159 (1972).
[80] P. M. Yavorsky, *EPRI Conf. Coal Catal., Santa Monica, California* (September 1973).
[81] C. W. Zielke, *EPRI Conf. Coal Catal., Santa Monica, California* (September 1973).

Chapter 2

Materials for Nuclear Fission Power Reactor Technology

MASSOUD T. SIMNAD

ADVANCED ENERGY SYSTEMS
GENERAL ATOMIC COMPANY
SAN DIEGO, CALIFORNIA

and

JOHN P. HOWE*

MATERIALS, ADVANCED ENERGY SYSTEMS
GENERAL ATOMIC COMPANY
SAN DIEGO, CALIFORNIA

* Present address: Energy Center, University of California at San Diego La Jolla, California.

31

I. INTRODUCTION [1–29]

A. Objectives and Approach

At the social level controversy appears to be the paramount characteristic of nuclear energy. What about the technological and scientific levels? Amid very sound progress in nuclear materials there are some questions on both counts due primarily to economic uncertainties and the narrowing of technological options, a trend partially corrected by the program plans [1] of the Energy Research and Development Administration (ERDA).

Energy from nuclear fission can supply a large portion of the needs of civilization. Together with coal it is one of two assured economical sources of energy that will be available beyond the next few decades. Various research and development programs that have been under way since 1942 have readied nuclear power systems barely in time to assume the major task of providing energy for a growing and developing world civilization. New developments promise more efficient utilization of our uranium and thorium resources, and less waste energy and fission products per unit of useful energy. In a large measure, study and development of materials systems such as fuel elements and other components for fission reactors will continue to set the pace for the use and advancement of nuclear fission energy technology.

A simple rubric under which to summarize items of progress in reasonable perspective is difficult to devise. In order to see the subject matter whole one must regard it from varying distances and from several points of view. The topics lie in the field of materials science and engineering recently characterized by a National Academy of Sciences study [2] directed by M. Cohen. The materials engineering is a part of the design, development, and operation of power producing systems and the processes that make up their fuel cycles. Present day objectives, trends, achievements, and difficulties have their historical background. In use, new kinds of material respond to somewhat unfamiliar forces. Each material has its own complex story that must be melded with others to describe the behavior and performance of components and systems as wholes. While materials science has contributed crucially to nuclear engineering, the reverse is equally true. Most of this interplay is healthy.

Thus, to conduct our discussions from the points of view and perspective that we have chosen requires some repetition. At the same time, some important items will not be apparent. We hope that the dimensions of the subject and some of its high points are made evident. We regret that the literature of the subject makes complete documentation and attribution sufficiently difficult that we have not attempted it uniformly.

To approach the objectives of this chapter the reader and the authors must agree on some boundary conditions. In 1974 at least 1500 important articles were summarized in *Nuclear Abstracts* under the general heading of materials. Research, development, and engineering involving nuclear materials is under way in over 50 laboratories located in 38 countries. We cannot introduce, define, and exhaust all topics. Neither may we stop at merely listing progress of interest to experts in the field nor simply give introductions to basic topics for general readers. Difficulties attend any balancing of these different objectives. We have attempted to state our views on the more durable themes and issues that characterize the field at the present time and substantiate them with data. But often we have found that the number of variables and the constraints and qualifications on much of the data may be too numerous to explain in a reasonable space. Further, many qualifications are known only implicitly to those working intimately with a given problem and may be brought out only in oral discussions, if at all.

B. The Status of Nuclear Fission Energy

Vigorous programs in research, development, engineering, construction, economic analysis, and environmental and safety studies have prepared the technology of nuclear fission energy for a major role in supplying the world's energy. This has occurred at a time when it has become apparent that reserves of petroleum and natural gas are limited relative to demand, and that clean fuel from coal, oil shale and other copious reserves will be several times as costly as hydrocarbon fuels have been in the past. Potentially, uranium and thorium can provide more than a hundred times as much energy more economically than can fossil fuels. The international aspects of nuclear power and its fuel cycle have been discussed in great detail in two recent conferences. The transfer of nuclear technology was the topic of an international meeting held at Persepolis, Iran, April 10–14, 1977. Nuclear power and its fuel cycle was the subject of an international conference organized by the International Atomic Energy Agency (IAEA) in Salzburg, May 2–13, 1977. The Proceedings of these two conferences should be available in 1978. Unfortunately, there are a number of cross currents that make the future course of nuclear fission energy systems very uncertain.

In reports prepared by the staff of the Atomic Energy Commission (AEC) and published prior to 1974, there are forecasts that in 1985 nuclear fission power plants will be supplying 280 GW(e) or approximately 44% of the total capacity for electric power generation in the United States. During 1974 the number of reactors in operation, under construction, or definitely planned to be in operation by 1985 would produce approximately 185 GW(e). By early 1975, policies, the very high cost of capital money, and other economic practices and uncertainties had led to the deferment of 77 of these units, 11 of them indefinitely. Thus, the potential foreseen in 1972 is unlikely to be fulfilled, and in 1985 fission reactors may supply no more than 20–25% of the nation's electric power. Additional uncertainties are introduced by campaigns in several states to legislate moratoria on the construction of nuclear plants. A vociferous, determined, and small number of individuals are emphasizing improbable nuclear disasters about which most people have little reliable knowledge. At the same time, the plan advanced by the executive branch of the government insists that a greatly accelerated program of using and developing nuclear energy sources approximately to the extent forecast in 1972 must be an essential part of the nation's plan for energy independence in order to avoid severe economic dislocation, unemployment, and pressures for war attending the high cost of imported petroleum.

Thus, large and in some cases imponderable forces must resolve themselves before the future will become clear. The debate involves tacit assumptions, beliefs, and feelings quite outside the arena of the argument. Many concerned and informed persons might agree that need and opportunity will bring about extensive and safe use of fission energy [3]. Competence and good will are required in order to benefit from nuclear energy. More of man's needs than merely energy are jeopardized by incompetence and ill will. Both opponents and proponents can work toward removing the causes of these ills.

Despite uncertainties, the need for and the momentum of the program seem likely to lead to the use of fission energy for at least 50 years. During this period of time, the availability of economical energy from other sources

TABLE 1

Status of World Use, Construction, and Planning of Electrical Power from Nuclear Fission [3a]

	World gigawatts GW(e)	Number of plants[a]	Largest single plant MW(e)	Number of countries
LWR	382	438	1300 U.S. & Germany	30
GCR	14.4	47	625 U.K.	5
PHWR	15.3	36	745 Canada	9
HTGR	0.6	2 demonstration	330 U.S.	3
		2 prototypes	(1500 in design) U.S.	
LMFBR	3.3	7 demonstration	1200 France	6
		6 experimental		

[a] Total plants operable, under construction, or planned (as of June 1976) are:

U.S.: 212 units, 207,200 MW(e);

Non-U.S.: 318 units, 208,000 MW(e); 33 countries;

World total: 530 units, 415,200 MW(e); 34 countries.

such as fusion reactions and solar radiation will be clarified while the accumulation of radioactive materials and the probability of hazards will not exceed easily manageable limits. Consequently, the summary in Tables 1 and 2 of nuclear power plants in operation or planned around the world will provide much of what we need to know about the future. The summary shows that, regardless of the decisions in any one country, that country's friends and enemies near and far are both likely to use power from fission. Major references [4–9] are given in which more details about nuclear fission reactors may be found. Power reactor concepts are defined in Table 3, and their characteristics are listed in Table 4.

C. Highlights in the Development of Nuclear Reactors and Materials

Many subjects [9] that would provide useful background for our discussion will have to be left stored in reports. Some indication of how the major themes arose and how firm these bases are may be sketched as follows.

Since the beginning of concerted development efforts in 1942, it has been realized that in order to make effective use of resources, nuclear fission reactors would be required to produce as much or more of one of the fissile nuclides, ^{233}U or ^{239}Pu, through *conversion* or *breeding* with the fertile nuclides, ^{232}Th or ^{238}U, as is destroyed by capture and fission. By 1944 in the United States, Zinn, Fermi, and others at the Metallurgical Laboratory[1]

[1] These are the wartime laboratories that became the Argonne National Laboratory (ANL), the Los Alamos Scientific Laboratory (LASL), and the Oak Ridge or Hollifield National Laboratory (ORNL), respectively, under contracts with the AEC and now the Energy Research and Development Administration (ERDA).

TABLE 2

World Nuclear Power Program [3b]

	1975	1980	1985	1990	1995	2000
Installed nuclear electrical generating capacity, GW	50	300	500	1,200	2,000	2,500
Cumulative nuclear electrical energy sent out, GW year	120	650	1,850	4,350	9,000	16,000
Spent fuel reprocessing load, t/year[a]	700	3,000	8,000	14,000	31,000	46,000
Cumulative fuel reprocessed, t	2,000	9,000	36,000	90,000	200,000	400,000
Cumulative electrical energy associated with reprocessed fuel						
GW year	60	270	1,100	2,700	6,100	12,000
kWh $\times 10^{12}$	0.5	2.5	10	25	50	100
Plutonium throughput, t/year	7	30	80	140	400	900

[a] All fuel quantities are based on the equivalent of high burn-up LWR fuel.

TABLE 3

Power Reactor Concepts

Reactor abbreviation	Reactor name and characterizing features
LWR	Light Water Reactor. Light water moderates neutrons and removes heat from slightly enriched uranium dioxide fuel rods.
PWR	Pressurized Water Reactor. Circulating liquid water at about 140 atm and 300°C transfers heat from fuel elements to steam generators.
BWR	Circulating liquid water transfers heat from fuel rods, boils at top of reactor, and furnishes steam at about 300°C, 70 atm directly to turbines.
LWBR	Light Water Breeder Reactor. Pressurized water provides cooling and moderation. Core consists of ThO_2 rods containing up to 6% ^{233}U, blanket consists of ThO_2 rods containing up to 3% ^{233}U.
HWR	Heavy Water (moderated) Reactors.
PHWR–CANDU	Heavy water in a calandria moderates neutrons; light water at 100–150 atm, circulating in pressure tubes, transfers heat from natural uranium metal or dioxide fuel rods to steam generators.
SGHWR	Steam Generating Heavy Water Reactor. Similar to PHWR and BWR. Light water at 100 atm boils in upper part of reactor supplying steam directly to turbines.
GCHWR	Gas Cooled Heavy Water Reactor. Similar to PHWR except that circulating carbon dioxide transfers heat from uranium metal fuel elements to steam generators.
HWOCR	Heavy Water Organic Cooled Reactor. Similar to PHWR except that a high boiling, stable organic coolant at a pressure of 3–4 atm transfers heat from fuel rods to steam generators at about 400°C.
GCR	Gas (carbon dioxide) Cooled Reactor.
Magnox	Graphite moderates neutrons, provides much of core structure. Circulating CO_2 transfers heat from natural uranium metallic fuel rods to steam generators at about 400°C. Name derives from magnesium–aluminum (0.8%)–beryllium (0.050%)–calcium (0.008%) alloy cladding developed for fuel rods.
AGR	Advanced Gas Cooled Reactors. Grapphite moderator and CO_2 heat transfer similar to Magnox. Stainless steel cladding for slightly enriched UO_2 fuel rods permits steam generation at over 500°C.
HTGR	High Temperature (helium) Gas Cooled Reactors. (HTR in U.K.; AVR, THTR in Germany.) Graphite moderator provides core structure. Circulating helium at 700–1000°C and 50–80 atm transfers heat to steam generators (550°C), or work to gas turbines or heat to chemical processes.
FBR	Fast (neutron) Breeder Reactor.

(continued)

TABLE 3 (*Continued*)

Reactor abbreviation	Reactor name and characterizing features
LMFBR	Liquid Metal Fast Breeder Reactor. Liquid sodium 400–600°C, 2–3 atm transfers heat from sealed uranium–plutonium oxide or carbide fuel rods to intermediate heat exchanger from which sodium transfers heat to steam generator. The system has reached the demonstration phase.
GCFR	Gas (helium) Cooled Fast (breeder) Reactor. Helium at 300–600°C transfers heat from vented and pressure equalized uranium–plutonium oxide (or carbide) fuel rods to steam generators. The system is in the development phase.
FTR or FFTF	Fast Flux Test Reactor. Sodium cooled, under construction at Hanford, Washington.
MSBR	Molten Salt Breeder Reactor. Molten Li, Zr, Be, U fluoride salt circulates through channels in structural graphite moderator and a Hastelloy N heat exchanger. Li, Zr, Be fluoride salt transfers heat at about 600°C to steam generators. Uranium bearing fuel and thorium bearing blanket salts processed continuously or in batches. The system is in the development phase.
TRIGA	General Atomic Training, Research and Isotope (production) Reactor. Cooled and shielded by light water circulating at ambient pressure. Core consists of rod containing Zr–U hydride clad with Al or SS.
Swimming pool reactor	A pool of water provides cooling, moderation, and shielding. Core consists of plates of aluminum–enriched uranium alloy clad with Al.
HIFR	High Flux Reactor. Similar to Swimming Pool but equipped with carefully designed water circulation and cooling system, ports for neutron beams, and other experimental features.
ETR	Engineering Test Reactor. Similar to above, but with regions for test apparatus. There are three generations of test reactors.
Water boiler research reactor	Solution of highly enriched uranium in light water, usually in a stainless steel container, often equipped with heat exchanging coils of tubing through which cooling water is circulated. (Use largely discontinued because of great care needed to confine radioactivity.)

TABLE 4

Typical Fuel Element Design Parameters

Characteristic	PWR	BWR	CANDU/PHW	HTGR	LMFBR	GCFR
Fuel material	UO_2 pellets	UO_2 pellets	UO_2 pellets	UC_2,ThC_2 particles	UO_2-PuO_2 pellets	UO_2-PuO_2 pellets
Fuel density, g/cm³	10.3 (94%)	10.3 (94%)	10.6 (96.7%)	0.58	9.90 (91%)	10.15 (93.3%)
Pellet diameter, mm	9.29	12.4	14.25	15.70	4.94	6.21
Fuel portion length, mm	3650	495	500	793	900	1360
Fuel initial enrichment, %	3.0	2.5	0.7 (natural)	93	15 (Pu)	15 (Pu)
Fuel peak temperature, °K	2530	2690	2270	1520	2470	2470
Fuel average burnup, MWd/kg	24.4	27.5	9.0	100	100	100
Fuel maximum burnup, MWd/kg	35	40	—	700	150	150
Fission gas release, %	1–4	1–4	1–4	10^{-3}	80–100	100
Fuel-clad gap (diametral), mm	0.18	0.25	0.08	0.076	0.18	0.15
Cladding material	Zr-4	Zr-2	Zr-2	Graphite	20% c-w 316 SS	20% c-w 316 SS
Cladding wall thickness, mm	0.62	0.81	0.38	0.17/0.16	0.375	0.370
Cladding peak temperature, °K	620	569	583	1520	920	998
Cladding fast neutron fluence, > 0.1 MeV	10^{22}	10^{22}	10^{22}	8×10^{21}	3×10^{23}	2.3×10^{23}
Fast neutron flux, > 0.1 MeV (average)	2×10^{14}	2×10^{14}	4.5×10^{13}	2.4×10^{14}	5×10^{15}	3.8×10^{15}
No. of pins per element	204	49	28	132	271	331
No. of elements per channel	1	1	12	8	1	1
Average core rating, W/g	34.8	22.0	22.1	76	170	132
Peak heat rating, W/cm²	171.2	134	126.1	42	230	105
Linear heat rating (peak), W/cm	577	600	603	210	475	500
Average power density, kW(t)/liter	90	50	10	8.4	300	250

of the Plutonium Project at the University of Chicago, and also Morrison and others at the Los Alamos weapons laboratory,[1] had begun the development of the fast (neutron) breeder reactors based on metallic fuel elements. This approach was attractive because (1) the yield of neutrons produced per neutron absorbed in ^{239}Pu had proved to be the largest of known fission reactions provided that the incident neutron had energies in the range 0.1–10 MeV, and (2) because compact reactor cores favor the smallest inventories. In 1945, Wigner and Weinberg and others associated with the Clinton Laboratories[1] at Oak Ridge, Tennessee, emphasized that fission of ^{233}U with neutrons having thermal energy also offered favorable neutron yields, conversion, and breeding with fertile ^{232}Th, and that fluid fuels with continuous on-line reprocessing promised minimum investment of fissile material and loss of neutrons to fission products.

Among other determinants of the program supported by the AEC, these two valid, long-ranged themes in nuclear reactor and materials development have persisted and spread from the three national laboratories to other organizations in many nations. Thus the development of solid fast reactor fuels, the chemistry of liquid sodium in steel conduits, molten fluorides of U, Th, Zr, Be, and Li and their containers, the effects of atom displacements in metals, graphite, and other materials, fission recoil and fission product effects in uranium alloys and compounds, and other indispensible topics have been under way since before 1944.

The commercial program that has led to the present extensive deployment of light water reactors (LWR) in many countries arose differently. Patents on light-water cooled reactors moderated with graphite, heavy water, or light water were written in the period 1942–1943 at the Metallurgical Laboratory, University of Chicago. The use of clean light water to provide adequate heat transfer from nuclear fuel elements, around 1 MW/m^2, was also developed and demonstrated in the 1942–1943 period, before the Hanford reactors began producing plutonium. But the dominant influence on commercial nuclear reactor technology in the U.S. was the naval reactors program initiated in 1946. Of the two reactor systems developed by Westinghouse and General Electric for submarine propulsion and derived from concepts originated at ANL and ORNL, the LWR system was more seaworthy than the sodium intermediate (neutron spectrum) reactor (SIR) largely because of the behavior of materials and components. Following this choice of LWR, Westinghouse Electric and General Electric Company extended the developments that had originated in the Argonne National Laboratory and the naval reactors program to the commercial PWR and BWR, respectively. Good fortune with three developments in materials made these programs possible.

Particularly fortunate were the two discoveries that hafnium-free zirconium had a low thermal neutron cross section of approximately 0.18 b and, sufficiently later to be suspenseful, that approximately 2–4% tin alloyed with zirconium improved corrosion resistance enough to permit its use for 1–3 yr in water at 300°C. These developments [10–12] served the naval program first. Also crucial was the Kroll process for producing pure zirconium. This earlier development, which came from outside the AEC, was adopted by the naval reactors project.

The third development came from a research program on intermediate neutron spectrum reactors supported by the AEC prior to the naval reactor project at General Electric Company. During 1945–1946 researchers in this program demonstrated that prospects of using uranium dioxide in stainless steel tubes to provide adequate fuel life and thermal output for power production. Unfortunately, because neutrons of intermediate energy, 10^{-1}–10^5 eV, are captured with relatively high probability by ^{239}Pu, the net neutron yield proved to be too low for breeding.

The development of UO_2 accompanied indications that metallic fuel elements would probably provide only a fraction of the 10% burnup required for an economical fuel cycle. Dilute alpha-uranium alloys had proved adequate at burnup of as much as 0.5% of the uranium atoms in water-cooled graphite-moderated reactors used to produce plutonium in the United States for the PHWR in Canada, and for GCR to produce both plutonium and power in the United Kingdom and France. But, in retrospect, it might have been anticipated even before 1947 that distortion due to atom displacements and the accumulation of fission products would limit the life of alpha-uranium base alloys too severely for economical power production. Well before 1944, uranium alloys, for example, the body centered gamma phase alloys of uranium containing about 10 atom % Mo and uranium–silicon alloys based on U_3Si, were known to be more stable tnan alpna phase alloys. While continuing development work has indicated that conditions under which these alloys might find use, apparent limits of 5% burnup and 500°C peak temperature have deterred their practical application. On the other hand, the development of the carbides of uranium and thorium, which began in 1942, is continuing.

Work on helium cooled reactors in the United States began almost independently of the AEC after 1955 and was strongly influenced by the European GCR technology. Designers of reactors in the United States had tended to emphasize high power density and expected heat transfer with circulating gas to be lower than desired at reasonable pressure, velocity, and pumping power. However, it has been demonstrated that circulating helium at pressures between 70 and 100 atm consumes less than 4% of the reactor thermal power, provides about the same heat flux as water, as it is typically employed, and matches the heat output of long lived fuel elements. Reactive impurities in helium may be maintained at sufficiently low concentrations that reactions at high temperature with graphite, iron, chromium and nickel-based alloys, refractory metals, and other materials of construction are within safe limits. These somewhat late entries into the list of options [13] in the United States require more thorough evaluation than they have received to date because the combination of high temperature heat and neutron economy conserves uranium and provides energy options not available from other nuclear systems.

D. An Overview of Materials of Principal Interest

Although early work in nuclear materials science and engineering led to other themes [14], e.g., the physical metallurgy of U, Pu, and their complex alloys, the foregoing were the more conspicuous results or trends attending the resurgency after 1952 of attention to the production of commercial electric power from nuclear fission heat. By the middle of that decade, exploration had shown that uranium resources in the United States were fairly large compared to the shortages that had been predicted in the earlier decade. Thus, in 1950 a group directed by Rickover initiated the Shippingport reactor project and in large measure predetermined the commercial development of LWR over at least eight other concepts that were carried toward or through reactor experiments in the period 1950–1965. The Rasmussen Report has assessed the safety of LWRs [15]. (See also [7].)

This work with alternative concepts revealed limitations and possibilities of the following sort. Of several fluid fuels, including uranyl salts dissolved in D_2O, uranium in bismuth, plutonium in magnesium, and other low-melting plutonium alloys, only the molten fluoride system can be contained reliably in available materials. Polyphenyl organic compounds may serve as both moderator and coolant up to 400°C if circulated and purified rapidly, conditions not readily attained in an organic moderated reactor but feasible in organic cooled (only) reactors. By itself pure sodium is inert, but readily penetrates stringers and cracks that may be associated with welded joints in commercial steel steam generators thus leading to caustic embrittlement. Anisotropic moderating solids such as graphite, Be and BeO cannot survive fast neutron exposure higher than about 4×10^{22} neutrons/cm^2, depending on temperature. Of these, graphite, because of its low cost and radiation stability, is the only practical solid moderator for commercial power systems. Although ferritic pressure vessel steels embrittle at neutron fluences around 10^{18} neutrons/cm^2, they may be used safely at 50°C above their ductile–brittle transition temperature.

The elimination of several alternative concepts has resulted in commercial demands being filled largely by water moderated reactors. Advanced effort is largely concentrated on **LMFBR** (liquid metal fast breeder reactor). A relatively smaller effort is under way on helium cooled reactor systems. Exceptions to the predominant use of LWRs occur as follows. The **PHWR** meets Canada's requirements for economical use of her supplies of uranium, can be adapted to the thorium–uranium cycle, and promises excellent resource conservation. India, Argentina, Pakistan, and possibly other countries not only want to use their own supplies of natural uranium and to be independent of isotope enrichment, but also appear to be interested in producing plutonium for weapons as well as electric power with the **PHWR**. To replace the GCR, the United Kingdom Atomic Energy Authority chose the **SGHWR** over the LWR and HTGR. While helium cooled reactors offer uniquely useful characteristics, the present-day scarcity of capital slows the evaluation and adoption of high temperature systems and the thorium–uranium fuel cycle. The United Kingdom has now abandoned the SGHWR in favor of the AGR.

The case for breeder reactors in the United States and the expectations of ERDA [1] were described in its 1975 program and in the environmental statement [16] for the LMFBR program. Chief among the basic assumptions for the case are the following:

1. The demand for electrical power will grow at a rate decreasing from approximately 6% to 4% by the year 2020;
2. consumption of primary energy by the utilities will increase from about 25% to 65% by 2020;
3. the capital costs of **LMFBR** electrical generating systems will, after a 10-yr learning period, be about the same as LWR systems ($368/kW 1974 $) and about 15% greater than fossil fuel systems;
4. a supply of plutonium from LWR systems will be available during a rapid introduction of LMFBR systems;
5. introduction of these systems will start in 1990 to 1995 and the number will double every 2 yr;
6. no other kinds of reactors will be constructed during the build-up period;
7. the industries essential to rapid introduction of **LMFBR** such as fuel fabrication, reactor construction, fuel reprocessing, shipping, and fissile material safeguards will be prepared in advance for this massive program;
8. a large program of fuel element development will provide the following capabilities: (a) heat output per unit length of approximately 1000 W/cm, (b) fuel life approaching 150 MW(t)d/kg, (c) fuel cycle costs less than 1 mil/kWh;
9. the AEC estimates of uranium supplies and costs will prove correct and coal prices will also trend upward.

A cost benefit analysis based on the above (and other subsidiary) assumptions and done by linear programming methods indicates a least cost through the year 2020 of filling a fixed projected demand for electric power from LMFBR compared to fossil fueled, LWR, and HTGR energy systems. Further, analyses of environmental and social effects of various energy alternatives purport to show that the LMFBR would have the least impact on the United States. Of the large economical, social and environmental benefits calculated to accrue from the LMFBR program by the year 2020 (some four to five times the $4–6 \times 10^9$ development cost if the discount rate is 10% per year) over 80% are due to the costs and hazards associated with resources, chiefly greatly reduced mining of coal and uranium, and to the elimination of isotope enrichment.

In very large measure, realization of these benefits will depend on a large program to develop U–Pu carbide or nitride fuel elements with low swelling cladding and on the reliability and cost of manufacturing and maintenance of sodium-to-water and sodium-to-steam heat exchangers. It is ironic that the tentative selection of oxide fuels, primarily because of their known performance, destroyed the basis for the choice of sodium as a coolant. A much higher breeding ratio is possible with the helium coolant.

Three other themes have also motivated work in nuclear materials science and engineering. These may be listed as: reactors for research and development emphasizing neutron physics and chemistry, irradiation effects, and materials testing; reactors for space applications including both electric power and propulsion for long duration space missions; and propulsion of military vehicles other than naval vessels. The latter two efforts have been discontinued for lack of an essential application. Nevertheless, they have contributed knowledge concerning the high-temperature, short-time conditions, and limitations on graphite, beryllium oxide, refractory carbides, molten salts, hydrogen-heat transfer and corrosion, thermionic emission, and several other high temperature materials and phenomena. High neutron flux in the range 10^{14}–10^{15} neutrons/cm^2/sec for materials testing, irradiation effects, and research with neutron beams is provided by water moderated and cooled reactors in which enriched (20–93%) uranium is dispersed in aluminum clad plates from which heat may be transferred most effectively without large temperature differences. To provide neutron flux in the range 10^{12}–10^{14} neutrons/cm^2/sec and pulses or bursts of millisecond duration yielding more than 10^{15} neutrons/cm^2, reactors using fuel elements based on $ZrH_{1.6}$, in which uranium and erbium (for better control and life) are dispersed, have proved to be very effective and safe. The third type of reactor that also may be pulsed uses uranium dioxide as the fuel and H_2O as the coolant moderator.

For the future, there are other attractive options in addition to those indicated by the foregoing prologue. Clearly, four principles of optimization must govern future reactor development and uses: optimum use of fuel resources, optimum use of heat from fission, optimum provision for nuclear safety, and management of radioactivity and other potential factors affecting the safety, health, and environment of living things and, finally, optimum development of and use of options. Each of these aspects relates directly to materials science and engineering through efficiency of use of neutrons and heat, through system integrity, and through the options for materials and components.

Approximate notions of the general arrangements, size, and shape of components inside reactors are essential to our discussions. However, in order to save time and space in this chapter we ask the reader to refer to more detailed descriptions in the referenced sources [3–5] or to picture reactor internals in his mind with the help of the following principles.

Most large nuclear fission reactors have similar topology. Hence, there are many formal similarities among the corresponding components of different power reactor concepts.

In all power reactors, coolant flow is axial along cylindrical fuel rods. Access for loading and unloading fuel rod bundles may be from either end although it is definitely preferable to insert fresh fuel at the low temperature end and remove spent fuel from the high temperature end. PHWR and GCR are able to use this advantage in fuel management. Instrumentation and control devices, which must be removable, are also inserted and withdrawn at one end. Thus vertical and approximately cylindrical symmetries prevail. GCR and pressure-tube reactors, like CANDU, may be horizontal. Except in the case of the FBR, the approximately cylindrical core has a ratio of length to diameter that minimizes surface area and neutron leakage. To avoid an unsafe positive temperature coefficient, fast reactor cores are considerably shortened to enhance axial leakage. Whenever possible, support and shielding structures that are heated by neutrons and gamma rays are cooled by incoming coolant. The size, the amount and distribution of fissile material within a reactor core, the amount and location of fertile material, the volume of moderator, the mass and flow of coolant, the size and location of control elements, the thickness of cladding, and the strength of pressure vessels or tubes are determined by means of optimization calculations based on coupled fundamental mathematical descriptions of neutron transport, nuclear reactions, heat and fluid flow, and mechanics of materials, all subject to boundary conditions and constraints selected by the engineer. The calculations are programmed for large digital computers. Many computer programs are widely available and are shared among members of the profession.

For reasons sketched above, and also for convenience and compactness in discussion, it is possible and useful to classify components, materials, and processes under loosely generic types. The most important classes and their associated classifications are:

Fuel and fertile materials, fuel elements, and their cladding materials.

Fuel systems and fuel cycles, which include materials preparations, fuel element fabrication, fuel reprocessing, fuel element refabrication, and waste management.

Moderating materials fall in two essential subclasses, liquid and solid, and for power reactors are at the present time restricted to H_2O, D_2O, and graphite, although $ZrH_{1.6}$ serves well in research reactors and could serve in compact power producers, and beryllium and its compounds have excited significant interest.

Coolants are often classified as gas or liquid. However, each of the useful fluids H_2O, CO_2, He, Na, mixed molten fluorides of Li, Be, and Zr, and the organic polyphenyls give rise to distinctive chemical and physical problems outside of those classed under the heading heat transfer and fluid flow. Thus, each coolant is best put in a class by itself. We shall not be able to do justice to the chemistry of coolants.

Control materials and components use several materials with high neutron absorption cross section, namely, boron, cadmium, indium, hafnium, samarium, and europium, often as a stable compound dispersed in a stable metallic matrix. They are employed in solid shapes, rods or blades, and in the case of boron compounds, as fluids to adjust neutron flux and to act as first, second, or third alternative shutdown devices.

Shielding materials turn out to constitute a rather indefinite class because structural, cooling, neutron absorbing properties, reasonably high electron density, and distances all enter the shielding function, which is the scattering attenuation and absorption of neutrons and gamma rays. Moderating materials and steel often provide the closest shielding; concrete is usually the outer shielding material.

Thermal insulation for helium and CO_2 cooled reactors requires special attention to limit heat transport from the reactor core to surrounding structural materials. Metal supported siliceous and related refractory materials often serve as insulators.

Structural materials that serve primarily to keep everything in its proper position include fuel element cladding, coolant channels or conduits, solid moderator blocks, control element matrices and cladding material, grid structures, shielding, reactor vessels, coolant piping, heat exchanger materials, and many others. Depending on temperature and calculated mechanical loads, well known materials usually serve this function. Strength, deformation, and corrosion supply familiar problems, but less familiar phenomena occur.

The naming and classification of reactors, reactor materials and components and the associated processes and cycles aptly illustrate Wittgenstein's thesis [17] that we must use overlapping and interconnected classes rather than distinct and disjoint categories if we are to name things at all.

The goal of the material scientist might be to describe a nuclear fission reactor as a multi-constituent-many-phase-system that evolves in time in response to physical–chemical forces. The initial state of the system is that specified by the design engineer. The goal of the engineer and operator is to regulate its evolution in time. The reactor core, the coolant, the reactor vessel, and the coolant loop and heat exchangers and, in some instances, the heat-to-power conversion equipment constitute the bulk of the whole system. Side loops that are used to adjust the composition of the coolant or of an atmosphere above a coolant constitute lesser coupled systems. Within the reactor some regions such as a sealed fuel element may constitute an isolated subsystem except for the flow of heat and the mechanical loads or constraints, unless failure occurs. As is shown in Section II, the power plant is the output terminal of an extensive fuel cycle.

The forces causing change are thermodynamic in nature. In principle, and frequently in practice, they may be represented by a gradient or difference of a potential energy. However, in addition to the usual flows of energy and matter, those due to nuclear recoils and atomic collisions must not be overlooked. And, in particular, recoil and collision processes must be included among the atomic and molecular mechanisms for change.

II. FUEL CYCLES FOR NUCLEAR FISSION ENERGY SYSTEMS [30–51]

A. Introduction

The fuel cycle includes the full circle of operations from mining to recycle of the spent fuel and management of wastes. Much social and technological questioning surrounds the several stages in this aspect of nuclear energy. Major steps and nuclides in the nuclear fuel cycle are depicted in Figs. 1, 2, and 3. The sequence of steps in a nuclear reactor fuel cycle consists of:

1. Mining and milling of uranium;
2. conversion of ore concentrate to uranium hexafluoride (UF_6);
3. enrichment in the fissionable isotope ^{235}U;
4. conversion of enriched uranium to fuel material (U, UO_2, UC, etc.);
5. fabrication of fuel elements;
6. burnup of fuel in reactor core;
7. shipping and reprocessing spent fuel;
8. disposal of radioactive wastes.

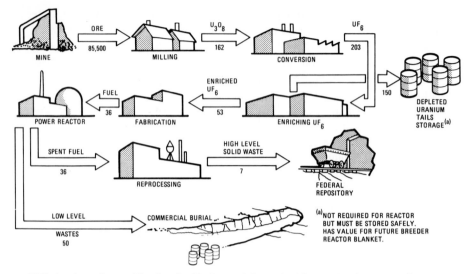

FIG. 1.   Annual quantities (tons) of fuel materials required for routine (equilibrium) operation of 1,000 MW(e) light water reactor [Source: ERDA].

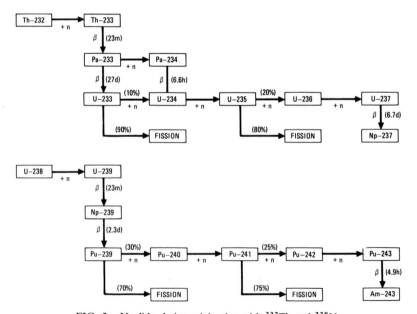

FIG. 2.   Nuclide chains originating with ^{232}Th and ^{238}U.

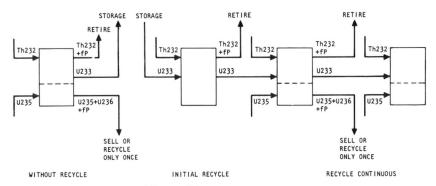

FIG. 3. $^{235}U/Th/^{233}U$ fuel cycle modes [234, p. 16].

The viability of the nuclear industry depends primarily on the lower cost of the nuclear fuel cycle compared with fossil fuels. The nuclear fuel cycle cost includes the costs of the unit operations of extracting, preparing, reprocessing, and disposing of nuclear fuels, and the credit allowed for reclaimed uranium and plutonium. The carrying charges during the residence time in the fuel cycle and the financing costs may reach 25% of the total cost. The fabrication costs of the fuel elements for LWRs correspond to about 20% of the total electricity generation cost. Comparative costs are shown in Table 5.

The limitation set by the fuel burnup at discharge is governed by the irradiation behavior of the materials and/or the reactivity characteristics. The natural uranium required to prepare the initial in-core fuel loading required in an 1100 MW(e) reactor is approximately 580 tons for an LWR, 450 tons for an HTGR, and 700 tons for a fast breeder reactor. The recycling of Pu in LWRs may decrease the uranium requirements by about one-third. Approximately 200 kg of Pu is produced per year in an 1100 MW(e) LWR.

Fuel management aims at the attainment of the most economical use of the fuel within the constraints set by the fuel design, the operation of the reactor, and the various cost items involved in the financing of the reactor. The cost of nuclear fuel must also include the unique requirements of safety and safeguarding of enriched and irradiated fuels in processing and reprocessing facilities, and precautions against illegal diversion.

The judicious selection of materials and fuels for the reactor cores must be based on considerations of the design, fabrication, and operation of the reactor. The effects of irradiation upon the physical, mechanical, chemical, and isotopic changes during operation of the reactor must be known in advance.

TABLE 5

Projected Cost of Electricity in New England, 1985 to 2000 A.D.[a]
[9a]

	Light-water nuclear	Low-sulfur western coal	High-sulfur eastern coal
Unit capital cost ($/kW)	863	565	697
Fuel cost ($/million Btu)	0.60	3.24	3.04
Cost of electricity (¢/kWh)			
Capital	2.690	1.798	2.374
Operation and maintenance	0.307	0.264	0.425
Fuel	0.625	2.902	2.880
Total	3.622	4.964	5.679

[a] Three types of generating plant are considered: a light-water nuclear fission plant, a coal-fired plant using low-sulfur western coal, and a similar plant using high-sulfur eastern coal. The figures are taken from a report by Arthur D. Little, Inc. and S. M. Stoller Corp. to New England Electric Co., "Economic Comparison of Base Load Generation Alternatives for New England Electric," Boston, Massachusetts, March 1975.

The primary objectives in fuel management are not only to minimize the fuel cycle costs, but also to allow optimum utilization of the fuel, and assure availability of fuel resources. Fuel management includes:

1. the out-of-core activities of planning, budgeting, purchasing, designing and fabricating;
2. the planning of the in-core utilization of the fuel;
3. the control of the fuel cycle.

The design of the fuel elements aims to achieve the goals of adequate heat transfer, nuclear reactivity, retention of fission products, inherent safety under accident conditions, and retention of structural and mechanical integrity. In addition, the important economic parameters must be considered, namely, reliability and high specific power and burnup, optimum fuel management, high neutron utilization, and realistic specifications for manufacture and quality assurance. The cost of fabrication of uranium dioxide pellets for LWRs was approximately $200 per kg of U in the early 1960s, $100 per kg in the early 1970s, and is expected to decrease to about $65 per kg by the late 1970s. A 1000 MW(e) LWR is fueled with about 9 million UO_2 pellets contained in approximately 150 km of fuel rods, with 40 fuel rods per MW(e).

There is a tradeoff between the specific power and burnup limits of the fuel and the influence of these factors on the capital cost in relation to the core dimensions, components, pressure vessel, and containment sizes. The design of the core reflects the need to optimize the critical parameters by suitable choice of fuel enrichment, distribution of fuel, control rod and burnable poison distribution, etc. In LWRs the peak rod power is about 19 kW/ft, and up to 25 kW/ft under power transients. The ratio of peak to average power falls in the range 2.8–3.0.

Fuel performance predictions are based on design modeling which includes both the results of experiments and analytical studies. The principal operating conditions governing fuel lifetime are the burnup (MWd/kgU) and specific rod power (kW/m). Also, the fission gas pressure, ratchetting between the fuel and cladding, and irradiation effects on the cladding (swelling, loss of ductility, irradiation-creep, fission product corrosion) influence the behavior of the stressed cladding. Allowance is made for the released fission gases by means of a large plenum volume in the fuel rod. In the gas-cooled fast breeder (GCFR) fuel rods, the fission gases are vented to external fission product traps.

An estimate [33] of 1981 fuel generation cost of LWR nuclear fuel for a 1000 MW(e) plant (in 1975 dollars) is given in Table 6. The Th–^{233}U cycle is the most advantageous of the fuel cycles in the thermal and epithermal regions. The thorium cycle also depends on the initial fissile charge of ^{235}U or Pu to generate the fissionable ^{233}U. The reactors that are based on the Th–^{233}U cycle include the HTGR and the thermal breeders: MSBR, HWBR, and LWBR. The thorium cycle is associated with significantly higher conversion ratios and longer reactivity lifetimes compared with the uranium cycle. However, the fuel inventory and fuel fabrication and processing costs are also higher for the uranium cycle. Hence, increased uranium costs and lower interest rates favor the thorium cycle.

B. Mining and Processing of Fuel

The techniques used for preliminary prospecting for uranium include airborne radiometric surveys, sampling water and soil sources for uranium or radon, gamma ray logging of drilled holes, and color photography of the terrain from the air. Open pit mining may be carried out down to depths of 100 m. The "western states" (Wyoming, Colorado, New Mexico) are the primary sources of uranium in the United States. The deposits in these regions range from 1 to 30 m in thickness. Over 10^7 m of drilling per year, at a

TABLE 6

Nuclear Fuel Costs 1965–1985 [33]

Factors (per unit uranium)	Range			Cost (mils/kWh)[a]	
	1965–1975	1974–1975	1984–1985 (estimated)	1975	1985
Ore cost ($/lb)	6–8	8–15[b]	15–25[a]	0.94	1.43
Conversion ($/lb)	1.0–1.25	1.25–1.40	1.40–1.80	0.09	0.10
Separative work ($/SWU)	28–38	38–43	50–100	0.58	1.24
Fuel fabrication ($/lb)	28–41	35–41	41–53	0.33	0.40
Shipping, reprocessing, and waste disposal ($/kg)[c]	20–40	40–80	100–150	0.15	0.28
Interest charges	12–15	12–15	15–18	0.61	1.23
Total				2.70	4.68
Coal (¢/10⁶ Btu)[d]	30–45	45–150	150–220	14.10	20.70
Incremental cost advantage, nuclear over coal				11.4	16.0

[a] Using highest value in range for each time period; all figures in 1975 dollars.

[b] Spot purchases over $35/lb have been recorded, but on a very thin market.

[c] Or cost of long-term fuel storage and safeguards without reprocessing and neglecting present worth of uranium and plutonium credits.

[d] Assuming a heat rate of 9000 Btu/kWh for coal versus 10,000 Btu/kWh for nuclear. Considerably higher spot-purchase prices have been recorded for both uranium and coal, but on thin markets. Oil prices are 30–50% higher than coal.

cost of about $4.00 per m (1974), is carried out in the United States. About 0.5 kg of U_3O_8 is discovered per meter drilled. An investment of about $30,000 has to be made per ton of new annual U_3O_8 production capability. It takes about 8 yr from exploration to production.

The major commercial uranium ores in the United States are uranite (pitchblende, uranium oxide) and coffinite (hydrous uranium silicate). Other sources of uranium ore are carnotite (U–vanadate), phosphate rock, and gold ores. Igneous rocks and shales contain a few parts per million, and sea water contains 3.34 parts per billion of uranium. The average uranium content of commercial ore is approximately 0.22% U_3O_8. The ground ore may be acid leached (sulfuric, nitric, or hydrochloric), or carbonate leached (sodium carbonate or bicarbonate). The uranium is precipitated from the purified solution by the addition of ammonia, H_2O_2, NaOH, or MgO. At this stage, the purity of the U_3O_8 is approximately 75–99%. The cost of uranium is $20/kg U when the commercial oxide (yellowcake) U_3O_8 is priced at $8/lb.

The total cost of raw materials investment for uranium in the United States is estimated will be about $10 billion by 1990.

Thorium is obtained from monazite sands (rare earth phosphates) which occur in large quantities in beach sands in India, Sri Lanka, Brazil, and Australia. It is also found in the United States (the Carolinas, Idaho, Southern California, Montana). The thorium oxide is dissolved and purified by solvent extraction.

The production of uranium hexafluoride (UF_6) for the enrichment plants is carried out in conversion plants. The cost of this process is approximately 4% of the fuel cycle cost. There are two commercial processes, namely:
1. the refining–fluoridation process (Kerr-McGee);
2. the dry fluoride volatility process (Allied Chemical Corporation).

The refining–fluoridation process consists of solvent extraction of uranium from a nitrate solution, which is washed with water to remove impurities. The uranium is then re-extracted into dilute nitric acid solution (0.01 N–HNO_3), and the uranium oxide formed is reduced with hydrogen to UO_2 which is converted to UF_4 (green salt) by reaction with HF gas, and then to UF_6 with fluorine gas. The dry fluoride process involves fluid bed reduction, hydrofluorination and fluorination of UO_2. The UF_6 is then double-distilled to produce the pure product. The enriched UF_6 is reacted with aqueous ammonia to yield ammonium diuranate (ADU), which is heated in an atmosphere of steam and hydrogen to yield UO_2.

The enrichment process involves the diffusion of UF_6 vapor through a series of porous membrane barriers. Since the maximum theoretical separation per stage is governed by the ratio of the masses of gas molecules in the UF_6, namely, 1.00429, a large number of stages extending several miles are required. For example, in order to attain an enrichment of 4% ^{235}U, a cascade of 1500 stages is required. At each stage the gas which diffuses from the tube through the barrier is fed to the next higher stage, and the remaining portion, about 50%, is recycled to the lower stage. The estimated demand for enrichment by 1980 is about 3.5×10^7 kg of separative work units (SWU).[2]

The separative work in enrichment entails about one-third of an average fuel cycle cost. The separative work unit is a measure of the work required to carry out the separation of feed into tails. For example, the production of 1 kg of 3% ^{235}U requires 4.306 units of separative work and uses 5.479 kg of uranium feed material (0.71% ^{235}U to yield tails of 4.479 kg having a ^{235}U content of 0.2%). The separative work costs are made up of power cost

---

[2] SWU is a measure of the expense of isotope separation (costs of pump capacity, power demand, total barrier area, etc.).

(49%), capital cost (35%), and operating, research, and development (16%). The three United States diffusion plants require 6000 MW(e) power capacity and consume 4.5×10^{10} kWh of electric power. At full capacity these plants have a capacity of 1.72×10^7 SWU per year. New enriching facilities will be required by 1983 to meet the increasing demand for enriched fuel.

Other enriching techniques include the centrifuge processes and laser separation. Both these methods are under intensive study and appear to have a potential for lower power requirements and capital costs and higher yields.

The characteristics of LWR-grade plutonium generated from three cycles of operation in a large reactor are as follows: the isotopic composition is 1% ^{238}Pu, 58% ^{239}Pu, 23% ^{240}Pu, 13% ^{241}Pu, and 6% ^{242}Pu. The isotopes ^{239}Pu and ^{241}Pu are fissile. The major sources of alpha radiation are ^{239}Pu, ^{240}Pu, and ^{241}Pu. The gamma-emitters are ^{241}Pu and the daughter products of ^{241}Pu (13-yr half-life), namely ^{241}Am and ^{237}U. Also, neutrons are emitted by spontaneous fission from ^{238}Pu, ^{240}Pu, and ^{242}Pu.

C. Reprocessing of Fuel

The reprocessing of spent fuel serves to reduce fuel cycle costs (by about 12% in LWRs) and diminishes the fuel requirements. The basic process used for LWR fuels is the solvent extraction process. The fuel pins are first disassembled (about 4 months after removal from the reactor core) in a chop-leach step to remove the fuel from the clad. The fuel is dissolved in nitric acid and the solution is then subjected to solvent extraction (Purex process) to strip first the Pu and then the U from the solvent. Following purification cycles by means of subsequent solvent extractions [tributyl phosphate (TBP) in kerosene], the Pu is recovered as the nitrate in aqueous solution and the U as UO_2 or nitrate in dilute nitric acid solution. The fission products in the waste solutions are stored for several years in cooled tanks to remove much of the decay heat, and are then solidified. About 100–300 gal. of fission product waste solutions are generated per ton of U fuel.

The four main solid fission product isotopes from spent reactor fuels are Sr, Cs, Ce, and Pm. By the year 1980, the total quantity of these fission products developed is estimated to be about 4×10^9 Ci. It has also been proposed that the elements Ru, Rh, Pd, Xe, Kr, and tritium may be produced economically as by-product isotopes from fission products. Fission product yields are shown in Figs. 4 and 5.

A 1000 MW(e) LWR generates approximately 200 kg of Pu annually, worth about \$1.5 million. By the year 1980 about 100 tons of Pu will be available for recycling. The fabrication of recycle Pu poses problems of shielding arising from gamma radiation from ^{241}Pu and the decay daughters

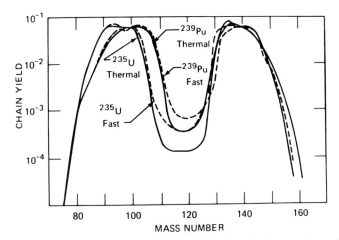

FIG. 4. Chain yields as a function of mass number of the chain for fast- and thermal-neutron flux spectra and for ^{235}U and ^{239}Pu [110, p. 173].

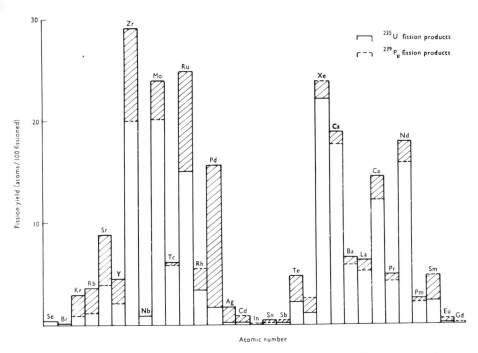

FIG. 5. Yield of major fission products for fast fission of ^{235}U and ^{239}Pu [143, p. 51].

^{237}U and ^{241}Am, as well as neutrons from the spontaneous fission of ^{238}Pu, ^{240}Pu, and ^{242}Pu. The licensing requirements for Pu fuel fabrication are under review by the National Regulatory Commission and no such fuel may be fabricated or shipped until the review is completed, perhaps by 1978.

The operational experience with mixed oxide (U, Pu)O$_2$ fuel in several thermal reactors (PRTR, Saxton, San Onofre) has been quite successful, and high burnups have been achieved (e.g., 48 MWd/kg at 19.5 kW/ft in Saxton). The only commercial LWR fuel reprocessing plants that will be operational by 1980 are the Allied/General plant (1500 tons U per year capacity) and the Nuclear Fuel Services plant (750 tons per year). A plant with 1700 tons fuel per year reprocessing capacity will save the equivalent of one million barrels of oil, 250,000 tons of coal, or 4,000 tons of uranium per day. It will provide the needs of 39 LWR power plants. The United States is now reassessing reprocessing.

D. Transportation, Safeguards, and Waste Disposal

The spent fuel from LWRs contains approximately 50% of fissile material that can be used for the reload batch. The fission product activity from a 3000 MW(t) core after 1 yr decay is approximately 3×10^8 Ci. Decay heat of the discharged fuel is lowered by storing at the reactor site for a period of 3–4 months before shipment. The amount of fission product activity shortly after shutdown is about 10 Ci per thermal watt of power. The shipping cask for LWR spent fuel consists of an annular stainless steel shell with depleted uranium or lead in between for shielding. The decay heat is removed by means of cooling fins. The spontaneous fast neutrons from the ^{242}Cm and ^{244}Cm are shielded with a neutron moderator several centimeters thick. The cask may weigh up to 100 tons, but highway transportation is limited to under 20 tons (about 0.5 tons U fuel capacity). Hence, railroad transportation is required.

The safeguards on the fuel cycle are based on four methods:

1. accounting for the materials balance on a continuing basis by means of a computer program, NMIS (Nuclear Materials Information System);
2. surveillance;
3. nondestructive assay by remote detection systems, e.g., detecting the gamma radiation from ^{239}Pu;
4. physical protection.

The assessment of the safety of nuclear reactors has been based on design basis accident (DBA) considerations. This approach identifies the events that can lead to the release of radioactivity and harm to people and property. The reactor designs incorporate safeguards against the worst physically possible chain of events. It has been estimated that almost half the effort in the design and operation of nuclear reactors is related to safety features. Recent analysis [15] of the probabilities of all events that can lead to release of radioactivity show that accidents with an LWR might kill ten people once in 350,000 yr.

The principal fission products and other radionuclides in the reactor effluents for waste disposal are: ^{3}H, ^{58}Co, ^{60}Co, ^{85}Kr, ^{89}Sr, ^{131}I, ^{133}Xe, ^{134}Cs, ^{137}Cs, and ^{140}Ba. By the year 2000 A.D., with 1000 gW installed capacity, about 19,000 tons of spent fuel would be reprocessed per year. The volume of liquid waste would be 5.8 million gal/yr and 77 million gal accumulated through the year 2000 A.D. When solidified, the waste volume would be about 580,000 ft^3/yr and 770,000 ft^3 accumulated by 2000 A.D.

III. NOTES ON URANIUM AND THORIUM RESOURCES [52–71]

More certain knowledge of uranium and thorium resources is badly needed to improve the bases of decisions and plans for the development and use of nuclear reactors. Data from estimates published by ERDA [57, 58], (Tables 7-9); by Searle [59, 60]; and by Brinck [61] show the spread in interpretation of the results of exploration and of geochemical knowledge concerning uranium. The AEC data on identified reserves are regarded by a few experts as optimistic, but appear to be accepted by a majority of those who

TABLE 7

U.S. Uranium Resources as of January 1, 1975 (Tons of U_3O_8)

Production cost (kg)	Proven reserves	Potential reserves			Total
		Probable	Possible	Speculative	
$17	200,000	300,000	200,000	30,000	730,000
$17–22 increment	115,000	160,000	190,000	80,000	545,000
$22	315,000	460,000	390,000	110,000	1,275,000
$22–33 increment	105,000	220,000	250,000	100,000	675,000
$33	420,000	680,000	640,000	210,000	1,950,000
$33–66 increment	180,000	460,000	700,000	200,000	1,540,000
$66	600,000	1,140,000	1,340,000	410,000	3,490,000
By-products[a] 1975–2000	90,000	—	—	—	90,000
Grand total	690,000	1,140,000	1,340,000	410,000	3,580,000

Source: Energy Research and Development Administration.
Note: Numbers may not add to totals because of rounding.
  [a] By-product of phosphate and copper production.

TABLE 8

Foreign Resources of Uranium (Tons of U_3O_8)

	Reasonably assured	Estimated additional	Total
$10/lb. U_3O_8			
Australia	340,000	50,000	390,000
Canada	240,000	250,000	490,000
South Africa and South West Africa	260,000	10,000	270,000
France, Niger, Gabon	140,000	70,000	210,000
Other[a]	60,000	60,000	120,000
$10 subtotal	1,040,000	440,000	1,480,000
$10–15/lb. U_3O_8			
Australia	80,000	50,000	130,000
Canada	160,000	280,000	440,000
South Africa	80,000	30,000	110,000
Sweden	350,000	50,000	400,000
France, Niger, Gabon	40,000	50,000	90,000
Other[a]	30,000	60,000	90,000
$10–15 subtotal	740,000	520,000	1,260,000
Less than $15 subtotal	1,780,000	960,000	2,740,000
$15–30/lb. U_3O_8			
Canada	100,000	300,000	400,000
South Africa	55,000	70,000	125,000
Sweden	150,000	200,000	350,000
Spain	15,000	250,000	265,000
Other[a]	70,000	140,000	210,000
$15–30 subtotal	390,000	900,000	1,350,000
Grand total	2,200,000	1,900,000	4,100,000

Source: Energy Research and Development Administration.
Note: Numbers may not add to totals because of rounding.
    [a] Argentina, Brazil, Denmark, Finland, India, Italy, Japan, Mexico, Portugal, Spain, Turkey, Yugoslavia, Zaire.

have examined the matter carefully. Granting the availability of these reserves, a great expansion of the uranium mining and milling industry will be required to supply the projected demand for uranium. At the same time, the reserves are scarcely adequate for the projected deployment of light water reactors.

On the other hand, because it is based on reasonable analogies with other resources, Brinck's [61] statistical analysis of the world-wide occurrence of uranium deposits must be taken into account. It suggests that exploration will eventually locate very large amounts of reasonably priced uranium. As he and others have shown for each of many ores, the proved

TABLE 9

Projected U.S. Nuclear Fuel Requirements[a] April 1, 1975

Reactor status	Number of reactors	Capacity [MW(e)]	Uranium oxide requirements to 2000 A.D. (tons)	"Lifetime"[b] U_3O_8 requirements (tons)
Operable	55	36,600	150,000	160,000
Being built	76	76,900	350,000	440,000
Planned	105	121,400	440,000	640,000
Total	236	234,900	940,000	1,240,000

Source: Energy Research and Development Administration.

[a] Assuming: Enrichment tails assay; 0.030% ^{235}U; no plutonium recycle; 30-yr reactor life.

[b] Period covered: 1975–2018 (43 yr).

distributions of deposits of certain sizes (i.e., mass of ore body) and grade (i.e., concentration of metal) may be very well represented by a log-normal distribution function.[3] Upon fitting this distribution function to known uranium deposits and calculating the costs of mining and refining ore bodies of various mass and grade, Brinck concludes that the world supply of uranium at reasonable prices is orders of magnitude greater than the identified deposits. While statistics do not predict, it is not unreasonable to postulate that when all resources are explored experience with uranium will prove to be similar to that which is well established for other minerals of comparable abundance.

On the basis of well-established data and a very modest extrapolation of the known statistics of uranium deposits in the United States and elsewhere, Searle [60] suggests that the uranium that will become available at costs less than $100/lb is considerably greater than the reserves identified by the AEC and may be 13×10^6 tons.

---

[3] As is customary, normal refers to the Gaussian probability distribution function, $A \exp[(x - x_0)^2/a^2]$, where A is the constant that normalizes the function, a related to the width of the distribution, and x_0 its center. In this case, the argument of the function is the logarithm of the concentration. Such a probability distribution may apply to situations wherein the final distribution is the result of many events, the probabilities of each of which are multiplied, that is to say, may be combined using *and* logic. Brinck has given examples of such processes for purposes of illustration. There are many additional possibilities going back to the origin of the universe. Because these events may be widely separated in time and space, multimodal distributions seem likely.

After reviewing Searle's study, Brinck's calculations, and the other information, Klein *et al.* [58] arrived at conclusions as follows. Up to the price level of $30/lb their estimates are consistent with other AEC estimates and not inconsistent with those by Searle. The experts involved in this critique of the statistical model offered by Brinck are not identified. While they are probably correct in rejecting certain assumptions that Brinck used to show how the log-normal distributions might arise, their arguments are outside the question of its applicability.

The uncertainties and controversy are associated largely with the question whether ores in the range of concentration recoverable at prices between $30 and $100/lb will be found in nature as statistically indicated. A major unknown appears to attend the number, extent, concentration, and types of deposits at varying depths below the surface of the earth and perhaps in localized areas [62], a matter deserving the attention of materials scientists and engineers. There is wide agreement that the amount of uranium potentially available from shales (average concentration approximately 60 ppm) and some granites (concentration approximately 15–50 ppm) is very large and represents a virtually inexhaustible supply of energy. There are contrary opinions on the environmental feasibility of mining such extensive masses of rock even though their net energy content is comparable to or larger than that of coal and oil shale. Sea water, which contains 3.34 parts per billion has also been considered an extremely large potential source that could be tapped if necessary [63, 64], in spite of the very difficult task of handling over 3×10^{10} tons of sea water per ton of uranium. At issue here is the energy consumed in processing raw material compared with that produced by its use and the rate of growth of the demand.

Because of relatively low interest and demand, thorium resources are not thoroughly identified. The U.S. Geological Survey [65] has assembled much of the information on which most statements are based. From such information it seems reasonable to conclude that thorium resources are at least as great as uranium resources, but differently distributed.

Wilkinson [69a] and Bellamy and Hill [69b] have provided concise summaries of the practical mineralogy and processing of uranium and thorium ores. Contributions to symposia [69c] have updated these subjects. Perhaps the following observations will illustrate the geochemical issues.

The abundances of uranium and thorium are 1.7–4 ppm and 5.8–13 ppm, respectively. Because of their charge and comparatively large ionic radii, 0.093 and 0.099 nm, respectively, tetravalent uranium segregates during the later stages of magma solidification, the more abundant and larger Th^{4+} entering isomorphous solid phases earlier than U^{4+}. Important primary minerals are solid solutions of thoria and urania, called thorianite or uranite depending on the dominant species, and of silicates, called thorite, in which the uranium–thorium ratio may range from very small to 1. The similar ions Ce^{4+} (0.089 nm), Zr^{4+} (0.074 nm), Ce^{3+} (0.11 nm) and other lanthanides may be present in these primary minerals. The youngest igneous rocks, such as pegmatites, frequently contain 10–100 ppm U and sometimes more.

Secondary minerals are the important known sources of both uranium and thorium because magmatic minerals are for the most part localized, not surface, extensive, rather difficult to find, and not concentrated. Further, because the oxidation states of uranium differ from those of thorium, the secondary mineralization of each is distinctive.

Oxygen at ambient chemical potential, prevailing since Precambrian times (about 2.3×10^9 yr ago), readily oxidizes U^{4+} to U^{5+} and U^{6+}. The fluorite structure of UO_2 and UO_2–ThO_2 solid solutions accommodates excess oxygen ion defects and the higher valent cation species. If U predominates, new phases form and changes in dimensions cause fragmentation of the crystals, facilitating oxidation. Crystallites having excess thorium may remain intact, housing U^{6+} and excess O^{2-}. Hydrothermal dissolution and surface water leaching can put U^{6+} in aqueous solution as the soluble uranyl ion, $UO_2{}^{2+}$ with naturally occurring anions and complexing ions such as carbonate, sulfate, and lignitic substances. Secondary minerals are formed by several concentration processes: principally, reduction to relatively insoluble U^{4+} by any of several naturally occurring reducing agents such as hydrocarbons, H_2S, Fe^{2+}, and the anaerobic living and dead organic matter at the bottom of bodies of water; secondly, adsorption and exchange in permeable carbonates, in phosphates of trivalent lanthanides wherein divalent ions compensate charge; intercalation in graphitoid solids like coal and complexing adsorption in lignite, formation of complex vanadates such as carnotite, and many others. Exchange reactions to give monazite type ores would appear not to require the oxidation–reduction steps or aqueous reactions. The important supply areas in the United States such as the Colorado plateau were formed by aqueous transport and concentration into porous sandstone by reduction. Many geologists believe such superficially well-explored areas to be unique and the only practical sources in the United States of ores of reasonable grade ($> 0.1\%$ U); thus, they have concluded that United States supplies are limited to those stated by the AEC, unless one is willing to process the Chattanooga shales and Conway granites. Some others believe that many additional primary and secondary ore deposits will be found as more widespread exploration proceeds to lower depths.

In any event, the known uranium and thorium provinces are clearly the result of erosion of ancient granitic shields and granitic and other intrusions, aqueous transport of soluble or particulate matter, and the many processes of concentration and segregation.

It would be most interesting to know more about the distribution of uranium prior to the oxygenation of the atmosphere. Researchers in France [52] have proved the occurrence of an extensive natural fission chain reaction in Gabon, Africa, around the Precambrian period when natural uranium contained about 2.3% ^{235}U. Going back to the time of the origin of the earth one finds that the amounts of ^{238}U and ^{235}U were about equal.

In contrast to uranium, which occurs in over 70 minerals,[4] thorium occurs primarily in six [69a]. The relative insolubility of the only naturally occurring state of oxidation, Th^{4+}, (that is to say, the stability of crystals containing Th^{4+}) limits the geochemical processes underlying ore formation of those in magma, hydrothermal ion exchange, erosion and classification of particulate matter during transport by flowing water. The last process is the primary reason for dense monazite sands constituting the major practicable sources and the occurrence of thorium in placer type deposits. However, for the North American continent, thorium occurring in, for example, uranothorianite in Canada and thorite in Idaho may, for a time at least, prove to be important sources.

With respect to the questions concerning the size of resources and reserves, we may conclude that the tasks ahead include the following: establishing what statistical distribution functions correctly represent the natural distribution of mass and concentration of accessible deposits of uranium; resolving Searle's probability of 0.5 for the availability of 13×10^6 tons of U_3O_8 at \$100/lb; and establishing commercial sources of thorium and learning all of the deposits accessible to each interested nation. McElvey and also Erickson [69] have put the problem of predicting ultimate mineral reserves in excellent perspective, by showing that reserves for many minerals such as lead have proved, after extensive experience, to be about $2–3 \times 10^2$ tons times the crustal abundance.

Two recent assessments of uranium supply in the United States address the near-term and long-term requirements and availability of uranium [70,71]. The report of the Edison Electric Institute, as summarized by Hogerton [70], points out that U.S. utility requirements for uranium will increase from the 10,000 tons (9.07×10^6 kg) U_3O_8 produced in 1975 to 30,000 tons (27.2×10^6 kg) per year by 1980 and to approximately 70,000 tons (63×10^6 kg) per year by 1990. The conclusion is that beyond the mid-1980s new uranium reserves will require accelerated development at an average rate of 60,000 tons (54.43×10^6 kg) U_3O_8 per year, which is twice the highest rate achieved in the past (in the 1950s). Hogerton states that "there continues to be a severe imbalance between the utility procurement appetite and the supply industry's contracting capability and/or readiness . . . . The lifetime

---

[4] These may be further differentiated; Wilkinson [69a, pp. 708–716] lists 151.

uranium requirements of the existing universe of United States reactors, including those for which commitments have been announced to date, amount to something under 1 million tons (907.2 × 10^6 kg) U_3O_8. By 1980 this lifetime requirements figure will have increased to about 1.4 million tons (1.27 × 10^9 kg) . . . . We have 45 years in which to discover and develop the "missing" ore. It does not seem unlikely that the ore can and will be found in the next 45 years."

A more pessimistic assessment has been made recently by Lieberman [71], who predicts that a critical shortage of uranium will occur in the United States by the late 1980s. He recommends a crash program to conserve energy, develop processes for low-grade uranium resources (e.g., shale, sea water), and by implication, to speed up the introduction of the fast breeder reactors.

IV. SOME PRINCIPLES OF NUCLEAR MATERIALS SCIENCE [72–110]

There has been increasing activity and reasonable progress in observing and understanding the behavior of materials suffering fast neutron damage in the fluence range of up to 1 × 10^{23} neutrons/cm^2, fission recoils and the accumulation and partial escape of 20–30% fission product, and large thermal gradients. These gradually maturing topics are perhaps paramount among those that extend kinetic and irreversible thermodynamic theories of matter and provide bases for the design of energy systems.

A. Irradiation Effects

Nuclear reactions yield several kinds of energetic particles that, through multifarious interactions with nuclei, atoms, electrons, and ions in a solid, provide both forces and mechanisms for change in materials systems. This field of research has been developed rapidly and extensively since 1942 when Wigner predicted and Neubert observed significant effects induced by fast neutrons from fission in the graphite used in the plutonium production reactors built and operated at the Hanford Engineering Works, Richland, Washington. As a result, thorough treatments of the subject have been published that provide both background and detail for our summary [72–109].

While we shall discuss most of these effects of irradiation under the behavior and performance of fuels, structural materials, and moderators, a brief summary of some definitions of concepts, phenomena, and models will support later discussions.

Irradiation has become a general term for the situation in which quanta or energetic particles interact with matter to produce changes in composition, structure, and properties.

Radiation damage has also become a widely accepted general term for the changes occurring in a solid when it is irradiated.

Particles. The input of energy and momentum into an irradiated material may be represented by a flux of electromagnetic quanta, electrons, fast neutrons, beams of protons and other accelerated ions, and of neutralized ions and recoil ions and atoms from nuclear reactions such as capture followed by fission and/or emission of gamma-rays or beta particles within the material.

Interaction potentials that are some reasonable function of the distance between colliding particles are used to give the fields, forces, and changes in momenta and energies of each particle in a collision. Quanta, electrons, protons, alpha particles, and atomic ions, the last at distances greater than their diameter, are accurately represented by electromagnetic fields, the Coulomb potential sufficing. A "hard sphere" potential function between neutrons and nuclei often suffices for elastic collision energies not exceeding the excited states of the nucleus. Ion–ion, ion–atom and atom–atom potentials may, in principle, be calculated from the fields between all interacting particles on quantum mechanical principles.

However, few dynamic calculations have been done with accurate potentials. Commonly, the repulsive potential between atoms, or ions and atoms, is represented by $V(r) = A \exp(-ar)$, first extensively used by Born and Mayer and bearing their names, where A and a are empirical constants that, however, may be estimated from the elastic constants of the solid. At shorter distances, that is, for close high energy collisions, the interaction involving both nuclei and electrons is sometimes represented by the screened Coulomb or Bohr potential $V(r) = (A/r) \exp(-ar)$, where A is the product of the charges on each nucleus and, again, a is an empirical constant that fits elastic constants. In order to represent interaction strengths at distances less than and equal to ionic radii, Brinkman proposed the function [88, p. 12]

$$V(r) = (BZ_1 Z_2 e^2) \exp(-br)(1 - \exp Br)^{-1}$$

where B and b are empirical constants that may be estimated from macroscopic properties of the crystal. This function interpolates between the Coulomb repulsion of the nuclei having charges $Z_1 e$ and $Z_2 e$, respectively and the electron cloud repulsions, reasonably represented by the Born–Mayer potential. Nelson [88] compares these functions with graphical representations of the interaction of copper atoms calculated by quantum mechanics based on the Thomas, Fermi, Dirac statistical representation of electrons in atoms.

Collisions between particles having an atomic mass of 1 or greater may be calculated using Newtonian mechanics, if an interaction potential is assumed. Three kinds of physical approximations must be kept in mind: pairwise interactions, a simplified analytical potential function and, usually, calculation only of initial and final states, not complete trajectories.

In near head-on collisions the hard sphere potential may represent the interaction of atoms reasonably well. Another physical approximation, the impulse approximation, serves to simplify calculations of higher energy collisions wherein the distance of closest approach is equal to or greater than the sum of the radii of the particles. The impulse approximation assumes that the struck atom does not move during the collision. Since the target area of these more distant collisions is larger than for close collisions, there are many more of them. The strength of the interaction at these larger distances is an important issue affecting the spacing between damage or displacements.

When all paths of the initial or incident energetic particles are considered, a spectrum of energies of struck or knock-on particles results. Also the primary or incident particles may have an energy spectrum. Primary, secondary, and successive *n*-ary collisions may be summed to give an energy spectrum of all *n*-ary particles participating in collisions resulting from a given incident particle. If the number, energies, and momenta of the primary particles can be described by spectra, summing or integrating over each spectrum can yield the energy spectrum for all participating particles. Such calculations are most conveniently done with numerical computer codes and are well systematized.

Displacements occur when a struck or knock-on particle is given suitable energy and momentum (direction). In the simplest model for recognizing the many-body mechanics of collisions in solids, one assumes a displacement energy E_d such that knock-ons receiving less energy do not leave their normal site in the solid. Atoms receiving more energy than E_d may move from their initial location and participate in a new state of the solid. As Sosin and Bauer [86] and also Nelson [88] show, neither this energy nor the probability and mode of displacement are simple quantities since the direction of displacement relative to neighboring atoms and their motion affects the probability and mode of displacement. For purposes of illustration, some observed displacement energies are tabulated in Table 10.

The data in the table are not inconsistent with the qualitative concepts that displacement energies of atoms are in the neighborhood of four times their sublimation energies, that heavier atoms have higher displacement energies, and that at lower temperatures longer lifetime for some of the more easily produced crystal defects is allowed.

Modes of motion, crystal lattice effects, and mechanisms of displacement

TABLE 10

Threshold Displacement Energies[a]

Material	Atom displaced	Temperature of irradiation (°K)	Threshold energy of atoms (eV)
Graphite	C	300	25–28
Diamond	C	300	80
Silicon	Si	260	14, 21
Aluminum	Al	4	13.5–19.5
Iron	Fe	< 25	17 (estimated)
Nickel	Ni	20	24
Copper	Cu	4	19
		80	24
Molybdenum	Mo	20	35–37
Tungsten	W	20	> 35
Gold	Au	4	33–36

[a] The minimum energy of a beam of electrons causing observable change, usually of electrical resistivity, is determined. From the electron energy the maximum energy of struck atoms or ions is calculated. The data are taken from A. Sosin and W. Bauer [86, Vol. 3, pp. 200–201].

have been inferred from experimental measurement of changes of properties caused by irradiation. Several simple mechanical models have been invented and analyzed. The important mechanisms have been demonstrated by the computer programmed calculations of the motions of assemblies of the order of 1,000 atoms and have been rather well confirmed by experimental observations of recoil atoms in and from crystals. These key motions and configurations of atoms in a crystal may be used for synthesizing a general picture of irradiation effects (Fig. 6).

The creation of interstitial–vacancy pairs is the first and simplest concept. The struck atom is considered to move between its neighbors to an interstitial position sufficiently far away that the stress field around it cannot cause intervening atoms to rearrange and fill the vacancy. Of course, in order for its effects to be observed the displaced atom must not migrate to a vacancy or trap during the time required for experimental measurement. This simple process of pair formation must be supplemented by others and by additional details.

Collision cascade is the term used to denote the events following the collision in which a fast neutron or other energetic particle gives an atom or ion in a solid enough energy to cause secondary, tertiary, etc., collisions and displacements in the course of dissipating the energy of the primary and all *n*-ary particles.

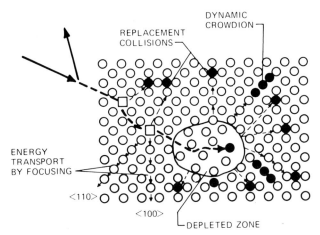

FIG. 6. A schematic two-dimensional picture of neutron-induced irradiation damage in a face-centered, cubic metal [110, p. 405].

Multiple displacements is a concept related to the cascade and conveys the idea that some collisions may be closely spaced, thus creating an aggregation of defects. Brinkman [88] has shown that the interaction potential between atoms or ions having a mass number greater than about 50 is such that collisions having energy between 100 and 5×10^4 eV can be closely spaced and that clusters of two or more vacancies surrounded by the displaced interstitials may occur. Such a configuration of several defects is unstable. Its collapse is probably accompanied by a rather large amplitude motion of all the atoms in the region and the formation of dislocation loops of both signs.

Thermal spikes provide another description of the consequences of the large multiple or closely spaced displacements mentioned above. The high amplitude motion of atoms of ions associated with the displacements might be described by a temperature and the transport of energy away from the region regarded as transport of heat. If sufficiently large numbers of atoms are involved or if an average is taken over a large number of events, the concept of the temperature is reasonable. This concept introduced by Brooks is well presented by Chadderton [76]. Further, the concept of a temperature or thermal spike is very useful in interpreting the effects of energy given to the electrons in a solid by energetic charged particles, particularly fission products. Over much of its range a recoil fission product is highly ionized. It loses most of its energy by Coulombic interactions with surrounding electrons of the solid. Electronic motion may transport heat more rapidly than atomic motion and spread the energy over distances of the order of 10^{-8} m surrounding the track of the particle.

If the distance for equilibration of electrons and ions is shorter than approximately 10^{-8} m, as it will be in nonconductors or highly defected solids, the region may be thought of as being heated to a high temperature. If electrons dissipate the heat over larger distances, as they probably do in a crystalline metal, probably few tracks will reach an extremely high temperature. Fission recoils traveling parallel to the surface of metallic films of thickness of the order of 10^{-8} m show tracks due to evaporated atoms. Thicker metallic films show no evaporation, only dislocation loops and vacancy clusters when observed by high resolution electron microscopy including diffraction, demonstrating that displacement mechanisms were predominant. Fission recoils in uranium dioxide heat local regions to very high temperatures, particularly near interfaces and other strong scatterers of electrons, causing sputtering, sintering, transport, bubble migration, resolution of gases, and related effects.

Collision sequences and focusing are names for characteristic motions of struck atoms having a kinetic energy of 50–500 eV in close-packed crystals, predicted first by Silsbee and others [89a]. A symmetrically placed ringlike group of atoms may act like a lens focusing atoms passing between them at a small angle with respect to a normal to the plane of the focusing group. The directions in a crystal along which focusing occurs are those of lines of atoms or ions. Thus, the struck atom immediately collides with its neighbor in the line and each successive motion approaches the direction of the line. If the lines of atoms in the direction in question are closely packed, the motion tends to assume a form of a compression wave sometimes called dynamic crowdion. If before dissipation the wave comes to a discontinuity in the line of atoms such as a stacking fault or twin or other boundary, the final atom may become an interstitial. Along lines of less closely spaced atoms, the first struck atom may displace the second and so on down the line. This mode is called a replacement collision. If the crystal is reasonably perfect and the amplitude of the vibration is not too great, as, for example, in an fcc crystal at or below room temperature, either mode may propagate many atom distances losing approximately 0.3 eV per atom distance. Collisions having energies above a few hundred volts are not focused in this manner because the effective collision diameter is appreciably less than the atom spacing.

Channeling occurs when an ion having a kinetic energy of many million electron volts travels thousands of atom distances without deflection along a channel usually parallel to directions of closest atomic packing in regions of perfect crystallinity. At these energies the effective collision diameter is small. Although of little importance in irradiation effects, the phenomenon is useful in determining crystalline disorder and in ion implantation.

Reactions of lattice defects provide the concepts and terms used to describe the processes that follow the dynamic events named above and by which changes in solids take place. Thus, interstitials and vacancies may move and combine with one another and with other irregularities in the crystal, e.g., minority components, loops, dislocations, stacking faults, twin boundaries, voids, and grain boundaries. Each of these structural features changes in turn in ways that may be described as reactions.

Vacancy motion and reaction have been inferred from the studies of diffusion, quenching, irradiation effects, the behavior of dislocations and, in ionic crystals, spectroscopy as well as from theory. For many solids the energies of formation and thermally activated migration are known [75]. The structure of vacancies is reasonably well known; the main questions concern the modes of motion of the immediately surrounding atoms. Sherby and Simnad [81] and others have shown that there are useful correlations between the thermodynamic functions for the activated state of motion of vacancies (as a mechanism of diffusion) and those for melting of the solid. Also, a number of authors has described the dynamics of vacancy motion in terms of the normal modes of the crystal.

The combination of vacancies with dislocations to cause climb is familiar. Clusters of vacancies due to multiple displacements may be partially resolved by electron microscopy. Aggregation on crystallographic planes through thermally activated motion of vacancies yields dislocation loops, which then may react according to the rules for dislocations influenced by line and fault energies and by interaction with constituents of the solid. The aggregation of vacancies into voids was unexpectedly discovered by Cawthorne and Fulton [83] in stainless steel cladding for FBR irradiated at $600–900°K$ at fluences above 10^{21} neutrons/cm^2. The factors contributing to the nucleation of this configuration having higher surface energy than loops are supersaturation of the vacancy concentration, presence of helium atoms, and stress. Stresses may arise from external forces, thermal gradients, dislocation climb, and the growth of loops. Once nucleated, growth by diffusion of vacancies to the void is easily understood. The swelling of the solid accompanying the formation of voids is, of course, due to the displaced interstitial atoms which add internal or superficial planes of atoms to the crystal. Interstitialcy behavior may be more important in swelling than that of vacancies.

Interstitialcy reactions have been inferred almost entirely from studies of irradiation effects and from calculations. Their energies of formation are too high to permit observable concentrations in most solids in thermal equilibrium. The structure of the interstitialcy is less clear than that of the vacancy. In graphite the concept of an atom located between slightly distorted graphitic planes is satisfactory. In fcc solids it is more likely that two symmetrically displaced atoms share a lattice site. The known activation energies for motion through a crystal lattice are considerably less than for migration of vacancies. The temperature range for thermally activated migration in copper is less than one-fifth the absolute temperature of melting. The aggregation of interstitials onto crystallographic planes to form loops readily occurs, as already stated. Close-packed planes having strongest bonding between atoms are preferred as is clearly indicated by the direction of growth of irradiated solids and the crystallography of the loops observed by diffractive microscopy. Interstitials also participate in dislocation climb. To us it seems likely that dilational stresses induced in specific regions by interstitial loop formation and dislocation climb can play a role in the nucleation of voids. The dynamical motion of interstitialcies and their interaction with solutes and other extrinsic defects play a role in determining the location and distribution of interstitialcy loops.

Irradiation creep probably results from the reactions listed above. Stress introduces crystalline anisotropy that can influence the planes for formation of loops and the capture of point defects by dislocations and, thus, the directions for climb and dilation as well as the direction of glide of dislocation segments.

Irradiation hardening is an obvious result of the tangles and other configurations of dislocations that result from the reactions of defects. Other kinds of crystalline disorder such as vacancy clusters, disordered alloys, redistribution of minor constituents, alteration of dispersed phases and possibly glassy amorphous regions may also play roles in altering the yield stress of solids.

Dislocation channeling is observed in several irradiated solids. We can speculate that dislocation glide through a region of tessellated stresses and strains resulting from irradiation-induced climb, and formation of loops and climb of dislocations can permit rearrangement and elimination of these features. However, other than the usual concepts about initiation of glide, what determines the region of the solid in which channels occur is less clear.

Fission recoils and fission product atoms or ions are a major cause of change in fuel materials. The kinetic energy of each recoil, 60–80 MeV, can displace 10^4–10^5 atoms or ions in displacement cascades, many of which are displacement spikes or thermal spikes as mentioned above. Some of the results are: disordering and phase changes in alloys [72], such as the transformation of the two-phase molybdenum–uranium alloys to a single gamma phase; conversion of crystalline oxides and other compounds and alloys to glass-like structures [77], as mentioned above, sintering of UO_2, if the particle size is less than 1–2 μm, and sputtering or evaporation of many atoms from surfaces. In addition, the introduction of new chemical species in a nonequilibrium state in a solid has a large number of physical–chemical consequences that will be taken up in the discussion of fuel performance. In Tables 11–15 are listed the major chemical classes of fission products. Dias

TABLE 11

Burnup Conversion Factors

Material	Fissions/cm^3	% burnup of heavy atoms	MWd/kg	
Uranium	10^{20}	0.209	1.810	
	4.78×10^{20}	1.0	8.650	
			U	UO$_2$
UO$_2$	10^{20}	0.411	3.560	3.130
	2.43×10^{20}	1.0	8.650	7.630
			U	UC
UC	10^{20}	0.305	2.640	2.500
	3.28×10^{20}	1.0	8.650	8.220

and Merckx [73] have reviewed the basic phenomena and mechanisms for change associated with the formation, precipitation, migration, volume changes, and escape of fission gases in and from fission fuel compounds. The consequences of these changes and of the irreversible thermodynamic processes associated with other fission products and their compounds are discussed in sections on reactor fuel elements.

TABLE 12

Chemical State of Solid Fission Products in Irradiated UO$_2$[a]

Fission products	Predicted chemical state	Remarks
Cs	Metallic	Volatile and
I	Metallic	insoluble in UO$_2$
Te	Metallic	
Mo	Metallic	Involatile and
Tc	Metallic	insoluble in UO$_2$
Ru	Metallic	
Rh	Metallic	
Ba	As oxides (BaO, SrO) and possibly	Insoluble in UO$_2$
Sr	zirconates (BaZrO$_3$, SrZrO$_3$)	
Zr	As oxides; some of Zr may	Soluble in UO$_2$
Ce	exist as BaZrO$_3$ and SrZrO$_3$	
Rare		
earths		

[a] From Bradbury and Frost [77, p. 264].

TABLE 13

Probable Chemical and Physical States of Fission Products in Near-Stoichiometric Mixed-Oxide Fuel [110]

Chemical group	Physical state	Probable valence
Zr and Nb[a]	Oxide in fuel matrix; some Zr in alkaline earth oxide phase	4+
Y and rare earths[b]	Oxide in fuel matrix	3+
Ba and Sr	Alkaline earth oxide phase	2+
Mo	Oxide in fuel matrix or element in metallic inclusion	4+ or 0
Ru, Tc, Rh, and Pd	Elements in metallic inclusion	0
Cs and Rb	Elemental vapor or separate oxide phase in cool regions of fuel	1+ or 0
I and Te	Elemental vapor; I may be combined with Cs as CsI	0 or 1−
Xe and Kr	Elemental gas	0

[a] Although the most common oxide of niobium is Nb_2O_5, the dioxide NbO_2 has been assumed to be stable in the fuel. The choice of niobium valence is not critical since its elemental yield is only 4%.

[b] Cerium has a 4+ valence state and may be stable as CeO_2 in fuels of high oxygen potential. This element has also been found in the alkaline earth oxide phase.

B. Temperature, Temperature Gradients, and Heat Flow

The gradient of temperature is a major factor in the potential energy gradients or thermodynamic forces causing changes in fuel elements that presents many questions. Composition gradients and changes give rise to other major components of the forces causing change. Finally, mechanical forces associated with gradients and differences in temperature (thermal expansion), pressure, composition, radiation swelling and growth, etc., produce mechanical displacements or strain in radial circumferential, longitudinal, bending, and other modes.

We may examine the matter of temperature in two stages. First, as mentioned above, we can estimate the temperature levels and gradients in fuel elements required to produce the demanded heat and power. Second, in subsequent sections we can examine the physical–chemical consequences of these temperature regimens and how these consequences, in turn, limit temperatures. Unfortunately, in practice with nuclear fuels, the relation between a flux and a gradient or a force and a displacement is not given by a simple coefficient or constitutive relation. The matter of thermal conductivity illus-

TABLE 14

Final Fuel Composition (in Atomic Percent) of $(U_{0.85} Pu_{0.15})O_2$ Fuel Element in a Fast-Fission Environment at 10% Burnup [77a]

Elements	State	Location	Concentration (atom %)
Y	Oxide, solid solution	Columnar[a]	0.07
La	Oxide, solid solution	Columnar[a]	0.17
Ce	Oxide, solid solution	Columnar, equiaxed	0.44
Pr	Oxide, solid solution	Columnar[a]	0.16
Nd	Oxide, solid solution	Columnar[a]	0.51
Pm	Oxide, solid solution	—	0.06
Sm	Oxide, solid solution	—	0.11
Eu	Oxide, solid solution	—	0.02
Ba	Oxide, solid solution	Columnar,[a] equiaxed	0.21
Zr	Oxide, solid solution	Columnar[a]	0.68
Sr	Oxide, solid solution	Equiaxed	0.14
Nb	Oxide, solid solution	—	0.02
Mo	Metallic phase	Columnar[b] (inclusions)	0.66
Tc	Metallic phase	Columnar[b] (inclusions)	0.19
Ru	Metallic phase	Columnar[b] (inclusions)	0.69
Rh	Metallic phase	Columnar[b] (inclusions)	0.16
Pd	Metallic phase	Columnar[b] (inclusions)	0.41
Cs	Metallic phase	Columnar[b] (inclusions)	0.60
Rb	Metallic phase	—	0.07
		Total concentration	5.37

[a] Primarily in columnar–grain matrix.
[b] Primarily near columnar–grain boundaries.

trates this circumstance. A panel convened by the IAEA reviewed [157] the thermal transport in UO_2 in 1966. Since that time a large number of publications [110] have extended and consolidated such studies for LWR, PWR, and FBR fuel elements. For the most part, working relationships are thought to be in hand for design and operation of UO_2-based fuels for water moderated reactors. The relationships for satisfactory FBR fuels are not well in hand. In the following summary we examine some qualifications on this sort of data.

1. Operational Values of Heat Flow and Temperature of Fuel Elements
   Temperature of Fuel Elements

In Fig. 7 we have plotted values of $W_l(T_a, T_0)$, the linear heat rating of oxide and carbide fuel elements as a function of center line temperature, T_0, keeping the cladding temperature, T_a, constant. These functions are based

TABLE 15

Elemental Fast-Fission Yields (in Pile 365 Days, out of Pile 180 Days) [72a]

Element	Fission yield (%)		Element	Fission yield (%)	
	^{235}U	^{239}Pu		^{235}U	^{239}Pu
Se	0.6	0.4	Sn	0.1	0.3
Br	0.4	0.3	Sb	0.1	0.2
Kr	3.7	1.9	Te	1.0	1.0
Rb	3.9	1.8	I	2.0	1.5
Sr	9.1	3.7	Xe	20.6	23.7
Y	5.7[a]	1.8[a]	Cs	19.4	21.0
Zr	30.8[b]	19.1	Ba	6.6	5.0
Nb	0.0	0.0	La	6.1	5.3
Mo	27.0	22.9	Ce	12.7	11.9
Tc	5.8	6.0	Pr	6.0	5.3[a]
Ru	11.8	23.1	Nd	18.7	15.0
Rh	3.3[a]	7.0[a]	Pm	2.4	2.1[a]
Pd	1.5	14.8	Sm	1.1	3.2
Ag	0.03	2.0	E	0.2[a]	0.8
Cd	0.2	1.0	Gd	0.8[a]	0.4
In	0.02	0.1			

[a] Values estimated from adjacent-chain mass yields.
[b] Thermal values from F. L. Lisman *et al.*, *Nucl. Sci. Eng.* **42**, 191 (1970).

on the preferred values given by Washington [159]. For uranium and uranium–plutonium oxides, the two upper curves represent the maximum linear rating of fully densified fuel pellets held tightly by their cladding. This functional relation may prevail in some regions in some elements over some portion of their lives. The two lower curves might represent the temperature dependence of W_l for material having 10% porosity and a loose fit in its cladding, thus allowing cracks and gaps to reduce conductivity. This situation may prevail early in the life of fuel elements having the maximum tolerable deviation from specified sizes or having gaps intended to tolerate swelling or accumulation of fission products. Also indicated for comparison are the melting ranges of the three fuel materials and the maximum recommended linear rating.

The uppermost curves in Fig. 7 compare the linear rating of elements containing uranium–plutonium carbides assuming no gap or contact resistance between pellet and cladding. This curve is calculated from the relation for the thermal conductivity recommended by Washington [159] to which he assigns an uncertainty of $\pm 15\%$. The expression is $k = 16 + 3.4 \times$

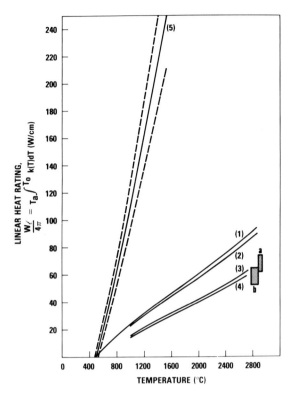

FIG. 7.   A graphical representation of likely values and variation of the linear power ratings, W_l, of oxide and carbide fuel elements as a function of the central temperature of the fuel with the surface temperature at 500°C (from A. B. C. Washington [159]). Curves 1 and 2 show smoothed values for completely densified UO_2 and $(U,Pu)O_2$, respectively, highly restrained by cladding. Curves 3 and 4 are similar for 90% dense oxides loosely held in cladding. The rectangles a and b illustrate the ranges of W_l and melting temperature observed for UO_2 and $(U,Pu)O_2$, respectively [166–169]. Curve 5 shows values for $(U,Pu)C$ calculated from relations given by Washington [159]. The dotted lines indicate the likely uncertainties.

$10^{-3}(T - 500)$ W/m °K, where T is in degrees centigrade. Assuming that surface heat flux is not limited by the coolant, true of Na but not of H_2O or He, elements containing carbides may provide about twice the linear rating of those containing oxides. Since we may conclude from the curve that this linear rating will require only that $T_0 - T_a \leq 500°C$, T_0 may be adjusted over a range up to perhaps 2000°C by putting an appropriate thermal resistance between the pellet and the coolant. Either the higher or lower ranges may prove advantageous. The curves given are only illustrative.

Variations in the resistance of the gap or contact between the fuel and the cladding and between fragments cause the largest uncertainties in estimates of the temperatures within the fuel element. Washington assigns an uncertainty of $\pm 33\%$ to the gap or contact resistance making the calculated values of these temperatures uncertain by $\pm 100°C$. These uncertainties exacerbate the difficulties of interpreting tests and evaluating fission reactor fuels. For operating reactors, safety factors applied to permitted values of W_t are believed to allow for these uncertainties.

The effective thermal conductivities of nitride, sulfide, and phosphide will be in the same ranges as the carbides, and the W_t will probably be comparable. However, the dimensional and compositional changes and temperature limitations have not been thoroughly studied.

For HTGR it is useful first to regard the fuel particle as the essential fuel element and inquire about its temperature gradients (see Fig. 38). Second, these variables are determined by the location of the particle in the carbonaceous fuel rod or stick, and the resistance to heat flow through the matrix, across a gap to the graphite moderator, through the moderator, and through the fluid boundary layer to the flowing helium. The average linear power rating is about 70 W/cm, the maximum about 200 W/cm. Again, the fuel rod–graphite and graphite–helium interfaces cause about half of the resistance and most of the uncertainty in estimating values for temperature. The maximum temperature of fuel particles having silicon carbide coatings is set at about 1350°C. Zirconium carbide coatings for particles may allow the maximum temperature to be about 1500°C. The temperature difference within a particle is at most a few degrees centigrade, but gives sufficient drive for mass transport to be one of the phenomena that limit temperatures.

2. The Thermal Conductivity of Fuel Materials

For reasons indicated above, a priori measurement of the thermal conductivity of fuel materials is only partially useful in predicting the temperature distribution and thermal performance of fuel elements. The dimensional and physical–chemical changes occurring during use alter the resistance and the paths for flow of heat. Nevertheless, in efforts to develop and understand fuel elements, knowledge of the mechanisms for conduction of heat and the phenomena that influence them is both instructive and useful.

Following Debye [159a] and Peierls [159b], the thermal conductivity of solids may be analyzed into three kinds of components: the energy levels and energy content or internal modes of motion of the system, the velocity of propagation of the modes of motion, and the scattering of each propagational mode. Electrons, holes, lattice and local oscillations or phonons, excitons, and photons serve as names and indications of the internal modes of motion that may transport energy in solids. These names or entities are often thought of as moving particles but in reality are terms representing one-body approximations of many-body dynamics. Other variations in the local structure of the solid such as atomic mass, dislocations and other lattice defects and twin, phase, and grain boundaries also scatter the energy carriers. Of all the carriers, electrons and holes in unfilled electronic energy bands ordinarily have the greatest mobility at ordinary temperatures and also share energy with most of the other modes of motion. Thus, if the number density of electrons and holes in unfilled energy states exceeds about 10^{17} per cm^3, they transport most of the heat. Since each electron may be thought of as carrying an average energy proportional to the absolute temperature as well as a charge, the ratio of the electronic component of the thermal and electrical conductivities is roughly proportional to temperature (Wiedeman–Franz law), which is the case for metals and alloys, carbides, nitrides, sulfides, and phosphides. Irradiation and the accumulation of fission products and components in solid solution alter the conductivity largely by reducing the mean free path for scattering of electrons; the electrical and thermal conductivities are decreased similarly. Although some additives undoubtedly alter the number of carriers of heat and charge no enhancement is readily achieved by adding solutes. Composites having a continuous phase with good conductivity may provide enhanced conductivity.

In contrast, uranium and plutonium oxides have no, or very few, free electrons or holes over the range of temperatures of use; thus heat is transported by phonons. The slope of the curve for oxide fuels in Figs. 8, 9, and

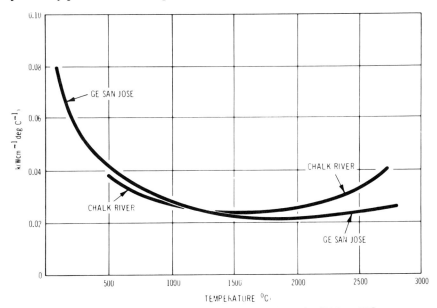

FIG. 8. UO_2 thermal conductivity to the melting point [115, p. 177].

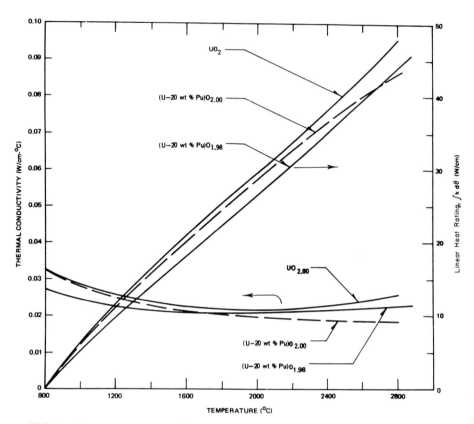

FIG. 9. Thermal conductivity (lower curves) and linear heat rating (upper curves) of $(U–20 \text{ wt}\% \text{ Pu})O_{2-x}$ fuel of two stoichiometries [176, p. 28].

10 is related to their thermal conductivity modified by any changes in the conductivities at interfaces in the fuel element. A decrease, minimum, and slight increase of thermal conductivity in the ranges up to, near, and beyond 1500°C, respectively, is indicated by the curves in these figures. Typical values of the measured thermal conductivity of unirradiated fuel materials are compared in Table 16.

In the fuel oxides phonons are scattered and the mean free path determined by grain boundaries, various defects, and other phonons. The decrease in conductivity as the temperature increases to about 1500°C is approximately proportional to the reciprocal of the absolute temperature and is characteristic of phonon–phonon scattering. In this temperature range the density of phonons is proportional approximately to Θ/T, where Θ is the Debye temperature and T the absolute temperature. The minimum conductivity corresponds to the shortest physically likely scattering mean

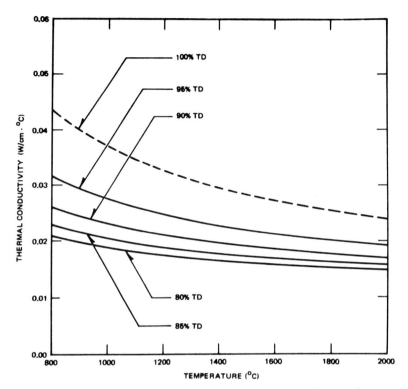

FIG. 10.   Thermal conductivity of $(U–20 \text{ wt}\% \text{ Pu})O_2$ as a function of density [176, p. 26].

free path, about equal to the size of the unit lattice cell of the fluorite type crystal structure of UO_2. Defects induced by irradiation and the accumulation of fission products will also scatter phonons. Below 500°C these defects will reduce the thermal conductivity significantly. As the temperature increases, this origin of lattice disorder becomes less important relative to phonon–phonon scattering, in part because of annealing.

The increase in thermal conductivity of UO_2 above approximately 1500°C is due to the increase in heat capacity. Some values of the heat capacity taken from Kerrisk and Clifton [160] are given in Table 17 in order to show the strong increase with little change in diffusivity (cm^{-1}, sec^{-1}) above 1500°C. The classical duLong and Petit value is approximately 75 J/°C mole and is exceeded below 500°C. The Debye temperature associated with the acoustic branch of the phonon spectrum is approximately 182°K [156]. A characteristic (Einstein) temperature for the optical branch is approximately 542°K [156]. Most probably, the increase in heat capacity is associated with local modes of motion of ions permitted by the defects

TABLE 16

Property	U	U–10% Mo	U_3Si
Melting point, °K	1405	1423	1203
Density, g/cm³	19.12	17.12	15.58
Heavy metal density, g/cm³	19.12	17.12	15.58
Crystal structure	a	b	bct
Thermal conductivity, W/cm-°K	0.35 (670°K)	0.29 (870°K)	0.2 (to 1170°K)
Thermal expansion, 10^{-6}/°K	19 (to 920°K)	12.3 (to 670°K)	16 (to 1070°K)
Electrical resistivity, ohm-cm	35×10^{-6} (298°K)	—	75×10^{-6} (to 1070°K)
Specific heat, cal/g-°K	0.026 (to 773°K)	0.035 (to 773°K)	0.043 (to 773°K)
Heat of fusion, cal/mole	4760	—	—
Vapor pressure, atm	5×10^{-6} (2300°K)	5×10^{-6} (2300°K)	—
Debye temperature, °K	200°K	—	—
Free energy of formation, kcal/mole	—	—	—
Heat of formation, kcal/mole	—	—	—
Entropy, cal/mole-°K	—	—	—
Poison ratio	0.21	0.35	—
Modulus of rupture, MPa	—	—	—
Modulus of elasticity, MPa	1.7×10^5	10^5	—
Shear modulus, MPa	0.85×10^5	3×10^4	—
Tensile strength, MPa	400	300	600
Compressive strength, MPa	—	—	2000
Thermal neutron fission cross section, barns	4.18 (natural)	4.18 (natural)	0.159 (natural)
Thermal neutron absorption cross section, barns	7.68 (natural)	6.68 (natural)	0.293 (natural)
Eta $(\eta)^d$	1.34	1.34	1.34

	UN	$(U_{0.8}Pu_{0.2})O_2$	$(U_{0.8}Pu_{0.2})C$
Melting point, °K	3035 (1 atm N_2)	3023	2758 ± 25
Density, g/cm³	14.32	11.04	13.58
Heavy metal density, g/cm³	13.52	9.80	12.3 (2.6 Pu)
Crystal structure	fcc (NaCl)	Cubic (CaF_2)	fcc (NaCl)
Thermal conductivity, W/cm-°K	0.2 (1023°K)	0.027 (1270°K)	0.18 (to 1270°K)
Thermal expansion, 10^{-6}/°K	9.3 (to 1270°K)	10.3 (to 1270°K)	12.2 (to 1670°K)
Electrical resistivity, ohm-cm	1.75×10^{-4} (298°K)	2×10^4	1.82×10^{-4}
Specific heat, cal/g-°K	0.049 (298°K)	0.10	0.047 (298°K)
Heat of fusion, cal/mole	12,750	—	10,920
Vapor pressure, atm	4.5×10^{-7} (2000°K)	—	8.1×10^{-9} (2000°K)
Debye temperature, °K	—	—	—
Free energy of formation, kcal/mole	−64.75 (298°K)	—	21.00
Heat of formation, kcal/mole	−70.70 (298°K)	—	21.18
Entropy, cal/mole-°K	15.0 (298°K)	—	14.80
Poisson ratio	0.263	0.28	0.295
Modulus of rupture, MPa	—	—	—
Modulus of elasticity, MPa	—	1.8×10^5	—
Shear modulus, MPa	1.01×10^5	0.53×10^5	0.78×10^5
Tensile strength, MPa	—	—	—
Compressive strength, MPa	—	—	—
Thermal neutron fission cross section, barns	0.143 (natural)	—	—
Thermal neutron absorption cross section, barns	0.327 (natural)	—	—
Eta $(\eta)^d$	—	—	—

[a] Orthorhombic (< 936°K), tetragonal (936–1043°K), body-centered cubic (> 1043°K).
[b] Orthorhombic plus tetragonal (< 838°K), body-centered cubic (> 838°K).

Properties of Fuels

U–Fs[c]	UO$_2$	UC	UC$_2$
1275	3138	2780 ± 25	2773
18	10.96	13.61	12.86
18	9.65	12.97	11.68
bcc (> 1000°K)	fcc (CaF$_2$)	fcc (NaCl)	fcc (CaF$_2$)
0.33 (820°K)	0.03 (1270°K)	0.216 (to 1270°K)	0.35 (to 1270°K)
17 (to 820°K)	10.1 (to 1270°K)	11.6 (to 1470°K)	18.1 (1970°K)
—	1 × 10^3	40.3 × 10^{-6} (298°K)	—
—	0.065 (700°K)	0.048 (298°K)	0.12 (298°K)
—	16,000	11,700	—
—	8.5 × 10^{-8} (2000°K)	1.7 × 10^{-10} (2300°K)	2.5 × 10^{-11} (2300°K)
—	< 600°K, 870°K	—	—
—	−218 (1000°K)	−23.4 (298°K)	
—	−260 (to 1500°K)	−23.63 (298°K)	−23 (298°K)
—	18.6 (298°K)	14.15 (298°K)	16.2 (298°K)
—	0.3	0.284	—
—	80	—	—
6 × 10^4	1.8 × 10^5	2 × 10^5	—
—	0.75 × 10^5	0.873 × 10^5	—
270	35	—	—
—	1000	350	—
	0.102 (natural)	0.137 (natural)	0.112 (natural)
	0.187 (natural)	0.252 (natural)	0.207 (natural)
1.34	1.34	1.34	1.34

(U$_{0.8}$Pu$_{0.2}$)N	Th	ThO$_2$	ThC
3053	2028	3663	2898
14.31	11.72	10.00	10.96
13.5 (2.7 Pu)	11.72	9.36	10.46
fcc (NaCl)	fcc < 1618°K < bcc	Cubic (CaF$_2$)	Cubic (NaCl)
0.19 (to 1270°K)	0.45 (923°K)	0.03 (1270°K)	0.28 (to 1270°K)
9.8 (to 1270°K)	12.5 (to 923°K)	9.32 (to 1270°K)	7.8 (to 1270°K)
—	15.7 × 10^{-6}	—	25 × 10^{-6} (298°K)
0.046 (298°K)	0.038 (970°K)	0.07 (298°K)	0.043 (298°K)
12,590	3300	25,000	—
2.1 × 10^{-6} (2000°K)	1.3 × 10^{-14} (1500°K)	5 × 10^{-9} (2000°K)	—
—	163.5°K	200°K	—
—	—	−279 (298°K)	−6.4 (298°K)
71.10	—	−293 (298°K)	−7.0 (298°K)
—	—	15.59 (298°K)	12.0 (298°K)
0.275	0.27	0.17	—
—	—	80	—
—	7 × 10^4	14 × 10^4	—
1.02 × 10^5	2.7 × 10^4	1 × 10^5	—
—	230	100	—
—	—	1500	450
—	—	—	—
—	7.56	—	—

[c] U containing 5% fissium (0.22% Zr + 2.5% Mo + 1.5% Ru + 0.3% Rh + 0.5% Pd). U–5% fissium is bcc above 10000°K, bcc + monoclinic U$_2$Ru between 825°K and 1000°K, and bcc + U$_2$Ru + tetragonal below 825°K.

[d] Number of fission neutrons released per neutron absorbed.

TABLE 17

Thermal Conductivity and Specific Heat of UO_2

Temperature (°C)	Thermal conductivity [116] K (W/cm-°C)	Specific heat [160] C_p (J/°C-mole)	$K/C_p \times 10^3$
250	0.053	57.8	0.92
500	0.041	76.2	0.54
1000	0.025	85.0	0.29
1500	0.021	89.6	0.23
2000	0.025	100.9	0.25
2500	0.035	131.1	0.27

having the structure that has been observed using neutron diffraction [156]. Local charge exchanges among uranium ions may be involved in these defect structures. The local modes of motion associated with these structures can propagate through the lattice from site to site carrying energy. An inspection of the mobility and mean free path for heat conduction would indicate little change from a distance of about one lattice spacing over the range 1500–2500°C. To determine this matter quantitatively, values for the crystal density and the velocity of phonons over the complete range of temperature would be required.

Because the change in heat capacity appears to be sufficient to account for the variation in thermal conductivity, speculation on photon transport seems unnecessary. Since the local modes suggested above may also involve movement of charge among uranium ions of valence 3, 4, 5, and 6, especially if excess oxygen is present, there may be exciton-like transport coupled with the lattice modes. At the highest temperature, intrinsic and, with some additives, extrinsic, electronic conductivity may play some role, but evidently a small one. Improved physical understanding would improve the self-consistency of the engineering relations used to model the performance of fuels.

In addition to the variation of temperature from center to surface of a fuel rod, the temperature varies over the length of a rod driving transport of the more mobile and volatile species primarily toward the cool end. Temperature and gradients also vary systematically over the reactor core. For the most part, discussion of the behavior of materials and the phenomena in reactor fuels focuses on regions of the reactor having the highest temperature and gradients. These occur in the central region of the core over perhaps the third quarter length of the element. However, the most drastic changes and potential failures may not always occur in these regions.

In conclusion, we note that the principles of irreversible thermodynamics recognize that a flux of heat is coupled to a flux of matter and provide the basic conceptual tools for a program of describing and predicting the performance of reactor fuel elements.

V. SOLID FUEL MATERIALS [110–209]

A. Introduction

The incorporation of the materials and phenomena associated with nuclear fuels presents materials science and engineering with an embarrassment of riches. It is necessary to narrow our attention primarily to relationships among composition, structure, properties, and behavior that are directly related to performance in reactors. Sources of information on crystal and microstructures, thermodynamic functions, and physical properties are listed as needed [110–209].

Thorium, uranium, and plutonium are grouped in the actinide series of chemical elements and display much analogous chemical behavior. Also, they exhibit strong differences because of the increasing stability and number of the 5f orbitals and electrons as one goes from $_{90}Th(6d^27s^2)$ through $_{92}U(5f^36d^17s^2)$ to $_{94}Pu(5f^66d^07s^2)$. Thus, the important chemical valence states are Th^{4+}, $U^{3+, 4+, 5+, 6+}$, and $Pu^{3+, 4+, 5+, 6+}$ with the $^{5+}$ and $^{6+}$ states of plutonium much less stable than those of U. These elements form analogous, stable, high-melting compounds with the nonmetals, boron, carbon, nitrogen, oxygen, sulfur, and phosphorous, those of thorium being the most stable. Of these compounds only the stoichiometric dioxides show no electronic conductivity at room temperature. Hydrides of these elements are moderately stable, uranium hydride being the least stable, yielding H_2 at 1 atm at 425°C.

The actinides also form many comparatively stable intermetallic compounds with a large number of metals due both to electron compound formation and to local electronic orbital overlap. The compound $^{238}PuBe_{13}$ provides stability and exposure of Be to the (α, n) reaction in beryllium–plutonium alloy neutron sources, but has found no use as a fuel.

Of these many stable compounds, the oxides and carbides are most easily prepared and are by far the most useful. The nitrides, although interesting because of their stability, are probably less useful than carbides. Metallic thorium is well behaved except for its rather ready reaction with oxygen and water and could find use in reactors.

Metallic uranium is less well behaved, but can be managed. Plutonium metal is hopeless. These systems of compounds and alloys provide a rich field for research and valuable potential applications. The properties of nuclear fuels are listed in Table 16.

B. General Requirements on Solid Fuels

Most useful fuel elements are cylindrical, as are reactor cores, and for similar reasons. The exception is the spherical element for the experimental AVR at Jülich, Germany. The overall active length of an element derives from the neutronic calculations that determine the size and fuel loading of the reactor. The diameter and number of the cylinders, usually called rods or pins, must satisfy two requirements: that the total volume of the pins contain the required mass of fuel and fertile material, and that the surface area multiplied by the local permissible heat flux and integrated over the reactor core equal the required power. A third requirement trades minimizing fabrication costs against maximizing specific power and forces the diameter into the upper range permitted by the first two requirements. Cladding thickness, strength, and neutron absorption enter these three requirements and are vital in a fourth, namely, fuel burnup and life. Prediction of fuel performance, life, probability, and mode of failure is the essence of nuclear fission fuel engineering. Implied in life and performance are stability of dimensions, tolerable corrosion, and confinement and management of fission products. For structural and handling purposes, fuel rods are grouped, spaced, and supported in conveniently sized bundles.

Physical–chemical–mechanical–metallurgical–mathematical models of fuel pin and element behavior [110] that incorporate many of the essential and interesting phenomena are being developed, programmed for numerical computation, and used to decide design parameters, development programs, and testing procedures and criteria. However, much instructive conceptual design and analysis of fuel elements may be done in one's head or on the back of an envelope.

A most important performance parameter of a fuel rod is the thermal power per unit length, which is directly related to the integral of the thermal conductivity of the fuel material from a permissible temperature at the center of the cylinder to the designed temperature at the edge. It may be shown that the heat output per unit length, represented by W_l, is

$$4\pi \int_0^a \frac{dr}{r} \int_0^r r'q(r')\,dr' = 4\pi \int_{T_a}^{T_0} k(T)\,dT = W_l$$

where $q(r)$ is the rate of heat production per unit volume (calculated from the neutron flux and fission cross section) in the fuel rod at a radial position r from the center of the rod, a the radius of the fuel cylinder, $k(T)$ the thermal conductivity of the material, which is a function of the temperature $T(r)$, T_0 the temperature at the center of the fuel cylinder, and T_a the temperature at the surface of the fuel cylinder.

Because $k(T)$ is often difficult to measure in a fuel element under conditions representing or equivalent to those of use, determinations of the thermal conductivity integral, W_l, has provided more useful design information than data on thermal conductivity.

The main point to be made about the thermal conductivity integral in the above equation is that the power per unit length of a fuel rod is a material property and independent of dimensions. The temperature dependence of the integral joins the behavior of this materials system with the thermal design of the reactor core. Given the thermal power required from the core, the total length of fuel rod of a given type is specified. Given the length of the core and hence the overall length of the rods, the number of rods is specified. Considerations mentioned above determine the diameter. The rods or pins obviously may be divided into lengths and grouped into bundles according to handling requirements, methods of spacing and support in the core, volume of moderator required, and cross section area and perimeter best for flow of heat to coolant and flow of coolant through the core.

If we examine the practical values of heat flux for the several reactor coolants listed in Table 3, we may conclude that the required diameters of solid fuel rods tend to be about 1 cm. Actually they may range from 0.5 to 0.8 cm for fast reactor uranium plutonium (mixed) oxide fuels through about 1 cm for LWR slightly enriched oxide fuels and HTGR composite uranium carbide–thorium carbide–carbon rods to as large as 1.5 cm for potential LMFBR carbide elements. For natural UO_2 rods, in PHWR for example, neutronic considerations rather than heat flow determines the 3.8 cm diam.

It turns out that moderation of neutrons in LWRs requires the ratio of volumes of H_2O to fuel to be between 3 and 4. For gas cooling, the optimum volume fraction for flow of gas is near 0.45. In principle, since sodium transfers heat most effectively of all reactor coolants, its volume fraction in an LMFBR core could be quite small as desired for LMFBR. However, as it has developed, reliability and safety considerations due mainly to neutron-induced dimensional changes tend to dictate comparable fuel rod spacings in both LMFBR and GCF[B]R (Gas Cooled Fast Breeder Reactors). For water moderated high flux research and testing reactors, obviously, the area for heat transfer is maximized by the plate type elements about 2 mm thick.

TRIGA research reactors usually do not have stringent requirements on heat transfer. As a result, the Zr–UH elements may be a convenient size, around 3.8 cm diam.

The determinants of the choice of cladding material and thickness include minimization of the absorption of neutrons subject to requirements on strength and fabrication costs. Corrosion and other interactions with constituents of the coolant and the fuel and effects of irradiation are major factors influencing the several properties having to do with strength. Practicable thicknesses of tubing used for cladding fall in the range from 0.025 cm for stainless steel to 0.05 cm for Zr, Al, and Magnox. Often the ratio of external-to-internal diameter of tubing used for cladding falls around 1.1–1.15.

C. Metallic Fuels

Uranium alloys have been used in the fuel elements in the carbon dioxide cooled first generation nuclear power reactors in Britain (Calder Hall or Magnox Reactors) and France, and in fast breeder reactor prototypes— Dounreay, EBR-I and -II (Experimental Breeder Reactor), and Fermi.

Metallic uranium is produced by the reduction of uranium tetrafluoride by magnesium or calcium in a pressure vessel. Uranium fuel rods are produced by casting, rolling, extrusion, machining, and heat treatment. Uranium undergoes three phase changes up to its melting point, namely, the alpha (orthorhombic) up to 666°C, beta (tetragonal) from 666° to 771°C, and the gamma (body centered cubic) from 771° to 1130°C (melting point). Anisotropic alpha-uranium is subject to dimensional changes under both thermal cycling and nuclear irradiation. The changes are governed by the structure and composition as well as temperature and burnup. Burke and Turkalo [110d] showed that the distortion during thermal cycling is due essentially to grain boundary sliding and relief of elastic stress–strain between adjacent crystals having different orientations and amounts of expansion at the higher temperature. On cooling, the differential stresses are relieved by slip and twinning of individual crystals. On the other hand, irradiation growth occurs within each crystal and is due to the aggregation of interstitial displaced atoms into new planes (loops) normal to [010] and the coalescence of vacancies into missing planes (loops) probably on (110) giving a contraction in [100]. A number of microscopic processes have been proposed that produce this net effect. However, the analogy with graphite (Section VI) seems adequate since the lattice spacing in the [010] direction is the largest, and atomic binding is strongest in the corrugated (010) planes.

Metallic uranium fuels are generally limited to operation below approximately 600°C maximum temperature, and to relatively low burnups of about 5 MWd/kgU because of irradiation damage. Swelling and growth become excessive primarily because of fission gas bubbles at high temperatures and the formation of lattice defects (vacancies, interstitials, dislocation loops, etc.) at low temperatures. Irradiation creep is also a problem at low temperatures. There is little swelling below about 400°C. The growth reaches a maximum in the range 400°–600°C. Above approximately 700°C fission gas swelling predominates. The Magnox reactors used uranium adjusted with iron (260 ppm), aluminum (650 ppm), carbon (800 ppm), silicon (20 ppm) and nickel (50 ppm), and the French EDF reactors used U–1% Mo in EDF-1, -2, -3, and -4, and " Sicral " alloy, uranium containing Al (700 ppm), Fe (300 ppm), Si (120 ppm), and Cr (80 ppm) in EDF-5. These minor alloying elements result in grain size refinement and very finely divided precipitates and the swelling diminishes by several orders of magnitude. These additions modify the $\alpha-\beta$ transformation and favor grain refinement and absence of preferred orientation upon quenching these alloys from temperatures in the beta range. Thus, heat treatment minimizes distortion of fuel elements due either to thermal cycling or irradiation growth, since induced intergranular stresses and strains are reduced, a typical grain size effect. These fuels were clad with magnesium alloys (Mg containing 0.8% Al, 0.002 to 0.05% Be, 0.008% Ca, and 0.006% Fe in the United Kingdom, and Mg containing 0.6% Zr in France).

Uranium alloyed with 10% by weight molybdenum has a metastable, single phase, body centered cubic structure, with higher strength, solidus temperature, and irradiation stability that is maintained if the fission rate is above a threshold. It was used as fuel in the Fermi fast breeder demonstration reactor. However, its irradiation stability was degraded by eutectoid decomposition of the γ-phase in the lower flux of Fermi, compared to its behavior in the high neutron flux test reactors in capsule irradiation tests (Fig. 11).

Metallic fuels for breeder reactors have been developed and studied at Argonne National Laboratory. An alloy of Zr–15%U–10%Pu has a solidus temperature of 1155°C. This fuel is satisfactory after irradiation to 6 atom percent burnup when adequate void space was provided between the fuel and cladding to accommodate 25–30% swelling and a plenum for fission gases. The fuel was sodium-bonded to the cladding and had a smear density of 75%. It attained a power rating of 15 kW/ft and expanded to touch, but not strain, the cladding. The swelling rate was $2\frac{1}{2}$% per atom percent burnup (the solid fission products accounted for 10% of the dilation, the fission gas bubble swelling for 22–66%, and 60–70% of the fission gases were released via cracks and fissures). There was a marked effect of external pressure on swelling. For example, the swelling per 1 atom percent burnup was 50% at 1 atm, and 10% at 67 atm. The cast alloy U–5 wt% fissium (2.5% Mo, 2% Ru, remainder Si, Rh, Pd, Zr, Nb) has been tested as a fuel in EBR-II.

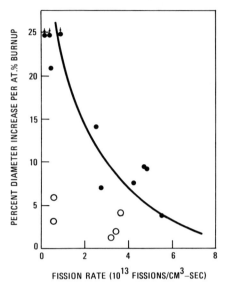

FIG. 11. Effect of the fission rate on the swelling of U–10 wt% Mo rods [110a] (● ~ 480°–590°C, ○ ~ 340°–450°C).

Swelling becomes marked above 500°C, and it can be exposed to a burnup of 1.2%. This fuel when made into rods by centrifugal casting developed a texture which resulted in excessive swelling. Over 4% burnup was achieved without rupture of the stainless steel cladding. Recent work on this fuel was presented by C. M. Walter *et al.* in an Argonne report (ANL-76-28).

Work at Chalk River has included extensive studies of U_3Si alloys that also may be used to perhaps 5% burnup at 500°C if a central void is provided to accommodate swelling [153, 154].

D. Oxide Fuels

The dioxides of Th, U, and Pu have the face-centered cubic fluorite structure. They are completely miscible in solid solution. Uranium dioxide can take up oxygen interstitially to form hyperstoichiometric UO_{2+x} where x may be as high as 0.25 at high temperatures. As the temperature is lowered a phase having the composition U_4O_9 precipitates. Hypostoichiometric oxides of uranium UO_{2-x} form under conditions of low oxygen partial pressure at high temperatures, and revert to stoichiometric UO_2 precipitating U on cooling. Unsintered, finely divided UO_2 powders oxidize to U_3O_8 at room temperature when exposed to air. The dioxides of Pu and Th form only the stoichiometric dioxides because of the stability of the Th^{4+} and Pu^{4+} noted earlier. In UO_2–PuO_2 solutions Pu^{4+} may be reduced to Pu^{3+}

Extensive studies [110] have elucidated the physical–chemical–mechanical behavior of the oxides, at the same time revealing mechanisms of diffusion, defect structures, and other features of fluorite type crystals. Here we can touch on only a few observations.

Uranium dioxide is the most widely used fuel material in nuclear power reactors, usually in the form of cylindrical, cold-pressed, and sintered pellets with densities in the range 92–97% of the theoretical. The properties which combine to make UO_2 such a unique fuel material are: (1) high melting point (2800°C), (2) chemical stability in water cooled reactors, (3) compatibility with cladding (Zircaloy and stainless steel), (4) excellent irradiation stability, and (5) ease of fabrication. The unit cell structure of UO_2 is shown in Fig. 12.

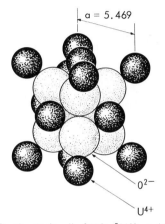

FIG. 12.   Unit cell of UO_2 [110b, p. 164].

Deviations of composition from stoichiometry have a profound influence on its properties, diminishing the already low thermal conductivity, lowering the melting point and strength, increasing creep, fission product migration and release, and altering the complex irradiation behavior. The increase in oxygen activity with burnup can be very significant in leading rods in LWRs (5% burnup) and in fast breeder reactor fuels (over 10% burnup) (see Figs. 13–20 for oxide fuel phase relations).

The allowable values of the thermal conductivity integral defined in Section IVB and the temperatures within pellets have been estimated from observation of microstructure, e.g., the melting point boundary corresponds to 2865°C, columnar grain growth to 1700°C, and equiaxed growth to 1500°C. The reported integral conductivity values from 500°C to melting range from 63 to 73 W/cm. The thermal conductivity of UO_2 decreases as the O/U ratio is increased (see Figs. 21–23 for thermal conductivity data).

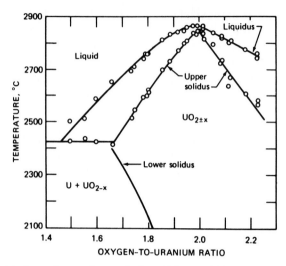

FIG. 13.   Partial phase diagram for urania from $UO_{1.5}$ to $UO_{2.23}$. The separation of the peaks of the liquidus and solidus curves at $O/U = 2.0$ is undoubtedly due to measurement errors. The UO_2 melts congruently; thus, the curves should coincide for $UO_{2.0}$. Similarly, the lower solidus curve should intersect the corner of the upper solidus and horizontal lines [110, p. 118].

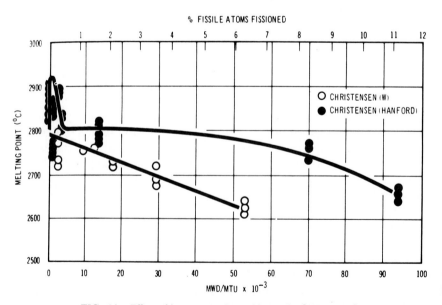

FIG. 14.   Effect of burnup on the melting point [115, p. 182].

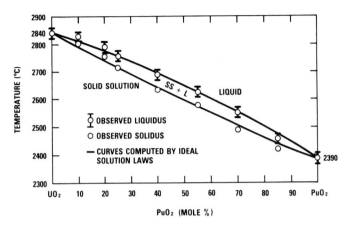

FIG. 15.   (U,Pu)O$_2$ fuel system [176, p. 4].

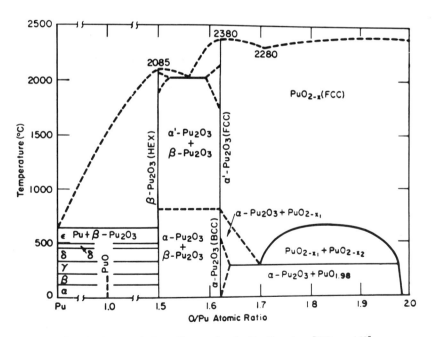

FIG. 16.   Proposed phase diagram for the Pu–O system [110c, p. 143].

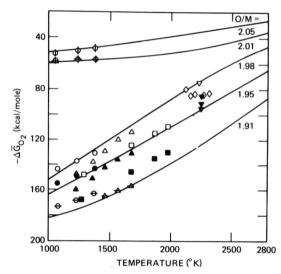

FIG. 17. Calculated and measured oxygen potential versus temperature curves for $U_{0.80}Pu_{0.20}O_{2\pm x}$ [188, p. 313].

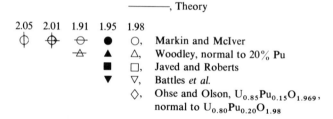

——————, Theory

2.05	2.01	1.91	1.95	1.98	
⊘	⊕	⊖	●	○,	Markin and McIver
		△	▲	△,	Woodley, normal to 20% Pu
			■	□,	Javed and Roberts
			▼	▽,	Battles *et al.*
				◇,	Ohse and Olson, $U_{0.85}Pu_{0.15}O_{1.969}$,
					normal to $U_{0.80}Pu_{0.20}O_{1.98}$

The melting point of stoichiometric UO_2 is $2865 \pm 15°C$. It drops to 2425°C at an O/U ratio of 1.68 and to 2500°C at an O/U ratio of 2.25. The lowering of the melting point to 2620°C at a burnup of 1.5×10^{21} fissions/cm³ has been reported. The melting points of ThO_2 and PuO_2 are 3300°C and 2400°C, respectively.

Particularly striking among the behavioral features of UO_2 is the large increase, as the O/M exceeds 2, in rate of creep, sintering, diffusion, and other processes depending on mobile defects. Creep data on UO_2 demonstrate this effect.

We illustrate and summarize the behavior and performance of oxide fuel in reactors using the following selected observations, emphasizing the effects of irradiation and the accumulation of fission products.

Irradiation-induced creep has been detected in UO_2 and UO_2–PuO_2 at temperatures as low as 250°C. The creep rate increase in-pile is proportional to the fission rate and is attributed to a Nabarro–Herring or diffusional

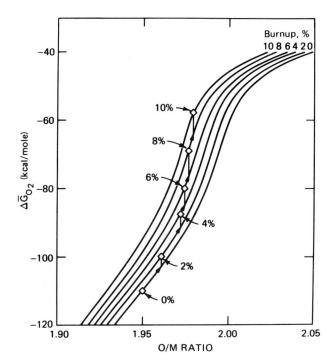

FIG. 18. Change in $\Delta\overline{G}O_2$ with O/M ratio for natural-urania-plutonia mixture at 2000°K [188, p. 318].

creep process involving the defect structure produced in the lattice during irradiation.

The densification [171–175] of UO_2 powder compacts and fuel pellets having particle sizes less than 2 μm takes place even at low temperatures where no significant thermal sintering would occur, permitting collapse of some PWR elements that contain improperly prepared pellets.

Solid fission products in UO_2 are found to segregate at grain boundaries or in precipitates. The UO_2 matrix usually contains the fission products Zr, La, Ce, Pr, and Nd, while the inclusions or precipitates consist of Mo and Ba containing Rh, Ru, Tc, Ce, Nd, and Sr. The occurrence of helium in thorianite to more than 10 cm^3/gm corresponds to an age of $> 500 \times 10^6$ yr and is equal to the concentration of fission gases at 30% or nearly 300 MWd/kg burnup of uranium. This is an indication of the high degree to which fission gases can be retained in oxide fuel at low temperature ($< 800°C$).

Little fission gas is released below 1200°C, and most of the fission gases are released above 1800°C. The low and constant release at low temperatures is by recoil from the surface region. Fission gas retention is "effectively

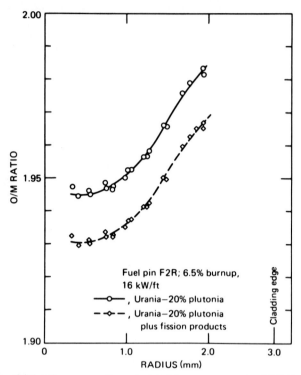

FIG. 19.   O/M ratio versus radius calculated from oxygen potential [188, p. 321].

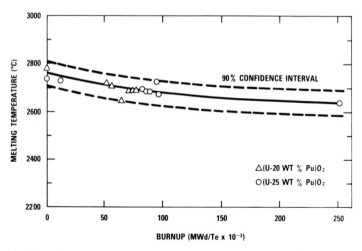

FIG. 20.   The effect of burnup on melting points of UO_2–PuO_2 compounds [176, p. 7].

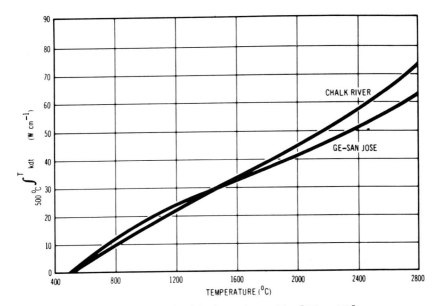

FIG. 21.   UO$_2$ conductivity integral to melting [115, p. 176].

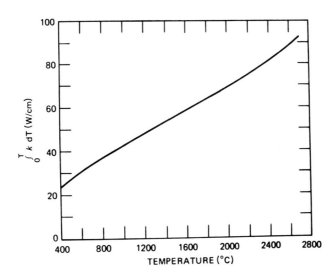

FIG. 22.   Minimum thermal conductivity integral for (U$_{0.8}$Pu$_{0.2}$)O$_2$ [110, p. 131].

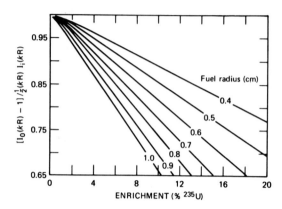

FIG. 23.   Effect of flux depression of the conductivity integral in thermal reactor flux rods
[110, p. 131].

complete" to 1.5% burnup (15 MWd/kg) up to 1200°C and to at least 10%
burnup (100 MWd/kg) below about 800°C.

As part of the development work [136] on the Shippingport core 2, about
60 Zircaloy-clad UO_2 fuel plates were irradiated to burnups as high as 14%
(140 MWd/kg). Interim examinations indicated initial volume changes of
approximately 0.045% per 0.23% burnup (1 MWd/kg) up to a threshold
exposure when the void spaces in the fuel were filled, beyond which the fuel
swelled at a rate of about 0.2% per 0.23% burnup (1 MWd/kg). The threshold burnups varied with heat rating, cladding restraint, and fuel density.
The critical burnup decreased with increasing heat rating and fuel density
and increased cladding restraint (Fig. 24). The fission gas release also increased with burnup.

Swelling in UO_2 becomes significant only after a burnup above approximately 2 atom %. For example, it was found that in UO_2 fuel pins clad with
stainless steel, 2% burnup was necessary before diametral expansions of the
cladding became significant. Below approximately 1400°C, swelling in UO_2
can be accommodated in porosity in the fuel pellets at burnups of at least
1% (10 MWd/kg).

Pressure from fission gases within a fuel rod during irradiation was
found to be directly proportional to the UO_2 fuel volume fraction operating
at temperatures above 1600°C, and negligible below this temperature.

For UO_2 fuels irradiated up to 2% burnup (20 MWd/kg), less than 1%
of the fission gas is released from sintered UO_2 at below about 1700°C.
Between 50% and 100% may be released from fuel which has recrystallized
during irradiation (above 1700°C).

Oxide fuel at temperatures below that required for gross structural
changes releases between 1% and 5% of the fission gases by a diffusional

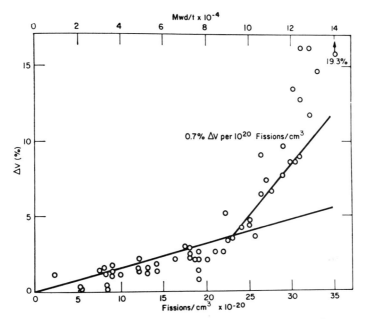

FIG. 24.   Swelling of 87–88% dense UO_2 plates [110c, p. 73].

mechanism. Fission gas release increases with burnup and decreasing fuel density.

At temperatures below approximately 1500°C, fission gas release is observed to take place by slow bubble migration and by diffusion. In the temperature range 1500–1800°C the fuel is plastic, equiaxed grain growth occurs, and the fission gases collect at the grain boundaries in voids or bubbles. The fission gas release increases with temperature. Above 1800°C, columnar grain growth takes place due to the voids sweeping up the temperature gradient, and the fission gas bubbles grow to a critical size and migrate to the center of the fuel with virtually total release (see Figs. 25–29).

E. Carbide Fuels

Thorium, uranium, and plutonium carbides are homologous but display differences. As already stressed, their stability decreases with increasing atomic number. Discussion of the carbides of uranium will illustrate most of the essentials.

There are three compounds in the uranium–carbon system: UC, U_2C_3, and UC_2. UC has the highest uranium density, has a face-centered cubic structure, and is stoichiometric at a 4.8 wt% carbon composition. At lower carbon contents, free uranium metal is present, generally at grain boundaries

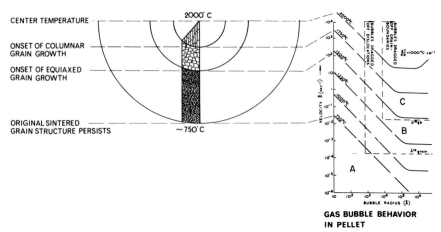

FIG. 25. Schematic representation of the way in which bubble radii and velocities change across the section of a UO$_2$ pellet [119, p. 234].

and as small particles within the grains. The hyperstoichiometric UC exhibits a Widmanstätten structure of UC$_2$ platelets in the UC grains. The monocarbides of Th, U, and Pu have the fcc NaCl structure and are completely miscible. ThC and UC are stable to their melting points. The tetragonal CaC$_2$ structure of UC$_2$ [191] transforms to a fluorite type lattice at about 1700°C. The arrangement of C$_2$ ions in UC$_2$ and ThC$_2$ is somewhat

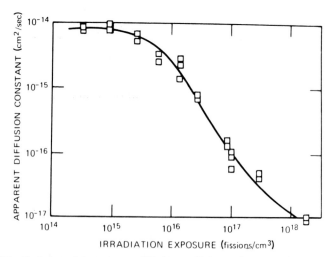

FIG. 26. Variation of the apparent diffusion coefficient of fission gases in single-crystal UO$_2$ with prior irradiation exposure (at low temperature). Postirradiation annealing conducted at 1400°C [110, p. 305].

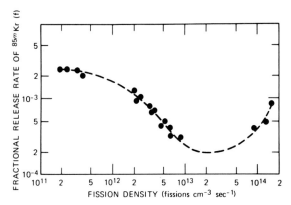

FIG. 27.   Fractional release rate of ^{85m}Kr from high-density UO_2 during irradiation at 1400°C [110, p. 305].

different [191]. Transformations in U_2C_3 have been studied but are not entirely clarified. Figure 30 shows the U–C system.

Compared with UO_2, UC has a higher uranium density, has at least five times greater thermal conductivity, and is almost as refractory. Selected data on the structure and properties of the monocarbides are given in Table XI.

The C and C_2 ions are relatively mobile in the carbides at temperatures of reactor interest. Prevention of transport of C from UC to cladding materials such as austenitic steels or Nb–1 % Zr may require control of the chemical potential of carbon. Thus, the compatibility of UC with stainless steel depends on the stoichiometry and whether the gap between pellet and cladding is filled with gas or sodium. The cladding acts as a sink for carbon, and sodium enhances the transport of carbon from the fuel to the cladding. The decarburized fuel tends to crack, and the carburized cladding loses ductility quickly, even at 600°C. With a gas gap, there is no significant interaction with stainless steel cladding below about 800°C.

The results obtained in compatibility tests of UC with sodium-bonded Nb–1 % Zr alloy are shown in Table 18.

TABLE 18

Compatibility of UC with Nb–1 % Zr Alloy

Stoichiometry	Increase in Time (hr)	C-content of Clad (%)	Thickness of NbC layer on clad (in.)
(hypo) 4.61 % C	5,700	0.08	0.0001
(hyper) 4.98 % C	11,000	0.20	0.00035

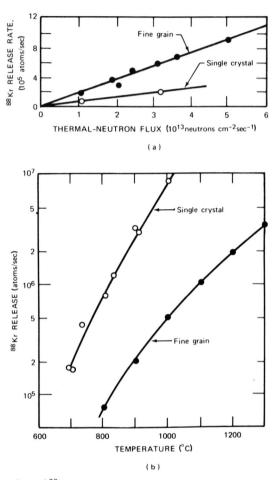

FIG. 28.   Comparison of ^{88}Kr release rates from single-crystal and polycrystalline UO_2 at (a) low temperature ($<600°C$) and (b) high temperature at 2.5×10^{19} fissions/sec. The specimens in (a) had the same geometric surface area, and the release rates in (b) were normalized to the same geometric surface area [110, p. 306].

With a gas gap, the stoichiometry had no consistent effect and the reaction zone thickness, x, at 870°C is given by $x = 0.0026t^{1/2}$ mils, where t is time in hours. For a reaction zone of 0.001 in., in 25,000 hr the temperature must be below 600°C.

Hypostoichiometric UC and stoichiometric UC are compatible with sodium up to 870°C, but the hyperstoichiometric carbide is attacked.

$UC_{1.02}$ in contact with tantalum, niobium, and molybdenum showed slight reaction during heat treatments up to several thousand hours at

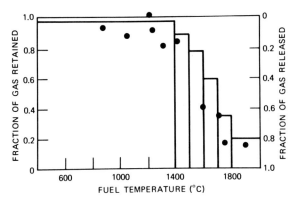

FIG. 29. Fission-product gas retention-release in UO_2 as a function of irradiation temperature; points represent measured gas contents of small samples; the histogram is the analytical gas-release function used in the computations [110, p. 326].

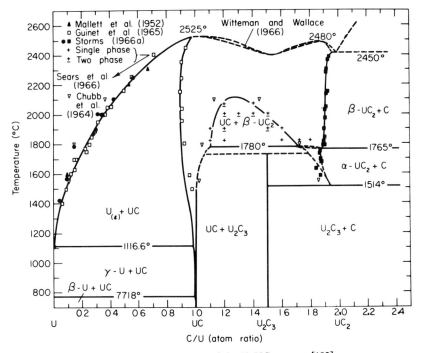

FIG. 30. Phase diagram of the $U–UC_2$ system [192].

1200°C. At higher temperatures the parabolic growth of the reaction layer proceeds as long as UC_2 is present, followed by a liquid phase formation in the case of niobium and tantalum, whereas the reaction does not continue in the case of molybdenum.

Mixed carbide fuels have also been studied in order to broaden the single-phase field in UC so that the undesirable second phases are not present or are rendered harmless. Alloying UC with ZrC appears to provide the most promising combination, since ZrC additions increase the melting point and lower the vapor pressure.

Both chromium and vanadium additions to UC have been reported to improve the compatibility of the carbide fuel with stainless steel cladding.

Irradiation tests [204] on sodium-bonded mixed carbide (UC–PuC) fuel modified with chromium and with iron were conducted in a thermal reactor to an average burnup of 50 MWd/kg. No evidence of incompatibility between fuel and cladding was observed. The chromium-modified fuel swelled less than the iron-modified and unmodified hypostoichiometric carbide fuels (maximum of 3% cladding deformation). The status of mixed (U, Pu)C carbide fuels for fast breeder reactors has been summarized recently [194–206].

Soviet experience with UC fuel (4.8–4.9% carbon) in the BR-5 fast reactor and the SM-1 reactor has been reported [249]. The carbide fuel pins were clad in stainless steel with helium and with sodium filled gaps. Irradiation conditions were as follows: fuel centerline temperatures of 880–1650°C and cladding temperatures of 460–640°C, at linear power levels of 254–605 W/cm, to burnups of 0.8–6.3 atom.% of the uranium. The smear density was 10.92. The swelling ranged from 1.2–3.5% per atom % burnup, and the fission gas released was from 0–20%. The swelling occurred at centerline temperatures above 1400°C (about 3.5% $\Delta V/V$ per atom % burnup). Interaction between the UC fuel and the stainless steel cladding with sodium bonding occurred at about 640°C in-pile, which was about 200°C below the threshold (800°C) for carburization in out-of-pile laboratory tests. The reaction between the UC fuel and the cladding was prevented by means of a chromium coating (25–35 μm thick) vapor-deposited inside the fuel pins by electron beam vaporization. Also, alloying the UC with 5–9% Cr improved the compatibility with the cladding [277, 278].

UC specimens containing 4.4%, 4.8%, and 5.2% carbon have been irradiated in capsules that were either tight fitting or had sodium filled gaps to fuel burnups of 14.3–40.1 MWd/kg at center temperatures of 1000–1350°C and surface temperatures of 625–1000°C [277]. The most stable specimens contained 5.2% carbon and withstood burnups to 36 MWd/kg at fuel center temperatures of 1100–1200°C, with 5% release of the fission gases. The diametral swelling rate was $1\frac{1}{2}$% per 1% burnup.

It has been established that a burnup of 5% of the uranium in UC can be achieved at temperatures up to 1150°C with a density decrease of 15% [195]. The lower thermal gradients in UC reduce the driving force for pore migration.

Arc-cast UC pellets irradiated at approximately 800°C to burnups of 2.5% (20 MWd/kg) swelled at a rate of approximately 0.4% $\Delta V/V$ per 10^{20} fissions/cm^3 [198]. Pellets of UC with a surface temperature of 650°C and a center temperature of 1300°C swelled at a predicted overall rate of approximately 1.8% per 1% burnup. It was also observed that the contribution of fission gas bubbles to the swelling in UC is negligible below 1100°C and 5×10^{20} fissions/cm^3, with good agreement between theory and experiment.

Swelling in hypostoichiometric UC increases with decreasing carbon content. In slightly hypostoichiometric and hyperstoichiometric UC the swelling is similar. For maximum fuel temperatures up to 1000°C, the diametral growth is approximately 1% per 10 MWd/kg in the range 5–40 MWd/kg [198], and at temperatures in the range 1300–1500°C, the swelling increases to 4–8%.

The release of fission gases from UC is less than 1% per 10 MWd/kg at temperatures below approximately 1100°C and increases with exposure and temperature. For example, a hyperstoichiometric carbide (5.2% C) released 5% of the fission gas at 40 MWd/kg at an average centerline temperature of 1060°C [198].

The "break-away" swelling in swelling rate curves occurs at a fuel surface temperature of about 1050°C for UC, at 1150°C for UO_2, and at about 1300°C for UN. However, this comparison is complicated by the fact that the thermal conductivity of UO_2 is much lower than that of UC or UN, resulting in a higher average temperature for the UO_2 specimens [195]. With this correction, the swelling rate of UO_2 would approximately equal that of UN in the temperature range from about 1200–1800°C. Bubble migration in a temperature gradient has been observed at about 1200°C [195].

The swelling rates found for pure $(UPu)_2C_3$ were unusually low: 0.1% $\Delta V/V$ per 10^{20} fissions/cm^3 at 1300°C. This is confirmed by results obtained in France [203] on the irradiation of U_2C_3 (10% UC) to a burnup of 7 atom % at 1400°C; no bubbles could be seen in these specimens at grain boundaries or within the grains. Similar results have been reported [191] on the unusually low swelling rates of the U_2C_3 composition in irradiated uranium carbides.

UC_2 at 100–1200°C and burnup around 100 MWd/kg apparently releases all of the volatile species.

Relatively little work has been carried out on mixed (UPu) carbide and mixed nitride fuels during the past decade in the United States, the Soviet Union, the United Kingdom, France, and Germany. However, during the last year a significant expansion has been initiated in the United States and abroad. The information available on the properties, fuel pin designs (helium-filled versus sodium-filled), and irradiation behavior of these fuels is still very limited and controversial. The recent completion of the high burnup (11.5 atom %) irradiation capsule test, which includes a variety of fuel pin designs, in EBR-II by Los Alamos is a significant milestone [208]. However, the post-irradiation examination is just beginning.

It will require at least another decade of high level effort before the requisite information on these fuels may be available. It would not be prudent to consider either of these fuels as an acceptable alternative to the reference oxide fuel any earlier time.

The major problems in the development of the carbide and nitride fuels are:

1. The feasibility and the relative merits of the two alternate fuel designs (helium-filled and sodium-filled);
2. the safety questions related to the transient overpower conditions.

Both of the fuel designs must accommodate the anticipated swelling of the fuel. Any significant mechanical interaction between the swelling fuel and the cladding must be avoided in order to prevent cladding failures.

In the helium-filled fuel pin the diametral gap is limited to under 0.3 mm in order to prevent central melting in the fuel. This results in fuel/clad mechanical interaction due to fuel swelling. The results of the United Kingdom and French irradiation tests suggest that failure of He-filled fuel pins will occur below the goal of 10 atom % burnup. However, there are no adequate data to determine the failure thresholds for different densities of fuel pellets.

In the sodium-filled fuel pins the initial fuel–clad diametral gap ranges from 0.4 to 0.8 mm in order to allow fuel swelling without fuel–clad contact. In effect, fuel with very low smear density may have to be used, which defeats a key advantage of these fuels, namely, a high metal density. Another problem with the sodium filling is the distinct possibility (observed in several irradiation tests) that the sodium between the fuel and the cladding will be displaced by the fission gases within 3 atom % burnup. This result may cause severe hot spot damage to the fuel with concomitant excessive swelling and/or fuel melting and failure of the pins.

The studies on the safety of carbide and nitride fuels address themselves to the problems of the smaller Doppler feedback to the reactivity, sodium void coefficient, and transient overpower operational problems.

The development of the carbide or nitride fuel would, if successful, offer such overriding advantages and marked improvements in the performance of the LMFBR that it merits a high level of support (see Fig. 31).

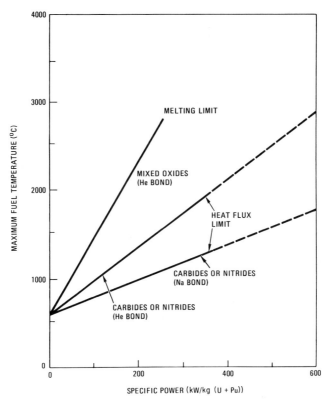

FIG. 31.   Fuel temperature as affected by specific power (0.64 cm dia. pin) [208a, p. 182].

The prospects for developing these fuels within the next 10 yr may be more confidently predicted after the results of the current tests are available. In addition, the safety problems of rapid over-power reactor transients, must be resolved before these fuels can be acceptable for use in commercial LMFBRs.

Compared with mixed oxide, the mixed carbide fuels have higher heavy metal density (13 versus 9.7 gm/cm^3), better neutron economics, greater thermal conductivity (10 times greater), higher linear heat rate capability (45 versus 15 kW/ft for oxide) and specific power (up to 500 W/gm U + Pu), improved breeding gain, and lower fuel cycle cost when compared with oxide fuel at the same burnup. The mixed nitride fuels also have higher metal density (13.6 gm/cm^3) and greater thermal conductivity (slightly higher than the carbide), and higher linear heat rate, specific power, and breeder gain, and lower fuel cycle cost compared with oxide fuels. The nitride fuel has a higher neutron penalty than the carbide because of the higher neutron absorption cross section of the nitrogen.

Compatibility of the advanced fuels with stainless steel cladding is good. However, with carbide fuels prevention of carbon transport from the fuel to the cladding material or vice versa requires control of the chemical potential of carbon in the fuel (e.g., by using stoichiometric UC composition, or by stabilizing the fuel with small additions of Cr, V, or Mo, or by Cr-plating the pellets). Both carbide and nitride fuels have good compatibility with sodium, but relatively poor oxidation resistance.

Compared with oxide fuels the carbide and nitride fuel elements have higher ($\sim 25\%$) costs of fabrication. However, the total fuel element fabrication costs for a core loading may be lower because fewer elements of larger diameter are required.

The irradiation behavior of the mixed carbide fuels depends on the composition, temperature, and burnup. The swelling rate of carbide is about twice that of oxide fuel and increases from about 1.7% to 3% per atom % burnup at temperatures below 1000°C to as high as 25% per atom % burnup at temperatures above 1300–1600°C (wide range of reported values). Fission gas release is much lower than from oxide fuel, being less than 5% at below 1000°C and up to about 40% at 1300°C and 10% burnup, compared to 50–100% release in oxide fuels. Fuel swelling and clad failure are the limiting failure modes in the helium-bonded carbide fuels because of the high creep strength of carbide fuel which can lead to clad failure by high stress/strain in interacting fuel. Fission product corrosion of stainless steel cladding has not been observed in carbide fuel because the corrosive fission products are tied up in the fuel and the oxygen activity is too low for corrosion by cesium.

The development of computer codes to predict the performance of mixed carbide fuels in breeder reactors has been under way by several groups during recent years [250]: for example, UNCLE (Argonne National Laboratory), SOCOOL II and NABOND (Combustion Engineering Company), and CYGRO-F (Westinghouse Atomic Reactor Division).

F. Nitride Fuels

Uranium mononitride has a face-centered cubic NaCl structure, a relatively high uranium density (13.6 gm/cm^3), and a slightly greater thermal conductivity than UC. The significant cross section of nitrogen for thermal neutrons reduces the breeding ratio. Thorium and plutonium nitrides are homologous with uranium nitrides. U_2N_3 phases are observed.

Both irradiation stability and chemical compatibility with claddings have been reported to be outstanding characteristics of nitride fuel [209]. The burnup of nitride fuel results in nitrogen release and also transmutation of nitrogen atoms by the (n, α) and (n, p) reactions. The distribution of fission product nitrides has been calculated on the basis of the thermodynamic potential of the various fission products for forming nitrides, and the conclusion was that UN would be stable with respect to most major cladding alloy components (niobium, tantalum, vanadium, chromium, iron, and molybdenum). Only minor additions of zirconium, titanium, or aluminum to cladding alloys can be tolerated before a tendency to reaction occurs. No fuel–cladding interaction was evident for mixed nitride fuel with a sodium filled gap in type 304 SS cladding after heating at temperatures of 700° and 800°C for 10,000 hr (Battelle–Columbus data). Inconel-625, Incoloy-800, and Nb–0.1 % Zr did not show any reaction when sodium-bonded to the nitride fuel. Thus, nitride fuel appears compatible in the presence of sodium or NaK and cladding. At a temperature of 1000°C, mixed nitride fuel exhibited surface oxidation when sodium-bonded to a variety of cladding materials. The nitride fuel appears to better oxygen from cladding materials at 1000°C.

The stoichiometry of the nitride fuel is more easily controlled than that of the carbides, and the vaporization rates of the nitride are lower. UN has a much higher stability in air than UC.

The evaporation of UN has been measured under various conditions at temperatures near 1700°C. Under dynamic vacuum conditions or sweep-gas conditions, the rate of evaporation of nitrogen approaches that of uranium, resulting in the evaporation of UN without leaving an accumulation of liquid uranium on the surface of the UN.

The radiation stability of UN was first studied in some detail by researchers at the Battelle Memorial Institute, Columbus, Ohio. Bare and stainless-steel-clad, isostatically hot-pressed UN pellets (99 % density) were irradiated to burnups up to 3.8 atom % of the uranium at surface and centerline temperatures up to 615°C and 1260°C, respectively, in NaK-filled capsules. The irradiated UN exhibited good dimensional stability (1.6 % diametral increase) and low (< 0.6 %) fission gas release. There was no reaction of UN with the NaK. These studies have included irradiation of UN at higher temperatures with tungsten–rhenium cladding at surface temperatures of 1500–1700°C. Fine-grained UN specimens at 1500°C surface temperature swelled up to 13 % at a burnup of 2×10^{20} (0.6 atom %) fissions/cm^3, whereas a cast large-grained UN specimen increased only 6 % in volume at the same burnup [209].

UN irradiated at ORNL at very high temperatures (cladding surface temperature about 1700°C) swelled 24 % at a burnup of 0.8×10^{20} fissions/cm^3 (0.25 % burnup). On the other hand, UN pellets with central holes to accommodate swelling have been irradiated at ORNL without any significant increase in fuel pin diameter [209].

UN single crystals and sintered specimens irradiated at ORNL have shown a striking difference in behavior. The single-crystal specimens did not swell because of the absence of grain boundaries as sites for trapping the fission gases. The polycrystalline sintered specimens swelled to a saturation level, and further burnup did not increase the swelling because of the formation of an interconnected network of fission-gas porosity along the grain boundaries.

The results of irradiation tests at Oak Ridge provide extensive information on the swelling behavior of UN at temperatures of 1105–2000°C to burnups of 1.6×10^{19} to 3×10^{20} fissions/cm^3 (0.04–0.9% burnup) at linear heat ratings to 8.5 kW/ft. The fuel pins were contained in tungsten–rhenium cladding. At temperatures below about 1400°C, there was little swelling or fission gas release. At higher temperatures, the polycrystalline (hot-press-sintered) specimens swelled, whereas single-crystal UN specimens did not. This difference is attributed to the absence of grain-boundary trapping sites for fission gas bubbles in the single crystals.

Studies at Livermore on UN have included irradiation tests at 1307–1417°C to a burnup of 2.3×10^{20} fissions/cm^3. The fuel pins swelled at different rates, depending on the grain size of the fuel. Fine-grained UN (30-μm-diam grains) swelled more than coarse-grained (105-μm-diam) UN, and single-crystal UN swelled least. The swelling increased with temperature. Most of the fission gases ($>98\%$) generated were retained in the UN in all specimens.

Thermal neutron irradiation tests of UN and (UPu)N at Battelle–Columbus have demonstrated outstanding fuel dimensional and structural stability at high power ratings (40 kW/ft) and to high burnups (~ 100 MWd/kg). More recently, the burnups on mixed nitride fuels (with helium- and sodium-filled gaps) have been extended to 18 atom % (50×10^{20} fissions/cm^3) in irradiations with thermal neutrons and to 3.6 atom % (10.1×10^{20} fissions/cm^3 or 32 MWd/kg) in irradiations with fast neutrons at 24.8 kW/ft (helium-bonded to type 304 SS cladding). No firm evidence of fuel–cladding reaction has been obtained for cladding temperatures of about 800°C. The thermal irradiation tests were conducted at fuel centerline temperatures of 1215–1455°C, and the fast irradiation tests were at fuel centerline temperatures of 1150–1760°C. The maximum swelling rates in the thermal irradiations were up to 0.48% per 10^{20} fissions/cm^3 (0.3% burnup) at temperatures below 1300°C and are ascribed to solid fission products. At temperatures above 1300°C, fission gas bubble precipitation and swelling begin.

The fission gas release rate in UN and (UPu)N increased from $<1\%$ above a burnup of about 90 MWd/kg to almost 10% at 150 MWd/kg. In the fast neutron irradiation tests, the fission gas release was 13.4%.

G. A Summary of the Effects of Irradiation in Solid Fuels

A reasonably coherent qualitative understanding of the behavior of solid
fuel materials has emerged from a largely phenomenological interpretation
of the results illustrated above. One combines the concepts of irradiation
effects (Section IV) with ad hoc statistical–thermodynamic models of atom
movements in solids, with some knowledge of defects in the specific solids in
question, with the chemistry of selected fission products, and with some
notions on the nonequilibrium thermodynamics in the temperature fields for
each kind of fuel element (Section IV and V). In addition to the microscopic
crystal and defect structural features of the specific substances, both fuel and
cladding, the macroscopic variables are neutron flux and fluence, fission rate
and distribution, heat flow, mass flow, their conjugate thermal and chemical
gradients and conductivities, temperature, chemical potentials, external
pressure and other forces, coefficients of expansion, elastic moduli, creep
coefficients, and other constitutive relationships. In spite of the complexity,
empirical numerical relationships among some of the variables and
processes have been devised and incorporated in computer programs that
afford some degree of correlation and prediction of the performance of fuel
elements that are of use in the design of reactors and of fuel element testing
programs.

The practical effects on the performance and life of the fuel element have
to do largely with mechanical deformation, corrosion and failure of the
cladding, and possible changes in the distribution of heat producing fissile
materials. Five major mechanisms may move the fuel radially toward the
cladding in an operating fuel element: thermal expansion, fuel swelling due
to the accumulation of fission products, thermal ratchetting, mechanical
ratchetting of cracked fuel, and thermal diffusion. These mechanisms are
interdependent and must be evaluated for the full service life of an element.
At the same time, the cladding undergoes changes in dimensions, ductility,
and strength due to fast neutron induced voids and loops, dislocation
tangles, helium bubbles at grain boundaries, and possibly grain boundary
attack by Cs_2O or Se or Te, this corrosion depending on the chemical
potential of oxygen. Diagnosis of tests of the life and output of fuel elements
must be done in terms of theories of the outcome of the several phenomena,
a situation dangerously close to circular reasoning.

As indicated in Section IV, recoiling fission fragments displace 10^4–10^5
atoms along their path. Particularly toward the end of their 4–6 μm range a
high density of crystal defects and rearranged atoms results. On coming to
rest at random many of the individual product atoms or ions become
strange constituents far from chemical equilibrium in their surroundings.
Random atom movements caused by displacements (spikes) can correspond

to an athermal diffusion coefficient of the order 10^{-16} cm^2/sec for a flux of 10^{15} neutrons/cm^2-sec. The longer-lived defects range from single vacancies and interstitials through clusters to loops, climbed dislocations, and voids. These defects can provide both traps and mechanisms for thermally activated migration of the strange species.

The simplest effect of fission product accumulation is the expansion of the solid due to the relative atomic volume of uranium and the fission products, which depends on the chemical state, e.g., cesium metal has a larger atomic volume than cesium ions. Thus, the expansion of UO$_2$ per atom percent burnup ranges from 0.13% to 0.23% if cesium segregates as Cs$_2$O, to 0.54% if as Cs metal. In more dense UC, swelling is at least 1.2% per atom percent burnup. The chemical state of fission products varies with the initial stoichiometry of the fuel and with burnup; that of some products is indicated in Tables 12–15. In the ranges of higher temperature and temperature gradients the products move toward some steady-state distribution [161–188] more or less in accordance with our incomplete notions of the stability and vapor pressure of various species. More systematic knowledge would help. But, the major volume changes are due to fission gases [161] and it is primarily their behavior that has been studied and modeled. We may sketch the picture as follows: about 15% of the fission product atoms are the fission gases xenon and krypton depending on neutron flux and spectrum; the fractions of the total fission gas atoms generated that are trapped and/or released depends on the many factors mentioned and governs the behavior of the fuel element.

At low temperature the large krypton and xenon atoms are relatively immobile and are metastably housed singly and in small aggregates within defect structures ranging from one or two vacant sites per atom to voids and bubbles containing many atoms. Surface energy accounts for the confining forces, even for single atoms, and up to bubbles of the order 0.1 μm in diameter where bulk yield and creep strengths prevail. Sufficient vacancies to permit equilibrium of the given bubble with surface forces are provided by the atom displacement mechanisms. In addition, displacement mechanisms, both knock-ons and spikes, disperse and move the aggregates, the former termed radiation induced resolution [163]. Thus, in a reactor even in low temperature regions atoms and small bubbles may move fractions of a micron per day resulting in some distribution of aggregate sizes.

However, the dominant effects are at temperatures where thermally activated motion of vacancies are higher. In this range, movement occurs of all sized aggregates of fission gases to traps such as dislocations, boundaries, etc., and thence, to more stable states, the most stable being a segregated gas phase.

Thus, recent studies of the effects of temperature, thermal gradients, and stress gradients on the nucleation and migration of bubbles have elucidated the influence of these factors on fuel swelling. The bubbles migrate up the thermal gradients toward the fuel center, being held and carried along by lattice defects (dislocations, grain boundaries, precipitates, and also defects formed by fission recoil damage) until they are large enough to escape from or migrate along the defects. The mechanisms of bubble migration have been postulated to be by Brownian motion for very small bubbles, by surface diffusion when they are larger, and by an evaporation–condensation process at elevated temperatures in very large bubbles (at $r \approx 10^4$ Å). The phenomenon of fission gas resolution in irradiated fuel has been incorporated in the picture. If enough bubbles collect on a grain boundary to touch one another, continuous paths result for escape from the solid. In addition, the operating history of the fuel in the reactor must be considered, since cracks form in the fuel (in grain boundaries at high burnups) during power changes and release fission gases to the fuel–cladding gap. These are some of the processes that have been incorporated in computer codes that have been developed to predict fission gas release and swelling.

A nonsteady-state (or ratchetting) mode of mechanical fuel–cladding interactions during power changes appears to be a primary cause of diametral increases in fuel pins and end of life failure of LWR elements [87, 110].

The BUBL computer program provides a model for fission gas swelling and gas release in UO_2, covering the full temperature range bubble-migration mechanism at elevated temperature ($> 1200°C$) and bubble resolution at lower temperatures [250]. The model accounts for the nucleation, motion, and coalescence of individual bubbles throughout their lifetime, assuming surface-diffusion-controlled migration along the thermal gradient at high temperatures ($< 1200°C$). The bubble re-solution mechanism may be important only at lower temperatures ($< 1000°C$). Cracking of the fuel pellets during power changes is also important in releasing fission gases, but is difficult to model.

A gas release and swelling model has been developed at Argonne for a computer code, GRASS, which has been written for inclusion in the LIFE code for fast breeder fuel elements [121]. It includes a treatment of bubble interaction and coalescence on dislocations and grain boundaries as well as those bubbles not associated with structural defects. This model requires a knowledge of the magnitudes of the physical parameters involved, namely, the diffusion coefficient for fission gas atoms and bubbles, the re-solution constant, the applicable gas law, the interaction energies, and the surface tension. Temperature gradients drive the bubbles toward the fuel center, and stress gradients near structural defects tend to pin the bubbles. Very small bubbles have much lower diffusion coefficients than are predicted by the theory of the surface-diffusion mechanisms.

Other computer codes that have been proposed to predict the behavior of oxide fuels include: BEHAVE (General Electric), OLYMPUS (Westinghouse), PROFIT (AI), SATURN (Karlsruhe), LIFE (ANL), and FMODEL (ORNL) [250].

Nichols and Warner [250] have made the cogent statement that "major disagreements exist concerning the dominant mechanisms for the release of fission gases from operating ceramic fuel elements. The multiplicity of phenomena occurring simultaneously, coupled with a dearth of gas-release data with accurate knowledge of the time–temperature history of the elements, makes the task of defining quantitatively the relative importance of specific mechanisms extremely difficult."

Neimark [250] laments that "one controversial area arose [in codes] over where to put the voidage in the fuel to accommodate fuel swelling. One code suggested we should put it in the central void; another code suggested we should put it in the colder fuel region ($< 1400°C$). I believe there is insufficient experimental evidence to support either view at this time . . . . The more adjustable parameters we have [in codes] the further we go from reality and the less useful the codes become in helping us."

Bailey [250] expresses the designer's viewpoint: "I think the time has come that we should stop looking at just the fuel pins and start looking at the subassemblies. I think some of the investigators too often get lost looking at the microstructures and forget to realize that the assembly can be the key to the success of the test."

Fuel rod design requires a knowledge of the fraction of fission gases that is released and the fraction that is retained in the fuel as gas bubbles. The principal stresses to which the cladding is subjected arise from released fission gas pressure, and from retained gas in the fuel in bubbles which are restrained by surface tension forces, the hoop strength of the cladding, and the creep strength of the fuel.

There are four competing mechanisms:

1. retention of the fission gases in the fuel lattice at low temperatures;
2. diffusion and release of the fission gases to the free surfaces of the specimen at very high temperatures;
3. growth of closed pores at intermediate temperatures leading to swelling;
4. fission gas resolution from bubbles into the matrix.

Hence, the tendency to swell is maximum at some intermediate temperatures.

Segregation of many of the solid fission products takes place in the irradiated fuel. The concentrations of the fission products vary, depending on the isotope, the thermal gradients, and the chemical activity.

The swelling due to solid fission products has been evaluated on a thermodynamic basis by assigning a chemical form to each of the fission products and a corresponding molecular or atomic volume. The nonuniform distribution of the fission products in the fuel does not allow a more quantitative assessment of the swelling. For example, cesium contributes a very large proportion of the swelling, and if it is present as cesium oxide at the fuel–cladding interface, the solid fission product contribution reduces to about 0.23% $\Delta V/V$ per 1% burnup.

In very high mixed oxide fuel at elevated temperatures, the migration of nongaseous fission products out of the fuel has been observed, thereby reducing the magnitude of the fuel swelling attributed to solid fission products [164].

Fission gas release from thermal and fast flux irradiations has been shown to be different because of the much larger flux depression in thermal flux irradiations. In a thermal flux, there is a higher local volumetric fission rate near the surface regions of the fuel than at the center, with the rate differing by a factor as high as two between the surface and the center. Hence, the fission gas concentration will be larger in the cooler surface regions of the fuel body. In a fast flux, on the other hand, there will be a relatively uniform volumetric fission rate and fission gas generation rate.

Another difference between thermal and fast flux irradiations is that the fission gas yield per fission is greater in a thermal flux than in a fast flux. In a thermal flux, the ^{135}Xe captures a neutron to yield ^{136}Xe, which remains gaseous instead of decaying to ^{135}Cs. Hence, there will be more fission gas present in fuel irradiated in a thermal flux. There is also a tendency toward lower fission gas release at lower linear power densities.

The effects of irradiation on the thermal conductivity of UO_2 and UC have been reported. The effects are significant at irradiation temperatures below $500°C$, and a small decrease in conductivity may occur at burnups in excess of about 2%.

VI. FUEL ELEMENTS [210–304]

A. Light Water Reactor Fuel Elements

As shown in Section IVA, the basic design limits for the fuel elements in LWRs are set by heat transfer, clad strain, center melting of the UO_2 fuel, and endurance of the fuel pin and element, including corrosion, fretting, vibration, and mechanical and metallurgical damage. Current BWR and

PWR peak local burnups in fuel assemblies are in the order to 50 MWd/kgU and peak steady state linear heat ratings of up to 630 W/cm, with a fission-gas release value of about 30% [210–219].

The most serious problem that has occurred in fuel element operational experience is fuel densification, which results in collapsed sections in fuel rods. This effect was observed during 1972 in several large PWRs [171–175]. The cladding collapses were found to have resulted from the occurrence of axial gaps in the fuel pellet columns within the rods. All of the rods with the flattened sections were of the unpressurized type. The inward creep of the cladding was not arrested where gaps in the fuel pellet column occurred until essentially complete flattening had taken place. Completion of the fuel densification process has required less than 2000 hr of reactor operation in the power range. The effects of fuel densification cause a decrease in the heat transfer and increase the linear heat generation rate of the fuel pellet, resulting in local power peaking and increased stored energy in the fuel rod.

In-reactor fuel densification is ascribed to irradiation-induced reduction of porosity with radiation-enhanced vacancy diffusion in the UO_2 fuel pellets, thermal sintering, and irradiation-enhanced creep of UO_2. Fuel structures (controlled pore size and grain size) have been identified which resist

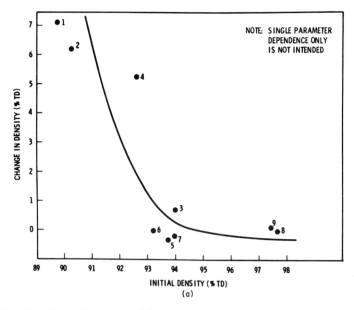

FIG. 32. The relationship between (a) initial density, (b) grain size, (c) pore diameter and observed change in density for pellets irradiated at the maximum burnup, fission rate, and temperature conditions [172, pp. 2-116, 2-117]. Numbers on the data points refer to type of pellets produced.

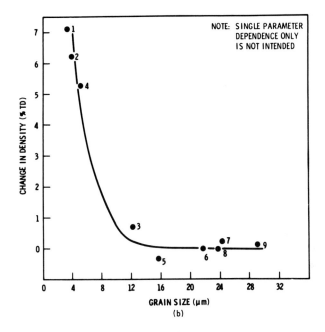

(b)

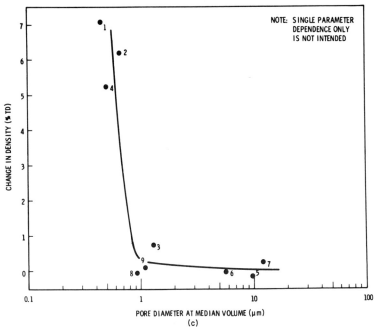

(c)

FIG. 32. *Continued.*

densification in-pile. Prepressurizing the fuel with helium also avoids clad flattening (see Fig. 32 for densification data).

In LWRs the fuel burnup at discharge is limited by reactivity considerations to approximately 30 MWd/kgU and up to 40 MWd/kgU peak exposures. In addition, the excellent corrosion resistance of UO_2 in water, good fission product retention, and low cost favor its use in LWRs. The BWR (GE) capsule tests achieved exposures of UO_2 pellets to over 80 MWd/kgU and specific powers of up to 970 W/cm successfully.

A gap is provided between the fuel pellets and the cladding to accommodate fuel expansion. Control of the pellet density near 90% and provision of concave or dished pellet ends, and possibly a central hole, also allow accommodation of fuel swelling.

The design limit for the plastic strain in the Zircaloy cladding is 1% (caused by swelling and thermal expansion of pellet). The linear power rating corresponding to this limitation is 930 W/cm.

In LWRs the types of fuel failures that have been reported include the following:

1. Clad failure by excessive strain from fuel swelling, and fuel–clad interactions;

2. internal corrosion of cladding resulting from presence of moisture, fluoride, or hydrogen in the fuel;

3. wear and fretting of clad by extraneous metallic pieces;

4. defects in the cladding or in the welds;

5. hot spots in the clad due to deposits of scale or poor heat transfer on corner rods.

It has been estimated that about 0.1–0.3% of the fuel rods have failed in LWRs (see Figs. 33–35 for LWR fuel element designs).

The costs of fuel may be reduced primarily by higher burnups, lower fabrication costs, and by increase in the maximum specific power output of the fuel rods. Fuel assembly prices have not risen because fabrication costs have dropped and balanced the rising labor and materials costs. The specific fuel costs have actually decreased with the higher burnups. Plutonium recycling in water-cooled reactors would further reduce fuel cycle costs and is considered to be desirable because of the delay in the introduction of breeder reactors. LWR fuel-cycle costs are shown in Fig. 36.

1. Zirconium Alloys

Zirconium alloys were developed for water reactors for fuel element cladding and pressure tubes because of their low neutron cross section, adequate strength in the operating temperature range, and satisfactory resistance to corrosion by water at high temperatures. In the United States

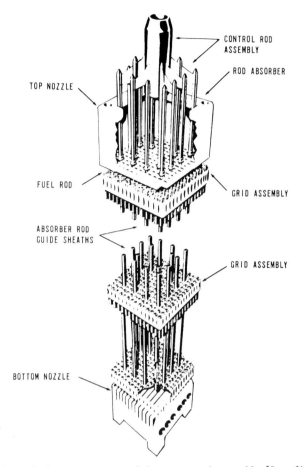

FIG. 33. Pressurized-water reactor rod-cluster control assembly [5, p. 329]. (Courtesy Westinghouse Electric Corp.)

over one million Zircaloy-clad UO_2 fuel rods have operated in LWRs. The compositions of zirconium alloys used in water reactors are shown in Table 4.

The Zircaloy cladding must withstand thermal, bending, and hoop stresses, and resist corrosion. The corrosion rate of the zirconium alloy and hydrogen embrittlement accompanying excessive corrosion limit the coolant temperature, and hence influence the thermal efficiency and capital cost of water reactors. The design criteria must make allowance for the degradation of heat transfer conditions and loss of ductility with time and temperature due to the buildup of the corrosion film, hydride formation (design limit

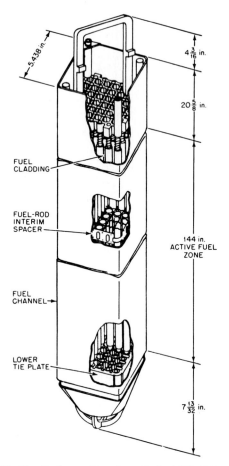

FIG. 34.   Fuel-assembly schematic for BWR [214, p. 69].

600 ppm), and crud buildup.[5] The maximum allowable strain in the zirconium alloy cladding is set at 1% throughout the core life. One cause of low ductility is the precipitation of hydride platelets normal to the stress direction. The fission gas pressure is limited by means of a plenum space to accommodate the gases released from the fuel pellets (about 0.116 ratio of plenum void space to volume of fuel).

The zirconium alloy claddings are also susceptible to hydriding on the internal surfaces from reaction with hydrogenous impurities in the fuel rod (e.g., adsorbed moisture, the presence of fluoride contamination, and hydrocarbons). These impurities can be eliminated by drying the fuel in the clad-

---

[5] Crud is loose corrosion product which circulates and deposits in water-cooled reactors.

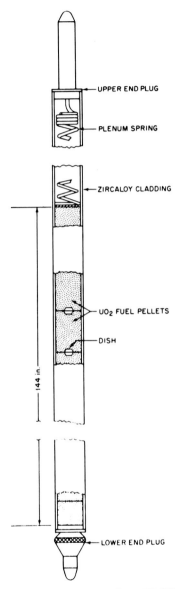

FIG. 35. Fuel rod schematic for BWR [214, p. 72].

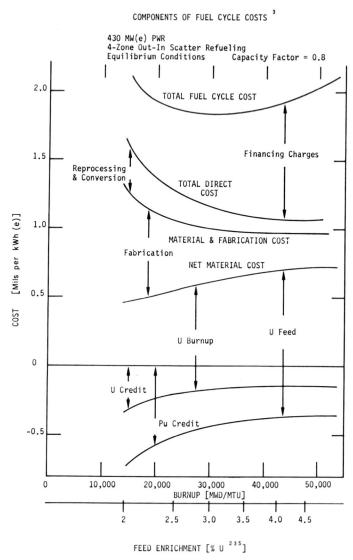

FIG. 36.   Components of fuel cycle costs [30, p. 18].

ding during fabrication (approximately 250°C for a day). In PWRs the corrosion rate (0.1–0.3 mg/dm²-day at 300–350°C) of Zircaloy is increased slightly by irradiation. In BWRs there is a significant increase (up to tenfold) in the corrosion rate of Zircaloys because of the radiolytic oxygen in the BWR coolant, but the rate of hydrogen pickup is similar for both types of reactors.

B. Heavy Water Reactor Fuel Elements

In PHWR–CANDU and SGHWR (see Table 4) reactors, the main structural materials are zirconium and aluminum alloys. The use of heavy water as a moderator provides good neutron economy and permits a wide range of possible fuel cycles (including Th–^{233}U or U–Pu) and fuel management schemes. The UO_2 fuel elements are positioned in Zircaloy-2 pressure tubes which pass through an aluminum calandria containing the heavy water moderator [220–223] (Fig. 37).

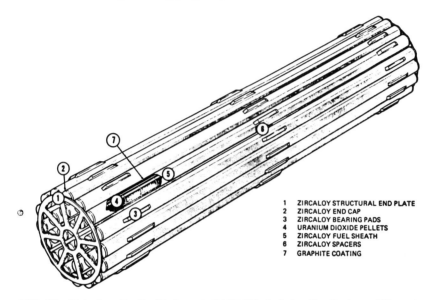

1	ZIRCALOY STRUCTURAL END PLATE
2	ZIRCALOY END CAP
3	ZIRCALOY BEARING PADS
4	URANIUM DIOXIDE PELLETS
5	ZIRCALOY FUEL SHEATH
6	ZIRCALOY SPACERS
7	GRAPHITE COATING

FIG. 37. Pickering G. S. 28-element CANLUB fuel bundle. It is a 500-mm-long, 100-mm-diam. bundle of 28 close-packed elements. Each 15-mm-diam. element contains high density UO_2 in 0.4-mm-thick Zircaloy-4 sheathing which depends on the support of the UO_2 to withstand heat transport system pressure (collapsible sheathing). A thin layer of graphite between fuel and sheathing acts as a lubricant to reduce stress concentrations and to impede stress corrosion cracking. The thin layer of graphite has been shown to be effective, and such fuel has been designated CANLUB fuel. These short, simple natural uranium bundles of basically a single type for all CANDU reactors has resulted in inexpensive fuel [217, pp. 1–71].

In the CANDU reactors the natural UO_2 pellet fuel is clad with Zircaloy-4 and the fuel rods are separated by Zircaloy-4 spacers brazed to the cladding. The fuel elements are made up of bundles of 28 rods. The maximum fuel rod rating is about 690 W/cm. The fuel temperature is 400°C surface and 2000°C center. The operational experience with the fuel elements in the CANDU reactor has been highly satisfactory [220–222]. A recent modification has significantly improved the fuel's performance. The new fuel rods include a thin graphite layer between the fuel and cladding, designated

CANLUB fuel. This has decreased friction and the strain concentrations in cladding expanded over fuel fragments resulting from power cycling. The mean burnup in heavy water reactors is about 10.5 MWd/kgU, and the maximum specific fuel rod rating is about 20 kW/ft.

C. Carbon Dioxide-Cooled Reactors

The first generation commercial nuclear power reactors in Britain and France were cooled by carbon dioxide gas. These reactors were graphite moderated and fueled with natural uranium metal rods clad in magnesium alloys the performance of which has been covered (Section VB). The first of these reactors (Calder Hall) started generating electricity in 1956. The second generation CO_2-cooled graphite moderated reactors in Britain (the AGRs) use slightly enriched UO_2 clad in stainless steel. These fuel elements can operate at higher temperatures, to much greater burnups, giving higher efficiencies and ratings [224, 225].

D. Helium-Cooled Reactors

The high temperature gas-cooled reactors (HTGR) use helium gas at about 800°C and 5 MPa (725 psi) as the primary coolant, graphite as the neutron moderator and fuel element structural material, and coated (Th–U) carbide or oxide fuel particles dispersed in a graphite matrix as the fuel [226–248]. Currently TRISO–UC_2 and BISO–ThO_2 (Fig. 38) are the candidate fissile and fertile fuel particles, respectively, for the large commercial HTGRs being developed by General Atomic. The manner in which advantage may be taken of high temperature materials deserves emphasis.

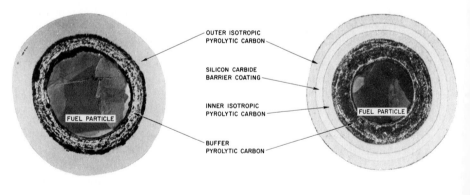

OUTER ISOTROPIC
PYROLYTIC CARBON

SILICON CARBIDE
BARRIER COATING

INNER ISOTROPIC
PYROLYTIC CARBON

FUEL PARTICLE

FUEL PARTICLE

BUFFER
PYROLYTIC CARBON

BISO   TRISO

FIG. 38. Typical coated fuel particles for HTGR [234, p. 8].

The choice of graphite as the moderator and core structural material is based on its unique high temperature chemical, physical, and mechanical properties and on its very low neutron cross section, satisfactory radiation stability, ease of fabrication, and low cost. The use of the graphite moderator as a diluent of the fuel permits much greater fuel dilution than would otherwise be possible and thereby minimizes radiation damage, increases specific power, and greatly extends the heat transfer surface.

The Th–^{233}U standard fuel cycle (Fig. 3) (with ^{235}U as the initial fissionable fuel) is used because of its potential for achieving a higher fuel utilization and lower power cost than any other thermal spectrum reactor system. The neutronic characteristics of ^{233}U are far superior to those of either plutonium or ^{235}U in thermal systems. A substantial portion of the power comes from fission of the ^{233}U converted from the fertile ^{232}Th. The carbon-to-thorium ratio is optimum at a value of 240. This concept promises a conversion ratio as high as 0.85, a steam-heat-power efficiency of about 39%, and a low fuel cost, even with high ore costs. The annual uranium requirements for the HTGR are 30–40% less than for a pressurized-water reactor (PWR) with plutonium recycle operation. Enrichment requirements for the initial core are relatively large for the HTGR, but the total separative work commitment over the life of the reactor is about the same for the HTGR and PWR. Because it can use plutonium as a fertile nuclide and provide a burnup of over 100 MWd/kg, the HTGR can also use plutonium more efficiently than LWRs.

The use of coated-particle fuel allows the high temperature operation of the core to very high burnup (80%) of the fissile fuel, with extremely high retention of the fission products. Also, the ^{235}U fissile particles are segregated from the ^{233}U bearing fertile particles, and thus, the neutron poisons ^{236}U and ^{237}Np can be separated during the fuel reprocessing operation. The average fuel burnup of 100 MWd/kg obtainable is by far the highest of all existing systems.

Inherent safety is achieved in the HTGRs by virtue of the single phase and inertness of the coolant, the high heat capacity of the fuel elements and moderator and their refractory nature, the negative temperature coefficient, which provides a safe shutdown mechanism, redundancy in the circulating systems and assured retention of 4% of the coolant. The fission product plateout activity is limited to low levels which permit direct maintenance.

The 1160-MW(e) HTGR reactor core consists of 3944 hexagonal nuclear-grade graphite block fuel elements of the same dimensions, 35.9 cm (14.2 in.) across the flats and 79.3 cm (31.2 in.) high. These fuel elements are arranged in columns eight blocks high. Internal coolant channels within each element are aligned with coolant channels in the elements above and below. The active fuel is contained in an array of small-diameter blind holes, which are parallel to the coolant channels and occupy alternating positions in a triangular array within the graphite structure. The coated-particle fuel, in the form of bonded rods, is placed in the fuel holes. The arrangement of the fuel element is shown in Fig. 39. The core is divided into 73 regions, each consisting of a central fuel-element column surrounded by up to six columns. Each central fuel-element column contains the vertical holes for the associated control rod pair and a hole for insertion of reserve shutdown material. Each region rests on a single hexagonal graphite support block and is located directly below a refueling penetration that houses a control-rod drive assembly.

The material flowsheet for the reference 1160-MW(e) HTGR is shown in Fig. 40. The fuel exposure, as measured in MWd/kg, is not an important constraint in the HTGR. The average exposure is about 95 MWd/kg, but the burnup in individual fuel particles in the HTGR reaches 0.75 fission per initial metal atom, or about 700 MWd/kg. In the HTGR core design, fissile

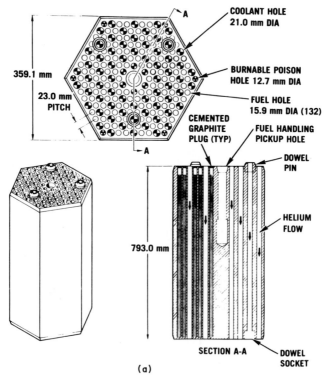

FIG. 39. (a) HTGR standard fuel element. (b) HTGR control fuel element. (c) HTGR fuel components [234, pp. 5, 6].

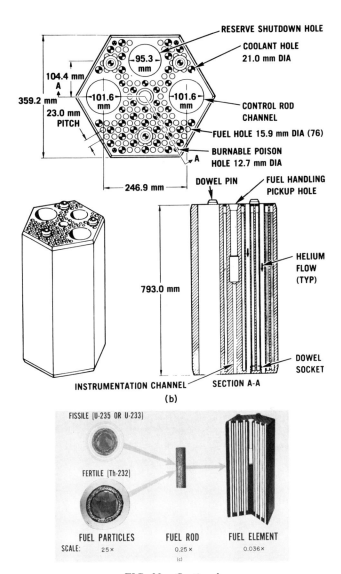

FIG. 39. *Continued.*

and fertile materials are zoned within the core in both the radial and axial directions. The zoning is achieved by using fuel elements containing different amounts of fissile and fertile material distributed uniformly within the fuel rods. The fuel rods in all fuel elements have the same dimensions. Varying amounts of high-conductivity granulated graphite are used as shims to make the volume of all the fuel rods equal. The shim is uniformly mixed with the

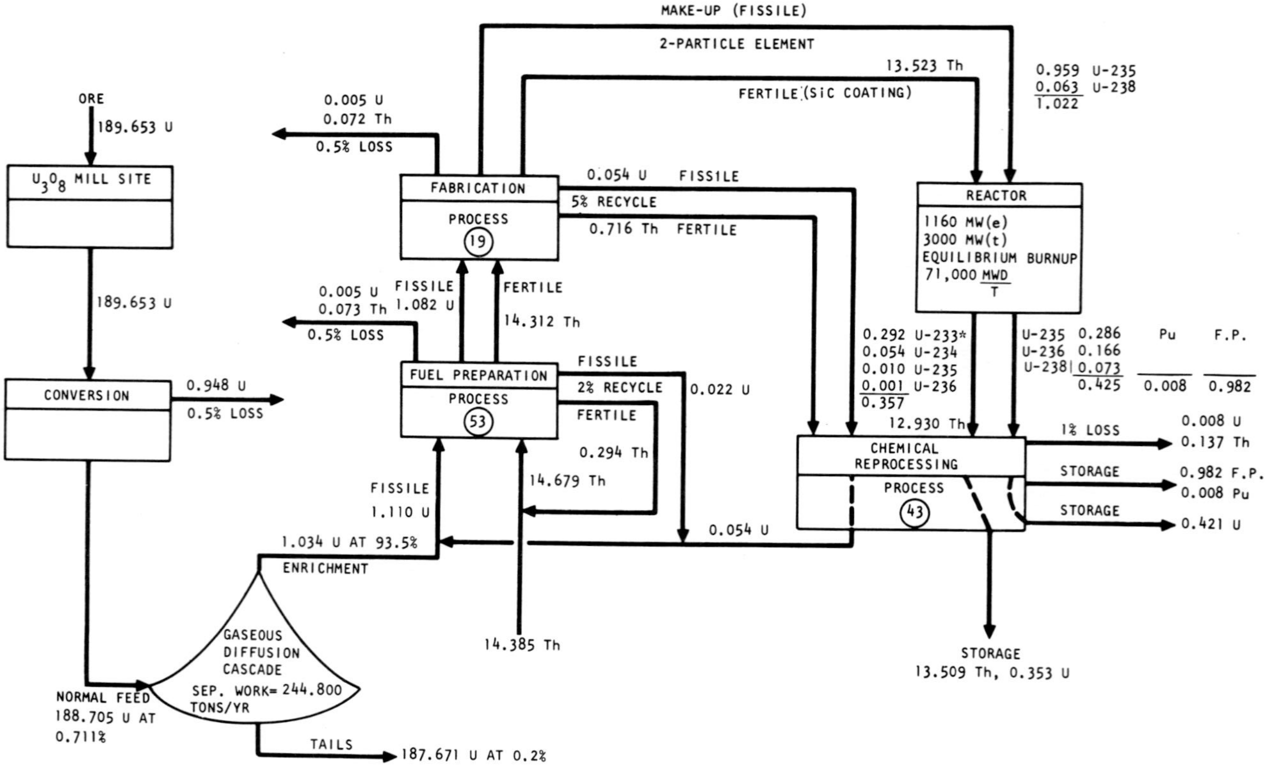

FLOW RATES IN METRIC TONS/YEAR
AT 80% REACTOR LOAD FACTOR

*INCLUDING Pa-233, SINCE ALMOST ALL Pa-233 DECAYS INTO
U-233 AFTER 250 DAYS OF COOLING AND SHIPPING
(Pa-233 HALF-LIFE $t_{1/2}$ = 27 DAYS)

FIG. 40. Equilibrium material balance flowsheet for reference HTGR at C/Th = 200 using TRISO/TRISO fuel with a 3-yr fuel cycle [248a, p. 6-33].

fissile and fertile particles before the rod is formed to produce a homogeneous mixture of particles.

The HTGR could make very effective use of ^{233}U manufactured in a GCFR with a thorium blanket. A high conversion ratio of 0.8 or more and a low fissile inventory requirement for the initial inventory as well as makeup fissile material would accrue from such a relationship. This would be a symbiotic arrangement in which the GCFR would be a fuel factory for the HTGR, but unlike the diffusion-plant alternative, it would produce electricity as a by-product rather than consuming it. One breeder could supply all of the fissile requirements of about three HTGRs if production of Pu for additional reactors would not be required. Figure 41 shows the He-cooled reactor economy.

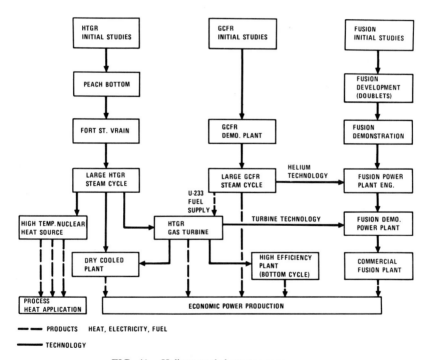

FIG. 41.   Helium-cooled reactor economy.

The fuel reprocessing and refabrication facilities for the HTGRs will be needed after the mid-1980s. The programs to develop and demonstrate the necessary technology are in progress at General Atomic, at Oak Ridge National Laboratory, and at the National HTGR Fuel Recycle Development Program in Idaho.

E. Fast Breeder Reactor Fuels

1. Introduction

The fast breeder reactors[6] will increase fuel use to over 70% of the uranium employed, compared to about 1% in present thermal reactors. The fuel cycle cost is also expected to be significantly lower in the fast breeder power reactors than in thermal reactors or fossil fueled power stations. The cost of uranium ore will not strongly influence the cost of power from fast breeder power reactors, whereas it strongly affects the cost of power from thermal burner reactors. The doubling time (i.e., the time necessary for the reactor to produce a surplus amount of fissile material equal to that required for its initial fuel loading) can be relatively short, about 10 yr, for fast breeder reactors, depending on breeding ratio, power rating, and recycle rate.

Although the technical feasibility and advantages of the fast breeder reactor have been demonstrated, the goal of the fast reactor programs at present is to develop the technology and data to build economically viable fast breeder power reactors during the next decade. Among the major problem areas are the already mentioned higher performance fuels and materials, fuel reprocessing and recycling, coolant technology (particularly steam generators), and the physics and engineering for the large fast breeder power reactors. The FBR fuel-cycle costs are shown in Fig. 42 and Table 19, and fuel processing flow sheet is summarized in Fig. 43.

The fuel in FBRs is a mixture of about 80% U–20% Pu, as oxides or carbides. As the fuel is recycled, there will be an increase in the heavier isotopes of plutonium until an equilibrium isotopic composition of fuel is reached. This actually improves the breeding ratio because of the more desirable nuclear properties of the heavier plutonium isotope, Pu-241.

The major factors of the FBR fuel cycle performance are the specific inventory, average burnup, breeding ratio, and doubling time. The specific inventory [Kg, Pu/MW(e)] is approximately equal to the breeding ratio times the doubling time, for a thermal efficiency of 40% and capacity factor of 0.8. The approximately 0.7 difference in breeding ratio between FBRs and LWRs is equivalent to a fuel cycle cost reduction for the FBRs of about 0.7 mils/kWh. By the year 2000, when the commercial FBRs could come on line, about 600 tons of Pu should be available from LWRs. This amount of Pu is sufficient fuel for as many as 200 FBRs. The value of Pu as a fuel is much greater in an FBR than in an LWR (one estimate is $5 million more

---

[6] The fast breeder reactor concepts are defined in terms of the coolants that are employed, namely, the liquid metal (sodium) cooled breeders (LMFBR) and the helium gas cooled breeders (GCFR).

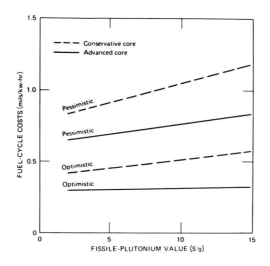

Economic criteria	Optimistic approach	Pessimistic approach
Working capital charge rate	10%/yr	13%/yr
Core-fabrication costs (excluding inventory costs of fissile materials), $/kg (U + Pu)		
Advanced	132	294
Conservative	175	355
Recovery costs (excluding inventory costs of fissile materials), $/kg (U + Pu)		
Advanced	72	140
Conservative	67	140

FIG. 42. FBR advanced- and conservative-design fuel-cycle costs determined from optimistic and pessimistic economic data [26, p. 431].

per ton of Pu). The high fuel inventory in the FBR cores calls for close control of the cost factor in the design of the core. There is also a strong incentive to shorten the cooling and reprocessing times.

2. Fuel Rods

The FBR cores require about five to six times higher fissile concentration in the fuel (15–20%) than LWR fuels. The higher burnup possible in FBRs (over 10 atom % burnup) helps to minimize the higher fabrication charges. Also, the power densities in FBR cores are much higher (about four times) than in LWRs, thereby reducing the inventory charges. The FBR with its high energy neutron spectrum has very small reactivity changes due to fission product poisoning, and the reactivity swings per cycle of operation is

TABLE 19

Energy Costs for a Reference 1000 MW(e) Liquid Metal Fast Breeder Reactor Core [26]

		Mills/kW-hr
Plant capital cost [$400/kW(e)]		8.00
Fuel cycle[a]		
Fabrication ($300/kg)		0.52
Capitalization (14%)		0.06
Inventory		0.48
Shipping and reprocessing		0.30
Losses		0.02
Plutonium credit		(0.41)
	Total fuel cycle	0.97
	Total energy costs	8.97[b]

[a] Plant factor of 90%; 100,000 MWd/tonne; plutonium; $10/g Pu.
[b] Excluding operation and maintenance costs.

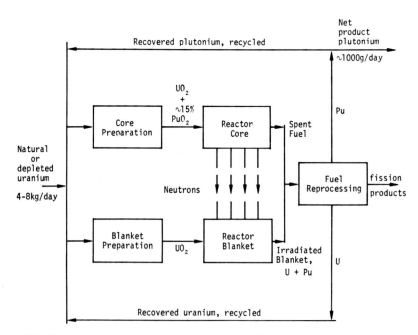

FIG. 43.   Fuel processing flow sheet for 1,000 MW(e) fast breeder reactor [30, p. 13].

about one-quarter of that in LWRs. FBR fuel rods consist of stainless steel clad mixed oxide, $(U, Pu)O_2$ fuel; however there is a real need to develop more stable alloys for cladding and in-core structural materials, and more efficient fuels for the LMFBR (carbide or nitride fuels). Pu-containing fuels require special shielded facilities for fabrication. Neutrons are produced in Pu by spontaneous fission and by alpha-n reactions with oxygen in the oxide fuel. Also gamma radiation is emitted from daughter products. In recycled or long-stored Pu significant amounts of Am-241 are present, which emits gamma radiation.

The limiting factors in the performance of the fuels include swelling and fission gas pressure, thermal conductivity, fuel density, low hot channel factors, fuel center temperature, fuel clad temperature, and loss of clad ductility (caused by neutron irradiation damage, the helium from (n, α) reactions, and the vacancy-void formation). Quality control is of crucial importance in order to minimize fuel failure (e.g., a large reactor will contain thousands of fuel rods having a total length of tens of miles). The relative design parameters of FBR and LWR are shown in Fig. 44.

In the fast breeder reactors the fuel configuration in the core consists of a highly enriched fuel surrounded both radially and axially by fertile blanket material. However, it is desirable to flatten the power distribution by means of zones of differing enrichment in the core. The fuel rods in the enriched region have small diameters in order to avoid excessive central fuel temperature with the high heat rating that is required. A high breeding ratio and low critical mass is achieved with the Pu–^{238}U cycle. The components of fuel cycle cost include the following items: fabrication, processing, breeding credit, inventory and capitalization charges (see Section II). Figure 45 shows the PFR element.

The breeding ratio in fast reactors is influenced by the type and composition of fuel and the size of the reactor. The ceramic fuels soften the neutron spectrum by elastic scattering by the light nuclei (oxygen or carbon), and thereby lower the breeding ratio. In the large cores the increased inelastic scattering in structural materials, in the coolant, and in the ^{238}U also softens the neutron spectrum further and decreases the breeding ratio. The accumulation of fission products in fast reactors does not have a very significant effect on breeding ratio. Moreover, the fuel plays a major part in fast reactors in determining the major reactivity changes and stability of control, namely, the Doppler, the sodium void, and the fuel movement effects.

The problem of accommodating the fission gas pressure [1 cm^3 gas/gm fuel/5% burnup, or about 25 cm^3 (NTP) of fission gas per cm^3 of fuel at 10% burnup or as high as 50 atm in a typical fuel element at 10% burnup] requires modifications in fuel design. There have been several approaches to fission gas accommodation, depending on whether pressurized fuel rods or

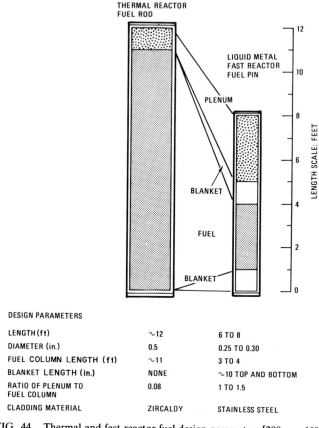

FIG. 44. Thermal and fast reactor fuel design parameters [208a, p. 468].

pressure-relieved fuel rods are used. In the pressurized fuel rods, the approach has been to use thick cladding and low density fuel vented to a plenum space in the fuel element. In the pressure-relieved fuel rods, the elements are vented either to an external fission product trap (as in the GCFR) or to the cover gas on the reactor core in LMFBRs through delaying paths in the fuel elements (such as capillary tubes, check valves, semipermeable porous plugs, or diving bells), in order to allow decay of the short-lived volatile fission products.

The composition and properties of the mixed fuel in a fast breeder reactor will change significantly with burnup. These changes include the ratio of uranium to plutonium, increase in the chemical potential of oxygen, and the production of as much as 20 atom % of fission products for a 10% burnup (100 MWd/kg) of the heavy metal atoms. The volume increase resulting

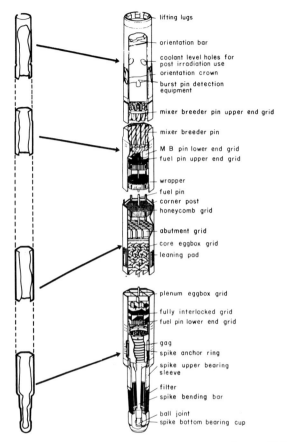

lifting lugs

orientation bar

coolant level holes for
post irradiation use
orientation crown
burst pin detection
equipment

mixer breeder pin upper end grid

mixer breeder pin

M B pin lower end grid
fuel pin upper end grid

wrapper
fuel pin
corner post
honeycomb grid

abutment grid
core eggbox grid
leaning pad

plenum eggbox grid

fully interlocked grid
fuel pin lower end grid

gag
spike anchor ring

spike upper bearing
sleeve

filter
spike bending bar

ball joint
spike bottom bearing cup

FIG. 45.   Arrangement of PFR fuel subassembly [254, p. 334].

from the formation of solid fission products may be approximately 5% for a 10% burnup of the heavy metal atoms in a mixed oxide fuel. The steep temperature gradient in mixed oxide fuels will also cause marked redistributions of the fission products, actinides, and oxygen.

A large FBR core contains several million UO_2 pellets. Ceramic fuels are prepared from carefully characterized powder. Mixed oxide—$(U, Pu)O_2$—fuel is blended mechanically or chemically. The powders are cold-pressed into pellets with binder and lubricant additives. The green compact is then sintered, centerless ground to the required dimensions, characterized, and clad. The fuel pins have to meet stringent quality control requirements and specifications. The mixed oxide fabrication process is shown in Table 20.

The highly complex behavior of fission products in fuel has to be examined in terms of the generation, motion, and release of the fission

TABLE 20

Highlights of Typical Mixed Oxide Fabrication Process (HEDL) [30]

PuO$_2$	10–20 m^2/gm, surface area
Calcining	675°–725°C soak, 60–75 min soak time, 200°C/hr rise, SS crucible
Screening PuO$_2$	−325 mesh, 6–12 m^2/gm surface area
Screening UO$_2$	−100 mesh, 8 m^2/gm surface area
Blending	(1) Hand premix through 20 mesh screen six times (2) V-blend 10 min
Ball milling	Low ash rubber lined mills; 12-hr cycle; tungsten carbide media; surface = 4–10 m^2/gm
Binder addition	3 wt% Carbowax 20M in H$_2$O (20 wt% solution)
Drying	3–4 hr at 70°C
Granulation	−20 mesh
Re-drying	3 hr at 70°C
Prepressing	$\frac{1}{2}$ in. diam die, 30–50K psi
Granulation	−20 mesh
Pellet pressing	20–30K psi, 53% theoretical density green density
Binder removal	
Atmosphere	Argon–8% H$_2$ 8 SCFH
Rate of temperature rise and cooling	120–140°C/hr (200°C/hr max)
Soak temperature	350–650°C
Soak time	4 hr
Batch size	3 kg max
Sintering	
Atmosphere	Argon–8% H$_2$ dried to < 1 ppm H$_2$O 1–6 SCFH
Soak temperature	1650°C
Soak time	4 hr
Batch size	6 kg max
Cycle	23 hr

(The furnace is evacuated at 850°C during the cool-down to reduce gas and
moisture content of sintered pellets.)

Gauging	Micrometers and dial indicators 0.001 in. accuracy
Grinding	Centerless (dry)
Pellet loading and fuel pin assembly	
Decontamination	
Closure welding	TIG weld, helium atm
Helium leak check	For cladding and weld integrity
Nondestructive testing	Gamma scan for fuel pellet placement and isotopic content
Cleaning and passivating	Caustic base cleaner and HNO$_3$ passivating
Surface contamination test	For removable and fixed alpha
Packaging and storage	

products. The migration and release occur by processes whose relative importance are shown to change with conditions and the properties of the fission products. For example, the variations in fission product yields for different fission isotopes lie principally in the position of the light element peak, which results in a significant effect on the constitution of irradiated oxide fuel. The temperature region of high fission product release is considered to be that of columnar grain growth, below which fission gas movement in oxide fuel is dominated by gas accumulation and release from the grain boundaries from interlinkage of gas bubbles. The distribution in size of the gas bubbles is correlated reasonably well by the calculation codes mentioned above that include competing processes of gas accumulation from the matrix by thermal diffusion and a dispersive radiation-induced resolution process. The activation energy for diffusion of fission gases shows an increase in the region of about 1,600°C, which is ascribed to a change of migration mechanism from radiation-enhanced to thermally activated diffusion.

Irradiation-induced resolution appears to be very important. This phenomenon has been observed in a number of experiments. The results of the Argonne work on irradiated UO_2 was carried out between 650° and 1400°C, using transmission- and scanning-electron microscopy and replication metallography. Irradiated specimens were also annealed out-of-reactor, and the fission-gas bubble distribution was compared with the as-irradiated distribution. It was concluded that the resolution process is operative over a substantial fuel volume of a typical LMFBR oxide fuel element [141].

The key areas of development for FBR fuel remain the following: fission product migration and behavior, fuel and oxygen migration, and fuel–fission product–cladding chemical interaction. Predicting the steady-state performance of the fuel pins and elements requires the setting up of design criteria and proof of performance in operating FBRs, particularly to validate the lifetime prediction methods. Problems of rod–bundle duct interactions must be assessed and tested in-pile. The performance characterization of the irradiation behavior of cladding includes studies of swelling, inelastic deformation, and other mechanical properties. The cladding compatibility tests with the coolant are needed in order to determine the effects of corrosion and exposure to the coolant on the properties of the alloys used. The loading on sliding contacts between fuel pin–grid interfaces results in friction and wear of the surfaces, and must be limited to acceptable levels by suitable choice of materials and surface treatments.

3. Cladding

The lack of reliable data on the strength, ductility, and dimensional stability (swelling) of the cladding material under irradiation precludes the accurate prediction of fuel element behavior in fast breeder reactors. However, the required data are being accumulated from the results of numerous irradiation tests, and these are being incorporated in the computer-based fuel element models. Most of the data have been obtained for the ranges of up to 15% burnup and fast neutron fluences of 8×10^{22} nvt ($E > 1$ MeV) for mixed oxide fuel clad with 20% cold worked type 316 stainless steel.

The rate of swelling of stainless steels and other alloys by fast neutrons is governed by temperature (maximum at about 575°C for 20% cold worked type 316 stainless steel), fluence, stress, and structure. Hence, inhomogeneous swelling in the duct, cladding, and structural parts will result in bowing and distortion. See Figs. 46–49 for irradiation data.

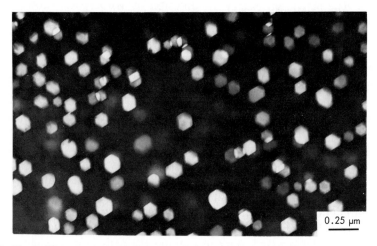

0.25 μm

FIG. 46. Void structures in high purity type 316 stainless steel irradiated to a fluence of 2×10^{22} n/cm² (>0.1 MeV) at about 600° [280a]. (Courtesy of J. Horak and P. Sklad, Oak Ridge National Laboratory, Tennessee).

Irradiation creep causes the maximum clad strain to occur in the region of maximum neutron flux rather than maximum temperature. Irradiation creep is linear with stress level and fluence and dependent on temperature. It is enhanced by swelling.

The main thrust of the results of the most advanced studies is the indication that the commercial austenitic stainless steels will exhibit excessive swelling at the high fast fluences required for the commercial fast breeder reactor, but that the alloys containing about 40% nickel and minor amounts of silicon and other elements and modified stainless steels will probably retain their dimensional integrity to the highest fluences. However, the irradiated alloys have very low uniform ductility and this factor will have to be

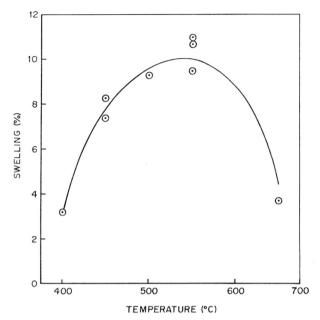

FIG. 47.   Temperature dependence of neutron irradiation induced swelling in high purity type 316 stainless steel for a fluence of 2×10^{22} neutrons/cm^2 (> 0.1 MeV). (Unpublished data by E. E. Bloom courtesy of J. Horak and P. Sklad, Oak Ridge National Laboratory, Tennessee).

considered in the core design. The maximum fast neutron fluences achieved in fast test reactors are still about a factor of two below the fluence required for the demonstration plant core structural materials and cladding, and a factor of four below that required for the commercial FBRs. In fact, a portion of the LMFBR program second only to the Clinch River Breeder Reactor (CRBR) in size is devoted to the development and testing of fuel elements in the Fast Flux Test Reactor (FFTF).

The simulation technique with heavy ion bombardment (nickel ions in particular) appears to be a powerful and effective screening test to evaluate the materials to the required fluences in terms of equivalent atomic displacements. The main differences are that the peak-swelling temperatures are somewhat higher in the simulation tests. Experiments by simulation tests furnish a striking demonstration of the propensity of the 300-series stainless steels to excessive swelling, no matter what their pre-treatments or composition, and reveal the encouraging information that the nickel alloys containing over 35% N [103], and modified stainless steels containing Si and Ti (ORNL) show relatively slight swelling [280a]. The low swelling alloys that have been identified include solid solution alloys as well as precipitation

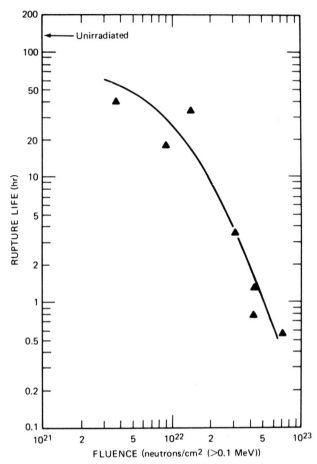

FIG. 48. Effect of neutron fluence on the postirradiation rupture life of type 304 stainless steel. The irradiation temperatures were between 370° and 430°C. The tests were performed at 600°C at a stress of 1.9×10^5 kN/m² [110, p. 445].

hardened alloys, showing that the presence of precipitation is not a necessary condition for low swelling. Swelling is strongly dependent on the major constituents, Fe, Cr, and Ni. The ferritic alloys also show negligible swelling. This provides the basis for selection of a compositional range in which special low-swelling alloys may be developed [80–106, 280–284].

The selection of cladding materials on nuclear grounds has been reviewed in some detail. A comparative study has also been made on the effect of cladding and structural materials on critical mass and breeding ratio in a fast breeder reactor with a core volume of 800 liters (see Table 21).

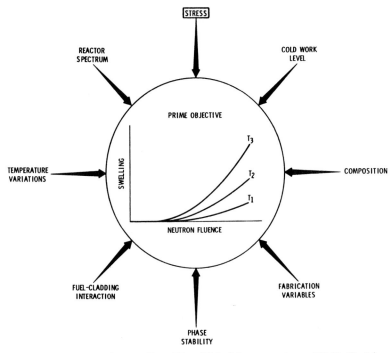

FIG. 49.   Factors influencing swelling. (Unpublished figure courtesy of T. K. Bierlein, personal communication, October 22, 1976).

The irradiation of nearly all materials by fast neutrons produces (n, α) reactions. At high temperatures (above 500°C) the helium that is produced can migrate and form gas bubbles at grain boundaries, and at dislocations, defects, or impurities within the grains. The striking discovery by Cawthorne and Fulton [83] in highly irradiated stainless steel is the presence of spherical

TABLE 21

Critical Mass and Breeding Ratio

Material	Critical mass, kg	Breeding ratio
Stainless steel	431	1.82
Ti	425	1.92
V	456	1.68
Zr	415	1.89
Nb	494	1.51
Mo	502	1.46
Ta	716	1.02

or polyhedral cavities (Fig. 46) ranging in size from the smallest observable to 1500 Å. These cavities were observed during electron microscope examination by the thin foil technique, on samples irradiated in the Dounreay Fast Reactor to total neutron fluences greater than 10^{22} nvt (fast) in the temperature range 400–610°C. The void population in this case was estimated to be too great to derive solely from helium [calculated from the Van der Waals relation $P(V - \kappa b) = nRT$ and the relation $P = 2\gamma/r$, where P is the gas pressure in the bubble, κ the number of moles of helium, r the bubble radius, γ the surface energy, V the bubble volume, b the Van der Waals constant, R the gas constant, and T the absolute temperature]. The number and size of voids is consistent with the idea of the condensation of vacancies, which are generated during irradiation in a fast neutron flux influenced by the helium atoms produced by (n, α) reactions, stress fields, and local alloy constitution.

The total void volume is peaked at 510°C. Annealing the specimens at 900°C for 1 hr completely removes the voids within the grains, leaving a residue of helium bubbles confined mostly to the grain boundaries. The maximum void volume observed in any specimen is 7%. This remarkable swelling of the clad was observed where the irradiation temperature was about 500°C, and the fast neutron fluence was 7.8×10^{22} nvt.

Clad wastage allowance of 8 mils (200 μm) in FBR fuel cladding includes consideration of reactions of fissions with the cladding, wall thickness variation in the as-fabricated cladding, mass transfer, corrosion by sodium, wear and fretting, and defects which lower the stress rupture life. There is an important incentive to determine the fuel endurance limit as a function of burnup and to establish the frequency, mode, consequences, and mean time to failure. More needs to be known about the effects of intergranular attack of the cladding by fission products [141–143] (particularly Cs, Te, I) and its influence on the strength and ductility of the cladding. The major problems requiring further study include mechanisms, kinetics, reaction product identification and characterization, effects of impurities (carbon, alumina, silica, calcia) in the fuel, the effectiveness and behavior of inhibitors, the evaluation of sol-gel fuels, the role of surface conditions, the change of oxygen potential and oxygen transport with burnup, and the oxygen potential thresholds for reaction in the presence of fission products. Also in need of further study are the effect of variations in cladding composition and microstructure and fabrication procedure. Illustration of the variation of cladding attack with temperature and stoichiometry of mixed oxide fuel is given in Fig. 50.

4. Analytical Codes

The LIFE code predicts diameters of restructuring zones in fuel, the gap between fuel and cladding, cladding strains, fission gas release and rod inter-

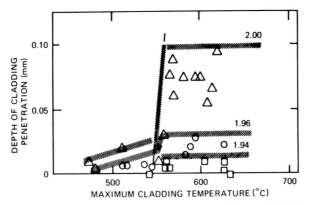

FIG. 50. Maximum depth of cladding attack of type 304 stainless steel cladding by mixed-oxide fuel. The data represent burnups ranging from 7% to 13%. O/M ratios: □, 1.94; ○, 1.96; and △, 2.00 [110, p. 190].

nal pressure [99, 104, 105]. The material property correlations used in the fuel rod design code LIFE are periodically reviewed by a national committee appointed by DOE and are used to predict the performance of the fuel rods for the LMFBR. The analytical expressions for material behavior develop from: (1) elasto-plastic equations, using (2) constitutive relations from correlations of empirical data. The best available data on 28 material properties of fuel, cladding, and contacts or gaps are incorporated into the code, such as thermal conductivity, swelling and creep, thermal expansion, gap conductances, fuel restructuring, fuel–cladding chemical reactions, and redistribution of plutonium, oxygen, and fission products. Property correlations that require continued major development include fuel swelling, fuel plasticity (hot pressing), and fuel–cladding friction coefficient. The difficulties in developing a fuel growth model from first principles stems from the admittedly complex relationships between swelling, gas release, hot pressing, sintering, creep, and cracking. For example, swelling of fuel generally precedes fission gas release. The creep of mixed-oxide fuel is enhanced by fission and affected by Pu content, O/M ratio, and impurities, and by the neutron flux spectrum.

The LIFE-III fuel element performance code user's manual has been prepared [123]. This document defines the code as follows:

The LIFE-III computer code was developed to calculate the thermal and mechanical response of mixed-oxide fuel elements in a fast-reactor environment. It incorporates a one-dimensional, steady-state heat-transfer analysis and a finite-strain-theory structural analysis based on generalized plane strain and the method of successive elastic solutions. An incremental approach is used so that temperature, stress, and strain can be calculated during any specified history of reactor power cycling. Fuel–cladding and sodium–cladding chemical attack is treated by a cladding wastage model. Up to seven axial sections are allowed to account for axial variations in power and coolant temperature. Fuel restructuring, migration of fabricated porosity; fuel and cladding swelling and creep, fission-gas release, and fuel cracking, crack healing, and hot pressing are included in the analysis.

The axial sections are divided into a maximum of 20 cylindrical shells. The thermal and mechanical conditions of the fuel element are calculated incrementally as a function of time. The reactor operating conditions are averaged over each time step. The code thus predicts the behavior of the fuel element as a function of the reactor operating history.

The thermal calculations are based on the assumptions that the heat flux is radial in the fuel and cladding and that the heat capacities of the fuel and cladding are negligible. The axial temperature distribution in the coolant is calculated from the coolant inlet and outlet temperatures and the axial power profile of the element. Then for every axial section the radial temperature distribution is calculated from the local coolant temperature, the local linear heat rating, the cladding–coolant heat-transfer coefficient, the cladding thermal conductivity, the fuel–cladding heat-transfer coefficient, and the fuel thermal conductivity. Encapsulated fuel elements may be treated at the user's option. All thermal properties vary with operating conditions and must be recalculated for every time step. The time steps are sufficiently short so that the thermal, restructuring, gas release and mechanical analyses can be decoupled within the time step. Once the temperature distribution is calculated for a time step, the incremental fuel restructuring and fission gas release, the thermal expansion of fuel and cladding, and the plenum pressure are calculated.

The mechanical analysis is based on generalized plane strain and the method of successive elastic solutions. The plenum pressure, the coolant pressure and the axial loads imposed by the core restraint system provide the boundary conditions. Until the gap closes, the boundary condition at the interfaces is the plenum gas pressure. When the gap is closed, the fuel–cladding interface pressure is calculated under the assumption the fuel and cladding displacement increments are equal at the interface. The incremental deformation is calculated for every time step. Fuel deformation mechanisms are thermal expansion, elasticity, restructuring, creep, fission product swelling, hot pressing, and cracking. Cladding deformation mechanisms are thermal expansion, elasticity, creep, and irradiation-induced swelling. Fuel–cladding gap closure and total cladding strain are determined by a combination of all the deformation mechanisms. Mechanical loading, fuel–cladding chemical interactions, and sodium corrosion are included in the calculation of cladding damage.

The development program for the SNR-300 fast breeder prototype has included comparisons between the results of 77-pin fuel bundles irradiated to 55 MWd/kg (fluence of 3.8×10^{22} nvt) in experiments in the Dounreay Fast Breeder Reactor (DFR) and the calculational results of the pin code SATURN [250]. The partial verification of the experimental results led to a number of conclusions. The surface temperature of the fuel increases steadily in the course of burnup, possibly because of reduced gap conductance caused by increasing impurities and irregularities in the gap and restructuring of the fuel. No contact pressures are evident since the clad may "swell away," leaving enough space for fuel growth. A diametral increase at the hot end of the pins is ascribed to a ratchetting effect which is not described by the code. It is concluded that the investigation of local failure modes instead of general limitations in the materials properties may generate some progress in the understanding of the behavior of a real pin.

A considerable amount of mathematical modeling of the fuel pin performance has been carried out which has served primarily to focus attention on
the important fuel and clad properties that influence the fuel pin performance. More accurate data are required on the critical properties to provide
engineering relationships mentioned above and portrayed in Fig. 51.

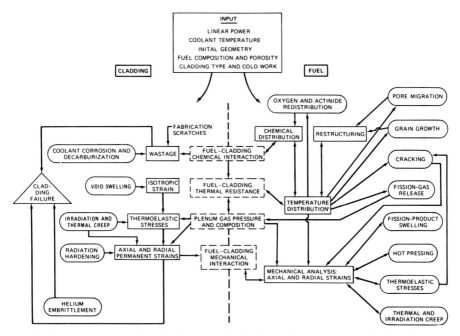

FIG. 51. Interrelation of mechanical, metallurgical, and chemical processes in fuel element
irradiation behavior [110, p. 567].

Irradiation creep studies in the DFR have included stress–relaxation
experiments on springs cut from Prototype Fast Breeder Reactor (PFR)
reference size tubing (20% cold worked M316SS, FV548, and solution
treated and aged PE16 [80,92]). The results indicate that the contribution
from a primary creep component is significant at temperatures as low as
300°C. This is of particular importance under conditions associated with
reactor power cycling imposing high stresses on the cladding, which will be
relieved at a rate determined by its stress–relaxation behavior. The PE 16, a
Nimonic alloy, exhibited a greater resistance to stress relaxation than the
austenitic steels.

A critical review of the core design literature was presented [280] in
which the effects of radiation-induced swelling and creep on fast reactor
design and performance were assessed. The design provisions for each core

component were presented together with the strategies to accommodate swelling and creep. This was followed by a period of silence in the literature on the design implications, " due in part to the traumatic and unsettling effect of this discovery on ' pre-swelling' approaches to core design " [250]. Both swelling and radiation-creep extrapolations were based on data at fluences at least a factor of ten below the target–fluence conditions in the FFTF and the United States demonstration plant. The models for this extrapolation are still not completely resolved and there is uncertainty concerning the prediction of fuel performance.

The significance of the swelling of core components in EBR-II has been reassessed by considering the deformation analyses of bowing of ducts, gap formation around spacer pads, fuel element bow, and duct-tube rounding from internal pressure. These ANL studies have evaluated fast neutron swelling and creep correlations, and analytical methods. An important conclusion is that " the fortunate occurrence of compensating fast-neutron creep is largely and coincidentally responsible for the satisfactory performance [of the core] " [280].

5. *Sodium Cooled Fast Breeder Reactors* (*LMFBR*)

The advantages of sodium as a coolant are its high boiling point, excellent thermal conductivity, large heat capacity, and low pressure operation [249]. Sodium can remove decay heat by convective cooling after shutdown if power is lost. The disadvantages of sodium are its chemical reactivity with air and water, corrosion of structural and fuel materials, opacity, prolonged radioactivity after reactor shutdown, void coefficient if boiling occurs or gas is entrained, and the requirement for heating on shutdown in order to retain fluidity. The characteristics of sodium-cooled fast breeder reactors are summarized in Table 4.

The development of sodium-cooled fast breeder reactors has been in progress for about 30 years. The first such reactor was the EBR-I, which operated from 1951 to 1963 in Idaho. It was replaced by EBR-II. Prototype demonstration size LMFBRs [about 300 MW(e)] are in operation in the Soviet Union (1972), France (1974), and the United Kingdom (1976).

The FFTR is under construction at Hanford, for testing fuels, cladding, and major components. The FFTR fuel element design parameters are shown in Table 4. About 40 km of cladding is required for the FFTR core [264–269]. Each fuel element consists of 217 fuel pins contained in a hexagonally-shaped flow duct. Spiral wire wrapping is used to give a pitch-to-diameter ratio of 1.24. The plutonium content of the fuel is varied radially across the core to flatten the radial power distribution. The first core contains about 600 kg of plutonium. The fuel quality assurance and vendor qualifications are based on the stringent requirements of meeting the

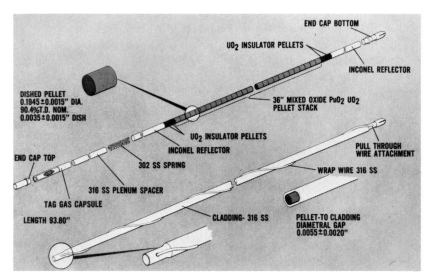

FIG. 52.   FFTF driver fuel pin [117].

specifications of composition (particularly plutonium content and distribution), density, O/M ratio, and impurity and moisture content. Computer-based process control, accountability, and materials flow control are employed. Statistical techniques are used in the procurement and fabrication of fuel (Figs. 52 and 53).

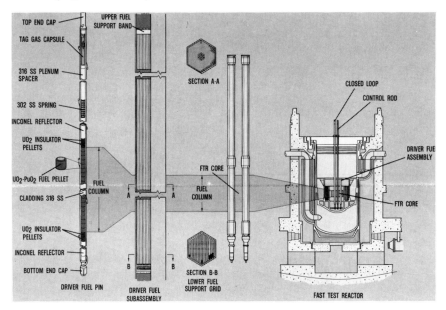

FIG. 53.   FFTF fuel system [117].

The 350 MW(e) Clinch River (Tennessee) LMFBR demonstration plant was under construction and was expected to be in operation by 1984. It was to provide information on the reliability, construction, operation, safety, and potential economics of LMFBRs for commercial power plants [261]. The reactor core consists of 198 fuel assemblies, each of which contains 217 wire-wrapped fuel rods, containing mixed uranium-plutonium dioxide fuel pellets. The axial and radial blanket regions are loaded with depleted uranium oxide pellets. The core region has inner (18.7%) and outer (27.1%) radial enrichment zones in the first core. Reflector and restraint assemblies surround the core. The fuel rod design is similar to the FFTR design, with the addition of axial blanket regions. There will be 150 radial blanket assemblies surrounding the core. In the blanket regions fertile ^{238}U is converted to fissile plutonium by capture of neutrons from the core. Table 4 summarizes the parameters of the CRBR and the commercial LMFBRs.

6. Gas-Cooled Fast Breeder Reactor (GCFR)

Because the first fast reactors had small high-power-density cores with very little room for cooling, the liquid metal fast breeder reactors were the type of breeders that received earliest and greatest attention in the United States and in most other countries. More recently, the economic and technological problems of sodium systems, and the rather modest breeding gain (0.2) (breeding ratio minus one) of the LMFBR with mixed-oxide fuel giving long doubling times, have led to increasing interest in gas-cooling for breeders in Europe, Japan, the United States, and the Soviet Union. The gas-cooled fast reactor breeder system presently under development in the United States is the metal-clad, oxide fueled, helium-cooled GCFR [295]. Except for the fuel, the power system resembles the HTGR: the primary system is integrated within a PCRV and will use similar components, such as the steam-driven helium circulators, helical-coiled once-through steam generators, helium purification system, etc. Thus, no new technology is required to build the GCFRs and, according to evaluations in Europe and the United States, the capital costs of GCFRs should be very close to those of HTGRs. The fuel elements of the GCFR can be made very similar to elements developed for the LMFBR, i.e., stainless-steel-clad (UO_2-PuO_2) rods. A high breeding ratio (1.4) is possible with the GCFR because of the very small interaction between neutrons and the helium coolant. Thus, the doubling time of the GCFR system can be made compatible with that of the current demand for electricity (8–10 yr).

A typical nuclear steam supply system for the GCFR is designed such that the entire primary system is integrated within a multi-cavity PCRV. The fuel elements, blanket (depleted UO_2 or ThO_2) elements, and shielding are contained in a central cavity, and the main and auxiliary cooling loops are located in cavities within the wall of the prestressed concrete reactor vessel (PCRV), similar to the arrangement in HTGRs. Steam-turbine-driven circulators circulate the helium coolant down through the reactor core, up through the steel and graphite shielding, and then through the steam generators back to the circulators.

The core of the 300-MW(e) GCFR consists of 91 fuel elements, 27 control elements, and 147 radial blanket elements. Although the active core is only 1 m high, the total length of the fuel elements is approximately 3.46 m, including top and bottom depleted UO_2 axial blankets, fission-product traps, and orifice (Fig. 54). Within the hexagonal duct, the fuel elements

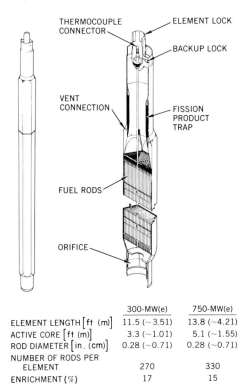

	300-MW(e)	750-MW(e)
ELEMENT LENGTH [ft (m)]	11.5 (~3.51)	13.8 (~4.21)
ACTIVE CORE [ft (m)]	3.3 (~1.01)	5.1 (~1.55)
ROD DIAMETER [in. (cm)]	0.28 (~0.71)	0.28 (~0.71)
NUMBER OF RODS PER ELEMENT	270	330
ENRICHMENT (%)	17	15

FIG. 54. GCFR fuel element.

contain 270 fuel rods, 7.2 mm in diameter, with artificial surface-roughening over three-fourths of the active core length to enhance heat transfer locally. The fuel is vented to the helium purification system. This provides management of fission gases and pressure equalization across the fuel-rod cladding and thus eliminates a major source of cladding stress, i.e., internal fission-gas pressure or external coolant pressure. In addition, the venting provides a means for locating and monitoring a leaking fuel or blanket element. The coolant outlet temperature from the core is measured by thermocouples installed in a tube in the center of each fuel element. Fixed, replaceable flow orifices are installed in the outlet end of the fuel and blanket elements. The fuel and blanket elements are held firmly in place in the grid plate by locks actuated from above the PCRV.

To provide control a number of fuel elements are modified by replacing the center 37 fuel rods by a guide tube that receives a 48-mm-diam moveable control rod containing boron carbide, which is vented directly to the helium coolant.

VII. MODERATOR AND CONTROL MATERIALS [305–335]

A. Moderator Materials

1. Nuclear Graphite and Pyrocarbons

a. Background   Much of the early work on reactors concerned applications in which the graphite was not above about 300°C. In this temperature range, irradiation by fast neutrons creates metastable interplanar atoms that move to traps or may react to form new hexagonal planes as the temperature is raised above the irradiation temperature, thus releasing stored energy. In the earlier studies emphasis was placed on this effect as well as on changes in dimensions, thermal conductivity, elastic moduli, coefficients of thermal expansion and strength. Later work has pertained to applications in which graphite is at 350–1500°C and stores little energy.

b. Origins of Graphite and Graphitoid Structures   A variety of very useful and interesting graphitoid and graphitic carbonaceous materials result from the pyrolysis of hydrocarbons. We differentiate them broadly into carbons and graphites according to the degree of hexagonal crystallinity, and also according to ranges of electrical, thermal, and mechanical properties. Both carbons and graphites are made up of very large planar (not necessarily flat) molecules within which most carbon atoms are bonded to three neighbors by strong aromatic carbon–carbon bonds, roughly 4/3 carbon–carbon single bonds per atom. Between planes relatively weak Van der Waal and electron exchange forces prevail that are influenced by the extent, curvatures, and registry of the planes. Graphitoid carbons show broad x-ray diffraction lines from which may be estimated a mean square distance of about 2–10 nm over which *a* and *b* stacking and hexagonal symmetry prevails. Graphites show much sharper x-ray diffraction patterns consistent with predominantly hexagonal stacking of thousands of graphitic planes. Crystallites contain varying amounts of faults, for example, rhombohedral stacking and also rotations about [002]. The interplanar spacing of graphitic crystals is 0.3354 nm while that of carbons may be as great as 0.3358 nm. Electron exchange bonding between planes apparently increases with the extent, flatness, and hexagonal registry of the two-dimensional macromolecules. The gap between the electronic valence and conduction bands narrows to zero, the number of electrons and holes increases, the number of traps for electrons decreases and the mean free patch for both charge carriers and phonons increases as the graphitic structure becomes more extensive and less convoluted.

Extensive rearrangement of the planar array of large, aromatic molecules in some coking processes results in the formation of needle cokes, so called because comminution yields particles that are elongated normal to the [001] direction. When mixed with a pitch binder, extruded, baked, and graphitized, logs formed from needle cokes have considerable preferred orientation of [001] directions normal to the direction of flow. These graphite bars have good longitudinal thermal conductivity [$i\,j\,o$] and behave relatively well as arc electrodes for steel making. Because of its commercial availability, needle-coke graphite has been studied extensively for reactor use.

Some starting materials and conditions of coking yield small particles that orient much less during flow. In addition, molding and other forming methods that cause less unidirectional flow than extrusion favor less orientation. Combinations of these effects may be employed to produce the nearly isotropic graphite that is preferable for nuclear reactor service. Smaller particles favor less extensive cracks in the material and hence greater mechanical strength and lower tesselated internal strains and stresses due either to shrinkage or to irradiation-induced growth of crystallites in their [002] direction. Naturally occurring Gilsonite cokes appear to yield graphites of this sort. Some recently developed commercial cokes are considerably better in this regard [315]. These graphites undergo moderate dimensional changes and retain adequate strength at neutron fluences approaching 4×10^{22} neutrons/cm^2.

c. Mechanisms of Irradiation Effects in Graphitic Materials   The classical and fairly simple picture of the primary displacement of atoms by fast neutrons followed by secondary, tertiary, and *n*-ary displacements by recoiling atoms gives a satisfactory description of the behavior of graphite and pyrocarbons in use. The minimum energy required to displace carbon atoms appears to be around 25eV. When averaged over all directions and kinds of collisions, of secondary carbon atoms or ions displaced by a fast neutron, 60eV appears to be the effective displacement energy [312]. The Kinchin–Pease theory of atom displacement predicts the approximate number of out-of-plane atoms as determined by neutron diffraction. Displaced atoms take up positions between the graphitic planes. Above about 120°K these interplanar atoms may move. Not all of their reactions are known, but it is clear that among the possibilities are the formation of successively higher aggregates of atoms and the growth of graphitoid new planes, the combination with vacant sites, and diffusion to edges of crystallites as mentioned above. Undoubtedly there are many other internal imperfections that trap atoms. In this discussion we are emphasizing the changes in graphite above about 700°C. Above this temperature not only may the interplanar atoms move freely either to aggregate or to add to crystal discontinuities, but also vacancies may rearrange. Above 1200°C, vacancies aggregate and result in pairs of dislocations in a single plane and thus cause its shrinkage. Thus, in the higher temperature range the predominant effect of neutron irradiation is to cause an extension in the [001] direction due to the growth of new planes and a shrinkage in the $[i\,j\,o]$ directions due to the coalescence of vacancies. It is interesting that these processes in pyrocarbons result in some increase in crystallinity suggesting that the new planes and the contracting planes tend, to some degree, to come into the graphitic registry [313]. At first, densification of both graphites and pyrocarbons occurs indicating that the extensions in the [001] direction tend to fill in the shrinkage cracks, while in some regions the shrinkage in the $[i\,j\,o]$ directions tends to pull the structure together. In any event, in all of these carbonaceous materials at fluences exceeding 2–4×10^{21} neutrons/cm^2 the resultant forces expand the material and above 1–2×10^{22} neutrons/cm^2 introduce new cracks and the material loses its integrity. Coarser crystalline needle coke graphites such as H-321 tend to crack badly by 2×10^{22} neutrons/cm^2. Low density pyrocarbons shrink so much that, if constrained, they crack at fluences $>2 \times 10^{21}$ neutrons/cm^2. Graphites having preferred orientation, such as the needle cokes, show expansion in the directions perpendicular to the direction of flow in manufacture and shrinkage in the other direction.

d. Commercial Nuclear Graphites Graphite has been used extensively in nuclear power reactors because it is an excellent neutron moderator material with good physical and mechanical properties at elevated temperatures, and is relatively low in cost [311].

The most stable commercial grade graphites are the near-isotropic materials, such as H-451 which is being evaluated for the large HTGRs (Fig. 55). There are no experimental data above 4×10^{22} nvt $(E > 0.1$ MeV) fluence range, the probable limit of usefulness of graphite in reactors [312–315].

The viscoelastic response of graphite materials in irradiation environments has been analyzed [307]. In this approach, the effect of irradiation-induced creep and dimensional changes are considered in the stress analysis.

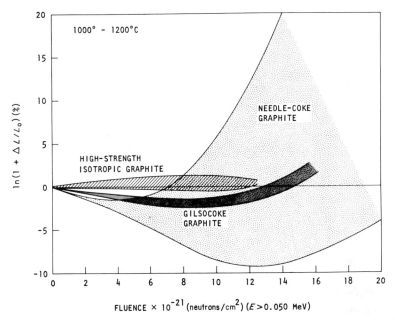

FIG. 55.   Dimensional changes versus fast neutron fluence at 1000–1200°C for needle-coke, near-isotropic, and high-strength isotropic graphites [316a]. The widths of the curves represent the scatter in data points.

The mechanical response of graphite is assumed to be viscoelastic and the constitutive relations are inferred from measurements. A computer program has been developed for analysis of plane strain, generalized plane strain, and axisymmetric problems, using the finite element method. The material properties are considered to be temperature dependent as well as neutron flux dependent.

   The dimensional-change behavior of nuclear graphites generally is in the pattern mentioned above. In extruded material there is first a contraction in the direction transverse to the extrusion direction and then a turn-around and rapid expansion. In the direction parallel to extrusion there is a contraction at an increasing and then decreasing rate, followed again by a turn-around and expansion. The rapid expansion is associated with porosity generation between filler particles; the isotropic and finer grained materials expand at a lower rate. The initial shrinkage rate is temperature dependent, decreasing up to 800°C and then increasing with increasing temperature up to 1200–1400°C [312–316].

   The thermal conductivity of the graphites is primarily by phonons and is markedly reduced by irradiation at low temperature. The rate of reduction declines and the conductivity approaches a saturation level, which increases as the irradiation temperature increases. Eventually, when irradiation-induced expansion starts, the conductivity again decreases probably because of internal cracking. The time constant for approach to saturation appears to increase linearly with irradiation temperatures, whereas the conductivity after saturation increases exponentially with irradiation temperature.

The irradiation-induced creep of graphite has been studied [314]. The transient creep strain and the steady-state creep constant increase with increasing irradiation temperature over the interval 500–1200°C. For different graphites, the transient creep strain and steady-state creep constant are both inversely proportional to Young's modulus. Creep strains up to 2.5% in tension and 5% in compression have been reported. However, there is some indication that in isotropic graphites compressive creep slows down or stops when the strain reaches 2–3%.

The new near-isotropic commercial graphites, which use isotropic petroleum coke as filler for improved radiation stability, have been fabricated in large sections and evaluated for use as core components in large HTGRs. These graphites are typified by grades H-451 (Great Lakes Carbon) and TS-1340 (Union Carbide Corp.). The mechanical properties and irradiation behavior (to about 8×10^{21} nvt fast fluence, and 1×10^{22} nvt for H-429 prototype) of the near-isotropic graphites have been determined [315].

The price of the new near-isotropic graphite is about $3.15 per kg. Machining this material into the required shapes in the shielding structures would cost approximately $0.65 per kg. The conventional graphite grades used in the HTGR core support region (PG-X) and permanent side reflectors (Great Lakes HLM) cost approximately $2.20 per kg, with the machining cost also about 20% of the unit price. However, less irradiation data are available on the new graphites.

2. Zirconium Hydride

The atomic density of hydrogen in many metal hydrides is greater than in liquid hydrogen or in water. Metal hydrides are efficient moderators and neutron shielding materials and are particularly suitable for minimizing the core shield volume [318].

Examples of the use of metal hydrides as moderators include the following reactor systems. In the gas-cooled Aircraft Nuclear Propulsion (ANP) reactor program (General Electric), yttrium hydride was in the form of large, hexagonal cross section rods which were metal clad and had central axial holes for fuel elements and coolant channels. These elements were capable of operation at 1000°C in air. The SNAP reactors (Atomics International) used uranium–zirconium hydride rods as a combination fuel–moderator element [320]. A similar uranium–zirconium hydride fuel element was developed for the TRIGA reactors (General Atomic) [319]. The sodium-cooled prototype reactor KNK (Interatom and Karlsruhe) contained metal-clad zirconium hydride as a moderator element for operation at temperatures up to 600°C.

The ZrH system is essentially a simple eutectoid, containing at least four separate hydride phases in addition to the zirconium and allotropes [321].

The hydrogen dissociation pressures of hydrides have been measured, and the equilibrium dissociation pressures in the ZrH system are given as follows. In the delta region, the dissociation pressure equilibria of the zirconium–hydrogen binary may be expressed in terms of composition and temperature by the relation [322]:

$$\log P = K_1 + (K_2 \times 10^3)/T,$$

where

$$K_1 = -3.8415 + 38.6433X - 34.2639X^2 + 9.2821X^3$$
$$K_2 = -31.2982 + 23.5741X - 6.0280X^2$$

and P is the pressure, in atmospheres, T the temperature, in °K, and X the hydrogen-to-zirconium atom ratio.

In the presence of thermal gradients in ZrH rods hydrogen composition gradients will be created in accordance with the phase–temperature–composition equilibria of the ZrH system. This will result in the unusual phenomenon of a negative thermal expansion [320] because "the hydrogen will redistribute, enriching the cold or concave side and depleting the hot or convex side of a rod (i.e., the lower temperatures correspond to a greater relative volume than higher temperatures). This will result (after a period of time depending on the hydrogen diffusion rate) in the cold side having larger dimensions than the hot side and, therefore, a reversal of the configuration with the cold side convex and the hot side concave. Furthermore, if during this redistribution and bowing change the thermal gradients are altered, as would be the case in an operating reactor, the conditions for sustained cycling are obtained." This phenomenon was verified experimentally.

Most of the irradiation experience to date is limited to the uranium–zirconium hydride fuels used in the SNAP and TRIGA reactors. The presence of uranium (about 8–10 wt%) complicates the situation because of the damage resulting from fission recoils and fission gases. Transmission electron microscope studies of irradiated samples indicated the presence of voids within the range of fission recoils, in the vicinity of the uranium fuel particles, with the regions far from the fuel particles retaining a microstructure similar to unirradiated material. The U–ZrH fuel exhibits high growth rate during initial operation, the so-called offset growth period, which has been ascribed to the vacancy–condensation type of growth phenomenon over the temperature range where voids are stable [320]. The voids are also associated with the delta–epsilon phase banding.

The behavior of unfueled delta and epsilon zirconium hydride under fast neutron irradiation has also been investigated [323]. In these experiments the samples (35 mm × 15 mm diam) were irradiated at temperatures up to 580°C to fluences of up to 1.15×10^{21} nvt ($E > 1$ MeV). The axial thermal gradients were between 520°C and 230°C. The post-irradiated tests include measurements of density and dimensions, metallography, x-ray diffraction, and hydrogen analysis. The results indicated that the epsilon phase had swelled about 0.51%, whereas the delta phase showed no growth. The growth was attributed to a radiation-induced change of the tetragonal epsilon structure.

3. Beryllium

Beryllium metal has been used as the moderator and reflector in a number of reactors, e.g., test reactors (MTR, ETR, ATR); the Oak Ridge research reactors (ORR and HFIR); EBR-II; research reactors in France, Japan, and the Soviet Union; and the SNAP reactors. Beryllium has many of the desirable nuclear properties for a moderator and reflector, such as low neutron-absorption cross section, high neutron-scattering cross section, a low atomic weight, high melting point, high specific heat, and fairly good corrosion resistance in water [324]. Its disadvantages are high cost, low ductility, toxicity, crystallite growth and swelling under irradiation at high temperatures. These features combine to eliminate its use as a structural material in commercial reactors.

Beryllium has a close-packed hexagonal crystal structure, alpha form, with a c/a ratio of 1.5671 and lattice parameters $a = 2.2866$ Å and $c = 3.5833$ Å. The alpha form transforms to the body-centered cubic form at about 1250°C, with a lattice constant of $a = 2.551$ Å.

There are a number of parameters that influence the mechanical properties of beryllium, such as the orientation of the test specimen, purity, iron and BeO content, grain size and anisotropy, strain rate, temperature, method of production, surface condition, and irradiation.

Commercial beryllium has good resistance to atmospheric corrosion. It shows variable behavior in high temperature water. The oxidation of beryllium in air or dry oxygen follows a parabolic rate at temperatures up to about 700°C. However, at temperatures above 750°C a "break away" reaction associated with the appearance of Kirkendal voids sets in after an induction period [325].

Chronic or acute poisoning results from inhalation of excessive amounts of beryllium dust. The permissible levels are 2 $\mu g/m^3$ of air for a 40-hr work week, and 0.01 $\mu g/m^3$ of air on a monthly average basis for nonindustrial environments. Very small acicular particles are the worst offenders.

Fast neutron transmutation reactions in beryllium yield helium and tritium through (n, 2n) and (n, α) capture, namely:

1. Be^9 (n, 2n)$2He^4$, threshold 1.85 MeV, 125 mb cross section
2. Be^9 (n, α)He^6, threshold 0.71 MeV, 25 mb cross section
   $He^6 \rightarrow Li^6$, $t_{1/2} = 0.82$ sec
   Li^6 (n, α)H^3, 950 b cross section
   $H^3 \rightarrow He^3$

For thermal neutrons, the reactions are:

1. Li^6 (n, α)H^3
2. He^3 (n, p)H^3

For each transmuted Be^9 atom, two atoms of He^4 are produced. Thus, for an exposure of 10^{22} neutrons/cm^2 about 22 cm^3 gas at STP, equivalent to 1 atom %, is present in the metal. The solubility of helium in beryllium is extremely low. Because of its high cross section the Li^6 soon reaches an equilibrium level, whereas the quantity of He^3 will increase with time. Both these isotopes have high neutron cross sections.

At temperatures below approximately 500°C the rate of growth of irradiated beryllium is about 0.2% per 10^{22} nvt ($E > 1$ MeV) up to a fluence of 10^{22} nvt. The helium at low temperatures is in enforced solid solution in the beryllium lattice, and no gas bubbles are observed. At elevated temperatures, the helium gas migrates and agglomerates into gas bubbles, which results in marked swelling. The gas bubbles themselves can migrate by a surface diffusion mechanism and coalesce into larger bubbles.

The stresses arising from the inhomogeneous growth of beryllium can result in cracking even in the low temperature range.

Bowing, cracking, and swelling occurred in the Materials Test Reactor (MTR) reflector blocks subjected to fast neutron fluences in the range 4–10×10^{21} nvt.

The design thermal stress for beryllium reflectors has been limited to 12,000–15,000 psi. Considerable swelling occurs when beryllium is irradiated to high fluences (above approximately 1×10^{21} nvt) at high temperatures (Fig. 56) or are annealed at high temperatures after irradiation at low temperatures. The results of the post-irradiation annealing studies are shown in Fig. 57. The swelling threshold (or breakaway) temperature decreases as the fast neutron fluence increases. The lowest threshold temperature reported is 550°C for a fast fluence of 1.6×10^{22} nvt. However, there is evidence from yield strength measurements which suggests that helium mobility is significant at temperatures above 300°C.

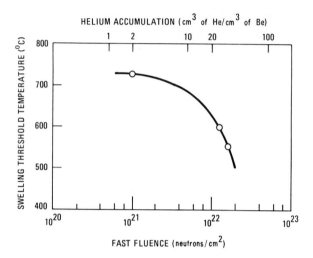

FIG. 56. Swelling threshold temperature of beryllium as a function of gas content and fluence [324].

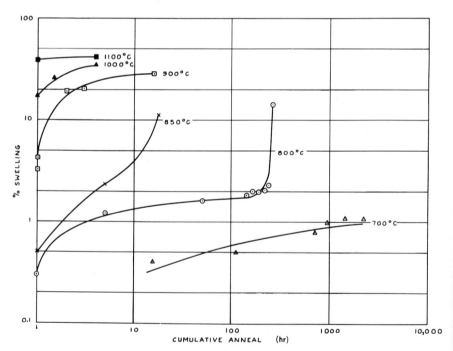

FIG. 57. Time dependence of swelling at various temperatures for beryllium irradiated at 280° and 480°C to a fast fluence of 2.75×10^{21} neutrons/cm^2 [325a].

4. Beryllium Oxide

There has been renewed interest in the use of beryllium oxide as a reflector and moderator material in nuclear reactors. In addition to excellent reflecting and moderating properties and low neutron capture cross section, BeO has favorable physical, mechanical, and chemical properties which allow its use at elevated temperatures. The relatively high cost and the deleterious property changes under irradiation have limited the applications of BeO in nuclear reactors [326, 327].

The properties of BeO bodies are governed largely by the nature of the starting materials and the methods of fabrication. The important variables in the fabricated piece include the composition, grain size, grain boundary composition, pore structure, and density.

At room temperature, BeO is a remarkably good thermal conductor, because of its large Debye Θ, about an order of magnitude better than most oxides and about equal to aluminum metal. With increasing temperature, the thermal conductivity of BeO decreases approximately as Θ/T, becoming about one-tenth the room temperature value above 1000°C. Above about 1600°C the thermal conductivity increases, possibly due to exciton transport.

BeO is very stable in hydrogen, vacuum, inert gas, and dry air to temperatures up to at least 2000°C. It is relatively inert to cold mineral acids and to aqueous alkalis, and is resistant to attack even at high temperatures. It will dissolve in hydrofluoric and phosphoric acids and in fused KOH. The troublesome reaction of BeO with water vapor at high temperatures has been discussed. It is generally agreed that the reaction is

$$BeO(s) + H_2O(g) = Be(OH)_2(g).$$

This corrosion reaction is rate limited by diffusion in the gaseous boundary layer, presumably by the slower moving $Be(OH)_2$ from the surface into the main gas stream.

After irradiation to 10^{20}–6×10^{20} nvt ($E > 1$ MeV) below 100°C, BeO single crystals exhibit an x-ray diffraction pattern characterized by duplex reflections, with one sharp and one broad component [326]. The sharp component is attributed to the crystal having been expanded by a random distribution of point defects. The broad component is attributed to the presence of interplanar clusters lying parallel to the basal plane. Annealing between 800° and 1600°C caused nearly complete recovery to the original line shape and position. These changes indicate removal of point defects and some growth of cluster size during annealing.

Dimensional changes during low temperature irradiations result from two processes: anisotropic lattice expansion and the formation of grain–boundary cracks. The resultant stresses across the boundaries between grains causes the latter process. The two processes lead to two approximately linear expansion rates. The initial rate of 0.2–0.3% per 10^{20} nvt corresponds to the lattice expansion, whereas the subsequent high expansion rate of 0.9% per 10^{20} nvt reflects the additional effect of grain–boundary cracks. Continued expansion at this higher rate leads to friability and powdering of the materials.

Processes that contribute to the expansion at elevated temperatures include lattice expansion, grain–boundary separation, and formation of defect clusters, and helium bubbles. Above $\sim 700°C$ the lattice expansion occurs primarily along the c-axis and becomes smaller as the temperature increases. The expansions permitted by grain–boundary separation above 700°C and at 100°C are comparable, except that total expansions range up to $\sim 1.0\%$ at higher temperatures compared to 0.4% at 100°C. Unlike the situation at low temperature, lattice expansion and grain–boundary separation do not account for all of the observed volume increase at 700°C. Helium bubbles and defect clusters probably contribute to the expansion.

Associated with the anisotropic growth during irradiation is the possibility of the formation of small coherent domains of the high temperature beta-phase under irradiation [327].

B. Control Materials

1. Boron Carbide

Boron carbide is used as the neutron absorber material in the control rods of many reactors [328–331]. The absorption of neutrons by ^{10}B results in the primary formation of ^7Li, helium, and tritium by the reactions:

$$^{10}_{5}B + ^1_0n \rightarrow ^4_2He + ^7_3Li + 2.3 \text{ MeV}$$

$$^{10}_{5}B + ^1_0n \rightarrow ^3_1T + ^4_2He + ^4_2He$$

The fast neutron capture cross section of the ^{10}B isotope is greater than that of any other isotope. The absorption cross section in a thermal neutron flux is much larger than in a fast reactor spectrum, resulting in considerable self shielding in a thermal reactor flux and a sharply decreasing reaction profile. Hence, it is difficult to extrapolate thermal reactor irradiation data to predict the behavior of boron carbide in fast reactor spectra, where there is virtually no self-shielding and the reaction rates and irradiation are homogeneous throughout the absorber material.

Boron carbide has a boron concentration of 85% of that of elemental boron. Natural boron contains 19.8% of the high cross-section isotope ^{10}B, and the content of ^{10}B in natural boron carbide is 14.7%. The thermal neutron absorption cross section decreases with increasing neutron energy by the $1/v$ relation for neutron energies below 100 eV. It remains fairly constant for energies between 100 eV and 0.1 MeV and has several resonances between 0.5 and 5 MeV. The cross section in a fast reactor spectrum is about 1 b.

Boron carbide pellets and structures can be produced by cold-pressing and sintering (70–80% density) or by hot pressing. In the hot pressing operation the B_4C powder is first cold pressed into pellet form and then hot pressed in graphite dies at temperatures from 2050° to 2300°C under pressures of 1500 psi. The density is controlled by varying the temperature and pressure. Some reactors have used boron carbide in the powder form, vibratory packed in 20% cold-worked type 316 stainless steel cladding.

The nuclear, physical, mechanical, and chemical properties of boron carbide have been determined. The compatibility of boron carbide with stainless steel cladding depends on the stability of a barrier layer.

There is extensive experience on the irradiation behavior of boron carbide, although much of this is from thermal reactor tests which are of limited value for the fast neutron irradiation in fast breeder reactors. Burnup levels to 80×10^{20} captures/cm^3 at 650–850°C have been achieved with a swelling of $\sim 3.5\%$.

2. Boronated Graphite for Control Rods and Shielding

The irradiation behavior of boronated graphite has been studied for the HTGR program [308]. Boronated graphites containing 23–43 wt% boron as B_4C were irradiated at 300–750°C, to fast neutron fluences up to 7×10^{21} nvt ($E > 0.18$ MeV). Irradiation-caused anisotropic dimensional change related to the preferred orientation of the graphite crystallites in the graphite matrix, a decrease in thermal conductivity, and an increase in thermal expansivity. The dimensional changes in the boronated graphites (less than 2% change in dimensions at 300–750°C and 7×10^{-1} nvt) were related both to the fast neutron fluence and alpha particles from ^{10}B (n, α). The damage increased with increasing ^{10}B isotope enrichment of the boron in the B_4C particles. The use of natural boron in the boronated graphite resulted in dimensional changes independent of the ^{10}B burnup gradient and correlated with the fast neutron fluence. The use of ^{10}B-enriched B_4C led to distortion directly related to the ^{10}B burnup gradient. The dimensional changes were not significantly influenced by swelling of the B_4C particles during the irradiation. (Earlier work at Hanford on boronated graphite has been reported [310].)

The boron carbide–graphite bodies are fabricated by either extrusion or warm pressing and then heat treated to 2000°C in an inert atmosphere. Heat treatment is limited to below 2200°C in order to prevent migration of boron into the graphite crystals, which would enhance radiation swelling of the matrix. The properties of the boronated graphite developed for the HTGR control materials (40 at %B) are described in reference [308].

3. Silver-Base Alloys

The combination of silver with 15 wt% cadmium and 5 wt% indium provides a control rod alloy with suitable neutron absorption properties over the spectrum of neutron energies present in pressurized light water reactors. This alloy clad in stainless steel or Inconel has been used as control rod material in PWRs [335]. However, with the rapid increase in the price of silver, alternate materials are under consideration.

4. Europium Oxide

Europium oxide has been under development and is being considered as a neutron absorber material for use in the control rods of fast breeder reactors in the United States, Britain, Germany, and the Soviet Union [332]. The BOR-60 fast breeder reactor (Soviet Union) has operated satisfactorily since 1972 with europium oxide in one of the control rods. Europium oxide control elements for the United States fast test reactor would be operable in the same way as the initial boron carbide rods, namely, pellets sealed in type 316 stainless steel tubing. The principal difference in design results from the absence of gas generation in europium oxide under irradiation with its (n, γ) reaction, and longer reactivity lifetime. This allows the use of thinner wall cladding with no gas plenum. With a pellet density of 93% and a diametral gap of 8 mils (200 μm) (to accommodate 1.5% swelling) the europium oxide assembly is expected to have at least the same nuclear worth (reduction of neutron multiplication) as the rods containing B_4C. The centerline temperature is calculated to be between 700° and 950°C, based on unirradiated thermal conductivity values. Lifetimes of at least 2 yr are predicted, based on assessment of the probable changes in the reactivity worth of Eu_2O_3 when exposed to fast neutrons because the nuclides resulting from transmutation also have large cross-sections. These studies indicate a rate of loss of reactivity worth with neutron exposure about one-third that of control rods containing B_4C. The main areas of concern for the FFTR application are determining actual nuclear worths, maintaining pellet dimensional stability, and accommodating decay heating. About 600 kg of europium oxide would be required for the FFTR if it is used in the control rods. The irradiation stability of europia is reported to be very good [332–334].

VIII. PRESSURE VESSELS [336–343]

Nuclear power reactors are contained in two types of pressure vessels, namely, steel pressure vessels (for most types of reactors) or prestressed concrete pressure vessels (for many gas cooled reactors). The exception to this is the use of Zircaloy pressure tubes in the heavy water moderated reactors (CANDU and SGHWR). The steels used are ferritic low alloy steels (Mn–Mo–Ni grades, type A302-B, A537B, and A533-B in the United States), which are lined with stainless steel in LWR applications.

The pressure vessel materials are subject to neutron irradiation during operation of the reactors. This results in significant effects on the mechanical properties of the steels used for the pressure vessels, primarily an increase of the yield strength, decrease of the ductility, and a rise in the brittle–ductile transition temperature and decreased fracture toughness. Available information on neutron irradiation embrittlement of reactor pressure vessels has been well reviewed recently in an IAEA monograph by L. Steele [336] and in the proceedings of a specialist's meeting on radiation damage units [338].

The neutron exposures on the LWR vessels range from about 5×10^{18} neutrons/cm^2 > 1 MeV to as high as 5×10^{19} n/cm^2 > 1 MeV. The important factors governing radiation embrittlement of pressure vessel steels are the sensitivity of the steel to embrittlement, the neutron fluence and energy spectrum, and the irradiation temperature. Much useful information is being accumulated from reactor vessel surveillance programs. The profound influence of minor constituents ("tramp impurities"), particularly copper and phosphorus, on the irradiation embrittlement of steel at elevated temperatures has been demonstrated at the Naval Research Laboratory. The temperature range of transition can be raised by as much as 300°C by neutron irradiation (Fig. 58).

The main criterion used in specifying the operational limitation of the pressure vessel steel is the nil-ductility transition temperature (NDT), which in the irradiated steel must not exceed 33°C below the lowest operating temperature. An important theoretical development in recent years has been the concept of the damage function, which evaluates the relative damage by a given neutron energy spectrum [338]. There is no significant temperature effect from room temperature up to approximately 230°C, above which the damage diminishes with increasing temperature.

The fine-grained vacuum deoxidized steels with low impurity content (Cu < 0.1%, P and S < 0.012%) provide a material with remarkably good resistance to irradiation embrittlement (Fig. 59) [337].

Recent assessments of the engineering damage cross sections for neutron embrittlement of pressure vessel steels have concluded that most (∼94%) of

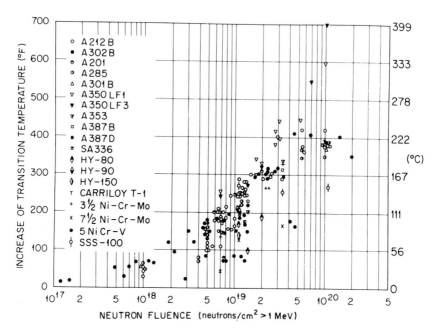

FIG. 58. Charpy-V-notch transition temperature increases of steels irradiated at less than 500°F (260°C) [336, p. 128].

the neutron embrittlement is caused by neutrons of energies > 0.1 MeV [341]. It is recommended that the threshold of > 0.1 MeV be adopted for use in assessment of neutron embrittlement of reactor pressure vessel steels and also that the computed "damage–fluence" values incorporating damage cross sections be used to account for the influence of neutron spectrum upon embrittlement.

Prestressed concrete reactor vessels (PCRV) have been used for the gas cooled reactors in France, Britain, and the United States. The PCRV consists of concrete reinforced with bonded, deformed steel bars and unbonded prestressing systems [342]. The main cavity, penetrations, and cross ducts are lined with a 20-mm-thick liner keyed to the concrete with anchors. The liner and closures form a leak-tight barrier for the primary coolant. The liner is cooled with water. Thermal insulation keeps the concrete within allowable temperature limits. The prestressing of the concrete acts to produce a net compressive stress on both the main cavity liner and the penetration liners, thereby making it highly unlikely that crack propagation could occur. The tendons that are in tension and that provide the confining strength are not irradiated. Moreover, they may be monitored and replaced if evidence of weakness is observed.

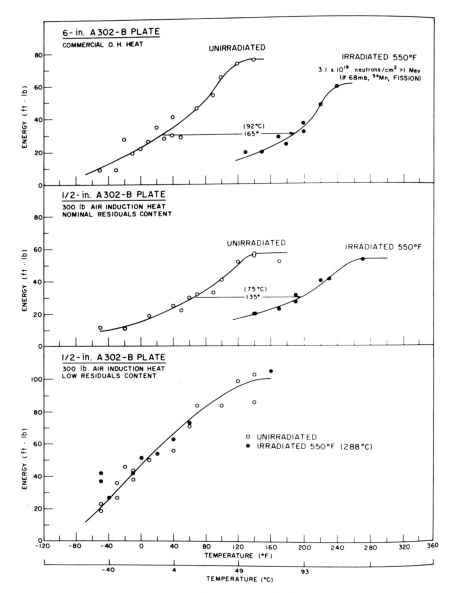

FIG. 59. Comparison of radiation embrittlement sensitivities of one large commercial heat and two air-induction heats of A302-B steel with nominal and low residual element contents based on Charpy-V-notch ductility following 550°C (288°C) irradiation [336, p. 116].

ACKNOWLEDGMENTS

The authors appreciate the support given by C. L. Rickard, R. H. Simon, and H. J. Snyder in the preparation of this chapter. They wish to express appreciation also for the permissions that were granted by authors and publishers to reproduce figures and tables from numerous publications. They are indebted as well to Joan Azar for her invaluable editorial assistance in the preparation of the manuscript.

REFERENCES

I. Introduction

[1]   A National Plan for Energy Research, Development and Demonstration: Creating Energy Choices for the Future, U.S. ERDA Rep. ERDA-48, Vols. 1 and 2, revised January 1976. Available from National Technical Information Service, U.S. Department of Commerce, Springfield, Virginia 22161.

[2]   Materials and Man's Needs, Summary Report of the Committee on the Survey of Materials Science and Engineering. The National Academy of Sciences, Washington, D.C., 1975.

[3]   32 scientists speak out! No alternative to nuclear power, *Bull. At. Sci.* **31(3)**, 4–5 (1975).

[3a] World list of nuclear power plants, *Nucl. News* **21(3)**, 49 (1978).

[3b] N. L. Franklin, *J. Brit. Nucl. Energy Soc.* **14**, 278 (1975).

[4]   "Directory of Nuclear Reactors," Vols. IV, VII, VIII, and IX. I.A.E.A., Vienna, 1962, 1968, 1971, 1970, respectively. Also current listing of reactors, I.A.E.A., Vienna, 1976.

[5]   M. T. Simnad, "Fuel Element Experience in Nuclear Power Reactors." American Nuclear Society, Hinsdale, Illinois, 1971.

[6]   H. Bethe, *Sci. Am.* **234(1)**, 21 (1976).

[7]   *Proc. Symp. Reliability Nucl. Power Plants, Innsbruck, April 14–18, 1975.* I.A.E.A., Vienna, 1975.

[8]   R. J. Campana and S. Langer, "Nuclear Power and the Environment: Questions and Answers." American Nuclear Society, Hinsdale, Illinois, 1976.

[9]   *Proc. Int. Conf. Peaceful Uses of At. Energy, 1st, 2nd, 3rd, 4th, Geneva, 1955, 1958, 1964, 1971.* United Nations, New York and I.A.E.A., Vienna, 1956, 1958, 1965, and 1972, respectively.

[9a] M. Benedict, *Technol. Rev.* **78**, 57 (1976).

[10] B. Lustman and F. Kerze, Jr., "The Metallurgy of Zirconium," National Nuclear Energy Series, Div. VII, Vol. 4. McGraw-Hill, New York, 1955.

[11] "Zirconium in Nuclear Applications," Special Tech. Publ. 551. American Society Testing Materials, Philadelphia, Pennsylvania, 1974.

[12] H. G. Rickover, *Met. Eng. Quart.* **3(1)**, 1 (1963).

[13] P. Fortescue, *Ann. Nucl. Energy* **2**, 787 (1975).

[14] H. M. Finniston and J. P. Howe, (eds.), "Progress in Nuclear Energy: Metallurgy and Fuels," Ser. V, Vols. 1, 2, 3, 4. Pergamon, Oxford, 1956–1959.

[14a] IMD Special Report series on, *Int. Symp. Nucl. Metall.*, conducted by the Metallurgical Society of the AIME, New York. Edwards Brothers, Ann Arbor, Michigan.

[15] N. Rasmussen (ed.), An Assessment of Accident Risks in U.S. Commercial Nuclear Power Plants, Nuclear Regulatory Commission Rep. WASH-1400, 1975. Available from National Technical Information Service, U.S. Department of Commerce, Springfield, Virginia 22161.

[16] Proposed Final Environmental Statement, Liquid Metal Base Breeder Program, USAEC Rep. WASH-1535, 1974. Available from National Technical Information Service, U.S. Department of Commerce, Springfield, Virginia 22161.

[17] T. S. Kuhn, "The Structure of Scientific Revolutions." Univ. of Chicago Press, Chicago, Illinois, 1970.

[18] G. Schleuter, Uranium demand of symbiotic reactor strategies during the transition period to fissile self-sufficiency, *Ann. Nucl. Energy* (in press).

[19] Potential Nuclear Power Growth Patterns, USAEC Rep. WASH-1098, 1970. Available from National Technical Information Service, U.S. Department of Commerce, Springfield, Virginia 22161.

[20] Nuclear energy maturity, *Trans. Am. Nucl. Soc.* **20**, 1 (1975).

[21] C. T. Rombough and B. V. Koen, *Nucl. Technol.* **26**, 5 (1975).

[21a] S. S. Penner, *Int. J. Energy* **1**(1), 45 (1976).

[22] R. M. Rotty *et. al.*, Net Energy From Nuclear Power, Rep. IAE 75-3. Institute of Energy Analysis, Oak Ridge, Tennessee, 1975.

[23] R. L. Murray, "Nuclear Energy." Pergamon, Oxford, 1975.

[24] V. S. Boyer, The economics of nuclear power, presented at the *Congr. Seminar Econ. Viability of Nucl. Energy, 3rd, Washington, D.C., June 7, 1976.*

[25] S. Glasstone and A. Sesonske, "Nuclear Reactor Engineering." Van Nostrand-Reinhold, Princeton, New Jersey, 1967.

[26] A. Sesonske, Nuclear Power Plant Design Analysis. Available as TID-26241 from National Technical Information Service, U.S. Department of Commerce, Springfield, Virginia.

[27] M. Benedict and T. Pigford, "Nuclear Chemical Engineering." McGraw-Hill, New York, 1976.

[28] *Proc. A.N.S. Nat. Top. Meeting Nucl. Process Heat Appl. 1st, Los Alamos, New Mexico, Oct. 1–3, 1974.* American Nuclear Society, Hinsdale, Illinois, 1975 (CONF-741032).

[29] M. T. Simnad and A. T. McMain, Jr., *in* "Energy Use and Conservation in the Metals Industry" (*Proc. Energy Mater. Conf. Metall. Soc., New York City, February 18, 1975* (Y. A. Chang *et al.*, eds.), p. 53. AIME, New York, 1975.

II. The Nuclear Fuel Cycle

[30] D. M. Elliott and L. E. Weaver (eds.), "Education and Research in the Nuclear Fuel Cycle." Univ. of Oklahoma Press, Norman, Oklahoma, 1972.

[31] E. A. Mason, *in* "Education and Research in the Nuclear Fuel Cycle" (D. M. Elliott and L. E. Weaver, eds.), p. 3. Univ. of Oklahoma Press, Norman, Oklahoma, 1972.

[32] R. G. Wymer, Nuclear Fuel Cycles: A Summary Review, ORNL Rep. TID-26071. Oak Ridge National Laboratory, Oak Ridge, Tennessee, 1971.

[33] E. Zebroski and M. Levenson, *in* "Annual Review of Energy" (J. H. Hollander and M. K. Simmons, eds.), Vol. I, p. 101. Annual Reviews Inc., Palo Alto, California, 1976.

[34] J. W. Clegg and D. D. Foley, "Uranium Ore Processing." Addison-Wesley, Reading, Massachusetts, 1958.

[35] R. C. Merritt, "The Extractive Metallurgy of Uranium." Golden Colorado School of Mines Research Institute, Golden, Colorado, 1971.

[36] J. J. Katz and W. Rabinowitch (eds.), "The Chemistry of Uranium," National Nuclear Energy Series, Div. VIII, Vol. 5. McGraw-Hill, New York, 1951.

[37] S. Villani, "Isotope Separation." U.S. ERDA, Washington, D.C., 1975.

[38] R. G. Wymer (ed.), *Proc. Int. Thorium Fuel Cycle Symp. 2nd Gatlinburg, Tennessee, May 2–6, 1966.* Available as CONF-660524 from Clearinghouse for Federal Scientific and Technical Information, Springfield, Virginia 22161.

[39] R. G. Wymer and A. L. Lotts (eds.), *Proc. Sol-Gel Processes and Reactor Fuel Cycles, Gatlinburg, Tennessee, May 4–7, 1970.* Available as CONF-700502 from Clearinghouse for Federal Scientific and Technical Information, Springfield, Virginia 22161.

[40] *Proc. Pacific Basin Top. Conf. Nucl. Develop. and the Fuel Cycle, Honolulu, Oct. 11–14, 1976.* American Nuclear Society, Hinsdale, Illinois, 1977.

[41] A. J. Feipot *et al.*, Pu fuel development and production, *Trans. Am. Nucl. Soc.* **20**, 596 (1975).

[42] O. Wick, "Plutonium Handbook," 2 Vols. Gordon and Breach, New York, 1967.

[43] N. L. Franklin, *J. Brit. Nucl. Energy Soc.* **14**, 273 (1975).

[44] *Proc. ANS Symp. Nucl. Safeguards, San Francisco, California, November 1973, Nucl. Technol.* **23**, 96 (1974).

[45] A. S. Kubo and D. J. Rose, *Science* **182**, 1205 (1973).

[46] *Proc. Symp. Combined Effects on the Environ. Radioactive Chem. Thermal Releases From the Nucl. Ind., Stockholm, June 2–5, 1975.* I.A.E.A., Vienna, 1975.

[47] *Proc. ANS Symp. Waste Management, Tucson, Arizona, April 1974, Nucl. Technol.* **24**, 265 (1974).

[48] R. G. Post and D. H. White, *in* "Education and Research in the Nuclear Fuel Cycle" (D. M. Elliott and L. E. Weaver, eds.), p. 321. Univ. of Oklahoma Press, Norman, Oklahoma, 1972.

[49] J. E. Lovett, "Nuclear Materials: Accountability, Management, Safeguards." American Nuclear Society, Hinsdale, Illinois, 1975.

[50] J. T. Long, "Engineering for Nuclear Fuel Reprocessing." Gordon and Breach, New York, 1967.

[51] A. S. Manne and O. S. Yu, *Nucl. News* **18(1)**, 46 (1975).

III. Uranium and Thorium Resources

[52] *Proc. Symp. OKLO Phenomenon, Libreville, Gabon, 23–27 June 1975.* I.A.E.A., Vienna, 1975 (Proceedings Series STI/PUB-405).

[53] D. M. Johnson, Some economic and technical aspects of uranium exploration, mining and milling, *Proc. World Energy Conf., 9th, Detroit, Sept. 23–27, 1974* (R. R. Ferber and R. A. Roxas, eds.), Vol. IV, p. 245. McGregor and Werner, Washington, D.C., 1975.

[54] A. Siegers, *Kerntechnik* **14**, 254 (1972).

[55] P. Hahn-Weinheimer, *Kerntechnik* **14**, 249 (1972).

[56] R. L. Faulker, Outlook for uranium production to meet further nuclear-fuel needs in the United States of America, *Proc. Int. Conf. Peaceful Uses of At. Energy, 4th, Geneva, Sept. 6–16, 1971* **8**, 23. United Nations, New York, 1972.

[57] *Proc. Am. Nucl. Soc. Special Session on Uranium Resources, New Orleans, June 1975, Nucl. Technol.* **30(3)**, 223 (1976).

[58] M. Klein *et al.*, Report on the LMFBR Program Review Group. Available from the National Technical Information Service, U.S. Department of Commerce, Springfield, Virginia 22161.

[59] M. F. Searle, Uranium Resources to Meet Long Term Uranium Requirements. Electric Power Research Institute, Palo Alto, California, 1974.

[60] M. R. Searle and J. Platt, *Ann. Nucl. Energy* **2**, 751 (1975).

[61] J. A. Brinck, *Euro-Spectra (Belgium)* **10**, 45 (1971).

[62] F. C. Armstrong, Uranium resources of the future: Porphyry uranium deposits, *in Proc. Symp. Format. Uranium Ore Deposits, Athens, May 6–10, 1974* p. 625. I.A.E.A., Vienna, 1974 (Proceedings Series STI/PUB-374).

[63] C. P. Haigh, The extraction of uranium from seawater, *Proc. CERR Symp. on Long-Term Stud.* U. K. Central Electricity Generating Board, London, 1974.

[64] F. E. Harrington *et al.*, Cost Commentary on a Proposed Method for the Recovery of Uranium From Seawater, ORNL Rep. ORNL/TM-4757, 1974. Available from National Technical Information Service, U.S. Department of Commerce, Springfield, Virginia 22161.

[65] W. I. Finch *et al.*, United States Mineral Resources, Nuclear Fuels, p. 455. U.S.G.S. professional paper 820. U.S. Geological Survey, Denver, Colorado, 1973.

[66] An ERDA-sponsored natural uranium resources evaluation (NURE) program begun in 1973 for completion in 1980.

[67] *Proc. IAEA Panel Uranium Exploration Geol., April 13-17, 1970, Vienna.* I.A.E.A., Vienna, 1970.

[68] *Proc. IAEA Panel on Nucl. Tech. and Mineral Resources Develop. Countries, Cracow, Poland, December 8-12, 1969.* I.A.E.A., Vienna, 1971.

[69a] W. D. Wilkinson, "Uranium Metallurgy," 2 vols. Wiley (Interscience), New York, 1962.

[69b] R. G. Bellamy and N. A. Hill, "Extraction and Metallurgy of Uranium, Thorium, and Beryllium." Pergamon, Oxford, 1963.

[69c] *Proc. Symp. Formation of Uranium Ore Deposits, Athens, May 6-10, 1974.* I.A.E.A., Vienna, 1974 (Proceedings Series STI/PUB-374).

[70] J. F. Hogerton, *Nucl. News* **19**, 73 (1976).

[71] M. A. Lieberman, *Science* **192**, 431 (1976).

IV. Radiation and Thermal Effects

[72] J. A. L. Robertson, "Irradiation Effects in Nuclear Fuels." Gordon and Breach, New York, 1968.

[72a] C. E. Johnson *et al.*, *Reactor Technol.* **15**, 305 (Winter 1972-1973).

[73] J. W. C. Dias and K. R. Merckx, *Ann. Nucl. Energy* **3**, 41 (1976).

[74] E. A. Aitken, Nuclear energy, In report of the *Conf. Thermodynam. Nat. Energy Prob. Warrenton, Virginia, June 10-12, 1974* p. 132. Available from the National Academy of Sciences, Washington, D.C.

[75] D. S. Billington and J. M. Crawford, "Radiation Damage in Solids." Princeton Univ. Press, Princeton, New Jersey, 1961.

[76] L. T. Chadderton, "Radiation Damage in Crystals." Wiley, New York, 1965.

[77] B. T. Bradbury and B. R. T. Frost, *in* "Studies in Radiation Effects in Solids" (G. Dienes, ed.), Vol. 2, Gordon and Breach, New York, 1967.

[77a] R. N. Singh, Theoretical Assessment of Some of the Effects of Fission-Induced Impurities on the Creep Behavior of Mixed-Oxide Fuel, ANL Rep. ANL-75-59, 1975. Argonne National Laboratory, Argonne, Illinois, 1975.

[78] A. Hoeh and H. Matzke, *J. Nucl. Mater.* **48**, 157 (1973).

[79] T. S. Elleman *et al.*, *J. Nucl. Mater.* **30**, 89 (1969).

[80] Irradiation effects in structural alloys, *Proc. Biann. Sympo. Committee E-10 on Radioisotopes and Radiat. Effects* ASTM Special Tech. Publ. STP-341, -380, -426, -457, -481, -529, -570, and DS-54. American Society Testing Materials, Philadelphia, 1963-.

[81] O. D. Sherby and M. T. Simnad, Prediction of atomic mobility in metallic systems, *ASM Trans. Quart.* **54**, 227 (1961).

[82] R. S. Barnes, *J. Nucl. Mater.* **11**, 135 (1964).

[83] C. Cawthorne and E. J. Fulton, *Nature (London)* **216**, 575 (1967).

[84] J. W. Corbett and L. C. Ianniello (eds.), *Proc. Int. Conf. Radiat-Induced Voids in Met., Albany, New York, June 9-11, 1971.* Available from National Technical Information

Service, U.S. Department of Commerce, Springfield, Virginia 22161 (AEC Symposium Series 26).

[85]   S. F. Pugh *et al.* (eds.), *Proc. Eur. Conf. Voids Formed by Irradiation of Reactor Mater., Reading, England, March 24 and 25, 1971.* British Nuclear Energy Society, London, 1971.

[86]   G. J. Dienes (ed.), "Studies in Radiation Effects in Solids," 3 Vols. Gordon and Breach, New York, 1966, 1967, and 1969.

[87]   *Symp. Theoret. Models for Predicting In-Reactor Performance of Fuel and Cladding Mater., Nucl. Appl. Technol.* **9(1,2,3)**, 3 (1970).

[88]   R. S. Nelson, *in* "Defects in Crystalline Solids" (S. Amelinckx *et al.*, eds.), Vol. 1. North-Holland Publ., Amsterdam, 1968.

[89a] M. W. Thompson, "Defects and Radiation Damage." Cambridge Univ. Press, London and New York, 1969.

[89b] W. N. McElroy *et al.*, Materials dosimetry, special issue of *Nucl. Technol.* **25(2)**, 177 (1975).

[90]   Fundamental aspects of radiation damage in metals, *Proc. Int. Conf., Gatlinburg, Tennessee, Oct. 6–10, 1975* Vols. 1 and 2. Available as CONF-751006-P1 and -P2 from National Technical Information Service, U.S. Department of Commerce, Springfield, Virginia 22161.

[91]   A. L. Bement, *Adv. Nucl. Sci. Technol.* **7**, 1 (1973).

[92]   *Proc. Eur. Conf. Irradiat. Embrittlement and Creep in Fuel Cladding and Core Components, London, Nov. 9–10, 1972.* British Nuclear Energy Society, c/o Institution of Civil Engineers, London, 1973.

[93]   *Proc. KTG/BNES Eur. Conf. Irradiat. Behavior of Fuel Cladding and Core Component Mater., Karlsruhe, West Germany, December 3–5, 1974.* British Nuclear Energy Society, c/o Institution of Civil Engineers, London, 1975.

[94]   J. A. Horak and T. H. Blewitt, Some implications of fast vs thermal neutron irradiation of metals, *Trans. Am. Nucl. Soc.* **16**, 73 (1973).

[95]   H. R. Brager *et al.*, Irradiation-produced defects in austenitic stainless steel, *Met. Trans.* **2**, 1893 (1971).

[96]   T. D. Gulden and J. L. Kaae, *J. Nucl. Mater.* **32**, 168 (1969).

[97]   Y. V. Konobeevski and A. V. Subbotin, *At. Energ.* (*U.S.S.R.*) **31**, 219 (1971).

[98]   D. Mosedale *et al.*, *Nature* (*London*) **224**, 1301 (1969).

[99]   R. J. Jackson *et al.*, Correlation of creep and swelling with fuel pin performance, *Trans. Am. Nucl. Soc.* **22**, 184 (1975).

[100] E. R. Gilbert *et al.*, Irradiation Creep Data in Support of LMFBR Core Design, Westinghouse-ERDA Rep. HEDL-SA-835 and -824 (1976).

[101] A. Hishinuma *et al. J. Nucl. Mater.* **55**, 227 (1975).

[102] H. R. Brager, *J. Nucl. Mater.* **57**, 103 (1975).

[103] W. G. Johnston *et al.*, Void Swelling in the Iron-Chromium-Nickel System, ASTM Rep. ASTM-STP-570,. p. 525. American Society Testing Materials, Philadelphia, 1976.

[104] J. A. Horak and T. H. Blewitt, *Nucl. Technol.* **27**, 416 (1975).

[105] F. W. Wiffen and E. E. Bloom, *Nucl. Technol.* **25**, 113 (1975).

[106] E. E. Bloom *et al.*, Temperature and fluence limitations for a type 316 stainless steel CTR first wall, *Trans. Am. Nucl. Soc.* **22**, 178 (1975).

[107] M. J. F. Notley and A. D. Lane, Factors affecting the design of rodded UO_2 fuel bundles for high power outputs, *Proc. Symp. Heavy-Water Power Reactors, Vienna, September 11–15, 1967* p. 773. I.A.E.A., Vienna, 1968 (Proceedings Series STI/PUB-163).

[108] A. B. Lidiard, Mass transfer along a temperature gradient. *Proc. Symp. Thermodynam. Nucl. Mater., Vienna, July 22, 1965.* I.A.E.A., Vienna, 1965 (CONF-650704-18).

[109] M. Bober and G. Schumacher, *Adv. Nucl. Sci. Technol.* **7**, 121 (1973).

V. Solid Fuels

[110] D. R. Olander, Fundamental Aspects of Nuclear Reactor Fuel Elements. Available as TID-26711-P1 from National Technical Information Service, U.S. Department of Commerce, Springfield, Virginia 22161.

[110a] J. H. Kittel *et al.*, Irradiation behavior of metallic fuels, *Proc. Int. Conf. Peaceful Uses of At. Energy, 3rd, Geneva, Aug. 31–Sept. 9, 1964* **11**, 230. United Nations, New York, 1965.

[110b] C. O. Smith, "Nuclear Reactor Materials," p. 164. Addison-Wesley, Reading, Massachusetts, 1967.

[110c] R. B. Holden, "Ceramic Fuel Elements." Gordon and Breach, New York, 1966.

[110d] A. N. Holden, "Physical Metallurgy of Uranium," p. 188. Addison-Wesley, Reading, Massachusetts, 1958.

[111] J. E. Harris and E. C. Sykes (eds.), "Physical Metallurgy of Reactor Fuel Elements." The Metals Society, London, 1975.

[112] D. R. Harries, *J. Brit. Nucl. Energy Soc.* **14**, 123 (1975).

[113] Nuclear Systems Materials Handbook, 2 vols., U.S. ERDA Rep. TID-26666. Hanford Engineering Development Laboratory, Richland, Washington, 1975.

[114] A. Pee and U. Schumann, *Nucl. Eng. Des.* **14**, 99 (1970).

[115] M. F. Lyons *et al.*, *Nucl. Eng. Des.* **21**, 167 (1972).

[116] *Proc. Int. Conf. Nucl. Fuel Performance, London, October 15–19*, British Nuclear Energy Society, London, 1973.

[117] E. A. Evans and W. F. Sheely, *in* "Education and Research in the Nuclear Fuel Cycle" (D. M. Elliott and L. E. Weaver, eds.), p. 153. University of Oklahoma Press, Norman, Oklahoma, 1972.

[118] O. L. Kruger and A. I. Kaznoff (eds.), *Proc. Int. Symp. Ceram. Nucl. Fuels, Washington, D.C., May 3–8, 1969* Special Pub. No. 2. American Ceramic Society, Columbus, Ohio, 1969.

[119] B. R. T. Frost, Studies of irradiation effects in ceramic fuels, *Proc. Int. Symp. Ceram. Nucl. Fuels, Washington, D. C., May 3–8, 1969* p. 225. American Ceramic Society, Columbus, Ohio, 1969.

[120] Y. R. Rashid, Mathematical modeling and analysis of fuel rods, From the *Int. Conf. Struct. Mech. in Reactor Technol., 2nd, Berlin, Sept. 10–14, 1973* CONF-730942, pp. vp. Paper D 1/1.

[121] V. Z. Jankus and R. W. Weeks, LIFE-I, A FORTRAN-IV Computer Code for the Prediction of Fast-Reactor Fuel-Element Behavior, ANL Rep. ANL-7736, 1970. Available from National Technical Information Service, U. S. Department of Commerce, Springfield, Virginia 22161.

[122] V. Z. Jankus and R. W. Weeks, *Nucl. Eng. Des.* **18**, 83 (1972).

[123] M. C. Billone *et al.*, The LIFE-III Fuel Element Performance Code User's Manual, Argonne National Laboratory Rep. ERDA–77–56 (July 1977).

[124] J. D. Stephen *et al.*, LIFE-III Fuel Performance Code: Evaluation of Predictions of Cladding Inelastic Strain, USAEC Rep. GEAP-13951-1. General Electric Company, Sunnyvale, California, 1973.

[125] J. Leary and H. Kittel (eds.), *Proc. Int. Symp. Advanced LMFBR Fuels, Tucson, Arizona, October 10–13, 1977.* American Nuclear Society, LaGrange, Illinois, 1977.

[126] S. Aas, *Nucl. Eng. Des.* **33**, 261 (1975).

[127] E. Rolstad, *Nucl. Technol.* **25**, 7 (1975).

[128] *Proc. Symp. Use of Plutonium as a Reactor Fuel, Brussels, March 13–17, 1967.* I.A.E.A., Vienna, 1967.

[129] J. M. Cleveland. "The Chemistry of Plutonium." Gordon and Breach, New York, 1970.

[130]   Plutonium: Physico-chemical properties of its compounds and alloys, *At. Energy Rev. Spec. Issue No. 1* (1966).

[131]   *Proc. Int. Conf. Plutonium 1970 and Other Actinides, Santa Fe, New Mexico* (W. N. Miner, ed.), Vol. 17. Metallurgical Society of the AIME, New York, 1970.

[132]   A review of plutonium utilization in thermal reactors, *Nucl. Technol.* **18**, 73 (1973).

[133]   Tables of Thermodynamic Data, IAEA Tech. Rep. Ser. 38. I.A.E.A., Vienna, 1964.

[134]   *Proc. Symp. Thermodynam. Nucl. Mater., Vienna, July 22–27, 1965.* I.A.E.A., Vienna, 1966.

[135a] *Proc. Symp. Thermodynam. Nucl. Mater., Vienna, May 1962.* I.A.E.A., Vienna, 1963.

[135b] *Proc. Symp. Thermodynam. Nucl. Mater., Vienna, September 8–14, 1967.* I.A.E.A., Vienna, 1968.

[136]   R. C. Daniel *et al.*, Effects of High Burnup on Zircaloy-Clad Bulk OU_2, Plate Fuel Element Samples, USAEC Rep. WAPD-263, 1962. Available from National Technical Information Service, U.S. Department of Commerce, Springfield, Virginia 22161.

[137]   *Proc. ANS Topical Meeting on Reactor Mater. Performance, Richland, Washington, April 23–26, 1972, Nucl. Technol.* **16**, 1 (1972).

[138]   B. R. T. Frost, *Nucl. Appl. Technol.* **9**, 141 (1970).

[139]   J. A. Turnbull and M. O. Tucker, *Phil. Mag.* **30**, 47 (1974).

[140]   J. R. Findlay, *J. Brit. Nucl. Energy Soc.* **12**, 415 (1973).

[141]   *Proc. Panel on the Behavior and Chem. State of Irradiated Ceram. Fuels, Vienna, August 7–11, 1972.* I.A.E.A., Vienna, 1974.

[141a] Fuel-cladding chemical interactions, *Trans. Am. Nucl. Soc.* **23**, 153 (1976).

[142]   J. H. Davies and F. T. Ewart, *J. Nucl. Mater.* **41**, 143 (1971).

[143]   W. Batey and K. Q. Bagley, *J. Brit. Nucl. Energy Soc.* **13**, 49 (1974).

[144]   Effects of environment of materials properties in nuclear systems, *Proc. Int. Conf. Corros., London, July 1–2, 1971.* British Nuclear Energy Society, c/o Institution of Civil Engineers, London, 1971.

[145]   Materials dosimetry, *Nucl. Technol.* **25**, 177 (1975).

[146]   *Proc. Symp. Neutron Std. and Flux Normalization, Argonne, Illinois, October 21–23, 1970.* Available as CONF-701002 from National Technical Information Service, U.S. Department of Commerce, Springfield, Virginia 22161.

[147]   *Proc. Conf. Corros. of Reactor Mater. Salzburg, 4–8, June 1962* 2 Vols. I.A.E.A., Vienna, 1962 (Proceedings Series STI/PUB-59).

[148]   J. C. Tverberg *et al.*, *Nucl. Eng. Int.* **17**, 1001 (1972).

[149]   J. F. Schumar, *in* "Education and Research in the Nuclear Fuel Cycle" (D. M. Elliott and L. E. Weaver, eds.), p. 187. Univ. of Oklahoma Press, Norman, Oklahoma, 1972.

[150]   W. D. Wilkinson, "Uranium Metallurgy," 2 Vols. Wiley (Interscience), New York, 1962.

[151]   A. S. Coffinberry and W. N. Miner (eds.), "The Metal Plutonium." Univ. of Chicago Press, Chicago, Illinois 1961.

[152]   J. F. Smith *et al.*, "Thorium: Preparation and Properties." Iowa State Univ. Press, Ames, Iowa, 1975.

[153]   M. A. Feraday *et al.*, Irradiation Behaviour of a Corrosion Resistant U-Si-Al Fuel Alloy, Atomic Energy of Canada Ltd. Rep. AECL-5028, 1975. Available from Scientific Document Distribution Office, Atomic Energy of Canada, Ltd., Chalk River, Ontario, Canada.

[154]   R. B. Matthews and M. L. Swenson, *Nucl. Technol.* **26**, 243 (1975).

[155]   J. Belle, (ed.), Uranium Dioxide: Properties and Nuclear Applications. Div. of Reactor Development, USAEC, Washington, D.C., 1961.

[156]   Report of the Panel on Thermodynamic and Transport Properties of Uranium Dioxide and Related Phases, Vienna, 16–20 March 1964. I.A.E.A., Vienna, 1965.

[157] Report of the Panel on Thermal Conductivity of Uranium Dioxide, Vienna, April 1965. I.A.E.A., Vienna, 1966 (Proceedings Series STI-DOC-10/59).

[158] R. A. Laskiewicz *et al.*, Thermal Conductivity of Uranium Plutonium Oxide, USAEC Rep. GEAP-13733. General Electric Company, Sunnyvale, California, 1971.

[159] A. B. G. Washington, Preferred Values for the Thermal Conductivity of Sintered Reactor Fuel for Fast Reactor Use, UKAEA RG TRG Rep. 2236. United Kingdom Atomic Energy Authority Reactor Group, Risley, Warrington, WA36AT, United Kingdom, 1973.

[159a] P. Debye, *in* "Vortrage über die Kinetische Theorie der Materie und Elektrizität" (M. Planck *et al.*, eds.). Tuebner, Leipzig, 1914.

[159b] R. Peierls, *Ann. Phys.* **3**, 1055 (1929).

[160] J. F. Kerrisk and D. C. Clifton, *Nucl. Technol.* **16**, 531 (1972).

[161] F. A. Nichols, *Nucl. Appl. Technol.* **9**, 128 (1970).

[162] F. Anselin, The Role of Fission Products in the Swelling of Irradiated UO_2 and $(U, Pu)O_2$ Fuel, USAEC Rep. GEAP-5583. General Electric Company, Sunnyvale, California, 1969.

[163] J. A. Turnbull and R. M. Cornell, *J. Nucl. Mater.* **41**, 156 (1971).

[164] J. R. Findlay *et al.*, *J. Nucl. Mater.* **35**, 24 (1970).

[165] J. I. Brammer and H. J. Powell, *J. Brit. Nucl. Energy Soc.* **14**, 63 (1975).

[165a] B. L. Harbourne and W. H. McCarthy, Axial fuel redistribution by vapor transport in LMFBR rods, *Trans. Am. Nucl. Soc.* **23**, 146 (1976).

[166] L. A. Lawrence and J. A. Christensen, *Nucl. Technol.* **12**, 367 (1971).

[167] R. D. Leggett *et al.*, Linear heat rating for incipient fuel melting in UO_2–PuO_2 fuel, *Trans. Am. Nucl. Soc.* **15**, 752 (1972).

[168] R. D. Leggett *et al.*, Influences of burnup on heat-rating-to-melting for UO_2–PuO_2 fuel, *Trans. Am. Nucl. Soc.* **19**, 235 (1974).

[169] J. L. Frankota and C. N. Craig, Melting point of Plutonia-Urania Mixed Oxides Irradiated to High Burnup, USAEC Rep. GEAP-13515. General Electric Company, Sunnyvale, California, 1969.

[170] M. G. Andrews *et al.*, Evaluating the solution to fuel densification, *Trans. Am. Nucl. Soc.* **19**, 140 (1974).

[170a] Y. R. Rashid *et al.*, Mathematical treatment of the mechanical densification of reactor fuel, *Int. Conf. Struct. Mech. Reactor Technol., 2nd, Berlin, Sept. 10–14, 1973* (CONF-730942, pp. vp, Paper C 1/8).

[171] W. P. Chernock *et al.*, In-pile densification of uranium dioxide, *Trans. Am. Nucl. Soc.* **20**, 215 (1975).

[172] M. D. Freshley *et al.*, The effect of pellet characteristics and irradiation conditions on UO_2 fuel densification, *Proc. Joint Topical Meeting on Commercial Nucl. Fuel Technol. Today, Toronto, April 28–30* p. 2–106. American Nuclear Society, Hinsdale, Illinois, 1975 (75-CNA/ANS-100).

[173] G. de Contenson *et al.*, UO_2 densification—Irradiation results, laws and models, *Trans. Am. Nucl. Soc.* **20**, 215 (1975).

[174] W. Chubb *et al.*, *Nucl. Technol.* **26**, 486 (1975).

[175] M. C. J. Carlson, *Nucl. Technol.* **22**, 335 (1975).

[176] B. F. Rubin, Summary of $(U, Pu)O_2$ Properties and Fabrication Methods, USAEC Rep. GEAP-13582, 1970. Available from Clearinghouse for Federal Scientific and Technical Information, National Bureau of Standards, U.S. Department of Commerce, Springfield, Virginia 22161.

[177] The Plutonium-Oxygen and Uranium-Plutonium-Oxygen Systems: A Thermochemical Assessment, IAEA Tech. Rep. Ser. 79. I.A.E.A., Vienna, 1967.

[178] M. H. Rand and T. L. Markin, Some Thermodynamic Aspects of (U, Pu)O₂ Solid Solutions and Their Use as Nuclear Fuels, UKAEA Rep. AERE-R5560. Atomic Energy Research Establishment, Harwell, England, 1967. Also in *Proc. Symp. Thermodynam. Nucl. Mater.*, Vienna, September 8–14, 1967, p. 637. I.A.E.A., Vienna, 1968.

[179] J. A. Christensen, Nonstoichiometry Effects on the Microstructure of Irradiated UO₂ − 25 W% PuO₂ Fuels, USAEC Rep. BNWL-SA-2030. Pacific Northwest Laboratory, Richland, Washington, 1968.

[180] N. A. Javed and J. T. A. Roberts, Thermodynamic and defect-structure studies in (U,Pu) mixed-oxide fuels, *Trans. Am. Nucl. Soc.* **15(1)**, 211 (1972). Also USAEC Rep. ANL-7901. Argonne National Laboratories, Argonne, Illinois, 1972.

[181] S. Guarro and D. R. Olander, *J. Nucl. Mater.* **57**, 136 (1975).

[182] D. R. Olander, *J. Nucl. Mater.* **47**, 251 (1973).

[183] D. R. Olander, *J. Nucl. Mater.* **44**, 116 (1972).

[184] M. Bober *et al.*, *Nucl. Technol.* **26**, 172 (1975).

[185] J. I. Bramman and H. J. Powell, *J. Brit. Nucl. Energy Soc.* **14**, 63 (1975).

[186] E. A. Aitken and S. K. Evans, *Nucl. Metall.* **17**, 772 (1970).

[187] R. O. Meyer *et al.*, Actinide Redistribution in Mixed-Oxide Fuels Irradiated in a Fast Flux, USAEC Rep. ANL-7929. Argonne National Laboratories, Argonne, Illinois, 1972 and 1973.

[188] C. E. Johnson *et al.*, *Reactor Technol.* **15**, 303 (1973).

[189] L. C. Michels and R. B. Poeppell, *J. Appl. Phys.* **44**, 1003 (1973).

[190] S. D. Gabelnick and M. G. Chasanov, Calculational Approach to the Estimation of Fuel and Fission-Product Vapor Pressures and Oxidation States to 6000°K, USAEC Rep. ANL-7867. Argonne National Laboratories, Argonne, Illinois, 1972.

[191] L. E. Russell *et al.* (eds.), *Proc. Symp. Carbides in Nucl. Energy, Harwell, England, November 1963* Macmillan, New York, 1964.

[192] E. K. Storms, "The Refractory Carbides." Academic Press, New York, 1967.

[193] The Uranium-Carbon and Uranium-Plutonium-Carbon Systems: A Thermochemical Assessment, IAEA Tech. Rep. Ser. 14. I.A.E.A., Vienna, 1963.

[194] K. Richter, Preparation methods for U-Pu non-oxide fuels (carbides, nitrides, oxycarbides, carbonitrides), *Trans. Am. Nucl. Soc.* **20**, 595 (1975).

[195] J. Board *et al.*, *Proc. Symp. Fuel and Fuel Elements for Fast Reactors, Brussels, July 2–6, 1973*, Vol. II, p. 95. I.A.E.A., Vienna, 1974 [Proceedings Series STI/PUB-346 (Vol. 2)].

[196] J. R. Routbort and R. N. Singh, *J. Nucl. Mater.* **58**, 78 (1975).

[197] A. J. Markworth, *Nucl. Sci. Eng.* **49**, 506 (1972).

[198] M. Montgomery and A. Strasser, Irradiations of (U, Pu) carbide fuel rods to 77,000 MWd/T, *Trans. Am. Nucl. Soc.* **21**, 180 (1975).

[199] H. J. Matzke, *J. Nucl. Mater.* **57**, 180 (1975).

[199a] J. F. Kerrisk, Experience Related to the Safety of Advanced LMFBR Fuel Elements, US ERDA Rep. LA-6028, 1975. Available from National Technical Information Service, U.S. Department of Commerce, Springfield, Virginia 22161.

[200] M. C. Billone *et al.*, UNCLE—A computer code to predict the performance of advanced fuels in breeder reactors, *Trans. Am. Nucl. Soc.* **19**, 96 (1974).

[201] J. T. Demant, *J. Brit. Nucl. Energy Soc.* **12**, 183 (1973).

[202] U. Benedict *et al.*, Study of advanced fuels in highly rated He-bonded pins, *Trans. Am. Nucl. Soc.* **20**, 293 (1975).

[203] R. Lallement *et al.*, Carbide fuel element for fast reactors, *Trans. Am. Nucl. Soc.* **20**, 296 (1975).

[203] R. Lallement *et al.*, Carbide fuel element for fast reactors, *Trans. Am. Nucl. Soc.* **20**, 296 (1975).

[204] A. Boltax *et al.*, Mixed-carbide fuel-pin performance analysis, *Trans. Am. Nucl. Soc.* **17**, 97 (1974).

[205] J. A. Vitti *et al.*, *Nucl. Technol.* **26**, 442 (1975).

[206] R. C. Noyes *et al.*, *Nucl. Technol.* **26**, 460 (1975).

[207] S. A. Casperson *et al.*, Economic and Nuclear Performance Characteristics of 500 MW(e) Oxide, Carbide, and Nitride LMFBRs. From American Nuclear Society Atlanta Section Topical Meeting on Advanced Reactors: Physics, Design, and Economics, Atlanta, Sept. 8–11, 1974. American Nuclear Society, Hinsdale, Illinois (TIS-4192).

[208] G. M. Nickerson and W. E. Kastenberg, *Nucl. Eng. Des.* **36**, 209 (1976).

[208a] S. Goldsmith *et al.*, Ceramic Fuels for Fast Reactors, USAEC Rep. BNWL-SA-1550. Pacific Northwest Laboratory, Richland, Washington, 1968.

[209] A. A. Bauer, *Reactor Technol.* **15**, 87 (1972).

VI. Fuel Elements

[210] M. F. Lyons *et al.*, *Nucl. Eng. Des.* **21**, 165 (1972).

[210a] Operating experience with thermal fuels, *Trans. Am. Nucl. Soc.* **23**, 254 (1976).

[211] R. Manzel and H. Steble, In-reactor experience with LWR fuel, *Trans. Am. Nucl. Soc.* **20**, 253 (1975).

[212] J. W. Beck and R. M. Grube, Vermont yankee and main yankee initial core fuel performance, *Trans. Am. Nucl. Soc.* **20**, 248 (1975).

[213] R. Traccucci, PWR fuel operating experience and performance evaluation program, *Trans. Am. Nucl. Soc.* **20**, 256 (1975).

[214] H. E. Williamson and D. C. Ditmore, *Reactor Technol.* **14**, 68 (1971).

[215] F. D. Judge *et al.*, A General Electric fuel performance update, *Trans. Am. Nucl. Soc.* **20**, 249 (1975).

[216] M. D. Freshley, *Nucl. Technol.* **18**, 141 (1973).

[217] *Proc. Meeting Commercial Nucl. Fuel Technol. Today, April 28–30, 1975, Toronto.* American Nuclear Society, Hinsdale, Illinois (75-CNA/ANS-100).

[218] P. Cohen, "Water Coolant Technology of Power Reactors." Gordon and Breach, New York, 1969.

[219] C. F. Cheng, *J. Nucl. Mater.* **57**, 11 (1975).

[220] A. S. Bain *et al.*, The performance and evolutionary development of CANDU fuel, *Trans. Am. Nucl. Soc.* **20**, 259 (1975).

[221] L. W. Woodhead and L. J. Ingolfsrud, *Ann. Nucl. Energy* **2**, 767 (1975).

[222] W. J. Penn *et al.*, CANDU fuel—Power ramp performance criteria, *Trans. Am. Nucl. Soc.* **22**, 209 (1975).

[223] D. O. Pickman *et al.*, Endurance of SGHWR fuel elements under steady and transient conditions, *Trans. Am. Nucl. Soc.* **20**, 263 (1975).

[224] B. Boudouresques *et al.*, 20 years experience in metallic natural U fuel cycle, *Trans. Am. Nucl. Soc.* **20**, 279 (1975).

[225] G. F. Hires, Magnox fuel element endurance studies: A review of CEGB experience, *Trans. Am. Nucl. Soc.* **20**, 278 (1975).

[226] G. Melese-D'Hospital and M. Simnad, *Energy: Int. J.* **2**, 211–239 (1977).

[227] H. B. Stewart *et al.*, Utilization of the thorium cycle in HTGR, *Proc. Int. Conf. Peaceful Uses of At. Energy, 4th, Geneva, September 6–10, 1971* **4**, 433. I.A.E.A., Vienna, 1972 (A/CONF.49/P837).

[228] P. R. Kasten, *At. Energy Rev.* **8**, 473 (1970).

[229] *Proc. ANS Nat. Topical Meeting on Gas-Cooled Reactors: HTGR and GCFBR, May 7–10, 1974, Gatlinburg, Tennessee.* Available as CONF-740501 from the National Technical Information Service, U.S. Department of Commerce, Springfield, Virginia 22161.

[230] *Proc. Int. Conf. High Temp. Reactor and Process Heat Appl., London, November 26–28, 1974.* British Nuclear Energy Society, London, 1974.

[231] *Proc. Nat. Topical Meeting on Nucl. Process Heat Appl., 1st, Los Alamos, New Mexico, Oct. 1–3, 1974* US ERDA Rep. LA-5795-C, 1974. Available from National Technical Information Service, U.S. Department of Commerce, Springfield, Virginia 22161.

[232] J. F. Schumar and M. T. Simnad, Materials for nuclear process heat applications, *Proc. Nat. Topical Meeting on Nucl. Process Heat Appl., 1st, Los Alamos, New Mexico, Oct. 1–3, 1974* p. 228, US ERDA Rep. LA-5795-C, 1974. Available from National Technical Information Service, U.S. Department of Commerce, Springfield, Virginia.

[233] D. J. Blickwede and T. F. Barnhart, Use of nuclear energy in steelmaking: Prospects and plans. *Proc. Nat. Topical Meeting on Nucl. Process Heat Appl., 1st, Los Alamos, New Mexico, Oct. 1–3, 1974* p. 169, US ERDA Rep. LA-5795-C, 1974. Available from National Technical Information Service, U.S. Department of Commerce, Springfield, Virginia 22161.

[234] R. C. Dahlberg *et al.*, HTGR Fuel and Fuel Cycle Summary Description, General Atomic Company Rep. GA-A12801, Rev. General Atomic Company, San Diego, California, 1974.

[235] M. T. Simnad and L. R. Zumwalt (eds.), "Materials and Fuels for High-Temperature Nuclear Energy Applications." M.I.T. Press, Cambridge, Massachusetts, 1974.

[236] M. T. Simnad, History of HTGR fuel element development at general atomic, *Trans. Am. Nucl. Soc.* **21**, 171 (1975).

[237] J. L. Scott, History of coated fuel particles, *Trans. Am. Nucl. Soc.* **21**, 170 (1975).

[238] H. G. MacPherson and W. P. Eatherly, History of graphite materials development for graphite-moderated reactors, *Trans. Am. Nucl. Soc.* **21**, 169 (1975).

[239] W. G. Homeyer and H. Huschka, Fabrication and Testing of Fuel for Large HTGRs, General Atomic Company Rep. GA-A13493. General Atomic Company, San Diego, California, 1975.

[240] D. P. Harmon *et al.*, HTGR Fuel Performance, General Atomic Report GA-A13506. General Atomic Company, San Diego, California, 1975.

[241] O. M. Stansfield *et al.*, *Nucl. Technol.* **25**, 517 (1975).

[242] T. Gulden *et al.*, Design and performance of coated particle fuels for the thorium cycle HTGRs, *Proc. Nat. Topical Meeting on Gas-Cooled Reactors – HTGR and GCFR, Gatlinburg, Tennessee, May 7–10, 1974* p. 176. Available as CONF-740501 from National Technical Information Service, U.S. Department of Commerce, Springfield, Virginia 22161.

[243] C. M. Hollabaugh *et al.*, *J. Nucl. Mater.* **57**, 325 (1975).

[244] R. F. Turner *et al.*, Economic Performance of HTGR Fuel Cycles, General Atomic Rep. GA-A13791. General Atomic Company, San Diego, California, 1976.

[245] E. Balthesen *et al.*, State and development of HTGR fuel elements in the FRG, *Trans. Am. Nucl. Soc.* **20**, 268 (1975).

[246] J. L. Kaae *et al.*, *Nucl. Technol.* (in press).

[247] J. M. Taub, Graphitic fuels for the nuclear rocket engine, *Trans. Am. Nucl. Soc.* **21**, 173 (1975).

[248] P. Hedgecock and P. Patriarca, Materials selection for gas-cooled reactor components, *Proc. ANS Topical Meeting on Gas-Cooled Reactors: HTGR and GCFBR, May 7–10, 1974, Gatlinburg, Tennessee* p. 525. American Nuclear Society, Hinsdale, Illinois, 1974.

[248a] High-Temperature Nuclear Heat Source Study, USAEC Rep. GA-A13158, p. 6–33. General Atomic Company, San Diego, California, 1974.

[249] *Proc. Int. Conf. Fast Reactor Power Stations, London, March 11–14, 1974*. British Nuclear Energy Society, London, 1974.

[250] R. Farmakes (ed.), *Proc. ANS Conf. Fast Reactor Fuel Element Technol., New Orleans, Louisiana, April.* American Nuclear Society, Hinsdale, Illinois, 1971.

[251] W. C. L. Kent and A. S. Davidson, Fast reactor fuel manufacture, From *Seminar Nucl. Energy, Barcelona, June 6, 1972.* Available as CONF-720662-2 from National Technical Information Service, U.S. Department of Commerce, Springfield, Virginia 22161.

[252] Fuel element fabrication, *Proc. Symp. Fuel and Fuel Elements for Fast Reactors, Brussels, July 2–6, 1973,* **II**, 359. I.A.E.A., Vienna, 1974.

[253] *Proc. Symp. Fuel and Fuel Elements for Fast Reactors, Brussels, July 2–6, 1973* Vols. I and II. I.A.E.A., Vienna, 1974.

[254] K. G. Eickhoff *et al.*, Theoretical and experimental studies supporting the design of fast reactor fuel elements, *Proc. Symp. Fuel and Fuel Elements for Fast Reactors, Brussels, July 2–6, 1973* **I**, 329. I.A.E.A., Vienna, 1974.

[255] J. A. G. Holmes, Design of oxide fuel for fast reactors, *Proc. Symp. Fuel and Fuel Elements for Fast Reactors, Brussels, July 2–6, 1973* **I**, 295. I.A.E.A., Vienna, 1974.

[256] D. Mosedale *et al.*, Cladding and structural materials for fuel components and assemblies, *Proc. Symp. Fuel and Fuel Elements for Fast Reactors, Brussels, July 2–6, 1973* **II**, 131. I.A.E.A., Vienna, 1974.

[257] Advanced FBR fuels: Programs and design; irradiation; and modeling and properties, *Trans. Am. Nucl. Soc.* **19**, 83 (1974).

[258] M. D. Donne *et al.* (eds.), *Proc. Int. Meeting on Fast Reactor Fuel and Fuel Elements, Karlsruhe, Germany, September 28–30, 1970.* Gesellschaft fur Kernforschung mbH, Karlsruhe, 1970.

[259] *Proc. Int. Conf. Irradiat. Exp. in Fast Reactors, Jackson Hole, Wyoming, September 10–12, 1973.* American Nuclear Society, Hinsdale, Illinois, 1974 (CONF-730910).

[260] Liquid Metal Fast Breeder Reactor Program Plan, USAEC Rep. WASH-1107, Rev. 1, Vol. 7, 1970. Available as TID-4500, UC 80 from National Technical Information Service, U.S. Department of Commerce, Springfield, Virginia 22161.

[261] Clinch River breeder reactor project, *Nucl. Eng. Int.* **19**, 835 (1974).

[262] G. W. Cunningham, *Nucl. Eng. Int.* **19**, 840 (1974).

[263] M. W. Dyos *et al.*, Economic factors affecting the choice of fuel for a liquid metal-cooled fast breeder reactor, *Proc. Am. Power Conf.* **31**, 150 (1969).

[264] E. A. Evans *et al.*, Status of engineering design and irradiation testing of the stainless-clad mixed-oxide fuel system for fast breeder reactors, *Proc. Int. Conf. Peaceful Uses of At. Energy, 4th, Geneva, September 6–16, 1971* **10**, 53. United Nations, New York, 1972 (A/CONF.49/P/077).

[265] S. Rosen *et al.*, Fundamental properties and behaviour of fuel materials in fast reactor environments, *Proc. Int. Conf. Peaceful Uses of At. Energy, Geneva, September 6–16, 1971* **4**, 325. United Nations, New York, 1972 (A/CONF.49/P/069).

[266] W. E. Roake *et al.*, The United States program for developing the manufacture of high-quality fast reactor fuels, *Proc. Int. Conf. Peaceful Uses of At. Energy, Geneva, September 6–16, 1971* **8**, 311. United Nations, New York, 1972 (A/CONF.49/P/063).

[267] C. M. Cox *et al.*, FFTF fuel pin design bases and performance, *Trans. Am. Nucl. Soc.* **20**, 313 (1975).

[268] W. F. Sheely, Materials considerations in the fast flux test reactor, *Meeting Am. Soc. Met., La Grange, Illinois, March 19* HEDL-SA-546. Hanford Engineering Development Laboratory, Richland, Washington, 1973.

[269] R. B. Baker and R. D. Leggett, Performance predictions of FTR fuel, *Trans. Am. Nucl. Soc.* **19**, 117 (1974).

[270] R. F. Hilbert *et al.*, High burnup performance of LMFBR fuel rods—Recent GE exper-
       ience, *Trans. Am. Nucl. Soc.* **20**, 315 (1975).
[271] F. Anselin *et al.*, Irradiation behavior of the mixed-oxide driver fuel elements of rapsodie
       and rapsodie-fortissimo, *Trans. Am. Nucl. Soc.* **20**, 318 (1975).
[272] J. Leclere and J. P. Marcon, Choice of the phoenix fuel element characteristics, and J.
       Leclere *et al.*, Super-phoenix fuel element design, *Trans. Am. Nucl. Soc.* **20**, 308 (1975).
[273] H. Mikailoff, The main problems raised by the in-pile behavior of mixed oxide, *Trans
       Am. Nucl Soc.* **20**, 285 (1975).
[274] G. Capart *et al.*, Design implications of material properties on (SNR-300) LMFBR fuel
       elements, *Trans. Am. Nucl. Soc.* **20**, 305 (1975).
[275] P. Murray, *Reactor Technol.* **15**, 16 (1972).
[276] K. Kummerer, General characteristics of fast reactor fuel pins, *Trans. Am. Nucl. Soc.* **20**,
       301 (1975).
[277] I. S. Golovnin *et al.*, *At. Energ.* (*USSR*) **30**, 216 (1971).
[278] A. I. Leipunskii *et al.*, Development of fuel elements for the BOR-60 reactor, *Proc. Int.
       Conf. Peaceful Uses of At. Energy, 4th, Geneva, September 6–16, 1971* **10**, 69. United
       Nations, New York, 1972 (A/CONF.49/P/460).
[279] W. Dienst and K. Ehrlich, Material problems in oxide fuel pins for fast breeder reactors,
       Trans. Am. Nucl. Soc. **20**, 283 (1975).
[280] P. R. Huebotter, *Reactor Technol.* **15**, 156 (1972).
[280a] J. M. Leitnaker *et al.*, *J. Nucl. Mater.* **49**, 60 (1973/1974).
[281] J. A. Board, *J. Brit. Nucl. Energy Soc.* **11**, 237 (1972).
[282] R. J. Jackson *et al.*, Swelling and creep effects upon fast reactor core structural design,
       Trans. Am. Nucl. Soc. **13**, 112 (1970).
[283] P. G. Shewmon *et al.*, Analytical methods to design and predict performance of fast
       reactor fuel elements, *Proc. Int. Conf. Peaceful Uses of At. Energy, Geneva, September
       6–16, 1971* **10**, 123. United Nations, New York, 1972 (A/CONF.49/P/838).
[284] J. R. Matthews, *Adv. Nucl. Sci. Technol.* **6**, 65 (1972).
[285] J. A. Christensen, Thermal Performance Limits for Fast Reactor Oxide Fuels, Hanford
       Engineering Development Laboratory Rep. HEDL-SA-89. Hanford Engineering
       Development Laboratory, Richland Washington, 1971.
[286] J. F. W. Bishop and J. A. G. Holmes, Evolution of fuel designs for sodium-cooled fast
       reactors, *Trans. Am. Nucl. Soc.* **20**, 311 (1975).
[287] J. F. W. Bishop, *J. Brit. Nucl. Energy Soc.* **10**, 643 (1971).
[288] J. F. W. Bishop, *J. Brit. Nucl. Energy Soc.* **10**, 321 (1971).
[289] J. F. W. Bishop, Performance development of the PFR fuel. . . . *Proc. Int. Conf. Fast
       Reactor Power Stations, London, March 11, 1974* p. 295. British Nuclear Energy Society,
       London, 1974 (CONF-740318).
[290] D. J. Foust (ed.), "Sodium—NaK Engineering Handbook," 5 Vols. Gordon and Breach,
       New York, 1972–1976.
[291] *Proc. Symp. Alkali Met. Coolants, Vienna, Nov. 28–Dec. 2, 1966.* I.A.E.A., Vienna, 1967.
[292] J. E. Draley and J. R. Weeks, "Corrosion by Liquid Metals." Plenum Press, New York,
       1970.
[293] W. E. Berry, "Corrosion in Nuclear Applications." Wiley, New York, 1971.
[294] R. H. Simon *et al.*, Progress toward a gas-cooled fast breeder demonstration plant,
       Trans. Am. Nucl. Soc. **20**, 127 (1975).
[295] R. H. Simon *et al.*, Gas-cooled fast breeder reactor demonstration plant, *Proc. ANS
       Topical Meeting on Gas Cooled Reactors: HTGR and GCFBR, Gatlinburg, Tennessee,
       May 7–10, 1974* p. 336. Available as CONF-740501 from National Technical Informa-
       tion Service, U.S. Department of Commerce, Springfield, Virginia 22161.

[296] *Nucl. Eng. Des.* special GCFR issue **40(1)**, 1–233 (January 1977).

[297] D. R. Vaughan, *J. Brit. Nucl. Energy Soc.* **14**, 105 (1975).

[298] A. F. Weinberg *et al.*, Gas-cooled fast breeder reactor fuel development, *Proc. ANS Topical Meeting on Gas-Cooled Reactors: HTGR and GCFBR, Gatlinburg, Tennessee, May 7–10, 1974* p. 355. Available as CONF-740501 from National Technical Information Service, U.S. Department of Commerce, Springfield, Virginia 22161.

[299] R. J. Campana, *Nucl. Technol.* **12**, 185 (1971).

[300] A. R. Veca *et al.*, *Nucl. Eng. Des.* (in press).

[301a] J. R. Lindgren *et al.*, *Nucl. Eng. Des.* (in press).

[301b] M. Peehs *et al.*, *Nucl. Eng. Des.* (in press).

[302] S. Langer and G. Buzzelli, Transport of volatile fission products in a vented fuel rod, *Trans. Am. Nucl. Soc.* **22**, 231 (1975).

[303] B. D. Epstein, A Review of the Literature Pertinent to Fission-Product Migration and Interaction in Fuel Rods, Rep. GA-A13423. Genral Atomic Company, San Diego, California, 1975.

[304] P. Fortescue, *Nucl. News* **15**, 36 (1972).

VII. Moderator and Control Materials

[305] J. C. Bokros, "The Structure and Properties of Pyrolytic Carbon," Rep. GA-8888. General Atomic Company, San Diego, California, 1968.

[306] R. J. Price, Property Changes in Near-Isotropic Graphites Irradiated at 300° to 600°C: A Literature Survey, Rep. GA-A13478. General Atomic Company, San Diego, California, 1975.

[307] T. J. Chang and J. R. Rashid, Visco-elastic response of graphites materials in irradiation environments, *Proc. Conf. Continuum Aspects of Graphite Design, Gatlinburg, Tennessee, Nov. 9–12, 1970* p. 307. Available as CONF-701105 from National Technical Information Service, U.S. Department of Commerce, Springfield, Virginia 22161.

[308] O. Stansfield, Irradiation Effects in Boronated Graphite, Rep. Gulf-GA-A12035. General Atomic Company, San Diego, California, 1972.

[309] L. A. Beavan, Control Materials in LHTGR Design, Rep. GA-A13260. General Atomic Company, San Diego, California, 1974.

[310] R. E. Dahl, Correlation of Radiation Damage in Boronated Graphite, USAEC Rep. BNWL-199. Pacific Northwest Laboratories, Richland, Washington, 1966.

[311] R. E. Nightingale (ed.), "Nuclear Graphite." Academic Press, New York, 1962.

[312] J. H. W. Simmons, "Radiation Damage in Graphite." Pergamon, Oxford, 1965.

[313] J. L. Kaae, *J. Nucl. Mater.* **57**, 82 (1975).

[314] G. B. Engle *et al.*, Status of graphite technology and requirements for HTGRs, *Proc. ANS Topical Meeting on Gas-Cooled Reactors: HTGR and GCFBR, Gatlinburg, Tennessee, May 7–10, 1974.* Available as CONF-740501 from National Technical Information Service, U.S. Department of Commerce, Springfield, Virginia 22161.

[315] G. B. Engle *et al.*, Development Status of Near-Isotropic Graphites for Large HTGRs, Rep. GA-A12944. General Atomic Company, San Diego, California, 1974.

[315a] W. R. Johnson and G. B. Engle, Properties of Unirradiated Fuel Element Graphites H-451 and TS-1240, USAEC Rep. GA-A13752. General Atomic Company, San Diego, California 1976.

[316] *Proc. Conf. Continuum Aspects of Graphite Design, Gatlinburg, Tennessee, Nov. 9–12, 1970.* Available as CONF-701105 from National Technical Information Service, U.S. Department of Commerce, Springfield, Virginia 22161.

[316a] G. B. Engle *et al.*, Status of Graphite Technology and Requirements for HTGRs, USAEC Rep. GA-A12885, p. 5. General Atomic Company, San Diego, California, 1974.

[317] W. K. Anderson and J. S. Theilacker (eds.), "Neutron Absorber Materials for Reactor Control." USAEC, Washington, D.C., 1962.

[318] W. M. Mueller *et al.* (eds.), "Metal Hydrides." Academic Press, New York, 1968.

[319] M. T. Simnad *et al.*, *Nucl. Technol.* **28**, 31 (1976).

[320] A. F. Lillie *et al.*, Zirconium Hydride Fuel Element Performance Characteristics, USAEC Rep. AI-AEC-13084, 1973. Available from National Technical Information Service, U.S. Department of Commerce, Springfield, Virginia 22161.

[321] K. E. Moore and W. A. Young, Phase Relationships at High Hydrogen Contents in SNAP Fuel System, USAEC Rep. NAA-SR-12587. Atomics International Div., Canoga Park, California, 1968.

[322] J. W. Raymond, Equilibrium Dissociation Pressures of the Delta and Epsilon Phases in ZrH-System, USAEC Rep. NAA-SR-9374. Atomics International Div., Canoga Park, California, 1964.

[323] P. Paetz and K. Lucke, *J. Nucl. Mater.* **42(1)**, 13 (1972).

[324] J. M. Beeston, *Nucl. Eng. Des.* **14**, 445 (1971).

[325] P. J. Spencer *et al.*, Beryllium: Physico-chemical properties of its compounds and alloys, *At. Energy Rev.* **11** (Spec. Issue No. 4), 1 (1973).

[325a] J. B. Rich *et al.*, *J. Nucl. Mater.* **4**, 288 (1961).

[326] R. Smith and J. P. Howe (eds.), Beryllium Oxide, *Proc. Int. Conf. Beryllium Oxide, 1st, Sydney, Oct. 21–25, 1963, J. Nucl. Mater.* **14**, 1 (1963).

[327] M. T. Simnad *et al.*, BeO—Review of properties for nuclear applications, *Proc. Symp. Nucl. Appl. Nonfissionable Ceram., Washington, May 1966* p. 169. American Nuclear Society, Hinsdale, Illinois, 1966 (CONF-660505).

[328] R. E. Dahl and J. W. Bennett (eds), *Meeting Absorber Mater. and Contr. Rods for Fast Reactors, Dimitrovgrad, U.S.S.R., June 4–8, 1973* USAEC Rep. HEDL-TME-73-91, 1973. Hanford Engineering Development Laboratory, Richland, Washington, 1973.

[329] A Compilation of Boron Carbide Design Support Data for LMFBR Control Elements, US ERDA Rep. HEDL-TME-75-19. Hanford Engineering Development Laboratory, Richland, Washington, 1975.

[330] A. L. Pitner and K. R. Birney, Performance analysis of LMFBR control rods, *Trans. Am. Nucl. Soc.* **22**, 225 (1975).

[331] W. F. Sheely and R. E. Dahl, Boron carbide performance in fast reactors, *Trans. Am. Nucl. Soc.* **20**, 291 (1975).

[332] A. E. Pasto *et al.*, Europium Oxide as a Potential LMFBR Control Material, USAEC Rep. ORNL-TM-4226. Oak Ridge National Laboratory, Oak Ridge, Tennessee, 1973.

[333] M. M. Martin *et al.*, Description of FTR Design—Oriented Irradiation Test of Eu_2O_3 in EBR-II, US ERDA Rep. ORNL-TM-4996. Oak Ridge National Laboratory, Oak Ridge, Tennessee, 1975.

[334] A. L. Grantz and E. L. Gluekler, An economic assessment of potential LMFBR neutron absorber materials, *Trans. Am. Nucl. Soc.* **22**, 513 (1975).

[335] W. K. Anderson and J. S. Theilacker (eds.), "Neutron Absorber Materials for Reactor Control." USAEC, Washington, D.C., 1962.

VIII. Pressure Vessels

[336] L. E. Steele, Neutron Irradiation Embrittlement of Reactor Pressure Vessel Steels, Tech. Rep. Ser. 163. I.A.E.A., Vienna, 1975 (STI/DOC/10/163).

[337] L. E. Steele, *Nucl. Eng. Des.* **27**, 121 (1974).

[338]   *Proc. IAEA Specialists Meeting on Radiat. Damage Units, Seattle, Washington, 30 Oct.–
        1 Nov. 1972, Nucl. Eng. Des.* **33**, 1 (1975). See also McElroy *et al.* [89b].
[339]   C. Z. Serpan and B. H. Menke, Nuclear Reactor Neutron Energy Spectra, ASTM Data
        Series Publ. DS 52. American Society for Testing Materials, Philadelphia, 1974.
[340]   Radiation Effects Information Generated on the ASTM Correlation Monitor Steels,
        ASTM Data Series Publ. DS 54. American Society Testing Materials, Philadelphia,
        Pennsylvania, 1975.
[341]   C. Z. Serpan, *Nucl. Eng. Des.* **33**, 19 (1975).
[341a] W. N. McElroy *et al., J. Test. Eval.* **3**, 220 (1975).
[341b] L. A. James, *J. Nucl. Mater.* **59**, 183 (1976).
[342]   Proceedings of the ICE/BNES Conference on Prestressed Concrete Pressure Vessels,
        London, March 1967. British Nuclear Energy Society, c/o Institution of Civil Engineers,
        London, 1967 (CONF-670301).
[343]   N. M. Schaeffer (ed.), Reactor Shielding for Nuclear Engineers, USAEC Rep.
        TID-25951. U.S. Atomic Energy Commission, Washington, D.C., 1973.

Chapter 3

Ceramics for Coal-Fired MHD Power Generation

H. KENT BOWEN

DEPARTMENT OF MATERIALS SCIENCE AND ENGINEERING
CERAMICS DIVISION
MASSACHUSETTS INSTITUTE OF TECHNOLOGY
CAMBRIDGE, MASSACHUSETTS

I. INTRODUCTION

In the late 1930s and early 1940s a concentrated effort was devoted to the development of two types of gas turbines, the mechanical turbine using moving blades, and the electrical turbine making use of the interaction between a conducting gas (plasma) and a magnetic field (Fig. 1). These new turbines showed promise of substantial fuel savings per unit of power produced. In our age of increasing environmental awareness, fuel savings also mean a reduction in thermal polution and conservation of natural resources. The mechanical turbine found rapid application in jet engines which today dominate the aircraft propulsion field. More recently, the gas turbine has found widespread use in the electrical power industry [1].

181

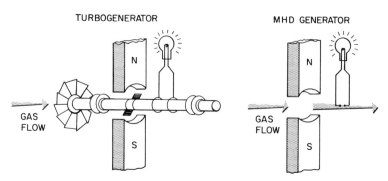

FIG. 1.   Comparison of the turbogenerator and the MHD generator.

The electrical turbine or magnetohydrodynamic (MHD) generator became a laboratory reality around 1960, through better understanding of the physics of high temperature gases and progress in high temperature technology. At that time, the economic and technological potential of MHD power generation was recognized by electrical manufacturers, by some utilities, and by military agencies. These different groups provided the early support for research in the field of MHD power generation. However, as a result of the construction of the first economical nuclear power plants, and of the slowing of the development of MHD power generation, the commercial interest of MHD began to wane [1].

In 1968 the Office of Science and Technology formed an MHD panel to evaluate the potential of MHD power generation. In its recommendations, the panel confirmed the soundness of the MHD power generation principles and underlined the valuable possibility of increased fuel utilization through higher efficiency and noted especially that if coal were used, fossil reserves could be extended. There have been two subsequent evaluations [1,2] which have recommended a national program including studies of materials.

The principle of the MHD channel is shown in Fig. 2, where the high velocity hot gas at velocity V interacts with the perpendicular magnetic field B to create a dc-current flux J in the third orthogonal direction. The active elements in the system are the electrode walls, the anode and the cathode, which transfer the electrons from the working gas to the external load (R_L).

MHD is the only advanced energy conversion technique that can directly use coal as a fuel. A schematic of the total system is given in Fig. 3. Pulverized coal is combusted with preheated air in the burner. To increase the electrical conductivity of the combustion products potassium compounds (e.g., K_2CO_3) are added. The "seeded" gas at 2800°K passes through the magnetic field where 10–20% conversion of thermal to electrical power occurs. On the low temperature phase of the combined cycle, a steam

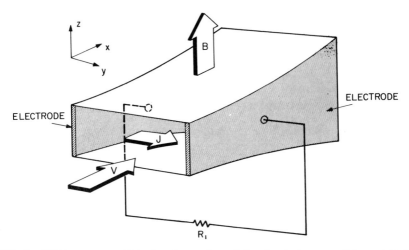

FIG. 2.   MHD generator: V is the plasma flow, J the electrical current flow, and B the magnetic field.

plant with about 40% conversion makes the overall process from 50–60% efficient.

The cooled gases pass through regenerative heat exchangers, necessary to preheat air to 1200°C in order to maintain high combustion temperatures; and finally the gases are cleaned of the seed, sulfur compounds and oxides of nitrogen. The potassium seed provides an efficient "getter" for sulfur in the fuel and condenses principally as K_2SO_4.

The extremely high plasma temperatures exiting from the combustor at nearly Mach 1 velocities are necessary to provide high electrical conductivity of the gas; but this temperature is also above the fusion and volatility temperature of most of the components in coal ash. Most U.S. coals have

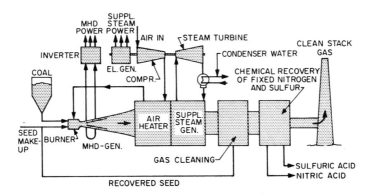

FIG. 3.   Schematic of MHD power plant.

ash contents of $\sim 10\%$. It is assumed that ash removal efficiencies in the combustor will not exceed 90%; thus thousands of pounds per hour of an iron–aluminum–silicate ash (slag) will be rushing down the channel of a 1000 MW plant.

II. MATERIALS REQUIREMENTS

First let us examine the important subsystems to determine the critical ceramic materials needs and describe the properties which are desirable for each of the ceramic components.

The wall of the MHD duct is constructed of alternating slabs of electrodes and insulators (Fig. 4). The electrode may be a metal, which is then

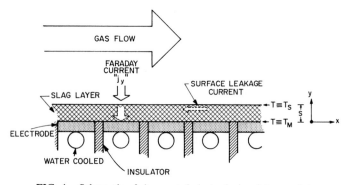

FIG. 4.   Schematic of slag coated electrode–insulator module.

generally operated at temperatures below 500°C, or a ceramic which is operated at much higher temperatures. In a clean fuel system there may be a coating of potassium sulfate seed on the electrode-insulator wall. For higher temperature systems operated in a clean mode, that is, not burning coal, the potassium compound would not condense on the walls. Since we are concerned only with coal-fired systems there will be a condensed layer of slag and seed except when the wall is operated at very high temperatures. At temperatures greater than about 1300°C the potassium content in the condensed layer will decrease and if operated at very high temperatures, in excess of 1700°C, the amount of condensate from the coal ash will be small and consequently a very minimal slag layer will result. The dc current arrives perpendicular to the wall, but there is also a field along the length of the generator which may result in breakdown through the insulator or current leakage along the surface or through the slag layer.

The electrodes and insulators must perform in an extremely corrosive and erosive environment. The high gas temperatures and near sonic gas velocities required for efficient MHD power extraction, combined with the presence of the coal slag and seed, place stringent requirements on the selection of these materials, especially if long duty life (> 1000 hr) is also a prerequisite. There are no electrode or insulator materials that can operate for long periods at the gas temperatures and therefore we must consider channel walls which would be operated at lower temperatures; however, wall temperatures below 1500°C result in significant sensible heat losses [3]. More specifically the electrode materials for the MHD channels should meet the criteria listed in Table 1.

TABLE 1

Properties of MHD Electrode Materials

1. Electronic electrical conductivity ($> 10^{-1} \Omega^{-1} cm^{-1}$) at current densities of more than 1 A/cm^2.
2. Small temperature dependence of the electrical conductivity.
3. Good thermionic (electron) emission.
4. Corrosion resistance: to potassium seed and coal slag.
5. Erosion resistance: to the high velocity gases, to liquid droplets and to particulates in the gas stream.
6. Good thermal conductivity: to provide for an adequate heat flux and to reduce thermal gradients and thermal stresses.
7. Oxidation (or reduction) resistance: stability in the MHD at oxygen partial pressures of 10^{-4}–10^{-2} atm.
8. Thermal stress damage resistance.
9. Compatability with insulator and combustion products.

Materials for use as insulators in MHD channels have been less extensively investigated than those for use as electrodes. However, the development of reliable insulation at very high temperatures will probably be a greater challenge; at very high temperatures all materials become highly conductive. The insulators must have the same mechanical, chemical and thermal stability as the electrodes while maintaining high electrical resistance (at least 100 times more resistive than the ionized gas phase and at least 10 times more resistive than the electrodes). For typical electrode–insulator design configurations there is an axial field of 30–50 V between electrodes and therefore the dielectric strength of the insulator must be sufficient at the high temperatures.

Preheating the combustion air is required to obtain the high flame temperatures and high electrical conductivity of the gas. For thermal efficiencies of 50–52 % the air must be preheated to 1100°C while a thermal efficiency of 60 % would necessitate preheat temperatures of about 1650°C. These high preheat temperatures would not be required if the combustion gas could be

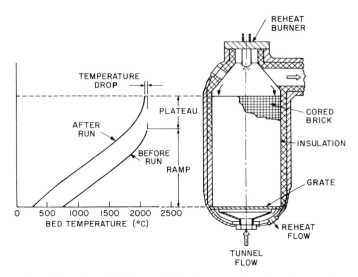

FIG. 5. Schematic of indirectly fired regenerative air preheater [4].

enriched with oxygen, but oxygen enrichment would be too costly for a full-scale plant. Figure 5 shows a schematic of a high temperature regenerative air preheater. Along the length of the heater there is a temperature gradient during each half of the cycle, for example, when the MHD exhaust gases are passing through the chamber. The air to be heated would then flow in the reverse direction. Initial heating of the air occurs in a lower temperature metallic preheater so that it enters the ceramic regenerator at a temperature of several hundred degrees and exits the preheater at 1000–1200°C. This schematic unit itself would represent one of probably six units that would be cycled during the continuous operation of the MHD system and thus will require large ceramic valves and ducts to direct the flow of the hot gases.

In selecting materials for regenerative air preheaters consideration should be given to the factors listed in Table 2. In tailoring materials to meet

TABLE 2

Requirements of Regenerative Preheater

1. Large air flow for 1000 MW station $\sim$ 700 kg/sec.
2. Thermal cycling of $\sim$ 200°C in 3–30 min cycles.
3. Air exit temperatures > 1100°C.
4. Duty cycle of several years.
5. Nonplugging configuration for large volume coal slag and seed (K_2SO_4) carry over.
6. Ceramics must be resistant to K_2SO_4 corrosion, to compressive creep and to thermal shock and thermal fatigue. The materials must be low in cost.

these requirements, trade-offs will frequently be necessary. For example, high mechanical strength and resistance to corrosion require high density bodies, but the resistance to thermal shock generally increases with increasing porosity and microcracks. For mechanical strength and resistance to erosion cored brick or stacked-plate checker configurations are preferred, while more efficient heat transfer is accomplished by the use of pebble beds. The most important consideration is the cost of the preheater because of the large volume of refractories which will be required.

III. PROPERTIES OF COAL SLAG

The most notable feature of the open cycle coal-fired MHD generator is the presence of coal slag. While the slag may affect strongly the performance of the materials as well as the performance of the generator, most of the past studies on MHD system components have been carried out under relatively clean conditions in the absence of coal slag. The overall content of slag or ash as well as its composition and properties vary considerably from one coal to another. The ash content covers a wide spread from 8–25% with values in the range of about 10% being fairly typical [5]. If it is presumed that 90% of the ash content of the coal is removed in the combustor, the slag content in the channel will be reduced to about the level of the seed content. Coal slags can be regarded as iron-rich, alkali–alkaline earth alumino silicates. The compositional range is: SiO_2, 40–53%; Al_2O_3, 26–33%; Fe_2O_3, 3–14%; CaO, 2–5%; MgO, 1–2%; Na_2O, 0–2%; and K_2O, 0–4%. There may also be several percent sulfur in coals in addition to a small concentrations of halides. Solution of the large fraction of the potassium seed into the slag may be anticipated whenever they come in contact in suitable ranges of temperature (above the glass transition temperature of the slag but below the vaporization temperature [3]). Consequently the composition of the slag in various positions through the MHD system will vary even more strongly than the variation of the composition of coal ash.

The viscosity data for two coal slags are shown in Fig. 6. The high iron concentration facilitates crystallization at lower temperatures and increases the measured viscosity (actually a multiphase system).

The electrical conductivity of the slags is relatively high because of the high iron content and also because of the high alkali content. It varies with the oxidizing potential (Fig. 7), and at 1400°C reaches the value required for an MHD electrode. Phase separation and precipitation of spinel compounds occurs between 1400–1550°K. Figure 8 contains data at 1300 and 1500°C for an alkaline-earth-rich alumino silicate slag for varying iron concentrations. At the low iron concentration, the conductivity is partly ionic

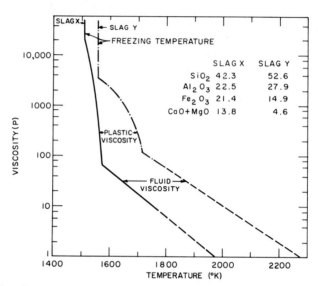

FIG. 6. Viscosity–temperature behavior of two coal slags containing no alkali oxides or sulfates [6].

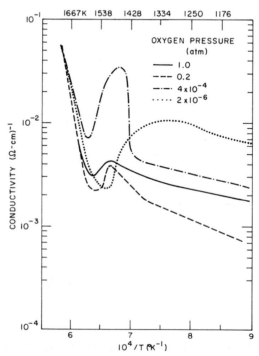

FIG. 7. Electrical conductivity of synthetic coal slag (in wt% 38 SiO_2, 32 Al_2O_3, 20 Fe_2O_3, 10 CaO) [7].

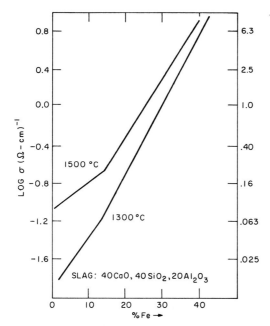

FIG. 8.   Effect of increased Fe-content on the conductivity of alumino-calcium-silicate [8].

and partly electronic, while at the higher iron concentrations the conductivity is nearly all electronic although ions can move due to the field.

The most volatile components of the seed slag mixture are the potassium and sulfur. The potassium will most likely enter the system as potassium carbonate and eventually form potassium sulfate. Below about 1200°C much of the potassium sulfate is liquid and some is soluble in the coal slag. However, at higher temperatures, most of the potassium and sulfur remains in the vapor phase. Other components will also preferentially vaporize out of the slag. The precise nature of what will be condensed and what will volatilize depends critically on the oxidizing–reducing conditions and the constituents in the gas phase; nevertheless, silica will begin to vaporize significantly in the range of 1550–1700°C, and the slag will become enriched with alumina, iron oxide, and alkaline earths. All of this will have notable effects on the physical properties.

IV. MATERIALS FOR THE MHD CHANNEL

The study and evaluation of electrode materials has involved the three major classes of materials: metals, intermetallics and ceramics [3,9]. Water-cooled metal electrodes promote high thermal and electrical losses as well as seed condensation, which is thought to cause erratic and degenerative electrical performance. The most deleterious effects arise from the low and nonuniform electron emission and large voltage drops across cold boundary layers, which in turn create severe discharge and arcing problems.

The electrical resistivity as a function of temperature of many of the potential electrode materials is shown in Fig. 9. The metals and highly conducting ceramic type materials (graphite, ZrB_2, SiC, doped $LaCrO_3$, and

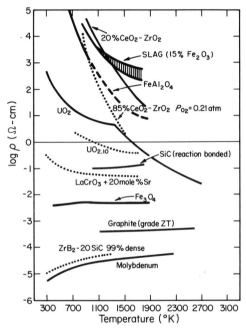

FIG. 9.   Resistivity of potential electrode materials.

Fe_3O_4) have very small variations in the conductivity (resistivity as a function of temperature); while the semiconducting and partially ionic conducting ceramics (UO_2, zirconia based oxides, $FeAl_2O_4$, and coal slag) have large (exponential) temperature dependencies. All of the materials in Fig. 9 have adequate conductivity at temperatures above 1900°K, but the large increase in resistivity with decreasing temperature because of the temperature gradient in actual electrodes will appreciably increase the Joule losses and may cause a thermal instability in the solid [10]. In addition, the zirconia based materials have significant ionic conduction. The transfer of ionic current from the electrode to a metal current lead must involve an oxidation–reduction reaction. Large chemical potential gradients may be required to drive these reactions at a rate sufficient to provide current densities on the order of 1 A/cm². These gradients decrease the efficiency of the current transfer process and can even lead to destruction of the electrode by electrolysis, as has been well documented for zirconia by Casselton [11]. Rossing [12] has developed a graded CeO_2–ZrO_2 electrode to alleviate some of these problems.

Electrical insulation materials for the insulating sidewalls or the inter-electrode insulators must have resistivities greater than $100 \ \Omega \cdot cm$ at the high operational temperatures. Those materials which have received most consideration have been Al_2O_3, MgO and calcium or strontium zirconates. The major problems are related to contamination and reaction product formation. Alumina reacts with potassium compounds to form

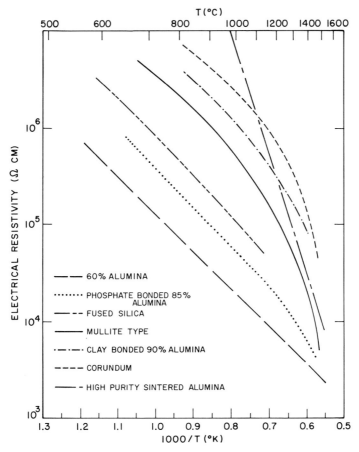

FIG. 10.   Resistivity of commercial refractories [13].

$K_2O \cdot 11Al_2O_3$ and all suffer from grain boundary attack and penetration of silicates and alkalis. Contamination from iron will lead to rapid degradation of the resistivity. Magnesia has good high temperature stability except that under coal-fired conditions and the presence of water vapor, the vaporization rate becomes too large above 1700°C.

It can be seen from the electrical property data in Figs. 10 and 11 that resistivity greater than 100 $\Omega \cdot$ cm would be difficult to maintain in most commercial materials at operating temperatures above 1700°C. Insulating cements are desired for ease of construction, but these may be too permeable to the slag–seed. The cements and controlled-porosity or pre-cracked sintered bodies do provide better thermal stress resistance [15]. The extremely large thermal gradients and nonlinear profiles make this a serious materials limitation.

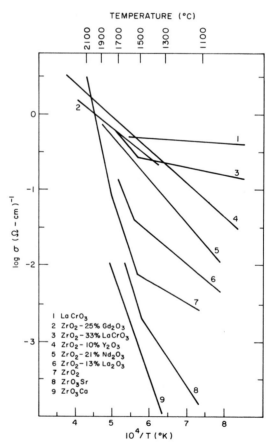

FIG. 11.   Resistivity of zirconia based materials [14].

V. MATERIALS FOR THE AIR PREHEATER

The high temperature regenerative air preheater is a critical element in the total system. Although preliminary units will be indirectly fired, i.e., not fired by the combustion products from the MHD channel, higher efficiencies are possible with directly fired units. The preheater consists of many subsections from low temperature metallic recuperators and finally high temperature regenerators constructed from ceramics. Experience in the Soviet Union has shown that for indirect gas-fired units (clean system) zirconia cored brick or pebbles can be operated up to ~1700°C. Because of costs and availability, magnesia and alumina refractories are the most extensively used.

Directly fired preheaters for coal fired MHD generators present serious problems of corrosion by the seed-slag and plugging of the air flow channels by the large volume of condensibles carried over from the combustor. Figure 12 shows the decrease in the modulus of rupture when polycrystalline alumina is exposed to the combustion products of heavy oil. Most of the

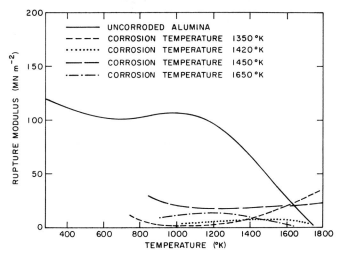

FIG. 12. Strength–temperature behavior for dense alumina fired in the combustion products of residual fuel oil for 500 hr [3].

degradation is a result of grain boundary attack. This is even more serious when porosity is present. Figure 13 contains corrosion rate data for alumina in molten K_2SO_4 at 1700°C [3]. As the porosity increases the logarithm of the corrosion rate increases. The logarithm of the corrosion rate was also shown to be proportional to the logarithm of the impurity content. Similar data were obtained for magnesia materials [3]. Zirconia-based materials have corrosion resistance which strongly depends on the processing conditions.

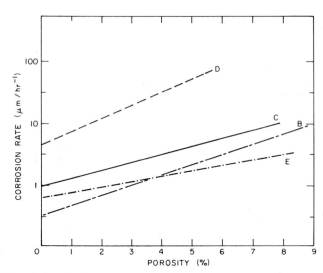

FIG. 13.   The relationship between corrosion rate and porosity for alumina: Line B—5 hr at 1700°K in K_2SO_4; lines C, D, and E are for corrosion at the equilibrium seed concentrations for 5 hr at 1900°K, and 25 hr at 1700°K, respectively [3].

The high silica content of most commercial zirconia refractories leads to preferential grain boundary attack by alkali sulfates [16].

The emphasis here is on alkali sulfate corrosion, because at the temperatures assumed for the first generation ceramic regenerators most of the silicate coal slag will be solidified but the K_2SO_4 will be liquid.

There are other important materials properties requirements listed in Table 2. Most critical are the creep strength and thermal shock resistance. Tailoring the microstructure by porosity, microcrack, or second phase control has been suggested [15,17], but this must be done in light of the corrosion problem. The creep resistance of commercial refractories was recently reviewed by Trostel [18]. Although the compressive loads are not large, the long time at temperature requirements (several years) in order to absorb the large capital costs will require creep rates $< 10^{-4}$ %/hr [3].

VI. MATERIALS DEGRADATION

Throughout this discussion, reference to the chemical stability of materials, e.g., corrosion, vaporization, etc., has been made. In this section, the degradation mechanisms for the most severe case, the MHD channel walls, will be examined. The conclusion which has been reached [16,19] is that only by control of the gas phase and condensed slag and seed phases will the degradation rates be held to a tolerable level. For example, a potassium sulfate saturated coal slag at 1800°K will penetrate and corrode most high density ceramics (ZrO_2, Al_2O_3, SiC, etc.) at rates of 0.1 mm/hr [16]. As an example of principles of stabilizing the system by making the electrode–insulator–slag chemically compatible, we choose the module consisting of a spinel electrode ($FeAl_2O_3$–Fe_3O_4) and alumina interelectrode insulator.

In general, the selection of any electrode–insulator pair which is to operate on a long term basis in coal-fired systems must allow for the minimization of chemical dissolution. The flux of electrode or insulator material into the slag is given by the diffusion coefficient and the concentration gradient, $J = -D \nabla C$, for the rate-limiting diffusion process. Dissolution will be tolerable either for a small ΔC ($\Delta C/\delta$, the gradient across the diffusion boundary layer thickness, δ) or a small diffusion coefficient. An approach relying on the minimization of the diffusivity will not be acceptable in that optimum slag properties (composition, conductivity, emission) only occur at elevated temperatures where D will be high. Suitable modules must therefore rely on minimization of the concentration gradient(s) for dissolution.

Taking MHD slag as essentially the four component Fe–Al–Si–O system and projecting liquidus equilibria onto the FeO · Fe_2O_3–Al_2O_3–SiO_2 plane will enable us to evaluate the conditions for equilibration of the spinel–alumina MHD module with coal slag. This chemical system was studied by Muan [20]. The region of interest is where ∇C will be the smallest for both crystalline phases (Al_2O_3 and spinel) and the slag. Thus, the goal is to control the slag composition by temperature and additives (e.g., Fe_2O_3) to equilibrate the slag with the electrode and insulator materials. Spinel and liquid compositions have been extracted from the Muan data and are given in Table 3.

The chemical compatibility and mode of stabilization of several electrode materials have been discussed in a recent publication [16] and are briefly

TABLE 3

Liquidus Data for the Spinel–Corundum Boundary, $P_{O_2} = 0.2$ [20].

Liquidus temperature (°C)	Approximate spinel composition (molar ratio, Fe_3O_4 : $FeAl_2O_4$)	Liquid composition (wt% SiO_2)
1460	52 : 48	20
1500	43 : 57	14
1550	36 : 64	10
1600	29 : 71	7
1650	25 : 75	4
1700	23 : 77	2

reviewed here. The use of thermochemical stability diagrams establishes the stable phases at the particular temperature and composition. For example, at 1800°K, SiC forms a cristobalite layer at oxygen pressures of $\sim 10^{-2}$ atm which offers some resistance to further oxidation. Unfortunately, in clean fired systems the oxide layer will lack efficient electron transfer as well as sufficient electron emission to meet the required 1 A/cm^2.

Zirconium diboride oxidizes to form a porous ZrO_2 scale filled with glassy B_2O_3. As temperatures increase the boron oxide vaporizes leaving the scale, which does little to prevent molecular transport of oxygen. By adding silicon carbide [22] the vaporization losses are reduced up to 1450°C. This is attributed to SiO_2 formation which has relatively small vapor pressures in this region. Using this in an MHD duct is subject to the same corrosion processes as in the SiC case. It also produces an oxide layer which is electrically resistive and which readily dissolves in coal slags. Slag saturation using additions of silica sand would stabilize SiC and ZrB_2–SiC electrodes. This would also require lower electrode–slag interface temperatures to keep the SiO_2 from vaporizing [16,19]. While doped zirconia has good high temperature stability and corrosion resistance, it degrades in slags due to grain boundary attack. Commercial material was shown to be no better than SiC or ZrB_2 [16].

Lanthanum chromite is stable for oxygen pressures greater than about 10^{-14} atm, but this assumes that it is simultaneously in equilibrium with chrome oxide gases. CrO_2 and CrO_3 have equilibrium vapor pressures of 6×10^{-8} and 8×10^{-7} atm, respectively, at 1873°K and 0.25 atm O_2. For these conditions, Meadowcroft and Wimmer [23] observed a vaporization rate of 5.4×10^{-5} gm/cm^2 hr. Evaporation resulted in a porous La_2O_3 scale. The presence of water vapor from fuel combustion will increase the rate of vaporization by an order of magnitude by forming volatile $CrO_2(OH)$.

There are some indications that UO_2 is quite resistant to chemical attack by seed and slag. Kravtsev *et al.* [24] found urania to be insoluble in potassium silicate melts at 1100°C and in addition, no reaction was observed between UO_2 and SiO_2 at 1800°C. These were based on the lack of the green tetravalent uranium ion in the melts.

Although they would otherwise be attractive as hot electrodes, the refractory metals and graphite are not generally useful because of their high oxidation rates at the typical oxygen pressures ($P_{O_2} \sim 10^{-3}$ atm) expected in MHD channels. However, for a slagging system the slag layer will slow the oxidation rate, perhaps sufficiently to allow these metals to be useful for short time applications in test cells or prototype facilities. It is well known, from experience in the glass industry, that in nonoxidizing conditions, molybdenum is compatible with commercial silicate glasses. Experiments have shown that at 1450°C the resistance of molybdenum to corrosion by coal slag with significant additions of K_2SO_4 is comparable to or better than many other candidate materials which normally have superior oxidation resistance [16].

In addition to the dissolution reactions discussed above, electrodes and insulators can degrade by forming reaction products and by penetration of contaminants. In the latter case, the degradation of the insulating and high temperature dielectric strength will be caused by interdiffusion of transition metal ions, particularly iron, and by interdiffusion of K_2O and SiO_2. The transition metal ions increase the electron transfer in insulators, while K_2O and SiO_2 may cause reaction products of low melting point liquids. Decreases in the electrical resistivity have been observed and related to the potassium seed [3].

While the interdiffusion coefficients for the particular interdiffusion couples of interest to MHD electrode modules are not known, we can infer some values. Generally, aliovalent impurities associate with the cation vacancies and increase the interpenetration. To estimate the magnitude of the diffusion coefficient which would be tolerable for 5,000 hr of operation, we assume a penetration distance of 0.25 cm, which yields a value of 4×10^{-9} cm²/sec. Tracer diffusion coefficients in, for example, Al_2O_3 are less than this value at 1700°C, but if point defect association occurs, higher penetration rates might be expected.

More serious penetration and degradation might occur down the electrode–insulator interface and at grain boundaries. The effects of potassium are the formation of low melting point liquids and accumulation of potassium compounds near the 1500°K isotherm of electrode modules [3]. Upon startup of the channel, nonseeding conditions should be used to coat the interfaces as a deterrent to potassium penetration.

The high slag temperatures (e.g., with the spinel electrode module) will reduce the potassium content of the slag; but in addition the lowering of the SiO_2 content further decreases the amount of seed incorporated in the slag. Plante *et al.* [25] observed the equilibrium potassium pressures over K_2O–SiO_2 solutions and compared their results to calculated pressures expected in the seeded plasma. Similar numbers are reported by Womack [26]; about 15 wt% K_2O is absorbed at 1900°K, but only 2 wt% at 2200°K.

VII. SUMMARY

The principal materials requirements and problems for ceramics in MHD power generators have been outlined. The physical properties of electrodes and insulators for open-cycle MHD have been specified; i.e., minimum electrical conductivity of electrodes ($\sigma \geq 0.1 \ \Omega^{-1}\text{cm}^{-1}$) and minimum dielectric strength ($40 \ \text{V/cm}$) and maximum conductivity of insulators ($\sigma < 0.01 \ \Omega^{-1}\text{cm}^{-1}$). The required properties have been related to compatibility with the operating environment; specifically, the large thermal loading of a 2600°C plasma, the corrosiveness of the oxidizing, potassium rich gas (clean system), or of the slag–potassium seed condensates (coal-fired system), and the compatibility of electrode with the insulator and electrode with the current extractor provide the additional design criteria. These also imply certain kinds of materials selections and solutions which appear to offer long term stability.

The advantages and disadvantages of several electrode–insulator/ electrode–current extractor systems were discussed including the electrode systems: Fe_3O_4–$FeAl_2O_4$, Mo, graphite, UO_2, $ZrB_2(+SiC)$, SiC and $LaCrO_3(+CaO, SrO)$. The magnitude of the heat flux from the plasma to channel walls suggests the following general operating parameters in a coal-fired system:

    1. Plasma-slag interface temperature >1600°C;
    2. slag layer thickness <0.5 mm;
    3. slag–electrode temperature ~ 1100°C;
    4. electrode thickness, <5 mm for ceramic and ~ 2 cm for metallic type electrodes;
    5. in all cases control of the slag properties by control of the composition and temperature is imperative.

The large thermal gradient in the electrode (~ 2000°C/cm) requires additional constraints on the conductivity. If the electrode material behaves as a semiconductor, $\sigma = \sigma_0 e^{-Q/RT}$, the magnitude of σ_0 must be large and Q must be small; i.e., the electrical conductivity should have a small temperature dependence. It is the integrated conductivity (conductance) which will to a large extent govern the system Joule losses.

The major constraints on refractories for the preheaters is their cost and resistance to corrosion by K_2SO_4. Thermal shock, thermal fatigue and long term creep resistance are also necessary.

REFERENCES

[1] J. F. Louis et al., Open Cycle Coal Burning MHD Power Generation, An Assessment and a Plan for Action. Published by the Office of Coal Research, Dept. of the Interior, June 1971.
[2] W. D. Jackson, MHD National Program. ERDA-Fossil Fuels Division.
[3] J. B. Heywood and G. J. Womack (eds.), "Open Cycle MHD Power Generation." Pergamon, Oxford, 1969.
[4] D. G. DeCoursin, FluiDyne Eng. Corp., Minneapolis, Minnesota.

[5] H. F. Feldmann *et al.*, Thermodynamic, Electrical, Physical, and Compositional Properties of Seeded Coal Combustion Products. U. S. Bureau of Mines Bulletin # 655, Washington, D.C., 1970.

[6] F. G. Ely and D. H. Barnhart, *in* " Chemistry of Coal Utilization " (H. H. Lowry, ed.). Wiley, New York, 1963.

[7] H. P. R. Frederikse and W. R. Hosler, *J. Am. Ceram. Soc.* **56**, 418 (1973).

[8] E. A. Pastukhov *et al.*, *Sov. Electrochem.* **2**, 209 (1966).

[9] L. L. Fehrenbacher and N. M. Tallen, Electrode and insulation materials in magnetohydrodynamic generators, *in* " Ceramics in Severe Environments " (W. W. Kriegel and H. Palmovr, III, eds.). 1971.

[10] H. K. Bowen, MHD channel materials development goals, *in* " Engineering Workshop on MHD Materials " (A. L. Bement, ed.). Massachusetts Institute of Technology, Cambridge, Massachusetts, 1974.

[11] R. E. W. Casselton, " Electricity from MHD," Vol. 5, p. 2951. Warsaw Symp., 1968.

[12] B. R. Rossing *et al.*, Febrication and properties of electrodes based on ZrO_2, *Proc. Int. Conf. MHD Elec. Power Generation, 6th*. ERDA and IAEA, Washington, D.C., June 1975.

[13] R. W. Wallace and E. Ruh, *J. Am. Ceram. Soc.* **50**, 358 (1967).

[14] A. M. Anthony and D. Yerouchalmi, *Phil. Trans. Roy. Soc. (London)* **261**, 504 (1966).

[15] B. R. Rossing and T. P. Gupta, The role of processing in the development of generator components, *in* " Engineering Workshop on MHD Materials " (A. L. Bement, ed.), Massachusetts Institute of Technology, Cambridge, Massachusetts, 1974.

[16] H. K. Bowen *et al.*, Chemical stability and degradation of MHD electrodes, *in* " Corrosion Problems in Energy Conversion and Generation " (C. S. Tedmon, Jr., ed.). The Electrochemical Society, New York, 1974.

[17] R. Goodof and D. R. Uhlmann, Thermal shock resistance of air preheater materials, *Symp. Eng. Aspects of MHD, 14th* (April 1974).

[18] L. J. Trostel, Deformation of oxide refractories, *in* " Deformation of Ceramic Materials " (R. C. Bradt and R. E. Tressler, eds.). Plenum Press, New York, 1975.

[19] H. K. Bowen, *et al.*, Chemical stability and degradation of MHD electrodes, *Proc. UTSI Symp., Aspects of Magnetohydrodynam., 14th. April 8–10,* 1974 (C. L. Wu, ed.), Tullahoma, Tennessee.

[20] A. Muan, Phase equilibria at liquidus temperatures in the system iron oxide—Al_2O_3–SiO_2 in air atmosphere, *J. Am. Ceram. Soc.* **40** (4), 121 (1957).

[21] T. O. Mason *et al.*, Properties and thermochemical stability of ceramics and metals in an open-cycle, coal-fired MHD system, *Proc. Int. Conf. MHD Elec. Power Generation, 6th*. ERDA and IAEA, Washington, D.C., 1975.

[22] W. C. Tripp, E. T. Rodine, and J. E. Stroud, Investigation of the Behavior of High Temperature Materials, Aerospace Res. Lab., AFL 71-0235, October 1971.

[23] D. B. Meadowcroft and J. M. Wimmer, The Volatile Oxidation of Lanthanum Chromite, Paper 87-B-73, 75th Annual Meeting of the American Ceramic Society, Cincinnati, Ohio, April 29–May 3, 1973.

[24] D. N. Kravtsev, E. A. Ippolitova, and Yu. P. Simanov, On the problem of formation of tetravalent uranium silicate in glass, Investigation in the Field of Uranium Chemistry, (V. I. Spitsyn, ed.), ANL-TRANS-33.

[25] E. R. Plante, C. Olson, and T. Negas, Interaction of K_2O with slag in open cycle, coal-fired MHD, *Int. Conf. Magnetohydrodynam. Elec. Power Generation, 6th, Washington* (June 9-13, 1975).

[26] G. J. Womack, " MHD Power Generation: Engineering Aspects." Chapman and Hall, London, 1969.

Chapter 4

Materials for Solar Energy Conversion

J. J. LOFERSKI

DIVISION OF ENGINEERING
BROWN UNIVERSITY
PROVIDENCE, RHODE ISLAND

I. INTRODUCTION

The sun is a nuclear "fusion reactor" in which energy produced by fusion of light nuclei into heavier nuclei heats the matter of which the sun is composed to very high temperatures. This heated matter emits electromagnetic radiation whose spectral distribution can be approximated reasonably well

201

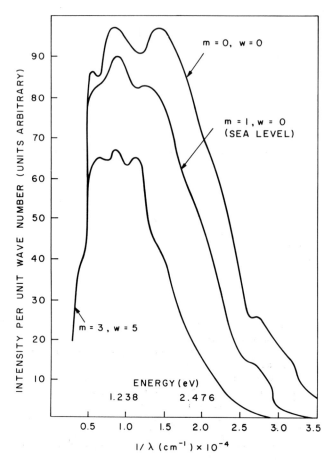

FIG. 1. The solar AM0 and AM1 spectra and an AM3 spectrum with substantial moisture content in atmosphere (from [70]). The parameter $m = 1/\cos\theta$ where θ is the angle between a line through the sun and a line through the zenith; w is the number of centimeters of condensible water vapor in the atmosphere and AM3 is the spectrum for $m = 3$.

by the spectrum characteristic of a 6000°K blackbody. Figure 1 shows the distribution of photons of various energies ·in the solar spectrum. The intensity of sunlight at sea level, normal incidence, with the sun at the zenith, is about 1 kW/m². It is the objective of solar energy conversion technology to transform—economically—the energy in sunlight into other useful forms of energy, namely, thermal and electric energy.

In this chapter we shall discuss the materials problems associated with the energy transformation processes underlying solar thermal and solar photovoltaic systems. In this introduction we first offer brief descriptions of the principles underlying solar energy conversion concepts. We shall then discuss certain problems common to both types of systems, namely, the area which must be covered by such systems if they are to contribute a significant portion of the United States and world energy needs, the problem of energy storage during periods of low insolation, and some of the economic aspects of solar energy conversion systems.

A. Solar Thermal Conversion

Solar thermal conversion is based on the " greenhouse " effect. A surface which is to be in contact with a heat transfer fluid (commonly water or air) is covered by a layer of sunlight-absorbing material. This layer can be produced by chemical dipping processes (e.g., Cu_2O on Cu), by evaporation processes to produce interference films on the surface or by painting the surface. The absorbing surface is then enclosed in a container covered by a material transparent to sunlight but opaque to the infrared radiation (beyond say 1 or 2 μm) emitted by the absorber after the absorption of solar energy increases its temperature. The equilibrium temperature achieved by the surface is determined by convective, radiative, and conductive heat losses and by the rate at which heat is transferred to the heat-utilizing or heat storage system. The equilibrium temperature provides a convenient basis for classification of solar thermal systems into those intended for low temperature and high temperature applications.

Low temperature systems are employed in the space heating of buildings and in producing hot water for residential and industrial applications. Heat transfer fluid temperature below about 90°C (about 200°F) suffice for these applications. Such systems commonly use flat plate collectors; a schematic diagram of such a collector is shown in Fig. 2. The transparent cover is glass or plastic. Various materials have been used or proposed as the basic element of the sunlight absorber; these include copper, steel, aluminum, and plastics. The relative advantages and disadvantages of various flat plate collector designs have been the subject of much discussion but no clearly superior solution has emerged, though many satisfactory solutions do exist.

The heat transfer fluid interacts with the materials used in the absorber and can affect its choice. For example, if water serves as the heat transfer fluid, the channel through which it flows must be constructed of material which will not corrode over the decades of operation demanded of such systems. The channel is commonly incorporated into the light-absorbing surface by extrusion, by joining pipes to the absorber surface, etc. As an example of the approach which is required, consider the problem of designing a solar thermal hot water heater system for a climate in which freezing temperatures are never encountered. The system can be simplified and the

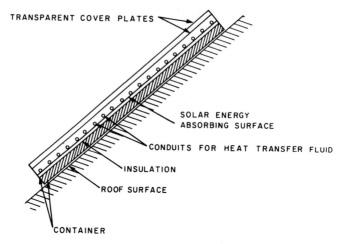

FIG. 2.   Schematic representation of a roof top flat plate collector (from [1]).

cost reduced if a heat exchanger is not required in the hot water storage
tank, i.e., if cold water from the mains can be fed directly into the solar
heater–storage system. Copper tubing or stainless steel channels would work
well in this application, because these materials do not corrode over very
long periods of use. If aluminum were used, a corrosion inhibiter would have
to be added to the water and this necessitates the incorporation of a heat
exchanger. If the solar thermal system is intended for a climate where freezing temperatures are commonly encountered, an antifreeze would have to be
added to the water or some fluid other than water would be used. In either
case, a heat exchanger is required and the materials corrosion problems are
entirely different; the choice of absorber material is not as severely restricted.

The absorbing surface of the flat plate collector must have a high absorptivity a for wavelengths present in sunlight (the value should be close to
unity) and a high a/e ratio, where e is the emissivity in the infrared over the
wavelengths emitted by the heated absorbing surface. In practice surfaces
with $a > 0.90$ and $a/e \sim 10$ are available.

High temperature systems are commonly intended for electrical power
generation, though, of course, the waste heat could be used for space condi-
tioning and industrial applications. The basic concept here is to substitute a
solar thermal converter for the fossil fuel or nuclear furnace. The thermal
energy would be used to drive a conventional turbine–generator system and,
since acceptable thermal-to-electrical energy conversion efficiencies require
hot reservoir temperatures of several hundred degrees centigrade, the
proposed high temperature solar systems would be intended to deliver ther-

mal energy at temperatures in excess of say 500°C to the turbine–generator system.

As we have stated, the ratio of the thermal energy extracted from the collector to the solar energy incident on it is a function of the heat losses in the system and of the optical properties of the heat absorbing system. These optical properties can be described by a figure of merit Xa/e, where a and e are the absorptivity and emissivity previously defined and X is the concentration ratio; its value is of course unity for flat plate collectors. Figure 3,

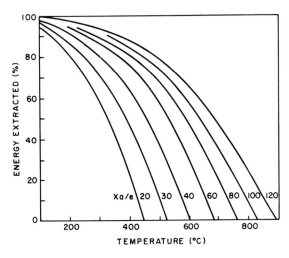

FIG. 3.   Percentage of energy that can be extracted from a selective absorber as a function of surface temperature and absorber characteristic Xa/e (from [2a,b]).

taken from a paper by Meinel and Meinel [2a,b], shows how the fraction of thermal energy extracted from the system depends on temperatures for various values of the figure of merit Xa/e. These curves are calculated on the assumption that $a = 1$ and that heat is lost only by reradiation from the hot absorber surface. Since Meinel and Meinel propose enclosing the absorber in an evacuated tube, convective losses would indeed be negligible but losses by conduction through supports, etc., need to be taken into account. Additional heat losses would change the shape of the curves in Fig. 3, and result in a lower fraction of energy extracted for a given Xa/e ratio. Therefore, the curves in the figure should be considered as only illustrative of the expected behavior.

With these caveats, it is evident from these curves that Xa/e must be equal to or greater than about 100 if at least 90% of the energy incident on the absorber is to be extracted at 500°C.

Let us examine the prospects for achieving such high values of Xa/e in flat plate collectors ($X = 1$); systems based on intermediate concentration ratios ($10 < X < 100$), and systems based on high concentration ratios ($X > 1000$).

Flat plate collectors are attractive because of their simplicity and their ability to transform solar into thermal energy even on cloudy days when direct sunlight is not available. In principle, a/e values in the required range (~ 100) are attainable in flat plate collectors covered by complex multilayer metal–dielectric interference film stacks. However, the high temperatures required for efficient solar-to-thermal-to-electrical energy conversion would promote interdiffusion, evaporation, and ultimately even self destruction of the films. Another approach to high a/e ratios involves the production of a fine "velvet" or "whisker" pattern on the absorbing surface, e.g., by means of photo-etching techniques; this concept is in an early stage of development and evaluation. As we have already noted, with currently available technology, the highest value of the a/e ratio attainable in simple (and therefore economically feasible), stable (up to say 500°C) thin film coated surfaces is around 10 with $a \sim 0.95$. Consequently, flat plate collectors are not practical as the basic building blocks of high efficiency solar-to-electrical energy conversion systems; some concentration is required if the high temperatures are to be achieved.

The linear concentrator system, one proposed solution to the problem, would work with concentration ratios between 10 and 100. The basic idea involves focusing the sun's rays on a pipe covered with a solar energy absorbing film through which a heat transfer fluid is flowing. This fluid transmits the heat to a thermal reservoir from which the heat is in turn transferred to the working fluid of a turbine generator system. Figure 4 shows a cross section of such a system in which a Fresnel lens and a cylindrical tube serve to concentrate the light; cyclindrical mirrors of parabolic or circular cross sections could also be used. The heat transfer fluid could be a liquid metal (Na or NaK), an organic fluid or a eutectic salt. Alternately, a heat pipe could be used to transfer heat from the linear array to the thermal reservoir. In such a system, energy in thermal form must be transported over rather long distances which introduces heat loss opportunities and increased complexity.

The "power tower" concept circumvents much of the heat transfer problems associated with linear systems by transporting the solar energy incident on a large area covered by mirrors via light beams to an absorber situated at the top of a tower. Such a system has a concentration ratio X in the 1000–2000 range. A specific proposal for such a system by Hildebrandt and Vant-Hull [3a,b] is a plant which would produce an equivalent of 200 MW continuously over the course of a year. This plant would have an area 3.3 km² covered by 2.6 km² of mirrors; the light would be focused on an

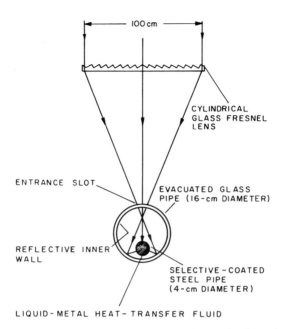

|← 100 cm →|

CYLINDRICAL
GLASS FRESNEL
LENS

ENTRANCE SLOT

EVACUATED GLASS
PIPE (16-cm DIAMETER)

REFLECTIVE INNER
WALL

SELECTIVE - COATED
STEEL PIPE
(4-cm DIAMETER)

LIQUID- METAL HEAT- TRANSFER FLUID

FIG. 4. Schematic representation of a cross section of a linear solar thermal collector system (from [2a,b]).

absorber placed on top of a 450 m high tower. Even if the a/e ratio were unity, the high value of X results in a high temperature at the absorber and ultimately in high system efficiency. It has been calculated [3a,b] that overall sunlight-to-electricity efficiencies in excess of 40% would result, whereas the efficiency would be about 26% in a linear system. In the "power-tower" system the power density at the absorber situated on the tower is in the range of 10^6 W/m^2 and liquid Na is therefore proposed as the heat transfer fluid. Durable mirror surfaces and a durable absorbing surface for the boiler are basic requirements of this kind of system.

In summary, solar-to-thermal energy conversion technology is in a highly advanced state. Systems can be designed for both high and low temperature applications. Economic analyses of both types of systems lead to the conclusion that they should be competitive with alternate energy sources and therefore it is reasonable to expect continually expanding utilization of this technology.

B. Solar Photovoltaic Conversion

The photovoltaic effect (PVE) is the generation of an emf as the result of the absorption of ionizing radiation. It can occur in gases, liquids or solids, but it is only in solids, specifically semiconductors, that useful efficiencies for conversion of sunlight into electricity have been achieved. A device in which the effect occurs is called a photovoltaic cell; a large number of such cells must be connected in series–parallel combinations to obtain high power outputs. Currently available cells have areas between 2 and 50 cm²; the efficiencies of "acceptable" cells lie above 5%. The highest reported efficiencies for conversion of sunlight into electricity in such solar cells are around 14.5% for air mass zero (AM0) and around 18% for air mass one (AM1) illumination. (AM0 illumination means exposure to a source whose spectral composition and intensity—1.39 kW/m²— are equal to those of sunlight above the earth's atmosphere; AM1 illumination means exposure to a source whose spectral composition and intensity—1 kW/m²—are equal to those of sunlight normally incident on a surface at sea level. The spectrum of AM1 sunlight differs from AM0 because of absorption in the atmosphere; in particular the absorption depletes photons in the "blue" end of the spectrum. The response of common photovoltaic cells is weaker in the blue than in the red and consequently the conversion efficiency for the AM1 can be higher than for the AM0 spectrum.)

The PVE is very commonly observed in semiconductors. It can be encountered across the junction between a metal and a semiconductor (a Schottky barrier); across the junction between two semiconductors which are in intimate contact (a heterojunction), and across the boundary between two different conductivity regions in a semiconductor (in the extreme case the semiconductor changes its conductivity type and a p–n homojunction is produced within the semiconductor). However, photovoltaic solar cells have been made on a relatively small number of semiconductors and in fact there is only one kind of photovoltaic solar cell available commercially. This is the single crystal silicon p–n junction cell whose only large scale utilization to date has been as the source of on board power for space vehicles.

The amount of silicon cells manufactured in the United States per annum is such that if all the cells manufactured during 1977 were deployed in space sunlight simultaneously, they would generate a maximum power of about 200 kW. The area covered by this annual production of silicon solar cells is about 1600 m². If a goal were set to install within a period of 10 yr a solar photovoltaic electric power generating capability equal to the 1977 United States electrical power generating capacity, about 1,500 km² of 10% efficient cells would have to be produced per annum during such a 10 yr period. (In arriving at this figure, the assumption is made that the cells will be deployed on the surface of the earth and that the total amount of solar energy incident on the cells can be approximated by an average annual insolation for the continental United States. Such an average takes into account the number of daylight hours during which the cells can operate, the effects of cloud cover, etc. In fact, the average annual insolation varies as a function of geography, but the ratio of the highest average annual insolation—southern Arizona—to the lowest—the northern states—is less than a factor of two.) This would require that the area production rate of 10% efficient cells would have to increase by a factor of 9.3×10^5. Such an increase would have a significant impact on the price of solar cells and indeed it has been suggested that, based on experience with other semiconductor devices, it should bring the price of solar cells down into the range required to make them competitive with currently used electrical power sources [4].

If the cells were to be deployed on a satellite solar power system (SSPS) like that proposed by Glaser [5] and described in more detail below, a somewhat smaller amount of solar cells would be required but the production rate would still have to be increased by an extremely large factor above the 1977 rate.

Figure 5 illustrates the SSPS system concept [1]. A satellite consisting of two large "wings" covered by several square kilometers of solar cells is

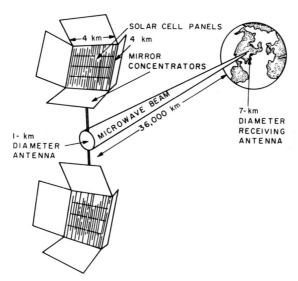

FIG. 5.   Schematic representation of a satellite space power system (from [5]).

connected to a microwave generator which transforms the dc energy produced by the solar cell panel into microwave energy. This, in turn, is transmitted to a receiving antenna in a ground station which converts the microwave radiation to dc and ultimately delivers it into the power distribution system. For a solar cell panel of 32 km² covered by 18% efficient solar cells, the useful power delivered to the power grid from one such SSPS would be 10,000 MW. The realization of satellite space power systems would require a major commitment of resources comparable to that required to land men on the moon. It has been estimated that about 20 yr would be required to place the first such satellite in orbit and that between 30 and 40 10,000 MW satellites would be needed to supply power for the United States by the year 2000; a few hundred would be needed to supply all the electrical power needs of the world.

C. Three General Problems Associated with Solar Energy Systems

There are a number of basic questions which arise when large scale utilization of solar energy is proposed; the answers to these questions are materials related.

The first such question is as follows: How large an area would have to be covered by solar energy converting systems before the energy derived from them would have a significant impact on the United States (and world) energy budget? Oviously, the area covered depends on the efficiency of the conversion process and therefore " high " efficiency is a basic requirement of solar energy conversion systems since the area covered by them must be minimized. Figure 6 shows the area of the United States which would have to be covered by solar energy conversion systems of various efficiencies to produce a given amount of energy. An appropriate average annual insolation was used in constructing this figure. The vertical lines show the total energy requirement and the total electrical energy requirement in the United States at various times in the future. (The total electric energy requirement is the amount of thermal energy needed to produce the electrical energy consumed at the indicated point in time. Since thermal-to-electrical energy conversion efficiency is about 30%, the actual amount of electrical energy used is about one third of the *total* electrical energy requirement.) From this figure it is seen that the area required, though large, is not unreasonably so. For example, if 10% efficient solar cells (like the currently available commercial silicon solar cells) were used to produce all the electrical energy consumed in the United States in 1977, about 5000 mile^2 (12,500 km^2) would have to be covered by the cells; this is 0.14% of the land area of the continental United States. Such solar energy conversion systems need not be concentrated in any one area (like Arizona); they could, for example, be dispersed on roof tops of existing or newly constructed buildings everywhere in the United States. The area required for the solar energy systems (1.25 × 10^4 km^2) is of the same order of magnitude as the area covered by single family residences in the United States (4 × 10^3 km^2) or covered by highways (> 5 × 10^4 km^2) [6].

The large areas required can affect the choice of materials used in solar energy systems simply on the basis of material availability at an acceptable

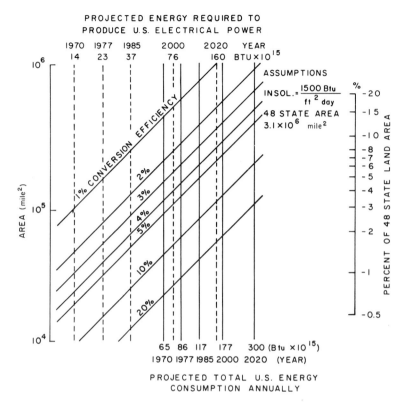

FIG. 6.   Area of United States which would need to be covered by solar energy converters of the indicated efficiency to satisfy projected total energy needs and electrical energy needs (from [1]).

cost. For example, about 5 million tons of 200 μm thick single crystal silicon cells (having 10% solar-to-electrical energy conversion efficiency) would be needed to cover the 15,000 km² area mentioned above. Since silicon is among the most abundant of the elements, production of this amount of silicon single crystal should not be a problem as regards material availability. Suppose, however, that these cells were to be made of gallium arsenide crystals of comparable weight per unit area. The current world production rate for gallium is about 20 tons per annum, and even if this could be increased by three orders of magnitude, about 250 yr would be needed to produce the required amount of gallium. Gallium arsenide is, however, a direct gap semiconductor and, therefore the thickness of GaAs required to produce a solar cell of a given efficiency is substantially smaller than the thickness of Si required for the task. For example, a 10 μm thick GaAs solar cell can absorb almost all the photons absorbed in a 200 μm thick cell. If such 10 μm GaAs cells were used as the building blocks of large scale solar energy systems, the amount of gallium required would be reduced by a factor of 20; and the amount of gallium available would not be insufficient.

As we have already pointed out, analyses show that solar-to-thermal-to-electrical energy systems could have overall conversion efficiencies between 20% and 40%. If such systems were used to produce the electrical energy referred to above, the area needed would obviously be reduced by factors between two and four in comparison with the areas required by 10% solar cells. The materials used in solar-to-thermal systems, i.e., Al, Cu, steel, plastics, etc., are very common and such large areas covered by such systems would not be limited by material availability problems.

A second basic question is as follows: How will the energy, be it thermal or electrical, be stored? Energy would be produced only when the sun shines but, presumably, it would have to be available on demand. The technology of large scale electrical or thermal storage is relatively undeveloped because currently used energy sources do not require such storage: energy is stored in the fossil or nuclear fuel. More fuel is consumed when more energy is needed. Consequently, large scale terrestrial solar energy utilization will require a parallel development of energy storage and transformation systems.

Possible methods of electrical energy storage include electrochemical storage batteries, storage in rotating masses and storage in the form of hydrogen (the hydrogen economy). A few comments are in order on these storage methods.

1. Electrochemical storage. Suppose the electrical energy were produced by solar cells deployed on the roofs of houses. Assuming average energy consumption, about 50 ft^3 of commercial lead–acid storage batteries would be needed to store a few days' electrical energy requirements for a 2000 ft^2 house. The volume required is not unreasonable. However, if lead–acid batteries were supposed to store a substantial fraction of all the electrical energy produced in the United States, it is questionable whether the supply of lead could meet the demand. Consequently, research and development on electrochemical systems based on more abundant elements is necessary. (See Chapter 9.)

2. Storage in rotating masses. Electrical energy can be transformed into mechanical energy and stored as kinetic energy of a rotating flywheel. In such a system, a motor-generator is operated as a motor when electrical

energy is available; the motor is connected to a flywheel which is set in rotation, ultimately at very high speeds. Energy is extracted by allowing the rotating flywheel to drive the motor-generator as a generator. The overall efficiency of such a system is potentially quite high. High density energy storage would require specially shaped flywheels composed of very high strength material capable of tolerating the high stresses developed at the extremely high speeds encountered in operation; here again is a materials problem.

3. Storage in the form of hydrogen. Hydrogen can be produced by electrolysis of water, inherently an extremely efficient process. The hydrogen can be stored in metal hydrides, as discussed in Chapter 8 or even as liquid hydrogen. The energy stored in the hydrogen could be extracted by fuel cells (see Chapter 7) or by combustion.

In summary, there are a number of large scale electrical energy storage methods which could potentially satisfy the storage requirements arising from large scale production of electrical energy from sunlight.

With respect to thermal energy storage, economic considerations have led to the use of water and crushed rock as heat storage media in most currently operating systems involving low temperature heat for the heating and cooling of buildings. Eutectic salts have also been used; they are attractive because a substantially smaller volume of such material is needed to store a given amount of heat. However, there do exist problems with the life of such salts; the cycling involved in storing and releasing the thermal energy can cause irreversible changes in their structure. High temperature solar thermal systems like those intended for the production of electrical energy by first converting the solar into thermal energy, storing the energy as thermal energy and using standard thermo-mechanical conversion have different thermal energy transport and storage problems. Salts like LiF have been proposed for heat transport and storage. Alternative systems involve the use of heat pipes for thermal energy transport and water for thermal energy storage.

The third question is the cost of the solar energy absorption–conversion and associated thermal or electrical energy storage systems. Obviously, if solar systems were economically competitive with alternate thermal and electrical energy generation systems (fossil fuel and nuclear energy systems), they would already be in use, since various technical solutions for the problems associated with solar energy conversion and storage already exist and they promise a lower impact on the environment than fossil or nuclear fuel systems. The gap between current estimates of the costs of solar systems and the costs which must be achieved to make the systems competitive is small for low temperature solar thermal systems like those needed to heat and cool buildings and large for photovoltaic electric power generating systems. In both instances, cost reduction is to a considerable extent a materials problem, and in both cases there is good reason to expect that at least one of the numerous possible technical alternatives available will satisfy the economic requirements. For example, in the case of photovoltaic cells, Wolf [7] analyzed the allowable cost per square meter of silicon cells for incorporation in systems deployed on roof tops of buildings and on space satellite power stations and concluded that these systems would become economically competitive if the cost of solar cells were in the range $20–$50/m^2. In 1977, the lowest price at which single crystal silicon cells were available was between $1000 and $2000 per square meter; thin film copper/cadmium sulfide cells promise availability at substantially lower prices but these cells require solution of certain important technical problems like long term (~ 20 yr) stability and higher yield of higher efficiency ($>5\%$) cells before they can be used in large scale solar energy conversion systems. Consequently in the case of single crystal silicon photovoltaic cells, new less expensive methods for producing thin single crystal plates and for introducing p–n junctions into the plates are under investigation. Processes such as pulling single crystal ribbons are being explored as alternatives to the cutting of single crystal platelets from large diameter single crystals grown by the Czochralski method [7, 8]. Processes such as ion implantation are being considered as alternatives to the diffusion process for producing p–n junctions in the single crystal plates. The possibility of developing polycrystalline silicon cells of acceptable efficiency and stability are also being actively investigated [9a,b]. Systematic investigations of the mechanisms responsible for the deficiencies in the behavior of thin film Cu/CdS cells are also underway [10a–c].

In the case of solar thermal systems, allowable costs for thermal collectors intended for heating and cooling buildings have been estimated to lie in the $20–$50 per square meter range [11]; in 1977, the range encountered lies between $60 and $200 per square meter. The relative costs of copper, steel, aluminum, and plastics as the base materials for thermal absorbers and of various coatings like cuprous oxide on copper, chromic oxide on steel, other kinds of interference films on steel, and various paints on aluminum are being explored. Experience has shown that flat plate collector panels can be produced and installed at prices such that the cost of thermal energy delivered by the solar system compares favorably with the cost of thermal energy produced from electricity but unfavorably with the cost of thermal energy produced by on-site combustion of gas or fuel oil. It is anticipated that parity with gas- and oil-produced thermal energy will be achieved as the production of solar panels increases.

II. MATERIALS FOR SOLAR THERMAL CONVERSION

A. Selective Absorbers

The solar energy absorber must have a high absorptivity a over the solar spectrum combined with a low emissivity e in the infrared over the spectral region where the absorber surface would be emitting after its temperature increases. Figure 7, taken from an article by Meinel and Meinel [2a,b],

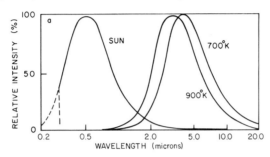

FIG. 7.   Relative intensity versus wavelength of solar spectrum and of black body radiation at 700° and 900°K (from [2a,b]).

compares the spectral distribution of photons in the solar spectrum (represented by a 6000°K blackbody spectrum) to the distribution of photons emitted by a blackbody at 900°K. About 90% of the solar spectrum is emitted with $\lambda < 1.3$ μm, while more than 90% of the photons emitted by a 900°K blackbody have $\lambda > 1.3$ μm. An ideal selective absorber characteristic for solar energy utilization would therefore be a " step function " with nearly 100% absorption for $\lambda \lesssim 1.3$ μm. Characteristics approximating this ideal to various degrees can be produced by a number of different surface treatments.

The simplest approach is to use a dull black paint. Paints having about 90% absorptivities over the solar spectrum and a/e ratios between 5 and 10 are available commercially. Thomason [12] has used such paints as the selective absorber in his corrugated steel roof panels. Löf has used a black paint to cover the aluminum panels which form the basis of the flat plate collectors used in the Colorado State University solar energy laboratory [13]. Black paints are adequate for low temperature systems but are not suited to high temperature systems because they cannot endure the elevated temperatures of such systems.

Another way to produce a surface approximating the ideal characteristic involves covering the absorber surface with interference films, since by means of such films one can achieve almost any desired absorber characteristic. Such an interference coating consists of a stack of thin metal and dielectric films whose thicknesses are so chosen that a phase balance is set up

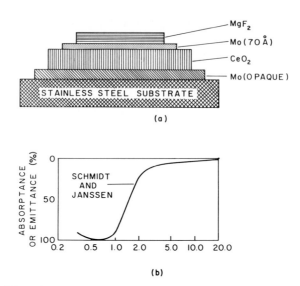

FIG. 8. (a) Schematic representation of interference stack designed by Schmidt and Janssen of Minneapolis Honeywell Company. (b) Absorption characteristic of this stack (from [2a,b]).

between the various layers. The layers are commonly produced by evaporating the constituents onto a surface. Figure 8a is a schematic cross section of an interference stack designed by R. Schmidt and J. E. Janssen of Honeywell Laboratories, while Fig. 8b shows the absorption characteristic of their interference film [2a,b]. The use of such interference films in high temperature systems poses a problem because of the possibility of degradation of the stack as a result of evaporation or interdiffusion of the metal and dielectric films. These effects may place an upper limit on the life of systems using interference films.

A third approach has been suggested by B. O. Seraphin of the University of Arizona. He proposed a system in which the two functions—high a and low e—are performed by two distinctly different components and in which the performance of these functions relies on intrinsic properties of the constituents rather than on their geometrical form as in the case of interference films [2a,b]. In his coating Seraphin uses a layer of Si between 10 and 100 μm thick to absorb the sunlight. (Figure 9 shows the fraction of photons in the AMl solar spectrum with energies in excess of 1.1 eV, the band gap of Si, which are absorbed in a given thickness of single crystal silicon.) In addition a noble metal (silver or gold) thin film is employed to suppress emittance from the stainless steel substrate on which the coating is deposited. The semiconductor absorbs the solar spectrum and heats the steel substrate by thermal conduction; the high infrared reflectance layer of Au or Ag prevents

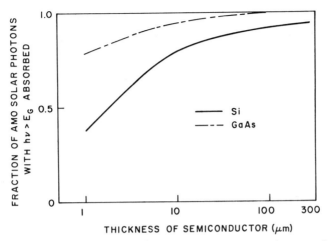

FIG. 9. Comparison of the fraction of AM0 solar photons with energy in excess of the energy gap E_G which are absorbed in comparable thicknesses of Si ($E_G = 1.1$ eV) and GaAs ($E_G = 1.40$ eV).

re-radiation from the substrate. Seraphin's proposed absorber has the structure shown in Fig. 10a; the SiO$_2$ film is intended to reduce reflection losses from the Si and the SiN film prevents alloying of Si and the noble metal substrate and provides a surface on which the Si can be deposited. The Si film is deposited by the chemical vapor deposition method commonly

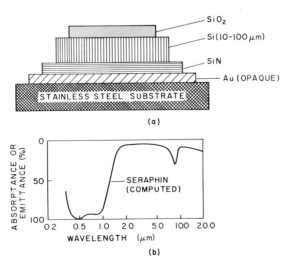

FIG. 10. (a) "Bulk" absorber stack proposed by B. Seraphin of the University of Arizona. (b) Wavelength dependence of absorption of such a bulk absorber stack (from [2a,b]).

employed to grow layers of Si on substrates by the semiconductor electronics industry. Figure 10b shows the absorber characteristic calculated by Seraphin for his structure. This type of coating is not tied exclusively to Si; any semiconductor having a band gap around 1.1 eV will serve. Indeed a direct gap semiconductor would have an advantage in this application because a thinner layer would suffice to absorb the incident sunlight. While coatings of this type are likely to be less subject to degradation at high temperature than interference stacks, there is still considerable uncertainty about their lifetime since interdiffusion of the layers comprising the film can change their characteristics.

A fourth possibility is to produce a layer on the substrate metal by chemically reacting a thin skin of the material. This procedure is used to produce Cu_2O layers on Cu plates used in solar energy absorbers [14]. Detailed studies of the physical characteristics of these layers have not been performed; it has been reported that the Cu_2O layer on " good absorbers " is between 15 and 20 μm thick. The value of a is in excess of 0.9 and the a/e ratio is around 10. The thickness of the layer makes it unlikely that its characteristics are based on interference effects. It is possible that it is an accidentally developed " Seraphin type " coating with Cu_2O as the absorber and the underlying Cu sheet serving as its own low infrared emittance material. The problem with such a proposed explanation is that the band gap of Cu_2O single crystals is about 2.3 eV. It is possible that the film produced by chemical dipping is heavily doped and that its properties are therefore different from those of Cu_2O crystals. Obviously, more investigation of this absorber system is necessary.

A fifth possible method for producing an absorber surface is to produce a structure covered with a roughened surface, e.g., a surface covered by microscopic cones or whiskers. The dimensions of the protrusions from the surface and the spacing between them should be comparable to the wavelength at which the transition from high to low absorption must occur. Such a surface can be produced on, say, aluminum by using photolithographic techniques [15].

B. Other Materials Used in Solar Thermal Systems

Flat plate collectors must be covered by a material transparent to sunlight and opaque in the infrared (see Fig. 2). The material should be inexpensive and durable. It should not be subject to damage by the elements or in the course of installation and repair of solar absorber panels. The obvious candidate materials are glass and certain plastics. Glass has superior optical properties and generally fits the requirements well [16]. It is heavy, however, and can be damaged. Consequently, interest continues to be directed toward possible plastic substitutes.

Mirrors for solar energy concentration systems do not generate any special materials problems. Like all components of the system, they must be made by simple, inexpensive processes.

The heat storage system can be based on sensible heat storage, i.e., the temperature of a substance having a high specific heat is raised during the storage process, or on latent heat storage, i.e., a substance with a high heat of fusion is alternately melted and solidified in the course of storage and recovery. Most solar–thermal space heating systems in operation rely on sensible heat storage and use either water or rocks to store heat. Water can store almost twice as much heat per unit volume as rocks can. Salts like sodium sulpha-decahydrate (Na_2SO_4–10 H_2O) have high heats of fusion and melt at temperatures below 200°C. They can store about six times as much heat per unit volume as can H_2O [17]. However, repeated cycling of these salts tends to degrade them. Furthermore, though inexpensive they cost more than the presently used alternatives. Finally, the volume occupied by H_2O storage tanks or rock storage bins is not unacceptable and, therefore, development of salts for thermal storage has not been pursued with great vigor.

III. MATERIALS FOR PHOTOVOLTAIC CONVERSION

A. Review of the Principles Underlying the Photovoltaic Effect

Three processes are involved in the photovoltaic effect. Firstly, excess positive and negative charges, at least one of which is mobile, must be generated in the semiconductor by the absorption of ionizing radiation. Secondly, the excess charges of opposite sign must be separated at some electrostatic inhomogeneity like a *p–n* junction, a metal–semiconductor barrier, etc. Thirdly, the mobile generated carriers must retain their mobility for a time long compared with the time they require to travel to the localized charge separating inhomogeneity. We shall examine each of these processes in detail.

1. Light Absorption in Semiconductors

When a monochromatic beam of light whose photons have an energy in excess of the forbidden energy gap of the semiconductor passes through a semiconductor, the flux of photons remaining in the beam after it traverses a distance x, [$N_{ph}(x)$] is given by

$$N_{ph}(x) = N_{ph}(0)e^{-\alpha(\lambda)x}$$

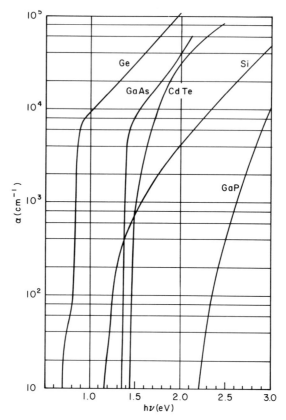

FIG. 11. Absorption constant α versus photon energy $h\nu$ for selected semiconductors of interest as photovoltaic solar cells.

where $N_{ph}(0)$ is the photon flux at $x = 0$ and $\alpha(\lambda)$ is the absorption constant for photons of this wavelength (energy).

Figure 11 is a plot of ln α versus photon energy $h\nu$ for semiconductors of interest for PVE solar energy cells. The threshold energy for the onset of the absorption corresponds to the forbidden energy gap E_G. It is, however, obvious from examination of the curves for various semiconductors that they are of basically two kinds: (1) those in which α rises very rapidly from very low values at $h\nu = E_G$ to values in excess of 10^4 cm^{-1} (the curves for GaAs and CdTe are in this category); (2) those in which α rises more gradually (the curves for Si and GaP are in this category). The rapidly rising absorption curves are characteristic of direct gap materials whereas the slowly rising curves are characteristic of indirect gap materials.

Because of their rapidly rising absorption curves, direct gap semiconductors are better suited for photovoltaic cells because the thickness of the material required to absorb all photons in excess of E_G for the material is smaller than for indirect gap materials; consequently, less material would be needed for the solar cells. Figure 9 compares the percentage of AM0 photons in excess of E_G which would be absorbed by various thicknesses of GaAs and Si. Whereas 1 μm of GaAs would absorb about 80% of the maximum number of AM0 photons absorbable in this material, about 10 μm of Si would be needed to absorb 80% of the number absorbable in Si.

2. Charge Separation in the Photovoltaic Cell

Charge separation requires that an electrostatic potential barrier be present in the photovoltaic cell. Excess carriers of opposite signs move in opposite directions at such an electrostatic potential barrier, which can be produced by a metal–semiconductor (Schottky) junction or a p–n junction. Two types of p–n junctions can be distinguished and are important in photovoltaic solar cells: (1) p–n homojunctions in which the junction exists in a single semiconductor; (2) p–n heterojunctions in which the n-side consists of one semiconductor and the p-side of a different semiconductor. Energy band diagrams of these types of junctions are shown in Fig. 12.

Metal–semiconductor photovoltaic cells can, in principle, be made using any semiconductor as the base material. For photovoltaic cells the barrier region must be of the depletion type. The height of the barrier and therefore the magnitude of the field in the depletion region depends on the work function difference between the metal and the semiconductor, and in principle diffusion potential barriers comparable in magnitude to those obtained in p–n junctions are possible. Such junctions are easy to make and should, therefore, be inexpensive. Metal–semiconductor cells based on Cu_2O and Cu were the first photovoltaic cells in commercial use. In spite of their favorable features and long history, metal–semiconductor photovoltaic cells have not been explored very intensely since the invention of the silicon p–n junction cell. The lack of interest in Schottky barrier cells can be attributed to the higher efficiencies that have been achieved in p–n junction structures and to the greater stability of p–n junction cells. Recent advances in understanding metal–semiconductor junctions resulting from the development of MIS devices have shown that stable, high quality junctions of this type are possible and that photovoltaic cells based on this simple system are worthy of re-examination.

In contrast to the metal–semiconductor situation, the number of semiconductors in which p–n homojunctions can be readily incorporated is quite limited. It includes the column IV semiconductors Ge and Si; all $A^{III}B^V$

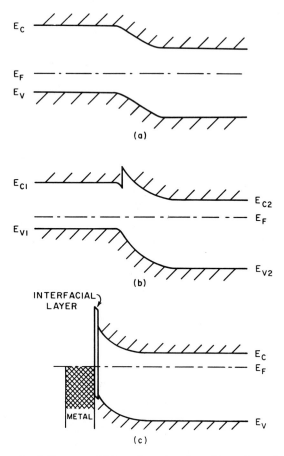

FIG. 12.   Energy band diagrams of (a) *p–n* homojunctions (b) *p–n* heterojunctions, and (c) Schottky barrier junction.

semiconductors like GaAs, InP, AlSb, etc; two $A^{II}B^{VI}$ semiconductors CdTe and ZnTe; certain $A^{I}B^{III}C_2^{VI}$ ternary semiconductors like $CuInS_2$, and a few other materials. The highest reported efficiencies for photovoltaic solar energy conversion have been observed in *p–n* homojunctions; values of about 14.5% for AM0 illumination and in the vicinity of 18% for AM1 illumination have been measured in Si and in GaAs homojunctions (in the latter case, a thin epitaxial layer of $Al_xGa_{1-x}As$ was incorporated to reduce surface recombination of light generated carriers).

On the other hand, the number of possible *p–n* heterojunction cells is limited only by the imagination since, in principle, any *p*-type semiconductor can form a heterojunction with any *n*-type semiconductor. Of course the matter is not quite so simple; problems of lattice mismatch, interface states, etc., are encountered and limit the number of feasible combinations. However, unpredictable combinations are possible, as witnessed by the important selenium photovoltaic cell which consists of a heterojunction between p–Se and n–CdSe, and by the important role in the development of large scale terrestrial solar energy development played by the thin film p–Cu_xS/n–CdS cell. In both these cases the two semiconductors have one element in common which suggests that in searching for other heterojunction structures those which have one or more elements in common deserve special attention.

Cells based on p–n homojunctions in a given semiconductor with a second semiconductor incorporated for some reason other than for the formation of the main p–n junction have also been investigated; these heterostructure cells should not be confused with heterojunction cells. An example of this type of structure has already been cited, namely, the p–$Al_xGa_{1-x}As$, p–GaAs/n—GaAs cell. As shown in Fig. 13, a barrier is produced between

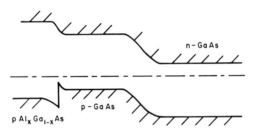

FIG. 13.   Energy band diagram of the p–$Al_xGa_{1-x}As$, p–GaAs/n–GaAs solar cell.

the p–$Al_xGa_{1-x}As$ and p–GaAs; it serves to reflect (minority) electrons generated in the p–GaAs away from the surface toward its junction with n–GaAs. Such a minority carrier mirror effects a substantial reduction in surface recomination losses.

3. Migration of Charge Carriers to the Charge Separation Site

The excess carriers generated as a result of the absorption of ionizing radiation must remain free until they reach the charge separation site, i.e., their recombination prior to separation must be minimized. Recombination can be intrinsic or extrinsic in origin; it can occur at the surface or in the bulk.

The magnitude of the intrinsic bulk recombination rate is determined mainly by the dependence of electron energy E on wave number k in the conduction and valence bands of the semiconductor, i.e., on whether the semiconductor is of the direct or indirect type. For low to moderate concentrations of excess carriers the recombination rate is high in direct gap semiconductors like GaAs, where its value leads to minority carrier lifetimes of the order of 10^{-8} sec; it is low in indirect gap semiconductors like Si where the intrinsic lifetime is about 10^{-2} sec. for material with a majority carrier concentration of about $10^{16}/cm^3$.

In real materials, intrinsic bulk recombination does not control the bulk recombination rate because of the presence of imperfections which act as recombination centers. The magnitude of the extrinsic bulk recombination rate depends on the concentration N_r and characteristics of the imperfections (their capture cross section and occupation probability). Specifically, the recombination rate is directly proportional to the product $N_r \sigma_c f$, where σ_c is the capture cross section for minority carriers and f is the occupancy factor. The semiconductor may contain a number of different species of recombination centers; that species with the largest $N_r \sigma_c$ product controls the recombination rate. Ultimately, it is the concentration of recombination centers N_r which will control the lifetime. In direct gap materials like Si and Ge, the direct recombination lifetime is long and therefore the lifetime encountered in the material and devices made from it is dominated by recombination centers present in very small concentrations: For example, in Si concentrations of centers N_r in the range of $10^{10}/cm^3$ control lifetime. In direct gap semiconductors, the band-to-band recombination lifetime is very small and consequently N_r would have to be considerably greater before it could affect minority carrier lifetime.

Light-generated excess carriers can also recombine at the free surfaces of the photovoltaic cell; the surface recombination process is commonly characterized by a surface recombination velocity s. The values of s can vary from zero—in which case the surface is a minority carrier mirror—to infinity—in which case the surface is a minority carrier sink. It is possible to treat semiconductor surfaces (with judiciously selected chemical etchants) to achieve acceptably low values of s. Sometimes the fabrication process is such that the value of s on the finished device is in an acceptable low range. Under any circumstances the surface recombination rate on the light receiving surface of solar cells must be minimized because strongly absorbed (short wavelength) photons generate carriers mainly near the surface and the fraction of carriers which reach the junction and contribute to the current is a strong function of the recombination rate. This problem can be solved in various ways; the use of a high band gap material as a minority carrier mirror has been successfully employed in the case of the previously mentioned p–$Al_xGa_{1-x}As$, p–GaAs/n–GaAs solar cell.

Short wavelength photons are also ineffective in contributing carriers to the short circuit current if the light receiving surface is covered by a " dead layer "—a layer in which the defect concentration is so high that the material can no longer be considered the same as the host material on which it was formed. Such a dead layer is produced as a result of the diffusion of dopants into silicon to form the solar cell *p–n* junction. The first few hundred angstroms are supersaturated with impurities and precipitates (e.g., if phosphorus is the diffusant, the precipitates are silicon phosphides) which so transform the material that it can hardly be considered an extension of the single crystal silicon wafer on which the solar cell was made. By eliminating this dead layer, Lindmayer and Allison [18] produced cells having increased blue response and therefore increased efficiency.

B. Optimum Properties of Single Semiconductor Solar Energy Converters

As we have pointed out, a photovoltaic cell contains an electrostatic potential barrier which can be either a *p–n* junction, a Schottky barrier, or an MIS barrier. Diagrams of the electron energy versus distance through such barrier regions are shown in Fig. 12. When voltages of opposite polarities are applied across these structures, the current flowing through them is high in one direction—the so-called forward direction—and low in the other—the so-called reverse direction. If the voltage is varied continuously, rectifier current–voltage characteristics having the form of curve (a) in Fig. 14 are obtained. When light is absorbed in the junction space charge region and/or in the adjacent material on either side of the junction, minority carriers flow toward the junction and increase the reverse current. The

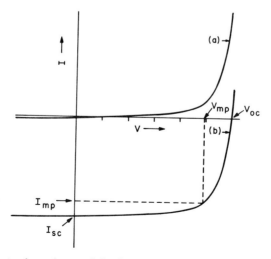

FIG. 14.   Current voltage characteristic of a solar cell. (a) in the dark, and (b) subjected to illumination.

current voltage characteristic of the illuminated cell has the form of curve (b) in Fig. 14. For curve (b), part of the *I–V* curve is now located in the fourth quadrant, where the current is negative and the voltage is positive; the *I–V* product—the electrical power—is negative, i.e., the device is generating power which can be delivered to a load connected across the junction.

1. Short Circuit Current

The magnitude of the light generated short circuit current I_{sc} can be determined from the following integral:

$$I_{sc} = q \int_{E_G = h\nu_G}^{\infty} Q(h\nu)N_{ph}(h\nu) \, d(h\nu) \tag{1}$$

Here q is the charge on the electron; $Q(h\nu)$, the minority carrier collection efficiency or the absolute spectral response of the cell, is defined as the fraction of carriers generated by absorption of photons of energy $h\nu$ which contribute to I_{sc}; and $N_{ph}(h\nu)$ is the number of photons/cm^2 sec with energy $h\nu$ incident on the solar cell. The collection efficiency $Q(h\nu)$ is a function of the absorption constant α, the bulk lifetime τ, the surface recombination velocities s_i on the exposed surfaces of the cell, and of the cell geometry. $Q(h\nu)$ can be and has been calculated for various models of a cell [19, 20]. Figure 15 shows such calculated $Q(h\nu)$ curves for silicon p–n junction cells

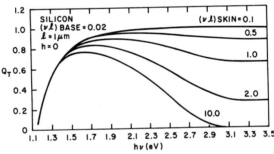

FIG. 15. Calculated total collection efficiency $Q_T(h\nu)$ versus $h\nu$ for Si. The parameter h is defined as $h \equiv s/D$, where s is the surface recombination velocity and D is the minority carrier diffusion constant (from [19]).

with a junction 1.0 μm below the surface; zero surface recombination for the light receiving surface; a minority carrier diffusion length in the base equal to 50 μm; a base thickness of 500 μm and various values of "skin" diffusion length between 10 μm and 0.1 μm. Figure 16 shows relative spectral response curves measured on a series of GaAs cells having various junction depths.

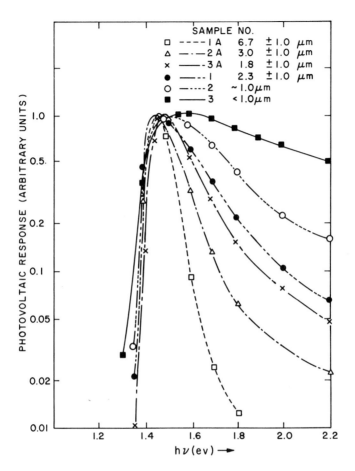

FIG. 16.   Relative $Q(hv)$ versus hv for GaAs cells having junctions at the indicated distance below the surface (from [19]).

It is an objective of solar cell technology to produce cells in which the absolute spectral response is close to unity for all solar photons having energies in excess of the forbidden energy gap. In the limit the response curve would have the form of a step function, i.e., $Q(hv) = 0$ for $hv < E_G$ and $Q(hv) = 1$ for $hv \geq E_G$. Figure 17 is a plot of I_{sc} versus E_G for the AM0 solar spectrum calculated from Eq. (1) assuming that $Q(hv)$ has this step function form. From this figure it is evident that for $1.0 \leq hv \leq 2.5$ eV, I_{sc} is an exponential function of energy gap; in this range of hv values

$$I_{sc} \approx I_{sc0} \exp[-k_1 E_G] \tag{2}$$

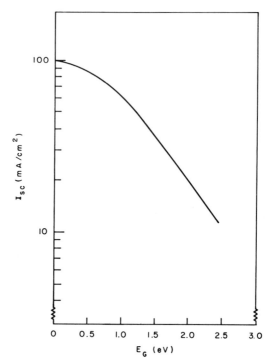

FIG. 17.   Maximum possible short circuit current I_{sc} versus energy gap E_G.

For the AM0 spectrum of Fig. 1, $I_{sc0} = 202$ mA/cm^2 and $k_1 = 1.2$. This approximate relation holds for solar spectra other than AM0 but the values of I_{sc0} and k_1 are different; k_1 ranges between 1.2 and 2 and $I_{sc0} \leq 200$ mA/cm^2. Clearly in real cells I_{sc} is less than the values calculated for the $Q(h\nu)$ versus $h\nu$ curve having the form of a step function and the value based on this $Q(h\nu)$ versus $h\nu$ relation constitutes an upper limit on I_{sc}.

2. Current Voltage Characteristic and Maximum Efficiency

The current through an "ideal" unilluminated p–n junction can be described by the expression

$$I_j = I_0[\exp(qV_j/AkT) - 1] \tag{3}$$

if internal series and shunt resistances are neglected. Here I_0 is the reverse saturation current of the junction, q the charge on the electron, V_j the voltage across the junction, k Boltzmann's constant, T the absolute temperature, and A a constant whose minimum value is $A = 1$ for an "ideal" p–n

junction but whose actual value is higher; its value depends on imperfections in the space charge region, etc.

For moderate illumination level, the load current–voltage characteristic of an illuminated photovoltaic cell (Fig. 14b) has the form

$$I_L = I_j - I_{sc} = I_0[\exp(qV_j/AkT) - 1] - I_{sc} \qquad (4)$$

The cell may also contain internal series and shunt resistances so that a lumped parameter model of a cell including these resistances would have the form shown in Fig. 18. For this circuit model the load current–voltage characteristic is described by the equation

$$I_L = I_0\{\exp[q(V_L + I_L R_L)/AkT] - 1\} - [(V_L + I_L R_L)/R_{sh}] - I_{sc} \qquad (5)$$

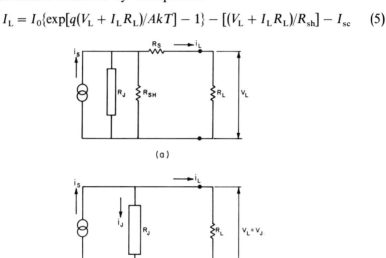

FIG. 18.   Circuit model of an illuminated solar cell (a) including internal series and shunt resistances (R_s and R_{sh}, respectively), and (b) without these resistances.

A more general model of the cell would treat the series and shunt resistances as distributed resistances and would decompose the series resistance into components contributed by the contacts to the two sides of the p–n junction, by the base region and by the diffused skin region; this last contribution is usually the largest. In a well-designed solar cell, R_{sh} must be large compared to the dynamic impedance of the junction at the operating point. The operating point is usually chosen so that maximum power is transferred from the cell to the load. The resistance of the matched load for which this occurs is calculated from the relation

$$R_{Lmp} = \partial V_j/\partial I_j \big|_{V = V_{mp}} = I_{Lmp}/V_{jmp} \qquad (6)$$

(The subscripts "mp" refer to the values of the indicated parameters under maximum power transfer conditions.) Now from Eq. (2)

$$R_j = \frac{\partial V_j}{\partial I_j} = \frac{AkT}{I_0 q} \exp\left[-\frac{qV_j}{AkT}\right] \tag{7}$$

For a 1 cm^2 Si cell with $I_0 \sim 10^{-8}$ A/cm^2, $A = 1.5$, and $V_{jmp} \sim 0.45$ V, $R_j \approx 23$ Ω. Consequently, easily achievable R_{sh} values of the order of a few thousand ohms suffice to make the effect of R_{sh} on the $I-V$ characteristics negligible. In the case of single crystal cells, the main contribution to R_{sh} arises from the periphery of the junction; in practice, it is found that no special treatments of the periphery is needed to satisfy the requirement $R_{sh} \gg R_{jmp}$.

The contribution to the series resistance R_s from the sheet resistance of the diffused skin can be reduced by forming a contact grid over the light receiving surface. In conventional diffused silicon solar cells, this sheet resistance is a few tenths of an ohm for cells with an optimized grid geometry.

Though the internal shunt and series resistances must be included in detailed analyses of the photovoltaic cell, for most purposes the simplified circuit model of Fig. 18(b) can be used to describe the solar cell because it is possible to design cells so that effects of R_{sh} and R_s can be neglected. The $I-V$ characteristic of this circuit is described by Eq. (4), from which parameters like open circuit voltage V_{oc}, current and voltage at the maximum power point, and the maximum efficiency can be calculated. For convenience the symbol $\Lambda \equiv q/AkT$ is introduced into Eq. (4) which then has the form

$$I_L = I_0[\exp(\Lambda V) - 1] - I_{sc} \tag{8}$$

The open circuit voltage V_{oc} is found by setting the load current $I_L = 0$, and

$$V_{oc} = (1/\Lambda) \ln(I_{sc}/I_0 + 1) \tag{9}$$

The values of current and voltage at the maximum power point can be found by setting the load impedance equal to the dynamic impedance at the maximum power point:

$$R_{Lmp} = \partial V/\partial I \,|_{V_{mp}} = \exp(-\Lambda V_{mp})/I_0 \Lambda = V_{mp}/I_{mp} \tag{10}$$

After substitution the following equations are derived for V_{mp} and I_{mp}

$$[\exp(\Lambda V_{mp})][1 + \Lambda V_{mp}] = I_{sc}/I_0 + 1 = \exp(\Lambda V_{oc}) \tag{11}$$

$$I_{mp} = [\Lambda V_{mp}/(1 + \Lambda V_{mp})](I_{sc}/I_0 + 1)I_0 \tag{12}$$

For a commercial silicon cell under sunlight illumination, $I_{sc} \sim 35$ mA/cm^2 and $I_0 \sim 10^{-8}$ A/cm^2; i.e., $I_s/I_0 > 10^4$, which simplifies Eqs. (9), (11), and (12).

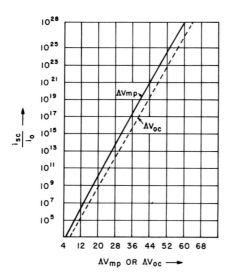

FIG. 19.   The products ΛV_{0c} and ΛV_{mp} where $\Lambda = q/AkT$; V_{0c} is the open circuit voltage and V_{mp} is the voltage at maximum power as a function of the ratio I_{sc}/I_0.

Figure 19 presents a plot of ΛV_{0c} and ΛV_{mp} as functions of $\ln(I_{sc}/I_0)$. From this figure, it is seen that for $(I_{sc}/I_0) > 10^4$, $\Lambda V_{mp} \geq 10$. From Eqs. (11) and (12), $I_{mp} \geq 0.9\, I_{sc}$ and $V_{mp} \geq 0.8\, V_{0c}$.

The maximum efficiency, i.e., the efficiency for matched load conditions, is given by

$$\eta_{max} = (\text{Power out})/(\text{Power in}) = I_{mp} V_{mp}/P_{in}$$

$$= \frac{\Lambda V_{mp}}{1 + \Lambda V_{mp}} \frac{\Lambda V_{mp}}{\Lambda V_{mp} + \ln(1 + \Lambda V_{mp})} \frac{I_{sc} V_{0c}}{P_{in}} \tag{13}$$

$$\equiv CF I_{sc} V_{0c}/P_{in} \tag{13a}$$

where CF is the curve factor, i.e., a measure of how closely the I–V curve approaches rectangular shape. As $V_{mp} \rightarrow V_{0c}$ and $I_{mp} \rightarrow I_{sc}$, $CF \rightarrow 1.0$. From Fig. 19 and the discussion presented above, for $I_{sc}/I_0 > 10^4$, $CF \gtrsim 0.72$, i.e.,

$$\eta_{max} \gtrsim 0.72\, I_{sc} V_{0c}/P_{in} \tag{14}$$

3. Open Circuit Voltage and Its Dependence on I_0

Let us now turn our attention to the parameter V_{0c} and explore its dependence on the properties of the semiconductor. According to Eq. (9), V_{0c} is determined by I_{sc}, I_0, and Λ. We have already discussed I_{sc} and its dependence on α, L, s, etc. The other two parameters appear in the expression for the I–V characteristic of a diode, i.e.,

$$I_j = I_0[\exp(\Lambda V) - 1] \tag{15}$$

Such an equation assumes that the I–V characteristic is a simple exponential curve. In fact this is not so; for example, the I–V curve of the silicon p–n junctions shown in Fig. 20 has two distinctly different slopes—one in the low voltage region and the other at intermediate voltages. Furthermore, at high voltages the internal series resistance begins to affect the shape of the

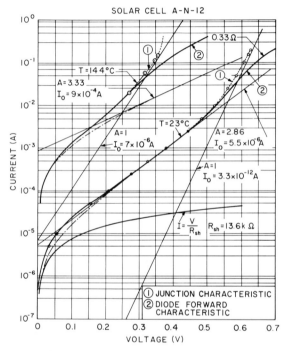

FIG. 20. Current voltage characteristics of a typical Si photovoltaic cell ($\eta_{max} \sim 12\%$) at room temperature (23°C) and at 144°C. Lines corresponding to $A = 1$ are shown along with I_0 values (from [20]).

curve and it departs from exponential behavior. Such an I–V curve can be fitted by an equation of the form

$$I_j = I_{01}[\exp(\Lambda_1 V) - 1] + I_{02}[\exp(\Lambda_2 V) - 1] + V/R_s \tag{16}$$

where the values of I_{01}, I_{02}, Λ_1, Λ_2, and R_s are determined by fitting the experimental I–V curve; R_s is the series resistance of the diode, whose magnitude could be a function of illumination level, i.e., it could be a photoresistor. In general, therefore, the $I-V$ characteristic for voltages and currents having values low enough so that internal series resistance can be neglected should be represented by an equation of the form

$$I_j = \sum_i I_{0i}[\exp(\Lambda_i V) - 1] \qquad (17)$$

Two basic mechanisms which can control the $I-V$ curves of a $p-n$ homojunction have been identified and analyzed theoretically. The first is based on the assumption that minority carriers move only under the influence of concentration gradients and that they cross the space charge region without any change in their concentrations. This is the classical $p-n$ junction diode and the $I-V$ expression has the form

$$I_j = I_{01}[\exp(qV/kT) - 1] \qquad (18)$$

and

$$I_{01} \sim n_i^2[1/\sigma_n L_n + 1/\sigma_p L_p] \qquad (19)$$

where n_i^2 is the concentration of carriers in intrinsic material; σ_n and σ_p are the dark conductivities of the two sides of the junction; L_n and L_p are minority carrier diffusion lengths in the two regions. Now

$$L \sim (D\tau)^{1/2} \qquad (20)$$

where D and τ are the minority carrier diffusion constant and lifetime, respectively. Also

$$n_i^2 = N_c N_V \exp(-E_G/kT) \qquad (21)$$

Consequently,

$$I_{01} \sim \exp(-E_G/kT) \qquad (22)$$

A second mechanism [21] is based on the assumption that generation and recombination occur in the space charge region. Under certain assumptions about the nature of the process [22], the $I-V$ relation assumes the form

$$I_j \sim (n_i/\tau)[\exp(qV/2kT) - 1] \qquad (23)$$

or

$$I_{02} \sim \exp(-E_G/2kT) \qquad (24)$$

For these two cases one can therefore write

$$I_{ji} = K_i \exp(-E_G/B_i kT)[\exp(qV/A_i kT) - 1] \qquad (25)$$

with $A_1 = B_1 = 1$, $A_2 = B_2 = 2$, and $K_1 \ll K_2$.

Recombination and generation in the space charge region can at least, in principle, be controlled by controlling the recombination center concentration. If this concentration is low enough, the I_j-V curve will be governed by the ideal diode equation. Now high efficiency requires high values of V_{0c}. The question arises: which of these two mechanisms would produce a higher V_{0c} for a given value of I_{sc}? The difference between V_{0c} values produced in the two cases is

$$V_{0c}^{(2)} - V_{0c}^{(1)} = (A_2 - A_1)(kT/q)\ln(I_{01}/I_{02}) \qquad (26)$$

From Eq. (26), $V_{0c}^{(1)} > V_{0c}^{(2)}$ provided only that $I_{02} > I_{01}$, which is always true. Because V_{0c}—and therefore η_{max}—is higher in cells in which space charge recombination and generation do not occur, it is a goal of solar cell technology to fabricate cells in such a way that their I–V characteristics conform to the ideal diode form, i.e., so that $A = B = 1$.

4. Metal–Semiconductor Barrier Characteristics

For metal semiconductor barriers, the I–V characteristic also has the form

$$I_j = I_0[\exp(qV/AkT) - 1] \qquad (27)$$

where I_0 is given by

$$I_0 = A^*T^2 \exp(-q\phi_B/kT) \qquad (28)$$

it being assumed that the reverse current originates from thermionic emission over the surface barrier whose magnitude is ϕ_B. The parameter A^* is Richardson's constant. As shown by Stirn and Yeh [23], the open circuit voltage of a Schottky barrier diode is given by

$$V_{0c} \sim (AkT/q)\ln(I_{sc}/A^*T^2) + A\phi_B/q \qquad (29)$$

Two solar cell parameters—the short circuit current I_{sc} and the series resistance R_s—impose conflicting requirements on the thickness of the metal film used to form the Schottky barrier. A thin film would minimize absorption losses and therefore maximize I_{sc}, while a thick film would minimize R_s and the losses associated with internal series resistance. A good compromise can be achieved between these requirements; films having a thickness of 100 Å or less covered with a contact grid have acceptable series resistance and the photon absorption in films of this thickness is not excessive although, of course, it reduces the value of I_{sc} below the values shown in Fig. 17.

The maximum efficiency of metal semiconductor cells is governed by the same equations as those previously deduced for p–n junctions [Eqs. (11)–(13)].

5. Dependence of η_{max} on E_G

The dependence of η_{max} on E_G can now be calculated starting with Eq. (14). The dependence of V_{0c} on E_G can be deduced by substituting the band gap dependence of I_{sc} on E_G [Eq. (2)] and of I_0 on E_G [Eq. (24) into Eq. (9)]:

$$V_{0c} \approx \frac{1}{\Lambda_i} \ln\left(\frac{I_{sc0} \exp(-k_1 E_G)}{K_i \exp(-E_G/B_i kT)}\right) \tag{30}$$

$$\approx \frac{1}{\Lambda_i}\left[E_G\left(\frac{1}{B_i kT} - k_1\right) + \ln\left(\frac{I_{sc0}}{K_i}\right)\right] \tag{31}$$

For the case of $A = B = 1$ (the ideal diode case) and $T = 300$ K, this becomes

$$V_{0c} = (kT/q)[38.8 \, E_G + \ln(I_{sc0}/K_1)] \tag{32}$$

For the case $A = B = 2$ (recombination–generation in the space charge region) and $T = 300$ K, this becomes

$$V_{0c} = (2kT/q)[18.8 \, E_G + \ln(I_{sc0}/K_2)] \tag{33}$$

In both cases, at a fixed temperature, V_{0c} increases as E_G increases. On the other hand, I_{sc} decreases with E_G; consequently, η_{max} passes through a maximum. Figure 21 shows η_{max} versus E_G for these cases based on convenient

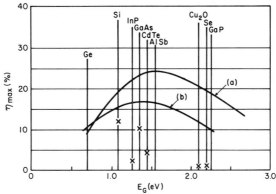

FIG. 21.   Maximum efficiency η_{max} versus E_G for AM0 illumination (a) for ideal junction ($A = 1$), and (b) for junction in which recombination and generation in the space charge region control I_0 ($A = 2$). The "X" marks indicate highest reported AM0 efficiencies for photovoltaic cells made from the indicated semiconductor. [70].

though rather artificial assumptions about I_{sc}, I_0, and Λ. It is assumed that the collection efficiency is unity and therefore that for Si ($E_G = 1.1$ eV), $I_{sc} = 56$ mA/cm^2. The values of the other important parameters in these equations are assigned in the following way:

1. The value of K_1 is determined by assuming that $A = B = 1$, i.e., $\Lambda = 40$ V^{-1} at room temperature, and that I_0 can be evaluated from diffusion theory using reasonable values for doping levels and lifetimes on the

two sides of the *p–n* junction. These assumptions lead to a value $I_0 = 2 \times 10^{-12}$ A/cm^2 for Si $(E_G = 1.1$ eV$)$. The value of K_1 is then calculated from Eq. (25).

2. The value of K_2 is determined by assuming that $A = B = 2$, i.e., $\Lambda = 20$ V^{-1} and that the value of I_0 corresponding to the $A = 2$ branch of the silicon *I–V* curve is $I_0 = 10^{-6}$ A/cm^2. The value of K_2 is then calculated from Eq. (25).

The curves in this figure are therefore approximate; they do, however, illustrate the general dependence of η_{max} on E_G. The maximum efficiency can be calculated more exactly for each semiconductor using measured or anticipated values of the various parameters which determine I_{sc}, I_0, R_s, R_{sh}, etc. Such calculations have been performed for silicon and for GaAs; the results will be discussed in subsequent sections of this chapter.

6. Dependence of η_{max} on Temperature

The dependence of η_{max} on E_G is a strong function of temperature T because of the strong temperature dependence of I_0. This leads to a linear relation between V_{0c} and T. From Eq. (30).

$$V_{0c} \approx (AkT/q)[(E_G/BkT) - E_G k_1 + \ln(I_{sc0}/K_i)] \tag{34}$$

$$\approx AE_G/Bq - (AkT/q)[k_1 E_G + \ln(I_{sc0}/K_i)] \tag{35}$$

Thus, the value of V_{0c} decreases as T increases. If $A = B$, the intercept of Eq. (35) is equal to the energy gap E_G/q. Now according to Eq. (13)

$$\eta_{max} = CF \ I_{sc} V_{0c}/P_{im} \tag{36}$$

The curve factor CF has a very weak temperature dependence because its numerator and denominator are both approximately linear functions of T. Therefore, the temperatures dependence of η_{max} is essentially the same as that of V_{0c}, i.e., η_{max} is a linear function of T. Figure 22 shows how η_{max} depends on E_G for various T [24]. These curves were computed using the assumptions similar to those used in calculating the curve of Fig. 21 for $A = B = 1$. They show that solar cells should be operated at low temperatures; that if cells are to be operated at higher temperatures—as is the case with combined thermal plus photovoltaic systems—they should be made from semiconductors with large bandgaps.

The η_{max} versus E_G of Figs. 21 and 22 are calculated for the AM0 solar spectrum. The solar spectrum on the surface of the earth differs from the AM0 spectrum because blue photons are more strongly absorbed in the atmosphere than red photons. As a result there can be significant differences in the efficiencies of cells exposed to these two spectra. Figure 23 shows how changes in the assumed solar spectrum affect η_{max} versus E_G curves for the case $A = B = 1$ [25]. The peak in the η_{max} versus E_G curve shifts toward

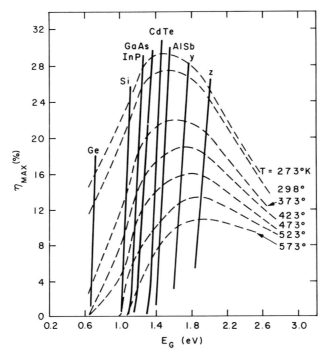

FIG. 22. Maximum efficiency η_{max} versus E_G for various temperatures. It is assumed that the junction I–V characteristic has $A = 1$ (from [24]).

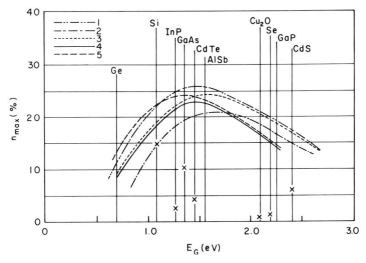

FIG. 23. Maximum efficiency η_{max} at 300°K versus E_G for various solar spectra and various assumption about I (from [25]). Curves 1 and 2 from [26] with two different values of K_1 from Eq. (25); curves 3 and 4 from Fig. 21 and curve 5 is from [27].

smaller energy gaps as the spectrum changes from AM0 to AM1 to AM2, etc.

In addition to this "fundamental" change in the η_{max} versus E_G dependence, it is also of course possible that the efficiency of a given solar cell changes as the spectrum changes because the collection efficiency of the cell Q is a function of photon wavelength. The curves in Fig. 23 are based on the assumption that Q is a step function, i.e., $Q = 0$ for $E_G > hv$ and $Q = 1$ for $E_G \leq hv$. In the case of commercial Si cell, the collection efficiency is higher for red photons than for blue photons; consequently the AM1 efficiency of Si cells is substantially higher than the AM0 efficiency.

C. Status of Candidate Materials and Photovoltaic Cells for Terrestrial Solar Energy Converters

As was pointed out, the calculations of η_{max} versus E_G curves described above are based on a number of simplifying approximations. They do show that the best semiconductors for solar cells have energy gaps $1.0 \leq E_G \leq 2.5$ eV; that the maximum possible efficiencies would lie between 20% and 25%, and that larger band gaps would be preferred if the cells are expected to perform at elevated temperatures. More accurate estimates of the ultimate efficiency obtainable from a given semiconductor can be calculated based on available information about its properties. Such calculations have been made for Si and GaAs and the results will be summarized in this section. We shall also discuss the current state of theory and experiment for solar cells based on other semiconductor systems; the most important of these is the thin film p–Cu_xS/n–CdS system.

1. Silicon

Silicon solar cells occupy a dominant position in discussions of large scale terrestrial photovoltaic power systems. Cells of this type have been the building blocks of power supplies on board earth satellites from the inception of the space program in the late 1950s. They have proven themselves to be reliable long lived devices. Standard 2×2 cm^2 commercial cells manufactured at the rate of more than a million cells per annum have AM0 efficiencies around 11% [28]; cells of new designs with AM0 efficiencies up to 14% have been fabricated in the laboratory [18] and analyses of the performance of silicon cells indicate that AM0 efficiencies between 18% and 20% are possible [29,30a,b]. These studies of Si solar cell efficiencies indicate that improvements in η_{max} would occur if the resistivity of the base wafer from which the cells are fabricated is decreased from 10 Ω cm to about 0.01 Ω cm while retaining a post fabrication minority carrier lifetime of about 1 μsec and if surface losses are reduced so that blue photons can be utilized more efficiently. Some of these improvements have already been realized in the so-called " Violet Cell " in which substantially increased blue response was achieved by altering the geometry and fabrication procedure of the cell [18].

Figure 24 (after M. Wolf) is a bar chart showing where losses occur in commercial 11% Si cells exposed to AM0 sunlight. The chart shows that 24% of the energy in the incident beam is lost because the photon energy is

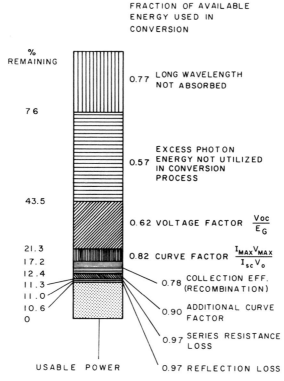

FIG. 24. Bar chart showing losses in Si solar cells (from [29]).

too low to produce ionization in silicon. In addition, 32.5% of the incident energy is lost because each photon having energy in excess of the band gap can generate only one hole–electron pair; the excess energy is degraded into thermal energy. These are basic physical processes and the losses associated with them cannot be reduced once the semiconductor (and its energy gap) have been selected. At this point, 44% of the energy in the incident beam remains. How much of this energy is actually delivered to the load, depends on the parameters of the material comprising the cell and on the properties of the cell, i.e., on such factors as the collection efficiency Q which, in turn, is a function of α, L, s, and the cell geometry; on the reflection losses which depend on the anti reflection coating used; on the doping levels on the two sides of the junction, on the $I-V$ characteristics of the junction, i.e., on I_0 and Λ.

If all the AM0 photons with energy in excess of the energy gap of Si (1.1 eV) produced a hole–electron pair and all the excess minority carriers contributed to the short circuit current, I_{sc} would be about 56 mA/cm². However, reflection losses $(1 - R = 0.97)$ and collection efficiency losses reduce I_{sc} to about 39 mA/cm² $(Q \sim 0.72)$. After including these loss mechanisms, 30.7% of the energy in the incident beam is left.

Diagnosis of cell performance can be aided by introducing certain parameters which define losses in the cell. The first of these is the voltage factor (VF), defined as follows

$$\text{VF} \equiv qV_{0c}/E_G$$

(In principle, one can conceive of a $p-n$ junction in which the doping levels are so high in both the p- and n-sides of the junction that the Fermi levels coincide with the respective band edges. The open circuit voltage in such a hypothetical cell could become equal to E_G if the illumination intensity, i.e., I_{sc}, were high enough.) For 10 Ω cm base wafers and I_{sc} of about 38 mA/cm², the values of VF in AM0 illuminated cells is 0.49. After including this loss, 15.0% of the incident energy remains. Thus we see that the VF must be significantly increased if η_{max} is to be increased.

The next loss factor occurs because at the maximum power point the current I_{mp} and voltage V_{mp} are less than I_{sc} and V_{0c}, respectively, giving rise to a fill factor FF defined so,

$$\text{FF} = I_{mp}V_{mp}/I_{sc}V_{0c} \tag{37}$$

If the measured values of these parameters are inserted in this equation, FF ~ 0.73 and the power remaining is 10.6%. The measured fill factor can be decomposed into three components:

1. A basic curve factor which would occur in a cell having $A = B = 1$ ($\Lambda = 40$ at 300°K). Substitution for parameters in Eq. (37) results in

$$\text{CF} = \frac{\Lambda V_{mp}}{1 + \Lambda V_{mp}} \frac{\Lambda V_{mp}}{\Lambda V_{mp} + \ln(1 + \Lambda V_{mp})} \tag{38}$$

with $\Lambda = 40$. In the case of Si the basic CF ~ 0.81.

2. An additional curve factor resulting from the fact that $\Lambda \neq 40$ at 300°K:

$$\text{CFA} = \frac{\text{CF}(\Lambda_m)}{\text{CF}(40)} \tag{39}$$

where Λ_m is the actual measured value of Λ. In the case of Si, CFA ~ 0.91.

3. A loss associated with internal series resistance which contributes to "softening" of the $I-V$ curve. In commercial cells, this loss factor has a value of about 0.97.

The product of these three loss factors is the fill factor and it has a value of about 0.73. These losses bring the efficiency down to 10.6% which is in the range of observed values.

The increased-efficiency COMSAT cell described by Lindmayer and Allison [18] is characterized by an increase in the $Q(1 - R)$ factor for AM0 illumination from 0.70 in standard cells to 0.78, in the fill factor from 0.72 to 0.83, and in the voltage factor from 0.50 to 0.53, which results in AM0 efficiencies of 14%.

To achieve even higher efficiencies, the loss mechanisms must be reduced further. Wolf [29] anticipates increasing the voltage factor to 0.71 by using more heavily doped base material; increasing the $Q(1 - R)$ product to 0.85 by reducing recombination losses in the bulk and at the surfaces of the cell, the curve factor to 0.86, and the additional curve factor to 1.0. If these values were achieved, the AM0 efficiency would be about 22%. Brandhorst [30a,b] anticipates that the maximum attainable value of $Q(1 - R)$ is about 0.82, of the voltage factor is 0.65, and of the fill factor is 0.80, and therefore that a practical upper limit on silicon solar cell efficiency is about 18%.

2. Gallium Arsenide

A detailed evaluation of the performance and potential of GaAs cells has been made by Hovell and Woodall [31] and Huber and Bogus [32]. The results of these studies are summarized in Table 1 and compared with Wolf's data on silicon. The table includes experimental results quoted by Hovel and Woodall [31] for a 12% $p-Al_xGa_{1-x}As$, $p-GaAs/n-GaAs$ cell; a $p-AlAs$, $p-GaAs/n-GaAs$ cell proposed by Huber and Bogus, and anticipated potential values of the losses based on the analysis of Hovell and Woodall. As is evident from this table, the main problem in GaAs is its low collection efficiency; the voltage factor and the fill factor are close to their potential maxima. The collection efficiency has been substantially increased by incorporating a higher band gap semiconductor layer over the GaAs $p-n$ homojunction. This layer reduces surface recombination losses normally encountered in GaAs cells. The high surface recombination velocity encountered in GaAs solar cell surfaces has an especially marked influence on collection efficiency because most of the solar photons are absorbed very close to the surface in this direct gap semiconductor. As shown in Fig. 13, the presence of this higher band gap material on the surface introduces a "minority carrier mirror" which repels minority carriers travelling toward

TABLE 1

Loss Mechanisms in Ordinary Silicon and Several Types of Gallium Arsenide Solar Cells Exposed to AM0 Illumination

Loss Mechanism	Ordinary 10 Ω cm n–p Si [29]	Observed values for ordinary p–n GaAs [32]	Calculated values for p–AlAs, p/n–GaAs [32]	Observed values for p–Al$_x$Ga$_{1-x}$As, p/n–GaAs [31]
1. Long photons not absorbed ($h\nu < E_G$) ≡ L	0.24	0.38	0.38	0.38
2. Excess photon energy not utilized ($h\nu > E_G$) ≡ E	0.43	0.34	0.34	0.34
3. Absorption in AlAs or Al$_x$Ga$_{1-x}$As window ≡ A	Does not apply	Does not apply	0.18	0.29
4. Voltage factor loss ≡ V	0.51	0.30	0.30	0.26
5. Ideal curve factor	0.19	0.10	0.10	—
6. Additional curve factor	0.09	0.09	0.09	—
7. Internal series resistance	0.03	0.03	0.03	—
8. Fill factor (function of 5 6, and 7) ≡ FF	0.71	0.79	0.79	0.79
9. Collection efficiency Loss $(1 - Q_T)$	0.28	0.54	0.20	0.21
10. Reflection loss R	0.03	0.03	0.03	0.03
11. Useable power/solar power input, η^a	0.11	0.12	0.14	0.13

[a] N.B. The efficiency η is calculated from the relation: $\eta = (1 - L)(1 - E)(1 - A)(1 - V)FF(Q_T)(1 - R)$

the surface and causes them to move toward the p–n junction where they can be collected. This concept, i.e., the incorporation of a large band gap material as a cover for a p–n junction cell to reduce surface losses, can be applied to any solar cell; it is especially beneficial in the case of direct gap semiconductors, since light is absorbed very close to the surface in such materials. The basic idea is to select an indirect gap semiconductor with a band gap substantially larger than that of the semiconductor material from which the cell is made and to join it to the host semiconductor. This "joining" works in the case of the $Ga_xAl_{1-x}As$ case in part because two constituents of the large band gap semiconductor are common with the host GaAs. Hovell and Woodall point out that this layer must be very thin—in the range of a few thousand angstroms—in order to decrease absorption losses in the $Ga_xAl_{1-x}As$ [31].

More recently, high efficiency ($\sim 10\%$) GaAs cells have been made by forming a thin (~ 30 Å) layer of oxide on GaAs and covering this oxide with a transparent metal coating forming an M–I–S structure [33]. Such cells may be less expensive to fabricate than conventional diffused cells and have the potential of comparable efficiencies.

Similar detailed analyses can be made for other semiconductors of potential interest as photovoltaic solar energy converters. In each case there is a basic loss mechanism associated with photons which cannot be absorbed by the semiconductor and with the energy absorbed photons have in excess of the forbidden energy gap. The other loss mechanisms are, to various degrees, dependent on material parameters and on device parameters which can be altered by changes in device fabrication procedures.

3. p–Cu_xS/n–CdS *Thin Film Cells*

Except for single crystal silicon cells, more effort has been expended on p–Cu_xS/n–CdS heterojunction thin film solar cells than on any other photovoltaic system. The active part of this solar cell is a thin (10–20 μm) polycrystalline film of CdS onto which an even thinner (0.1–0.2 μm) layer of a copper–sulfur compound is incorporated. Many thousands of large area (7.5×7.5 cm^2) cells of this type have been fabricated in various laboratories over the past decade. Occasional cells have exhibited solar energy conversion efficiencies up to about 9% at 25°C; a high yield of cells having efficiencies between 5 and 6% has been reported in a pilot plant operation. However, the cells do have stability problems; their output can decrease with time. Accelerated life tests on recently fabricated cells have, however, indicated that lifetimes in excess of 15 yr may be possible [10a]; furthermore, some cells fabricated almost 8 yr ago show no apparent degradation over this period of time. There is, therefore, reason to believe that the output can be stabilized and indeed a recent publication reports a substantial increase in stability resulting from modifications in cell fabrication procedure [10b].

The main reason for the continued high interest in these solar cells is the promise that they should cost less to manufacture because the processes involved in producing polycrystalline film cells are simpler than those required to produce single crystal cells. While much has been learned about the complex electronic and chemical processes occurring in the cell, gaps in the understanding of this heterojunction cell still prevent the degree of control over the fabrication process which is a prerequisite to effective commercial exploitation of the cell.

The photovoltaic effect in the Cu–CdS system was discovered and first reported by Reynolds *et al.* in 1954 [34] who deposited Cu films on CdS crystals and observed a significant photovoltaic effect in the system; they reported solar energy conversion efficiencies of a few percent. Also in 1954, Nadjakov *et al.* [35] reported observation of a weak photovoltaic effect in cells made from thin evaporated layers of CdS on which Cu was evaporated. Subsequently Middleton *et al.* [36] produced thin film cells with efficiencies (from 3 to 5%) comparable to those reported in the single crystal cells. These early thin film cells were fabricated on molybdenum substrates. Subsequent generations of cells have been made on plastic (kapton) and other metal (e.g., copper) substrates. In the case of kapton, metal films (e.g., chromium and gold) are first deposited on the plastic and the CdS evaporated thereon. This configuration is especially attractive for cells intended for use in space where a high power-to-weight ratio is of paramount importance. Such plastic substrate cells have been supplanted for terrestrial applications by cells made on copper foil substrates onto which a thin layer of zinc is deposited. The zinc provides a good ohmic contact to the CdS film (10–20 μm thick) which is deposited on the substrate by standard vacuum evaporation methods.

In the "standard" type of cell, the Cu_xS layer is formed via the chemical displacement reaction

$$2Cu^+ + CdS = Cu_2S + Cd^{2+} \tag{40}$$

by dipping the CdS into a hot ($\sim 90°C$) acidic solution of Cu_2Cl_2 [37]. Thermodynamic considerations indicate that the reaction proceeds toward the right. Thus, ideally two Cu^+ ions displace a Cd^{++} ion to produce Cu_2S. However, the Cu–S system is extremely complex; compounds stable at room temperature include chalcocite (Cu_2S), djurleite ($Cu_{1.96}S$), digenite ($Cu_{1.8}S$) and covellite (CuS). Each modification exists over a certain composition range as shown in the phase diagram generated by Cook [38].

The University of Delaware Group [10a] has shown that in their thin film cells the composition of the Cu_xS layer varies as a function of distance from the surface, i.e., the Cu/S ratio changes. According to Palz *et al.* [10b], the highest short circuit current, highest efficiency and highest stability is observed in cells in which the layer is "pure" Cu_2S. In the best cells, the thickness of the Cu_xS layer lies between 1000 Å and 2000 Å.

Over the years, many models have been proposed to explain the photovoltaic effect in the Cu–CdS system. The band model shown in Fig. 25 shows the main features on which agreement currently exists [10a, 39]. The most

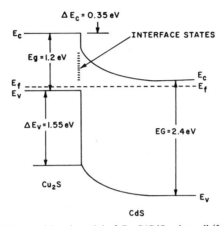

FIG. 25.   Proposed band model of Cu_2S/CdS solar cell (from [10a–c]).

firmly established point is that sunlight is absorbed mainly in the high conductivity Cu_xS layer. The chemical dipping process produces a layer which is mainly Cu_2S in the best cells. Subsequent heat treatment improves the $I–V$ characteristics of the cell. Of the various compounds in the Cu–S system only Cu_2S is known to be a semiconductor. It has a band gap of about 1.2 eV and is probably a direct gap semiconductor [40]. When excited by electrons, Cu_2S and the Cu_xS layer emit a luminescence band consistent with a band gap of 1.2 eV as shown in Fig. 26 [41]. The conductivity of this Cu_xS layer is high; measurements indicate carrier concentrations between 10^{17} and 10^{20} per cubic centimeter. The conductivity can vary over the surface of the cell [10a]. It is not clear at this time what the optimum conductivity is; indeed the process commonly used to form the Cu_xS layer does not allow good control over the parameter.

The $I–V$ characteristic of these cells can be fitted by an equation of the form

$$I_j = I_{01}(e^{\lambda_1 V} - 1) + I_{02}(e^{\lambda_2 V} - 1) \qquad (41)$$

It is found that I_{01} is an exponential function of the inverse temperature of the form $I_{01} \sim \exp(-0.45/kT)$ and $\lambda_1 \equiv 1/A_1 kT$ with $A_1 \approx 2$; I_{02} is also an exponential function of the inverse temperature of the form

$I_{02} \sim \exp(-1.1/kT)$ and $\lambda_2 \equiv 1/A_2 kT$ has $A_2 = 1$. The first pair of these parameters (I_{01} and λ_1) has been attributed to the presence of states at the interface between the Cu_xS and the CdS. According to one model [42] tunneling into these states from the CdS conduction band determines I_0,

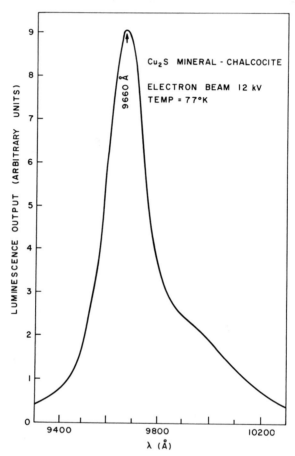

FIG. 26.   Luminescence spectrum of Cu_2S (from [41]).

and λ_1, while according to another model [10a] recombination through the interface states controls I_0.

In summary, the thin film Cu–Cds cell continues to hold promise for large scale terrestrial applications. More work must, however, be expended on this photovoltaic system in order to exploit its potential as fully as possible.

4. Cadmium Telluride Cells

Cadmium telluride is another II–VI semiconductor which has been and continues to be explored as the basis of a photovoltaic solar cell. Cadmium telluride is a direct gap semiconductor which can be made *n*- or *p*-type by doping with appropriate impurities; its 1.5 eV band gap indicates that CdTe homojunction cells should have efficiencies close to the peak in the η_{max} versus E_G curve. Both single crystal and thin film cells have been made from CdTe but the greater effort has been expended on thin film cells. Cusano first described such a thin film cell in 1963 [43] and published a more detailed report in 1966 [44].

Cusano's cells were made by depositing (by vapor reaction of Cd, CdI$_2$, and Te) about 10 μm of *n*–CdTe onto a sheet of molybdenum 0.003–0.005 cm thick and producing a layer of Cu$_2$Te on the surface by the chemical dipping method used to produce p–Cu$_x$S/n–CdS thin film cells; i.e., he formed heterojunction cells between p–Cu$_2$Te and n–CdTe. The band gap of Cu$_2$Te is 1.04 eV at 300°K [45]; it is an indirect gap semiconductor. According to Cusano, the Cu$_2$Te layer was 80–200 Å thick. The cell efficiency was as high as 6% and $V_{OC} \sim 0.55$ V. Cusano also reported work on single crystal CdTe cells made on zone refined *n*–CdTe. The barrier was formed by dipping into a CuCl solution at 85°C for about 10 sec. Ohmic contact to the base material was provided by In solder; based on measured values, he estimates that with an appropriate antireflection coating, the single crystal cells would have had an efficiency of 9–10%. The thin film cells were unstable; their output degraded with time.

In 1970, Lebrun [46] described CdTe cells in which the carrier concentration in the *n*–CdTe is controlled by adjusting the reaction temperature and the "Cu$_2$Te" layer is formed by flash evaporation. His "standard" cells have an area 8×8 cm^2, efficiencies around 4.8%, $V_{OC} \sim 0.590$ V, and $I_{sc} \sim 13$ mA/cm^2. Areas up to 320 cm^2 with efficiencies in excess of 4% have been fabricated. According to Lebrun, these cells are more stable than earlier types though they still exhibit some degradation with time.

These investigations have demonstrated that good photovoltaic solar cells can be made from CdTe. Recent work on this semiconductor has shown that single crystals of much greater perfection are now being grown and it may be possible to make improved CdTe photovoltaic cells from this material.

5. Other Semiconductors

As we have already pointed out, the criteria for selecting semiconductors for large scale photovoltaic solar energy conversion are rather few:

1. the forbidden band gap should lie between 1.1 and 2.5 eV;

2. the material should be a direct gap semiconductor;

3. it is preferable if both p- and n-type conductivity is possible in the semiconductor, but this is not essential since heterojunctions, Schottky barrier, and MIS cells can be made from almost any semiconductor;

4. the elements comprising the solar cell, its metallic contacts, etc., should be sufficiently abundant to allow manufacture of large amounts of cells.

Many semiconductors, besides those already discussed, satisfy these criteria and in this section we briefly describe some of them. For example, Table 2 (from the paper by Tell et al. [47]) tabulates the electrical properties of a group of ternary $A^I B^{III} C_2^{IV}$ semiconductors whose energy gaps are in the range required for good solar energy conversion. The values of conductivity, carrier concentration, and mobility do not represent maximum or minimum values since these parameters can be made to vary over a wide range by annealing in vacuum or under pressure of the chalcogenide constituent. The values of the energy gap and absorption edges are approximate since the quality of single crystals of these materials has not been comparable to that of the $A^{III} B^V$ or $A^{II} B^{VI}$ semiconductors and the work expanded on them has been only a small fraction of that expended on other better known semiconductors. Note that two of them ($CuInS_2$, $CuInSe_2$) can be produced as either n- or p-type materials, i.e., homojunctions can be made from them.

Wagner et al. [48a,b] have reported p-$CuInSe_2/n$-CdS heterojunctions having solar energy conversion efficiencies around 12%. The best of these cells were made by growing an expitaxial layer of CdS onto $CuInSe_2$ single crystals. They were illuminated through the CdS layer whose thickness was between 5 and 10 μm. The spectral response of the cell was reported to be constant between 0.55 and 1.25 μm (the absorption edge of $CuInSe_2$) and with a collection efficiency of about 70% over this wavelength region. Others of these ternary compounds should also be of interest for solar energy converters. However, to be of practical interest, the cells would have to be based on thin films rather than single crystals because of the availability of In whose world production is only about 100 tons per annum. As in the case of gallium such a production rate would suffice for thin film cells but would be too small for large areas of thick (say 200 μm) cells.

Another group of semiconductors of potential interest for solar energy conversion are $Zn_3P_2(E_G \sim 1.2$ eV$)$; $Zn_3As_2(E_G \sim 1.0$ eV$)$; $ZnP_2(E_G \sim 1.4$ eV$)$; $ZnAs_2(E_G \sim 0.92$–1.2 eV$)$; $CdP_2(E_G \sim 1.93$ eV$)$; $CdAs_2(E_G \sim 1.94$–1.0 eV$)$ [49a]. The Zn compounds in this group have always exhibited p-type conductivity while the Cd compounds have been n-type. Consequently, these materials can only be used in heterojunction or Schottky barrier cells.

TABLE 2

Comparison of Electrical Properties of I–III–VI_2 Compounds [47]

Compound	Energy gap (eV)	Absorption edge (μm)	Annealed under maximum S or Se pressure				Annealed under minimum S or Se pressure			
			Type	ρ (Ω-cm)	P (cm^{-3})	μ (cm^2/V-sec)	Type	ρ (Ω-cm)	n (cm^{-3})	μ (cm^2/V-sec)
CuAlS$_2$	3.5	0.37	p	10^2–10^3	$>3 \times 10^{15}$	<2	—	$>10^5$	—	—
CuAlSe$_2$	2.7	0.50	p	10^2–10^3	$>1 \times 10^{16}$	~ 1	—	$>10^5$	—	—
CuGaS$_2$	2.5	0.50	p	1	4×10^{17}	15	—	$>10^5$	—	—
CuGaSe$_2$	1.7	0.77	p	0.05	5×10^{18}	20	—	$>10^5$	—	—
CuInS$_2$	1.5	0.81	p	5	1×10^{17}	15	n	1	3×10^{16}	200
CuInSe$_2$	0.8	1.5	—	0.5	1×10^{18}	10	n	0.05	4×10^{17}	320
AgGaS$_2$	2.7	0.46	—	$>10^5$	—	—	—	$>10^5$	—	—
AgInS$_2$	2.0	0.62	—	$>10^5$	—	—	n	10	4×10^{15}	150
AgGaSe$_2$	1.8	0.69	—	$>10^5$	—	—	—	$>10^5$	—	—
AgInSe$_2$	1.2	1.0	n	10^1	8×10^{11}	750	—	0.02	5×10^{17}	6

Schottky barrier cells based on Zn_3P_2 exhibiting $\eta_{max} \sim 6\%$ have been reported by a group at the University of Delaware in 1978 [71].

The copper compounds Cu_2S and Cu_2O have been used in photovoltaic cells, $Cu_2S(E_G \sim 1.2$ eV) as part of the p–Cu_2S–CdS cell (efficiencies up to 8% have been reported [49b]) and $Cu_2O(E_G \sim 2.4$ eV) in the Schottky barrier Cu_2O/Cu cell (efficiencies of about 1% have been reported [50]). Efficiencies up to 5% have been reported for a p–Cu_2S/n–Si solar cell [72].

Other $A^{III}B^V$ compound semiconductors, InP and AlP, have also been explored as photovoltaic cells. Efficiencies up to 14% have been reported for p–InP/n–Cds solar cells by Shay *et al.* [53]. Aluminum phosphide $(E_G \sim 1.5$ eV) has not received much attention because it undergoes slow decomposition when it is exposed to the atmosphere. Of course, since solar cells must be hermetically sealed, this may not be a "fatal" defect for AlP.

Recently significant AM1 efficiencies $(\sim 5.5\%)$ have been reported in a noncrystalline photovoltaic cell specifically in amorphous silicon [54]. These cells are made by forming a Schottky barrier between platinum and a very thin $(1\ \mu m)$ film of amorphous silicon.

D. Increasing Efficiency By Means of Tandem PV Cell Systems

The maximum AM0 solar energy conversion efficiency which is theoretically expected from a p–n homojunction operating at room temperature is around 25% as shown in Fig. 21 and the accompanying discussion. From Fig. 24, it is evident that a substantial portion of the losses in a p–n homojunction are fixed once the semiconductor (and its E_G) are selected. One such loss is associated with long wavelength photons which cannot be absorbed because $hv < E_G$. Another substantial loss occurs because the absorbed photons have energy in excess of E_G; this excess energy $(hv - E_G)$ is degraded into heat. The magnitude of these losses can be decreased and the overall sunlight to electricity conversion efficiency can be increased by stacking homojunction cells of different energy gaps in series, i.e., by using tandem cell systems [55].

Consider, for example, the arrangement shown in Fig. 27 which depicts three p–n homojunctions made from semiconductors having different values of E_G arranged in such a way that the sunlight is incident on the material having the largest band gap E_{G1}. Those photons which are not absorbed in this cell are transmitted to the cell with the intermediate band gap E_{G2}, and the photons not absorbed in this cell are in turn transmitted to the cell with the smallest band gap E_{G3}. Let the values of E_{G1}, E_{G2}, and E_{G3} be chosen in such a way that each of them will absorb the same number of photons per unit area per unit time and therefore—on the assumption that they have the same collection efficiencies—they will have equal values of short circuit

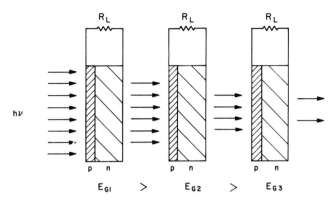

FIG. 27.   Schematic representation of a three-semiconductor tandem cell system.

current. This means that E_{G3} fixes the magnitude of the loss associated with photons whose energy is so low that they cannot be absorbed in the stack. Furthermore, if I_{sc}' is the short circuit current which would be produced by absorption of the spectrum in the E_{G3} material acting by itself, then for equal currents in all three semiconductors,

$$I_{sc1} = I_{sc2} = I_{sc3} = \tfrac{1}{3}I_{sc}' \equiv I_{sc} \tag{42}$$

Now assume that we are dealing with the AM0 spectrum for which I_{sc}' is shown as a function of E_G in Fig. 17. Assume further that $E_{G3} = 0.71$ eV (corresponding to the band gap of germanium); then from Fig. 17, $I_{sc}' \approx 79$ mA/cm^2, $I_{sc} \approx 26.3$ mA/cm^2, $E_{G1} = 1.75$ eV, and $E_{G2} = 1.10$ eV. To simplify calculations, we assume that the solar cells made from these three semiconductors exhibit ideal junction characteristics ($A = B = 1$) and that in Eq. (25), the parameter K_1 has the same value for the three homo-junctions. Further we assume that the value of I_0 in the material with $E_{G2} = 1.10$ eV (silicon) is $I_{02} = 2 \times 10^{-12}$ A/cm^2, the value used to calculate the curves in Fig. 21. The values of the I_{sc}/I_0 ratio for each of the three cells would be as follows: for E_{G1}, $I_{sc}/I_{01} \approx 2.6 \times 10^{21}$; for E_{G2}, $I_{sc}/I_{02} \approx 1.32 \times 10^{10}$; and for E_{G3}, $I_{sc}/I_{03} \approx 2.2 \times 10^3$. With the help of Fig. 19 and Eq. (13), we calculate $\eta_1 \approx 21.1\%$, $\eta_2 \approx 11.3\%$, and $\eta_3 \approx 2.4\%$. Consequently the total theoretical efficiency of this three semiconductor tandem structure would be about 34.8%. Note that in this example leaving out the smallest E_G material would result in a two semiconductor system having an efficiency of 32.4%.

Even higher efficiencies can be calculated for tandem cell structures based on more than three semiconductors. For example, a seven semiconductor system with E_G values of 2.5, 2.0, 1.75, 1.55, 1.35, 1.10, and 0.71 eV would have a total theoretical efficiency of 42.8%.

In these calculations, the assumption is made that I_0 decreases as an exponential function of E_G and that the pre-exponential coefficient in Eq. (25) is independent of E_G; if this were so, the value of I_0 for a p–n junction in a material with $E_G = 2.5$ eV would be 4.7×10^{-36} A/cm^2 and for $E_G = 2.0$ eV, $I_0 \sim 1.3 \times 10^{-27}$ eV. It is unlikely that such low values can be achieved in p–n junctions because other mechanisms, like generation and recombination in the space charge region of the semiconductor, will set a lower limit on the attainable value of I_0 and therefore, an upper limit on the value of V_{oc}. Current experience with semiconductors indicates that in GaAs AM0 efficiencies up to 15% are achievable and that improved cells with efficiencies up to 20% may be possible; from this one concludes that tandem cell structures based on semiconductors with $E_G \leq 1.35$ eV should work in the way described above. The effort expended on p–n homojunction solar cells in materials having values of $E_G > 1.35$ eV has not been sufficient to demonstrate what their full potential is or what ultimate efficiencies may be achieved in tandem cell structures, but the potentially high efficiencies which can be achieved in such systems warrant further investigations.

E. Materials for Ohmic Contacts to Cells, Substrates, and for Protecting the Cells from Their Environment

Materials problems of solar cell systems do not end with the selection of the semiconductor from which the solar cell is made. Materials for ohmic contacts, for substrates, and for sealing the cells to protect them from a possibly injurious atmosphere must also be carefully selected. In this section, brief descriptions are offered of the problems encountered in one such area, namely, finding an acceptable ohmic contact for single crystal silicon cells.

The first ohmic contacts used on single crystal Si cells were made by electrodeless plating a thin film of Ni in the required geometrical pattern on both faces of the cell. The light-receiving surface was covered by a "finger" pattern; the opposite face was completely covered by Ni. The cell was then dipped into a hot solder bath. The solder wet only those areas covered by Ni. Wires, bars, etc., could then be soldered to those areas. These Ni-based contacts were acceptable as ohmic contacts, but their mechanical properties were not satisfactory since a fraction of them did not adhere to the cell. Improved contacts were required.

In 1962, the Bell Laboratories introduced an evaporated Ti–Ag contact for the Si cells used on the Telstar Communications satellite [56]. The new contact was superior both electrically and mechanically to the Ni contact. Preparation of the contact required four steps. First a layer of Ti about 1000 Å thick was evaporated onto both surfaces of the cell through suitable masks. The function of the Ti was to "break through" the oxide layer always present on the Si cell. The second step involved evaporation of an Ag layer 3-5 μm thick over the titanium covered areas. The Ag covered region was a bit larger in area than that covered by Ti, so that the edges of the Ti–rutile layer were partially sealed from the atmosphere. In the third step, the Ti-Ag structure was sintered in an H_2-N_2 atmosphere for about 5 min at about 600°C. The sintering process promotes the formation of TiH_2, the bonding of Ag to Ti and of Ti + TiH_2 to Si. Finally, the cells were dipped into molten solder which adhered only to Ag covered areas. These contacts became standard for solar cells used on artificial earth satellites.

Now one of the problems associated with space vehicles is minimizing the weight of everything used in the system and especially eliminating dead weight. It was known that the amount of solder adhering to the Ag represented a significant fraction of the weight of a finished Si cell. Therefore, "solderless" Ti-Ag contacts, in which the last step—dipping in molten solder—was omitted, were proposed; the solder would be applied only at a few contact points between the bars used to connect cells in series-parallel combinations. A substantial number of such cells were manufactured; solar cell power supply systems were assembled from them and a satellite using a power supply with solderless Ti-Ag contacts was eventually launched. After a while it was observed that the power output from this system was degrading with time and tests were designed to determine the cause of the degradation. Eventually it was found [57] that when cells with solderless Ti-Ag contacts were exposed to so called high temperature ($\sim$85°C), high humidity ($\sim$95% relative humidity) conditions, the Ti-Ag contacts exhibited increased resistance and that after 10-30 days in such an atmosphere some contacts even separated from the Si surface. Of course, the cells on the satellite were not operating in such an atmosphere; however, during the many months which elapsed between the manufacture of the solar cells and the launching of the satellite, the cells and the assembled power supply were stored in the "normal" atmosphere of the manufacturing plant, etc., i.e., no special precautions were taken to protect the cells from moisture. Furthermore, sometimes in the course of testing the solar cell panels they may have been exposed to an elevated temperature under moderate humidity conditions.

An explanation of the failure of the solderless Ti-Ag contact was provided by Gereth *et al.* [58] who showed that an electrochemical reaction can occur at the Ti-Ag interface in the presence of H_2O. The reaction causes the Ti to dissolve anodically thus forming a titanium oxide at the interface between the Ti and Ag. This oxide layer increases the contact resistance and when its growth reaches a certain stage, the mechanical strength of the contact becomes so low that it can be easily separated from the silicon. Gereth *et al.*, proposed a solution to this problem. They showed that the Ti–Ag electrochemical cell could be passivated by interposing a thin layer of a noble metal (they used palladium) between the Ti and Ag. Such Ti(Pd)Ag contacts are considerably more resistive to corrosion in the presence of H_2O. However, they too undergo complex changes when exposed to "high temperature–high humidity" conditions as reported by Becker and Pollack [59]. Porosity of the Ag layer which allows H_2O to reach the Ti–Ag interface plays an important role in the degradation as shown by Bishop [60]. The pores can be sealed by dipping in solder.

An alternate path to durable contacts is based on aluminum contacts [61]. However, aluminum cannot be conveniently interconnected by soldering and therefore other methods like thermal diffusion or ultrasonic bonding would need to be used. Such methods may be acceptable for large scale manufacturing like that required for terrestrial utilization of solar energy but have not been attractive in space applications.

It should be pointed out that all of the contacts described above (Ni, Ti–Ag covered with solder, solderless Ti–Ag, and Al) are basically satisfactory. Improvements in durability and yield, and reduction in cost of the cells and systems produced from them have been the motive for continued investigation of contacts.

F. Special Materials Problems Associated with the Proposed Satellite Space Power Systems (SSPS)

We have already alluded to special materials problems associated with the SSPS. In this section, we discuss one of these problems, namely, the effect of radiation damage on solar cells in space.

The SSPS is intended to operate at synchronous altitude so that its position with respect to the microwave receiving antenna on the surface of the earth is fixed. Measurements of particle radiation levels at synchronous altitude lead to the conclusion that the fluxes of high energy electrons and protons are such that the radiation damage caused by these particles sets an upper limit on the useful life of a power supply based on silicon solar cells and indeed on solar cells made from any other semiconductor. Consequently, the mechanism of such radiation induced changes in photovoltaic cells has been the subject of intensive study since 1953 [62]. It was shown that as a result of the interaction between radiation (high energy electrons, protons, neutrons, alpha particles), the minority carrier lifetime decreases. This decrease results from an increase in the conncentration of deep-lying energy levels which act as minority carrier recombination centers. Under certain conditions (low minority carrier injection level, low concentration of recombination centers, no interaction between centers), the excess carriers introduced by absorption of light will exhibit a lifetime τ given by the expression

$$1/\tau = 1/\tau_0 + N_{rB}\,\sigma_{cB}\,f_B\,v \tag{43}$$

where τ_0 is the pre-irradiation lifetime, N_{rB} the concentration of recombination centers introduced by bombardment, σ_{cB} the minority carrier capture cross section of these centers, f_B the probability that the center is occupied by a majority carrier, and v the thermal velocity of the carriers. For low values of integrated particle flux (fluence) ϕ, N_{rB} is proportional to ϕ and we can write

$$1/\tau = 1/\tau_0 + \lambda\phi \tag{44}$$

where the definition of λ is obvious from Eq. 43.

Changes in τ affect the output of the cell because, as we have already pointed out, both I_0 and I_{sc} are functions of τ [see Eqs. (1), (19), (20), and (23)]. The radiation can change τ in both the base region and the diffused skin region. In the skin region of a Si cell the initial lifetime τ_{0s} is very low; its value corresponds to recombination center concentrations between 10^{15} and 10^{17} per cubic centimeter. This large recombination center concentration is introduced during the diffusion process. In the base region, the initial lifetime τ_{0b} is determined by recombination center concentrations between 10^{12} and 10^{13} per cubic centimeter. Consequently, radiation, which penetrates into the semiconductor to a depth comparable to the sum of the minority carrier diffusion lengths in the skin and the base, will first begin to reduce lifetime in the base region because the ratio of the defect concentration introduced by radiation to the initial recombination center concentration is higher in the base than in the skin. This occurs even though λ [from Eq. (44)] is about ten times as large in n–Si as it is in p–Si. Furthermore, most of the photons from the solar spectrum are absorbed in the base region of the semiconductor and, therefore, most of the contribution to I_{sc} arises from the base region. Consequently, the "red" response, i.e., the response to penetrating photons, is affected more strongly by penetrating radiation [63].

Figure 28 (from Baicker and Faughnan [63]) shows the measured total response of a silicon cell before and after irradiation by 8.3 MeV protons whose range in Si is about 500 μm. Curves for the skin response and the base response before irradiation were calculated using the formulation of Loferski and Wysocki [19]. The value of the base lifetime which enters into these equations was measured on the finished cell. The values of parameters in the diffused skin region were deduced on the basis of reasonable assumptions about the skin material. The post-irradiation curves were constructed as follows: (1) it was assumed in accordance with the considerations presented above that the skin response did not change as a result of irradiation,

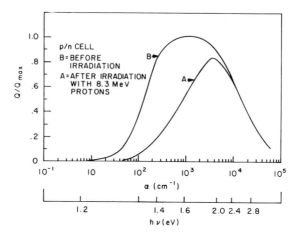

FIG. 28.   Total response of a Si solar cell before and after irradiation (from [63]).

and (2) the post-irradiation value of the base lifetime τ_b was measured and the base response corresponding to this value was calculated. It is evident from this figure that the decreased lifetime in the base adequately explains the observed reduction of red response.

If the range of the particles is comparable to the depth of the diffused skin (this would be true in silicon solar cells for protons having energy ~ 100 keV), then the damage would be confined to the skin region and a reduction in blue response would result [64]. Blue response also decreases if the surface recombination velocity increases as a result of irradiation [65].

There exists abundant evidence that the radiation defects stable around room temperature (satellite solar power systems are designed to operate around room temperature) in silicon (and by induction this can be expected in other less intensely studied semiconductors) are complexes consisting of either the vacancy or interstitial atom produced as a result of the primary radiation event and an impurity atom present in the crystal. For example, in electron-irradiated, phosphorus-doped n-type Si crystals grown by the Czochralski method (crystals pulled from the melt), Watkins and Corbett [66], using EPR techniques, showed that the dominant deep center is an oxygen vacancy complex, the so-called A-center. In phosphorus-doped n-type Si crystals grown by the floating-zone method, they showed that the dominant center is a phosphorus–vacancy complex, the so called E-center. As is well known, there is a large difference in the oxygen concentrations of silicon crystals grown in these two ways; the pulled crystal has oxygen concentrations in the range between 10^{17} and 10^{18} per cubic centimeter, whereas the floating-zone crystal has an oxygen concentration of about 10^{15} per cubic centimeter. Oxygen is not an electrically active defect in silicon;

energy levels associated with it affect neither resistivity nor lifetime. The vacancies introduced by irradiation activate the oxygen. However, in the case of Si, eliminating oxygen does not reduce the rate of degradation of lifetime since the vacancies produced by irradiation form complexes with the next most abundant impurity, e.g., phosphorous, which cannot be left out of the crystal because it is essential in the fabrication of the device. Although the atomic natures of the A- and E-centers are quite different, their effectiveness as recombination centers in *n*–Si is comparable. Attempts to make more radiation resistant cells by substituting other donor impurities like As or Sb for P in Si were frustrated because the complexes formed between these other Column V impurities and the vacancies introduced by irradiation are equally effective as recombination centers.

Studies of minority carrier degradation rates in irradiated *p*–Si showed that the complexes stable at room temperature are less effective as recombination centers than the A- and E-centers in *n*–Si. Consequently, if a solar cell is made by diffusing an *n*-type impurity into a *p*-type base (an *n*–*p* cell) its radiation resistance is about one order of magnitude greater than that of a cell made by diffusing a *p*-type impurity into an *n*-type base (a *p*–*n* cell).

From this experience with silicon, it is reasonable to expect that complexes involving the irradiation produced defects (vacancies and interstitial atoms) and impurities incorporated either deliberately or accidentally into the semiconductor are likely to play a dominant role in the minority carrier recombination process in the irradiated semiconductor and therefore in the radiation resistance of solar cells made from it.

This important role of impurities in radiation damage in Si led to the concept of the lithium-doped, "self-healing" solar cell. Vavilov [67] had shown that if Si doped with Li is irradiated, the lifetime degrades but it recovers partially after irradiation.

In 1966 Wysocki *et al.* [68] described a Li-doped solar cell which exhibited self-annealing after irradiation by electrons. Their cells were made by diffusing Li from an oil slurry or a tin alloy into *p*–*n* Si solar cells, cells in which the junction had been formed by diffusing boron into an *n*-type Si wafer. This configuration was chosen because based on Vavilov *et al.* [67] self-annealing was known to occur in *n*–Si. Since most of the damage by high energy particles occurs in the base region of a cell, treating this region in a way which could neutralize defects introduced by radiation could substantially increase the radiation resistance of the cell. Indeed, Wysocki *et al.* found that Li-doped *p*–*n* cells after recovery by self-annealing had higher residual efficiencies than standard *n*–*p* Si cells.

Wolf and Brucker [69] have reviewed the work on the Li doped *p*–*n* Si cell. Figure 29 (from Wolf and Brucker [69]) shows the relative maximum power output from various kinds of Li doped Si cells. It is found that the

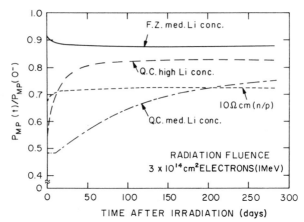

FIG. 29.   Room temperature recovery of various kinds of Li doped Si solar cells (from [69]) after irradiation by 1MeV electrons. The symbol FZ refers to cells made on base wafers cut from Si crystals grown by the "floating-zone" method; the symbol QC refers to cells made from "quartz crucible" or Czochralski-grown Si single crystals. A "medium" Li concentration is about 10^{15} per cubic centimeter while a "high" Li concentration is around 10^{16} per cubic centimeter.

recovery rate depends on a number of factors. The first of these is the oxygen content of the silicon. Crystals pulled from a quartz crucible by the Czochralski method have a substantially higher oxygen concentration (10^{17} to 10^{18} per cubic centimeter) than crystals grown by the floating zone method ($< 10^{16}$ per cubic centimeter). The lithium interacts with the oxygen as well as with the vacancies and interstitial atoms introduced by irradiation; the result is a reduction in the recovery rate as the Li/O concentration ratio decreases. Secondly, the recovery rate is a function of Li concentration; the higher the concentration the more rapid the recovery. Thirdly, the recovery rate is a strong function of the Li ion concentration gradient within and in the immediate vicinity of the cell's space charge region; the higher this gradient, the greater the recovery rate. Figure 29 shows the relative output power for these various types of cells. When the absolute value of output is plotted as a function of time, it turns out that the recovered power output after a given amount of radiation by electrons does not differ very much from the output of an *n–p* cell exposed to the same electron fluence. Thus, although the electron irradiation resistance of *p–n* cells is substantially increased by the incorporation of lithium, it is not much better in this regard than an ordinary *n–p* cell. When comparisons of performance after irradiation by neutrons and protons were made, it was found that the Li cells could tolerate substantially higher fluences of these heavy particles than ordinary *n–p* cells. Thus, the radiation resistance of Li cells is also a function of the nature of the incident particles. The Li doped cell has not replaced the *n–p* cell in space applications because the advantage it produces is not sufficient to compensate for its anticipated higher cost. Further, a different way to achieve higher radiation resistance in the space environment was demonstrated by the "Violet" cell [18]. This is an $n-p$ cell which has a higher efficiency and output under AM0 illumination than ordinary $n-p$ cells. The increased output is achieved by utilizing a greater fraction of the "violet" photons of the AM0 spectrum, i.e., the photons which are absorbed in the thin diffused skin of the cell. The initial minority carrier lifetime in the diffused skin is so low that irradiation to fluences which cause substantial decreases in the minority carrier lifetime in the base material produces virtually no change in the "skin" lifetime. Consequently, the enhanced "violet" response is virtually immune to irradiation effects.

Radiation resistance is important in cells intended for use on satellites and will, therefore, be an important factor in selection of cells for a satellite space power system [5]. Solutions other than those described above are also possible. One alternate path is to use other semiconductors. For example, GaAs solar cells have been shown to be more resistant to radiation than Si [64]. It is also possible that further changes in the design of cells could lead to improved radiation resistance.

G. Concluding Comments

From this brief examination, we see that materials problems are encountered in many aspects of solar energy utilization technology. The basic science underlying the functioning of solar energy converters and the relation between conversion efficiencies and material parameters are reasonably well understood. In many solar energy converting devices, adequate control over the important parameters has not been achieved and, therefore, the full potential of the solar energy converters has not been realized. The most important impediment to large scale utilization of solar energy is, however, the cost of the conversion devices. Because solar energy is diffuse, the converters must cover very large areas. Consequently, thin films play a large role in such devices. The materials problems encountered in thin films are in general difficult to handle in part because good methods for characterizing thin films are not as well developed as those available for characterization of bulk specimens. The challenge of economic solar energy conversion is and will continue to be an important spur to research in materials science and engineering.

REFERENCES

[1]   This Figure is from the report, Solar Energy as a National Energy Resource, NSF/NASA Solar Energy Panel, Dept. of Mechanical Engineering, Univ. of Maryland, College Park, Maryland (1972).

260   *J. J. Loferski*

[2a] A. B. Meinel and M. P. Meinel, *Phys. Today* **25**, No. 2, 44 (1972).
[2b] A. B. Meinel and M. P. Meinel, "Applied Solar Energy: An Introduction." Addison Wesley, Reading, Massachusetts, 1976.
[3a] A. F. Hildebrandt and L. L. Vant-Hull, *Mech. Eng.* 23 (September 1974).
[3b] A. F. Hildebrandt and L. L. Vant-Hull, *Science* **197**, 1139 (1977).
[4] E. L. Ralph, *Solar Energy* **14**, 11 (1972).
[5] P. E. Glaser, *Power* (August 1974).
[6] M. Wolf, *Energy Conversion* **14**, 49 (1975).
[7a] H. E. Bates, D. N. Jewett, and V. E. White, *Conf. Record IEEE Photovoltaic Specialists Conf., 10th, Palo Alto, California* p. 197, IEEE Cat. No. 73 CH 0801-ED (November 1973).
[8] Also see *Conf. Record IEEE Photovoltaic Specialists Conf., 12th, Baton Rouge, Louisiana* (November 1976).
[9a] P. H. Fang, L. Ephrath, and W. B. Nowak, *Appl. Phys. Lett.* **25**, 583 (1974).
[9b] T. L. Chu, K. Y. Duh, and H. I. Yoo, *Conf. Record IEEE Photovoltaic Specialists Conf., 12th, Baton Rouge, Louisiana* p. 74 (November 1976).
[10a] K. W. Boer *et al.*, *Conf. Record IEEE Photovoltaic Specialists Conf., 10th, Palo Alto, California* p. 77, IEEE Cat. No. 73 CH 0801-ED (November 1973).
[10b] W. Palz, J. Besson, T. N. Duy, and J. Vedel, *Conf. Record IEEE Photovoltaic Specialists Conf., 10th, Palo Alto, California* p. 69, IEEE Cat. No. 73 CH 0801-ED (November 1973).
[10c] Also see *Conf. Record IEEE Photovoltaic Specialists Conf., 12th, Baton Rouge, Louisiana* (November 1976).
[11] R. A. Tybout and G. O. G. Löf, *Natur. Resources J.* **10**, 268 (1970) republished in *Solar Energy* **16**, 9 (1974).
[12] H. E. Thomason, *Solar Energy* **10**, No. 1, 17 (1966).
[13] G. O. G. Löf and R. A. Tybout, *Solar Energy* **14**, 253 (1973); **16**, 9 (1974).
[14] J. T. Gier and R. V. Dunkle, *Trans. Conf. Use of Solar Energy, Univ. of Arizona* Vol. II (1955).
[15] C. N. Watson-Munro, *in* Report on U.S.-Australian Workshop on Solar Energy. Sponsored by U.S. National Science Foundation and Commonwealth Scientific and Industrial Research Organization, Australia (February 1974).
[16] E. A. Farber, *Solar Energy* **14**, 243 (1973).
[17] M. Telkes, *in* "Solar Energy Research" (F. Daniels and J. A. Duffie, eds.). 1956.
[18] J. Lindmayer and J. Allison, *COMSAT Tech. Rev.* **3**, 1 (1973).
[19] J. J. Loferski and J. J. Wysocki, *RCA Rev.* **22**, 38 (1961).
[20] M. Wolf, *Proc. IRE* **51**, 674 (1963).
[21] C. T. Sah, R. N. Noyce, and W. Shockley, *Proc. IRE* **45**, 1228 (1957).
[22] A. Jonscher, "Principles of Semiconductor Devices." Wiley, New York, 1960.
[23] R. J. Stirn and Y-C. M. Yeh, *Conf. Record IEEE Photovoltaic Specialists Conf., 10th, Palo Alto, California* p. 15 (November 1973).
[24] J. J. Wysocki and P. Rappaport, *J. Appl. Phys.* **32**, 371 (1960).
[25] J. J. Loferski, *Acta Electron.* (July 1961).
[26] W. Shockley and H. J. Queisser, *J. Appl. Phys.* **32**, 510 (1961).
[27] M. Wolf, *Proc. IRE* **48**, 1246 (1960).
[28] See for example, L. J. Goldhammer and B. E. Anspaugh, *Conf. Record Photovoltaic Specialists Conf., 8th, Seattle, Washington* p. 201 (August 1970) or D. J. Curtin and A. Muelenberg, *ibid.* p. 193.
[29] M. Wolf, *Energy Conversion* **11**, 63 (1971).
[30a] H. Brandhorst, *Conf. Record IEEE Photovoltaic Specialists Conf., 9th, Silver Spring, Maryland* p. 37 (May 1972).

[30b] P. N. Dimbar and J. R. Hauser, *Conf. Record IEEE Photovoltaic Specialists Conf., 12th, Baton Rouge, Louisiana* p. 23 (November 1976).

[31] H. J. Hovell and J. M. Woodall, *Conf. Record IEEE Photovoltaic Specialists Conf., 10th, Palo Alto, California* p. 25 (November 1973).

[32] D. Huber and K. Bogus, *Conf. IEEE Photovoltaic Specialists Conf., 10th, Palo Alto, California* p. 100 (November 1973).

[33] R. J. Stirn and Y. C. M. Yeh, *IEEE Trans. Electron. Devices* **ED-24**, 476 (1977).

[34] D. C. Reynolds, C. Leies, L. L. Antes, and R. E. Marburger, *Phys. Rev.* **96**, 533 (1954).

[35] G. Nadjakov, R. Antreitchine, and M. Borrison, *Izv. Bulg. Akad. Nauk* **4**, 10 (1954).

[36] A. E. Middleton, D. A. Gorski, and F. A. Shirland, *Prog. Astronaut. Rocketry* **3**, 275 (1961).

[37] E. R. Hill and B. Keramidis, *Rev. Phys. Appl.* **1**, 189 (1966).

[38] W. R. Cook, Ph.D. Thesis, Case Western Reserve Univ. (1971).

[39] P. F. Lindquist and R. Bube, *Conf. Record Photovoltaic Specialists Conf., 8th, Seattle, Washington* (May 1971).

[40] A. E. van Aerschot *et al.*, *IEEE Trans. Electron. Devices* **ED-18**, 471–482 (1971).

[41] S. Mittleman, Ph.D. Thesis, Brown Univ. (1974).

[42] A. L. Fahrenbruch and R. H. Bube, *Conf. Record Photovoltaic Specialists Conf., 9th, Silver Springs, Maryland* p. 118 (May 1972).

[43] D. A. Cusano, *Solid State Electron.* **6** (1963).

[44] D. A. Cusano, *Rev. Phys. Appl.* **1**, 195 (1966).

[45] G. P. Sorokin, Yu. M. Popshev, and P. T. Oush, *Sov. Phys.—Solid State* **7**, 1810 (1966).

[46] J. Lebrun, *Conf. Record IEEE Photovoltaic Specialists Conf., 8th, Seattle, Washington* p. 33 (August 1970).

[47] B. Tell, J. L. Shay, and H. M. Kasper, *J. Appl. Phys.* **43**, 2469 (1972).

[48a] S. Wagner, J. L. Shay, and P. Migliorato, *Appl. Phys. Lett.* **25**, 434 (1974).

[48b] S. Wagner, J. L. Shay, K. J. Bachman, E. Buehler, and H. M. Kasper, *Appl. Phys. Lett.* **26**, 229 (1975).

[49a] For a review of the properties of the $A^{II}B^{VI}$ semiconductors, see W. Zdanowicz and L. Zdanowicz, *Annu. Rev. Mater. Sci.* **4**, 000 (1974).

[49b] M. Wolf, *Conf. Record IEEE Photovoltaic Specialists Conf., 10th, Palo Alto, California*, p. 5, IEEE Cat. No. 73 CH 0801-ED (November 1973).

[50] D. Trivich, E. Y. Wang, R. J. Komp, and F. Ho, *Conf. Record IEEE Photovoltaic Specialists Conf., Baton Rouge, Louisiana*, p. 875 (November 1976); W. A. Anderson, J. J. Kim, and A. E. Dehahoy, *ibid.* p. 87.

[51] A. G. Stanley *in* "Applied Solid State Science," Vol. 5, p. 251. Academic Press, New York. This is a review of work on the Cu-Cd-S solar cell up to late 1974.

[52] A. Rothwarf and A. M. Barnett, *IEEE Trans. Electron. Devices* **ED-24**, 381 (1977).

[53] J. L. Shay, S. Wagner, K. J. Bachman, and E. Buehler, *J. Appl. Phys.* **47**, 614 (1976); J. L. Shay, S. Wagner, M. Bettini, K. J. Bachman, and E. Buehler, *IEEE Trans. Electron. Devices* **ED-24**, 483 (1977).

[54] D. E. Carlson, *IEEE Trans. Electron. Devices* **ED-24**, 449 (1977).

[55] J. J. Loferski, *Conf. Record IEEE Photovoltaic Specialists Conf., 12th, Baton Rouge, Louisiana* p. 957 (November 1976).

[56] K. D. Smith, H. K. Gummel, J. D. Bode, D. B. Cuttriss, R. J. Nielson, and W. Rozenzweig, *Bell Syst. Tech. J.* **42**, 1765 (1963).

[57] W. Luft, C. C. McCraven, and L. A. Aroian, *Conf. Record IEEE Photovoltaic Specialists Conf. 7th, Pasadena, California* p. 214 (May 1968).

[58] R. Gereth, H. Fischer, E. Link, S. Mattes, and W. Pschunder, *Energy Conversion* **12**, 103 (1972). This work was first presented at the Seventh IEEE Photovoltaic Specialists

Conference, Pasadena, California, May 1968.

[59] W. H. Becker and S. R. Pollack, *Conf. Record IEEE Photovoltaic Specialists Conf., 8th, Seattle, Washington* p. 40 (August 1970).

[60] C. J. Bishop, *Conf. Record IEEE Photovoltaic Specialists Conf., 8th, Seattle, Washington* p. 51 (August 1970).

[61] See for example K. Lui and R. K. Yasui, *Conf. Record IEEE Photovoltaic Specialists Conf., 8th, Seattle, Washington* p. 62 (August 1970).

[62] J. J. Loferski and P. Rappaport, *RCA Rev.* **19**, 536 (1958).

[63] J. A. Baicker and B. Faughnan, *J. Appl. Phys.* **33**, 3271 (1962).

[64] J. J. Wysocki, P. Rappaport, E. Davison, and J. J. Loferski, *IEEE Trans. Electron Devices* **ED-13**, No. 4, 420 (1966).

[65] J. J. Loferski, W. Giriat, I. Kasai, and H. Flicker, *Proc. Colloq. Int. Action Rayonnements sur Les Composants Semicond., Toulouse, France* paper A19 (March 1967).

[66] G. D. Watkins and J. W. Corbett, *Phys. Rev.* **121**, 4 1001 (1961).

[67] V. S. Vavilov, V. M. Patskevich, B. Ya. Yurkov, and P. Ya Glazunov, *Sov. Phys.-Solid State* **2**, 1301 (1961).

[68] J. J. Wysocki, P. Rappaport, E. Davison, R. Hand, and J. J. Loferski, *Phys. Rev. Lett.* **7**, 44 (1966).

[69] M. Wolf and G. J. Brucker, *Energy Conversion* **11**, 75 (1971).

[70] J. J. Loferski, *J. Appl. Phys.* **27**, 777 (1956)

[71] A. Catalano *et al., Conf. Record IEEE Photovoltaic Specialists Conf., 13th, Washington, D.C.* (June 1978).

[72] J. Shewchun *et al., Conf. Record IEEE Photovoltaic Specialists Conf., 13th, Washington, D.C.* (June 1978).

Chapter 5

Materials for Geothermal Energy Utilization

HOWARD L. RECHT

RESEARCH AND TECHNOLOGY
ENERGY SYSTEMS GROUP
ROCKWELL INTERNATIONAL CORPORATION
CANOGA PARK, CALIFORNIA

I. NATURE OF GEOTHERMAL SOURCES

A. Incidence

Throughout the world, where fracture and shifting of the crustal plates has permitted it, the hot underlying magma has moved near to the surface. This has caused regions of high thermal gradients near the surface. It has also led to volcanoes, hot springs, geysers, and other manifestations of this heat energy.

Regions where this has occurred include the large circum-Pacific ring, noted for earthquakes, which runs through the coastal regions of western South America, western North America (including Alaska), eastern Siberia, Japan, and New Zealand. A similar belt traverses Europe and extends into the Caucasian Mountains. A third such belt runs north and south through the mid-Atlantic, making geothermal effects manifest in Iceland.

B. Forms

There are many forms or types of geological activity which appear through the presence of these near-surface hot magma masses. These include, as indicated above, volcanoes, geysers, and hot springs. By drilling beneath the surface, other sources may be tapped. There are four principal types of geothermal sources currently used or considered suitable to supply geothermal energy. They are: dry steam, water-dominated systems, geopressured systems, and hot, dry rocks.

Dry steam (or vapor-dominated) systems are of relatively infrequent occurrence but have provided the bulk of current geothermal power generation. The principal locations are at the Geysers (California, USA), Lardarello (Italy), and Matsukawa (Japan). These sources are characterized by a cap rock which prevents continued replenishment of the reservoir with surface water. When these are tapped, the water boils so that steam, rather than brine comes to the surface. From their nature, there is a fall-off in production with time, necessitating the drilling of additional wells to maintain production.

The water-dominated geothermal systems are much more common. With these, the subterranean structure is porous so that there is continued entry of surface (meteoric) water into the reservoir. When tapped, the flow is a mixture of water and any steam that has flashed. As discussed below, these waters differ widely in temperature and composition from site to site.

Geopressured sources, which have been found at the north end of the Gulf of Mexico, in Wyoming, and in California, are characterized by insulating impermeable clay beds above high porosity deposits in which confined water has accumulated the earth's heat flow.

There are a number of regions in the world which are characterized by very high thermal gradients just below the surface of the earth. It is believed that these are due to hot magmatic intrusions that have penetrated near to the surface. The structures between the magma and the surface are relatively nonporous. There has been no intrusion of meteoric or other water to transmit the heat by convection, nor to act as an energy transfer medium. Heat transfer is by conduction through the rock structure.

Recently, several test drillings have been made in some of these high thermal gradient areas [1, 2]. In many cases, the temperature increase with depth is not maintained, so that temperatures which were projected to be usable (desirable) for power generation ($\geq 200°C$) are not obtained. Also, some porous intermediate layers with water content were encountered, complicating the drilling operation. These sources may better be described as warm, moist rocks. Nonetheless this type of source is the most widespread of geothermal sources and may provide much of our future energy needs.

In all but the hot (impermeable) dry rock sources, an *in situ* aqueous fluid (steam and/or brine) is available to transfer the thermal energy to the surface for utilization. The characteristics of these fluids will be discussed in the next section, while the bulk of this chapter will be devoted to discussion of processes and materials (and materials problems) for utilizing these fluids.

The hot, dry rock sources present a unique situation. As they have not as yet been exploited, consideration of them will be limited to a short review of current ideas on their utilization. To obtain the heat from the hot, dry rock sources, current thinking requires that the structure be opened up to provide flow paths and heat transfer surfaces, and the heat recovered by use of a heat transfer fluid. The hot rock structure may be fractured by hydraulic pressure, or by use of explosives, either conventional or nuclear. With luck, nearly all the fluid introduced into this structure will be localized and recoverable. Water is the most often mentioned fluid, but with minimal losses other fluids of lower boiling point, high enthalpy of vaporization, and high vapor density may be used. For example, isobutane or the freons which are suggested and/or employed for indirect power generation cycles should be suitable. Also, gases under pressure may be used. These nonaqueous fluids would not pick up salts, silica, and other materials which complicate use of brines for geothermal power generation, and, in the case of condensible fluids, provide the opportunity of relatively low cost, efficient use of " not-so-hot " sources.

C. Chemical and Physical Characteristics of Geothermal Fluids

The chemical and physical make-up of the brines and steam recovered from geothermal heat sources set the materials requirements for their handling and dictate the power generation process or processes whereby their energy may be utilized. All the geothermal aqueous fluids have or are associated with water with a content of dissolved salt and silica, and contain varying amounts of generally acidic noncondensible gases. The two key materials problems in geothermal energy production are scale deposition and corrosion in producing wells and power generation equipment. In this chapter, the nature and causes of these problems will be examined and proposed solutions will be considered.

The dry steam issuing from sources such as the Geysers, or Lardarello, contains noncondensible gases and other impurities in amounts which vary with time and location. (The Lardarello source is noted for its boron (borate) content. Indeed, the first use of this geothermal source was as a source of boric acid.) The compositions of several of these steam sources are given in Table 1.

Carbon dioxide is a principal component of all the gases, with varying amounts of H_2S, NH_3, N_2, and lower molecular weight hydrocarbons being also present. Hydrogen present is taken to indicate some reduction process liberating this gas has occurred in recovery.

The water-dominated geothermal sources vary widely in temperature (and pressure), salinity, nature of the dissolved components, and nature of

TABLE 1

Composition of Geothermal Steam from Various Sites[a]

	Larderello	The Geysers	Wairakei	Buttes–Salton Sea
Total noncondensibles (mole %)	2.0	0.3	0.06	0.25
Noncondensible gas composition (mole %)				
CO_2	92.8	69.3	91.7	94
H_2S	2.5	2.0	4.4	0.5–2.0
HC	—	11.8	0.9	0.25–4
H_2	—	12.7	0.8	0.5–4
NH_3	1.7	1.6	0.6	—[c]
H_3BO_3	0.45	—	0.05	—
N_2	—	—	—	0.5–2.0

[a] The above represent approximate average values from a number of drill holes. The data in first three columns are from A. J. Ellis, Quantitative interpretation of chemical characteristics of hydrothermal systems, *Geothermics, Spec. Issue 2* 516–528 (1970).

[b] Gases after condensation of total brine flow.

[c] 400 to 500 ppm NH_3 in condensate.

the (noncondensible) gases issuing from wells along with the brine and steam. Some idea of the range of characteristics to be encountered may be gained from the analyses of several sources listed in Table 2. There has been some attempt to classify sources into from four to sixteen groups by temperature and salinity; namely, from high temperature (> 240°C), high salinity

TABLE 2

Composition of Some North American Geothermal Brines

Component or Property	Brine Source/Content (ppm)				
	Shell No. 2 IID (Ca, USA)	Shell No. 1 State (Ca, USA)	Cerro Prieto (Mexico)	Mesa 6-1 (Ca, USA)	Mesa 6-2 (Ca, USA)
Sodium	53,000	47,800	4,450–6,100	7,960–10,939	704–918
Potassium	16,500	14,000	504–1,860	1,047–1,412	65–83
Lithium	210	180	12–19	—	—
Barium	250	190	~ 12	—	—
Calcium	28,800	21,200	210–390	1,020–1,635	3–40
Strontium	440	—	~ 10	—	—
Magnesium	10	27	6–33	10–40	0–1
Boron	390	290	4–21	—	—
Silica	400	—	151–770	100–220	140–301
Iron	2,000	1,200	0.2	—	—
Manganese	1,370	950	0.64	—	—
Lead	80	80	4	—	—
Zinc	500	500	—	—	—
Copper	3	2	—	—	—
Silver	< 1	< 1	—	—	—
Rubidium	70	65	—	—	—
Cesium	20	17	—	—	—
Chloride	155,000	127,000	7,420–11,750	14,300–20,730	665–793
$\sum CO_2$	500	5,000	0–1,600	44–318[a]	—
$\sum S$	30	30	0–700[b]	—	—
Total Dissolved Solids	259,000	219,500	—	26,200–35,110	2,130–2,830

[a] HCO_3
[b] SO_4

(> 100,000 ppm Total Dissolved Solids—TDS) brines to low temperature (< 90°C), low salinity brines (< 2000 ppm TDS). The gradations in temperature and salinity encountered in actual practice are so varied that while such a classification does have utility, each location and source will require unique consideration of potential materials problems to be encountered in energy utilization.

As far as high efficiency and simplicity of materials problems for handling these brines, a high temperature, low salinity source would be best. Unfortunately, the really high temperature (> 240°C) sources have attendant high dissolved silica and usually high content of dissolved salts. Prime examples of this are the brines from the Salton Sea area of the Imperial Valley (see Table 2), and some reported Italian brines [3].

It is instructive to consider in turn the major parameters of importance to energy utilization which characterize geothermal brines: (1) temperature, (2) salinity, (3) silica content, (4) noncondensible gas content and composition, and (5) mineral content. As was noted regarding Table 2, each brine has distinctly different composition. It is expected that there also may be considerable variation in brine composition from a single well as a function of operating time. The influence of brine composition is discussed below as it relates to major brine parameters.

1. Temperature

The downhole temperatures of geothermal brines are found to vary from 90°C or less to over 350°C in some cases. This temperature is found to correlate with a number of compositional parameters which may serve as an indirect indication of the magnitude of the downhole temperature [4]. These indicators and comments on them are given in Table 3. In addition, there is a direct correlation between downhole temperature and salinity (TDS).

2. Salinity (Total Dissolved Solids)

The magnitude of this parameter, and its make-up vary widely, generally being greater in the hotter brines. In all brines, the principal cation is Na^+ while Cl^- is the principal anion.

3. Silica Content

This parameter correlates well with downhole temperature. The level often corresponds to that of quartz solubility at this temperature. The importance of this parameter will be made clear in the discussion of scale formation later in this chapter.

4. Noncondensible Gases

All geothermal brines contain some dissolved gases which are released when the pressure is reduced and which remain as gases after the system cools. The reported compositions of these gases in the steam from various geothermal sources are given in Table 1. It may be seen that the principle noncondensible gas is CO_2 in all cases. This accounts for the low pH values (3–5) reported for many brines.

TABLE 3

Brine Compositional Indicators of Downhole Temperatures [4]

Indicator	Comments
1. SiO_2 content	Best of indicators; assumes quartz equilibrium at high temperature with no dilution or precipitation after cooling.
2. Na/K	Generally significant for ratios between 20/1 to 8/1 and for some systems outside these limits.
3. Ca and HCO_3 content	Qualitatively useful for near-neutral waters; solubility of $CaCO_3$ inversely related to subsurface temperatures.
4. Mg:Mg/Ca	Low values indicate high subsurface temperature, and vice versa.
5. Cl dilution	Assumes dilution of lower-Cl springs by cold water, permitting calculation of subsurface temperatures from required mixing ratios with highest Cl waters.
6. Na/Ca	High ratios may indicate high temperatures but not for high-Ca brine; less direct than Indicator 3 above.
7. $Cl/(HCO_3 + CO_3)$	Highest ratios in related waters indicate highest subsurface temperatures and vice versa.
8. Cl/F	High ratios may indicate high temperatures but Ca content (as controlled by pH and $CO_3^=$ contents) prevents quantitative application.
9. H_2/other gases	High ratios qualitatively indicate high temperatures.
10. Sinter deposits	Reliable indicator of subsurface temperature (now or formerly) > 180°C
11. Travertine deposits	Strong indicator of low subsurface temperatures unless bicarbonate waters have contacted limestone after cooling.

The noncondensible gases, as well as being principal potential contributors to corrosion throughout the geothermal processing system, act to develop back-pressure in turbines using steam directly, thereby decreasing power output. This is important in selection of a power generation cycle.

5. Mineral Content

Aside from the salinity content, certain of the brine components are of special significance as (1) scale formers, or (2) valuable minerals. Some of the scale formers are Ca^{2+}, Fe(II), Mn, Pb, Ag, and other so-called heavy metals. The valuable minerals include K, Li, and Ag. Their value depends on their concentration in the brine. In addition, recovery of the gas components (e.g., CO_2, NH_3) may prove of value. In California's Imperial Valley, carbon dioxide from geothermal sources was once used for producing dry ice.

There are several key materials-related characteristics of these various brines worth noting. First, they contain no oxygen until or unless air enters the reservoir or recovery system, but generally do contain CO_2 and H_2S with lesser amounts of NH_3. Thus, they are generally acidic and often reducing. These factors, together with those already noted define the scaling and corrosion behavior encountered in handling the geothermal fluids.

D. Geothermal Exploration and Source Assessment

To round out the geothermal power picture, mention should be made of the techniques used to locate geothermal sources and to assess their extent. For more information, the reader is referred to the published literature [5–7].

Surface manifestations such as geysers, fumaroles, and so on, are a good guide to sites for geothermal exploration. Aside from these, however, knowledge of the various factors such as the geology, weather (as source of water), and presence of groundwater and its composition serve to identify promising sites. A variety of geophysical techniques for measurements at depth have proved of value in locating geothermal sources. These include temperature gradients, electrical conductivity, and passive seismic methods (e.g., microearthquake measurements). Gravitational and magnetic surveys have also proved useful.

The existence of a usable hydrothermal (vapor- or liquid-dominated) or dry geothermal system can only be proved by drilling to full depth, generally below several thousand feet. This is expensive. To justify drilling, more than one geophysical indicator should show promise.

II. PROCESSES FOR GEOTHERMAL ENERGY UTILIZATION

A. Introductory Remarks

There are two, perhaps three, principal forms in which geothermal energy may be utilized. The heat in the geothermal fluid may be used directly for heating buildings, for agriculture, for process heat and similar applications. With fluids at temperatures below 180°C, this is the most attractive form of use. Such use is currently underway throughout the world, notably in Hungary [8], Russia [9], Iceland [10], New Zealand [11], and in Oregon in the USA [12], for example. Such heat must be used near the source. As the costs of fossil and nuclear fuels rise, such applications of geothermal energy will probably increase.

With fluids of higher enthalpy (temperatures above 180°C and preferably 200°C), electric power generation utilizing some form of turbomachinery is the method of choice. The various techniques in use or planned for such conversion will be discussed in more detail below.

A third mode of utilizing the heat in geothermal fluids is to produce potable (or generally low-salinity) water by condensation of steam from the geothermal fluid either from an electric power generation cycle or as a use in itself of the heat in the geothermal brine in conjuction with one of the available thermal desalination processes. The direct use of this heat for self-desalination of a geothermal brine has been investigated by the U.S. Bureau of Reclamation at its site at the Mesa region of the (California) Imperial Valley [13]. The materials problems for such a process are essentially the same as for handling the brine for electric power generation and will be treated together.

B. Electric Power from Geothermal Sources

Just about all methods of generating electricity from goethermal fluids (steam and brine) involve some form of turbomachinery driving generators. The fluid may be used directly, or the energy to be converted may be transferred to a second fluid for power generation (binary cycle). Those schemes where the geothermal fluids are used themselves for power generation are called direct cycles; those utilizing other fluids are known as indirect cycles.

1. Direct Cycles

Where the geothermal wells produce steam alone, such as at the Geysers in California, the steam is used directly to power turbines. Entrained solids must be removed, for example by using cyclone separators. It may be necessary to remove from the steam brine carryover or certain or all of the noncondensible gases (H_2S, NH_3, CO_2, etc.). These gases may lead to corrosion problems; their presence will lead to a relatively high back-pressure in the turbine condensers, reducing turbine efficiency. While the enthalpy of the steam may be transferred to a secondary fluid, the energy losses and costs involved have not warranted such action.

The energy of geothermal brines may be converted directly by (1) use in the so-called "total flow" devices, or (2) conversion to steam which is used to drive the turbogenerators.

Total flow schemes [14] involve converting the energy of the brine (brine and steam mixture as the pressure is reduced) into kinetic energy using nozzles plus turbines, helical screw expanders, or similar devices. The theoretical efficiency of such devices is greater than for other conversion schemes. However, much remains to be done in developing the devices and dealing with problems of brine handling, such as erosion, before they can be used in practice. A major effort is currently underway at the Lawrence Livermore Laboratory [14] (California) and at the Jet Propulsion Laboratory [15] (Pasadena, California) to develop such total flow devices.

Current direct cycles using geothermal brines (Cerro Prieto, Mexico; Wairakei, New Zealand; etc.) involve flashing of part of the brine to steam and using the steam to drive generating turbines. A direct steam cycle involving two stages of flashing is shown schematically in Fig. 1. As the brine flows up the well, the pressure decreases permitting partial flashing to steam within the well. The steam and brine mixture flows into a flash evaporator and/or separator where the steam to be used is flashed and separated from the brine. Additional steam is produced at a lower temperature and pressure in a second flash stage. The steam (after cleaning, if needed) is directed to the turbines and then to a condenser. Cooling of the condenser is done using either a wet or dry cooling tower. The steam condensate itself may be used in the cooling cycle if sufficiently clean.

A one-stage direct steam cycle is used at Cerro Prieto (Mexico). A higher efficiency of conversion at the price of more elaborate and expensive equipment may be obtained with two or more stages of flashing.

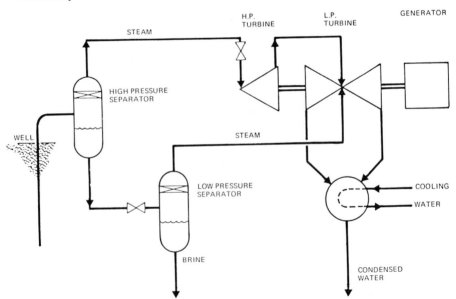

FIG. 1. Two stage brine–steam separator arrangement with steam turbine power generation.

2. Indirect Cycles

In indirect cycles, one or more fluids other than the brine and/or steam from the geothermal well is used for heat transfer and power generation. Fluids such as n-butane, isobutane, Freon, and ammonia have been considered for such use. Even use of pure water has been suggested as a secondary fluid. It would offer the advantages of minimal, controllable corrosion in the turbine and condenser subsystems with no scaling problems to deal with.

In operation, an indirect or binary system requires that the enthalpy of the geothermal fluid be transferred to the secondary or binary fluid in some heat exchanger. This may involve the total geothermal brine, the flashed geothermal steam alone, or both the steam and residual brine.

In principle, use of a binary fluid with the total well flow is most efficient if the brine is prevented from flashing. For this, a downhole (in-well) pump is required. A novel method for transferring downhole heat to the surface using a heat pipe has recently been patented [16]. As the residual enthalpy in the flashed brine is a relatively small fraction of that in the original stream, it is not likely that direct flashed brine–binary fluid heat exchange will be used. Rather, two or more stages of flashing, with the flashed steam exchanging heat to the binary fluid, is contemplated. The flow diagram for such a cycle with one stage of flashing is shown in Fig. 2.

The properties desired in such a binary fluid are (1) suitable normal boiling point, (2) high vapor density, (3) high enthalpy of vaporization, (4) chemical stability, and (5) low corrosion tendency of the fluid and its decomposition products. Candidates are discussed in Section III,B,2.

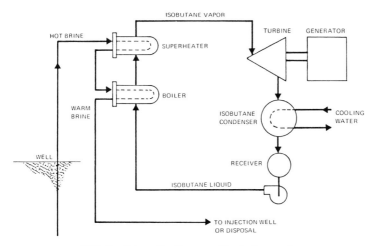

FIG. 2.   Schematic diagram of iso-butane cycle.

TABLE 4

Comparison of some Alternative Geothermal Power Options

Option	Advantages	Disadvantages	Development potential
1. Steam turbine operating on steam flashed from the brine.	Simplest flow arrangements. High cycle efficiency. Low capital cost. Flash evaporators less susceptible to brine fouling and corrosion than heat exchanger surfaces.	Additional steam purification equipment required to protect turbine from corrosion and deposition. Turbine design modifications may be required to accommodate impurities in the steam. Multi-flash staging requires a multi-stage admission turbine.	High development potential with low development risks.
2. Steam turbine operating on clean steam generated by a secondary brine-to-steam heat transfer system.	Turbine sees only clean steam. Conventional steam cleanup equipment is adequate.	Fouling and corrosion presents a problem in tubular steam generator. Lower turbine throttle temperature and pressure than Option 1. Increased capital cost due to secondary steam generator loop. Surface condenser required. Larger capital cost and lower capability than Option 1.	High development potential with low development risk. Development risk is probably lower than Option 1, but economics are less favorable.
3. Isobutane, n-butane, or ammonia turbine; flashed steam to sec. fluid heat transfer system.	Potentially smaller (and lower cost) turbine than steam cycles.	All of Option 2 (except turbine throttle press) plus: Larger steam generators due to poor thermal conductivity of secondary fluid. Large sec. fluid pumps and pump power. High working pressures increase heat exchanger costs. Development of large secondary fluid turbine required. Potential fire hazard.	Low. Has all of the development problems of Option 2 plus unavailability of sec. fluid turbines in the desired size range.
4. Freon-turbine, flashed steam-to-freon heat transfer system.	Potentially smaller turbine than butane or steam cycles.	All of Option 3 except no potential fire hazard. Freon decomposition may lead to corrosion.	Low.

Concept	Advantages	Disadvantages	Development potential
5. Total flow concept	Eliminates separators, flash evaporators, and steam generators. Uses rugged turbo-machinery with replaceable parts. High efficiency. Potentially a simple, low cost system.	More severe corrosion/erosion fouling problems than Options 1 and 2. Requires development of novel turbo-machinery concepts and development of basic analysis methods for high-velocity, two-phase brine flow.	Development potential uncertain. Requires further analytical evaluation.
6. Isobutane turbine, flashed steam-to-isobutane plus brine-to-isobutane heat exchangers.	Relatively simple flow arrangement, single flash vessel and no downhole pump.	All of Option 3 plus scale deposition on tubes of brine heat exchanger.	Low development potential due to scale deposition problems.
7. Isobutane turbine and downhole pump, total well flow to isobutane heat transfer system.	Simple flow arrangement, steam operator not required. High efficiency possible. Scale deposition in well is minimized.	All of Option 3 plus scale deposition on heat exchanger tubes. Large expensive alloy heat exchanger required due to low coefficients and corrosiveness of brine.	Low development potential due to scale deposition problems.
8. Downhole heat exchanger.	Turbine sees only clean steam. May ease heat exchanger corrosion and fouling problems.	Thermal refluxing problem with concentric tube in well casing, i.e., cold downstream must be thermally insulated from hot upstream. Uncertain knowledge of heat extraction mechanism from geothermal deposits. Outside scaling of casing and plugging of formation may be a problem.	Development potential is uncertain. Further analysis is required.
9. Inject hydrocarbon into hot brine and pass vapor to mixed vapor turbine.	Simple flow arrangement. Higher hydrocarbon temperature than 3. May ease well casing corrosion and/or scale deposition problems.	Removal of brine droplets from mixed hydrocarbon–steam vapor presents purification problems. Thermal refluxing in well. Larger fluid flow requirements in well. Requires special turbine development. Loss of hydrocarbon with brine, condensate, and noncondensible gas streams.	Requires further evaluation to establish development potential.

The boiling point should be such that it condenses near the condenser temperature so that condenser pressure is low. Also, the vapor pressure at the highest temperature reached should not be too high as equipment costs become excessive. The high vapor density reduces the sizes of required turbine and other components; high enthalpy of vaporization reduces the flow rate and consequent pumping requirements. The significance of the stability and low corrosion potential are self evident.

In order to avoid the scaling and corrosion problems (discussed at length in Section III, below) likely with geothermal steam and/or brine heat exchangers, direct fluid–fluid contact between the steam and/or brine and the binary fluid is under consideration. The fluids must have low mutual solubility, as significant binary fluid losses to the brine or condensate would be expensive, and probably environmentally unacccceptable. An alternate scheme has been put forth [17], whereby an intermediate fluid is used between the geothermal fluid and the turbine-driving binary fluid. There are more options open in selecting the required fluid parameters so that the overall system performance may be optimized.

Various options for direct and indirect cycles for geothermal power generation are given in Table 4, together with a summary of advantages, disadvantages, and the development potential at the present time of each option. Advances in technology in the near future may lead to changes in the assessment of these options.

The choice of a power generation cycle will depend on all the factors listed, plus the nature of the geothermal fluid available, in particular its temperature, salinity, corrosion and scaling characteristics, and amount of noncondensible gases contained.

The most significant materials problems in geothermal power utilization are scaling and corrosion (with erosion included). The nature of these problems and current methods for prevention or control constitute the heart of this chapter.

III. MATERIALS PROBLEMS IN GEOTHERMAL SYSTEMS

A. Introduction

In this section, the materials which have been and are being used in preparing geothermal wells, and in utilizing the geothermal energy are described, together with the nature of the materials problems encountered. Emphasis has been placed on corrosion and scale formation as these are the major problem areas.

B. Production and Conversion Materials

1. Well Production [18]

Where feasible, petroleum production technology has been applied to geothermal sources. Usually API casings are used [19]. In drilling, a recirculated mud is used. Because of the high temperatures (> 140°C) involved, conventional muds of a bentonite or clay base deteriorate in terms of their flow and filtration properties. In Italy, use of lignosulfonates of iron and chromium, together with added chromolignine, in a controlled composition with proper pH has yielded products of good flow and filtration properties with good stability.

In Japan, a chromite mud, a water-based system containing chromolignite and chromolignosulfonate with caustic soda, has proved effective and stable. A mixture of ferrochromolignosulfonate and chromite in a 3 : 1 ratio (neochromite) has also given satisfactory results. In New Zealand, good results were reported with a combination of chromolignosulfonate and lignite. In these formulations, the lignosulfonic and luminic acids form chelates with the cations (iron and chromium), with a resulting colloidal (gel) system of proper viscosity and stability.

Cooling towers are usually used in the field to reduce the temperature of the recirculating mud. Temperatures are lowered by 20–30°C per cycle. In drilling, structures are often encountered into which the mud flows and is lost. A variety of materials (mica, organics, cellophane, etc.) are often used to plug the open structures.

The casing, used to insure the integrity of the well and localize the source of geothermal fluid, is cemented in place. An inner liner is also sometimes employed, with cementation also used in the annulus. The key problems in selecting a suitable cement are: (1) very high temperatures, (2) contamination with drilling mud, and (3) contamination with well fluid. Aside from choice of the correct type of cement, special techniques are needed in the cementing operation.

Use of silica flour, to the extent of 20–50%, mixed in with the base cement has proved satisfactory at temperatures above 110°C [20]. Milled sand in portland cement has shown promise to 300°C in field condition tests [21]. Various additives are used along with the silica to control curing time, function, foaming, etc.

2. Binary Cycle Working Fluids [22]

In power generation from lower temperature geothermal sources, and from high temperature sources with high salinity, use of a second fluid is often considered. In operation, the enthalpy of the geothermal fluid (brine, steam, or both) is transferred to a second fluid which in turn gives up part of this enthalpy to a turbine or similar converter. The choice of working fluid is made based on physical and thermodynamic properties, stability and corrosiveness, availability and cost. The principal materials of choice are the lower molecular weight saturated hydrocarbons and halogenated (with fluorine and chlorine) hydrocarbons (Freons).

The saturated hydrocarbons are relatively inexpensive, stable to temperatures of $\sim 400°F$, and generally compatible with the usual metals of construction. They are subject to isomerization or cracking reactions at higher temperatures. These processes can be catalyzed so avoidance of contact with catalytically active materials is required.

Freons are relatively expensive. They are stable to 400°F. Their decomposition products (for example HCl) can be quite corrosive. Some metals will react with the Freons over extended periods at elevated temperatures. It is reported that in increasing order of reactivity some common metals are: Inconel $<$ 18-8 stainless steel $<$ nickel $<$ mild steel $<$ aluminum $<$ copper $<$ bronze $<$ brass $<$ silver.

C. Corrosion and Erosion in Geothermal Systems

1. Introductory Remarks

a. The Corrosion/Erosion Environment As described previously, certain properties of the geothermal fluid (and its source) will define the potential for corrosion and erosion. Among these properties are: (1) temperature, (2) composition, (3) entrained solids (and liquids in steam), (4) quantity and nature of noncondensable gases, (5) pH and redox potential, (6) flow and phase change parameters, and (7) scale formation. The significance of each property and the ensuing corrosion or erosion effects will be elaborated on below.

b. Materials for use in Geothermal Systems By and large, metals and alloys, because of their engineering properties, are and will continue to be the principal construction material for geothermal systems components. However, for cost and corrosion factors, other materials have been considered, and may well be used more and more, where feasible. These include ceramics, various coatings, composites, and plastics.

2. Nature of Metallic Corrosion

a. Electrochemical Nature of Aqueous Corrosion Aqueous corrosion, in whatever form it occurs, is almost universally an electrochemical (specifically, a galvanic) process.[1] This means that in the system undergoing corrosion, there are regions where anodic and regions where cathodic processes are occurring, these regions or sites being separated by both an electrolytic and an electronic conductive path.

Anodic processes are ones where electrons are lost—for example, iron forming ferrous ion, zinc dissolving to form zinc ions, or either chlorine or oxygen gas being formed from ions in solution. Cathodic processes are ones where electrons are gained; examples are hydrogen gas evolution, metals plating out of solution, and oxygen gas forming hydroxide ions. The electrons involved travel from the anodic to the cathodic region through the electronic conductive path. To enable the anodic and cathodic reactions or processes to proceed, the circuit must be completed by an electrolytic path also connecting the two (anodic and cathodic) regions. This may be seen more clearly in a simple example by referring to Fig. 3.

In Fig. 3a, an entire corrosion process involving dissimilar metals is shown. The anodic process is zinc-forming zinc ion in solution with loss of two electrons (e^-) per atom:

$$Zn \rightarrow Zn^{2+} + 2e^-$$

The cathodic process is hydrogen evolution at the nickel–solution interface; two electrons are gained for each hydrogen gas molecule formed:

$$2H^+ + 2e^- \rightarrow H_2(g)$$

The electrons given up at the anodic region travel to the cathodic region through an electronic path, involving the zinc, the nickel, and their contacting interface. In order to maintain electrical neutrality, two negative charges move in to pair up with the zinc ions, and two positive charges move in to replace the lost hydrogen ions, or an equivalent set of charge transport processes occur. This takes place, for example, as shown in Fig. 3a, by migration of Cl^- to the anodic region, and Na^+ to the cathodic region.

If there is no salt solution, as shown in Fig. 3b, nothing happens. Likewise (forgetting about other possible corrosion processes), if there is no contact between the zinc and the nickel, as in Fig. 3c, nothing happens.

It might be helpful to think of the corrosion process in terms of a short-circuited "dry" cell or battery (of cells). In this common cell, the zinc undergoes anodic dissolution—essentially corroding away. The cathodic reaction involves reduction of manganese dioxide to a lower oxide (rather than hydrogen evolution as in the example above). The electrolyte is a damp

---

[1] High temperature corrosion in the absence of an aqueous phase also is an electrochemical process involving transport of metal or oxygen ions through an oxide film with separate anodic (metal ion formation) and cathodic (oxide ion formation) processes. Molten salt corrosion is entirely analogous to aqueous corrosion.

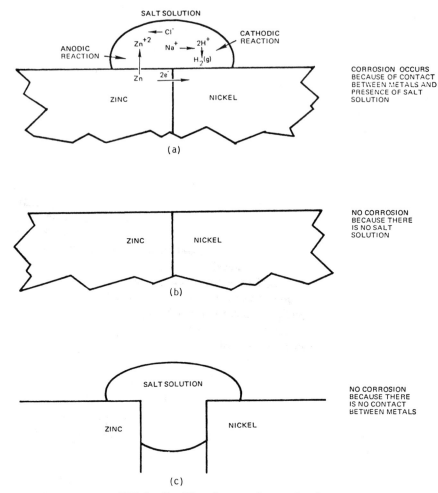

FIG. 3. Conditions for corrosion to take place.

paste of ammonium chloride. By short-circuiting, the cell's reactions proceed and corrosion of the zinc occurs.

In the example shown in Fig. 3, if the zinc and the nickel, separated as shown in Fig. 3c, were connected by an electronic conductor, corrosion would occur. If this conductor were, for example, a low-voltage light bulb, part of the energy of the process could be made to do useful work.

The rate at which a corrosion process occurs is determined by the rate at which the slowest of the various transport or transfer processes occurs. Each step in the current-carrying loop must proceed at the same net rate. Many corrosion processes are limited by the cathode reaction rate (i.e., the rate of hydrogen evolution or oxygen reduction). Others are limited by the electrolytic conductivity of the aqueous phase. Sea water corrosion is often of the severity encountered because of the (relatively) high conductivity of the 3.5% brine involved.

b. Forms of Corrosion   "It is convenient to classify corrosion by the forms in which it manifests itself, the basis for this classification being the appearance of the corroded metal" [23]. The eight forms are: (1) uniform or general attack, (2) galvanic or two-metal corrosion, (3) crevice corrosion, (4) pitting, (5) intergranular corrosion, (6) selective leaching, (7) erosion corrosion, and (8) stress corrosion including corrosion fatigue.

Uniform attack is the most common type of corrosion. The corrosion process proceeds uniformly over a large area, the metal becomes thinner and eventually fails. This form of corrosion, while it is of importance to consider for all applications, can usually be accurately estimated on the basis of simple tests and does not present the possibility of unexpected catastrophic failure. For geothermal systems, these tests are not so simple. This is especially true for downhole use, where test measurements of corrosion rates have never, to the author's knowledge, involved specimens subjected to flowing geothermal fluids. The only "tests" under these conditions have been of the well casing itself. However, downhole static tests of specimens have been carried out.

Two-metal (galvanic) corrosion occurs, as the name implies, when two different metals are in contact in the presence of an electrolyte. The example discussed in the previous subsection (Fig. 3) is of this type. This form of corrosion can be prevented by avoiding such metal couplings.

Selective leaching, as exemplified by removal of zinc from brass, may be avoided by proper materials choice.

Erosion corrosion results when flow velocities of fluids in a system result in much enhanced corrosion due to protective film breakdown, impingement, turbulence, or other related factors. It may be avoided by proper design and materials selection.

The four remaining forms of corrosion—crevice corrosion, stress corrosion cracking, intergranular, and pitting corrosion—are of special importance in materials selection for geothermal applications. They each are strong functions of the nature of the material and environment, and, most importantly, can lead to catastrophic failure without outside warning. These are discussed in some detail in the following sections.

c. Crevice Corrosion   Crevice corrosion, a localized intense attack, may result when a region of metal is covered or shielded while allowing some penetration by the aqueous fluid. This allows the covered region to become anodic resulting in its dissolution. While selection of the proper alloy may minimize such attack, the only sure cure is to avoid such coverage.

The threading of pipe sections together will produce such a crevice unless the overlapping region is completely sealed from brine entry. Bolted flanged joints likewise present the possibility of crevice corrosion.

Another type of crevice attack, often called deposit corrosion, may occur in geothermal systems where scale formation occurs. A hard glassy scale has been encountered in a number of geothermal wells in the Salton Sea KGRA. In some cases, this coating has been claimed to protect against corrosion. If such a coating is incomplete, however, anodic dissolution may occur below such deposits.

d. Stress Corrosion A wide variety of metals exhibit stress corrosion cracking (SCC) under certain environmental conditions.

Some form of tensile stress must be present for SCC to occur. This may be an applied stress, a residual stress from welding or shaping, or a combination of these. No cracking will occur if the stresses are compressional only. Shot-peening, which produces such compressional stresses, has been used to prevent SCC.

Initiation of SCC requires surface nonuniformity; it will initiate at points of high stress, such as at corrosion pits, depressions, or gouges. In addition to these, there must be some environmental factor which results in a corrosion cell being established with anodic and cathodic processes occurring. The anodic process is established at the crack tip; the resultant metal removal causes crack propagation which may be coupled with mechanical fracture at the point of failure.

Some of these metals and some of the environmental factors which can lead to SCC are listed in Table 5. There is such a variety of environments which can cause SCC in a given material that the only way to be certain that this will not occur is to test a material under operating conditions.

While carbon steel has been used in most geothermal systems, the high corrosion rates often encountered prompts one to consider more corrosion resistant materials. The most readily available and inexpensive of these are the austenitic stainless steels (e.g., Types 304, 316, and 347). However, austenitic stainless steels are subject to SCC. While caustics may cause such corrosion (actually through hydrogen embrittlement), chlorides are generally present in environments resulting in SCC failure of these steels.

It has been found that there is no threshold chloride level for attack; the attack is more rapid at the higher Cl^- levels. While elevated temperatures ($> 200°F$) are required for SCC to occur, such is the case in all geothermal systems.

TABLE 5

Environments Which May Cause SCC in Metals and Alloys

Material	Environment
Aluminum alloys	NaCl solutions, seawater
Copper alloys	Ammonia vapor and solutions
Gold alloys	$FeCl_3$ solutions
Ordinary steels	Acidic H_2S solutions, seawater
Stainless steels	H_2S, seawater, condensing steam from chloride waters
Titanium alloys	Seawater, methanol–HCl

With chlorides, the cracking is generally transgranular. The appearance of a 304 SS surface which has undergone SCC is shown in Fig. 4a. The same surface in cross section at higher magnification is shown in Fig. 4b; the transgranular nature of SCC may be seen as the cracks cross grain boundaries in their penetration of the specimen. With sensitized steels, the SCC mode may be intergranular. It should be mentioned that the specimen shown in Fig. 4 was stressed beyond the yield point and exposed to boiling 42% $MgCl_2$; this exposure is used as a standard, accelerated testing method for evaluating SCC in austenitic stainless steels.

Figure 5 shows the synergistic effect of chloride and oxygen on stress corrosion cracking in austenitic (18-8; Type 304) stainless steel. The presence of dissolved oxygen or other oxidizing species is critical to the cracking of austenitic stainless steels in chloride solutions. If the oxygen is removed, cracking will not occur. The author has found this to be the case in a number of laboratory tests involving high salinity geothermal brines, U-bend 316 SS specimens, and > 200°C test temperatures.

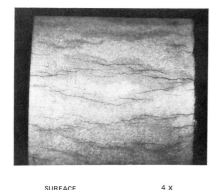

SURFACE   4 X   WALL   50 X

FIG. 4.   Stress corrosion cracking in austenitic 304 stainless steel (stressed specimen tested in boiling magnesium chloride).

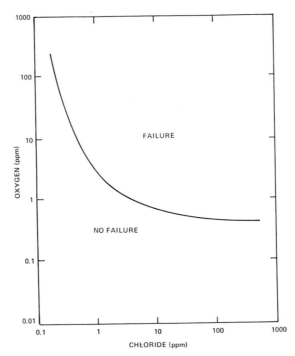

FIG. 5. Synergistic effect of chloride and oxygen on stress corrosion cracking of 18-8 stainless steel [W. L. Williams, *Corrosion* **13**, 359t (1957)].

It should be noted that at low pH values, high temperature, or both, hydrogen evolution becomes an alternative to oxygen reduction as a rapid cathodic process. In this case, SCC can occur in the absence of oxygen.

It might be thought that in surface equipment where temperatures are low enough, exclusion of oxygen would permit austenitic stainless steels to be used. However, the widespread incidence of corrosion failure in seawater desalination plants due to "poor process control" [24] leads one to select a material not as subject to SCC as are the 18-8 stainless steels.

Another route to avoid chloride SCC is to use a ferritic stainless steel. A number of such steels, usually of high Cr content, have been suggested for geothermal service [25].

Time to failure from stress corrosion cracking versus nickel content for iron–chromium–nickel alloys is shown in Fig. 6. It may be seen, first, that the *greatest* susceptibility and most rapid failure occurs at about 8% nickel, the nickel content of the most common stainless steels (e.g., Type 304, 316, 347, 310, etc.). Second, it can be seen that by going to much higher nickel content or by going to low nickel content, SCC may be avoided.

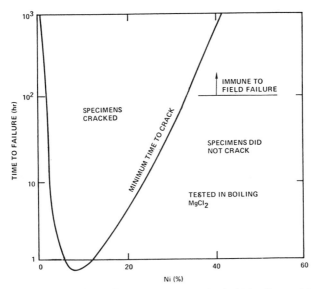

FIG. 6.   Ni alloy stress corrosion [H. R. Copson, *in* " Physical Metallurgy of Stress Corrosion Fracture" (T. Rhodin, ed.). Wiley (Interscience), New York, 1959].

Corrosion fatigue is often considered a special case of SCC. Fatigue is the tendency of a metal to fracture under repeated cyclic stresses. This can be purely mechanical. However, the fatigue resistance can be reduced by the presence of a corrosive medium, leading to corrosion fatigue. This has apparently been encountered frequently with geothermal turbine blades.

e. Intergranular Corrosion   This type of corrosion is encountered with alloys, particularly the austenitic stainless steels. The attack occurs at grain boundaries only; when it has progressed far enough, the grains may fall away from the bulk metal. The results of such attack are shown in Fig. 7.

The origin of intergranular corrosion is well understood. All austenitic stainless steels have some carbon dissolved in them. In the sensitizing temperature range 850–1650°F, carbon is rapidly precipitated out in the grain boundaries where it combines with chromium and forms chromium carbide ($Cr_{23}C_6$). This depletes the boundary region of chromium, rendering it much less corrosion resistant than the interior of the grains. This is shown schematically in Fig. 7. In a corroding environment, the system acts as a two-metal couple with the boundary region being anodic and dissolving away. The cathode role is apparently served by the chromium carbide.

There are three approaches for preventing this form of attack. First, by annealing the steel at the proper temperature (e.g., about 2800°F) and quenching rapidly, the carbon and chromium are redissolved and corrosion re-

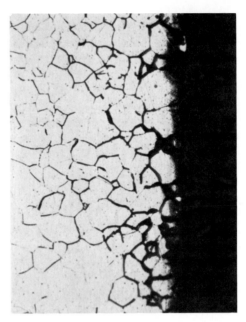

FIG. 7.   Intergranular corrosion in austenitic 304 stainless steel (material sensitized for 1 hr at 650°C).

sistance restored. A second approach is to limit the concentration of carbon in the steel. A value of less than 0.03% C is considered adequate to avoid sensitization during welding. The low carbon grades of stainless (e.g., 316 L) meets this requirement.

Where annealing after welding is impractical or the materials requirements prevent use of a very low carbon steel, a third approach can be used. This involves addition of a carbon "getter," for example, niobium, to give stabilization by alloying. The niobium combines with the carbon, allowing the chromium to remain in place. Such alloys can be welded without being sensitized.

f. Pitting   Pitting corrosion, as the name implies, involves the formation of a number of small-area, deep zones where corrosion proceeds. Pitting starts in regions which, through imperfection, impurities, and other factors, are anodic to the rest of the surface. If allowed to proceed, the attack penetrates deeper below the surface, leading to build-up of corrosion products which reduce access of air (or oxygen) thereby stabilizing the anodic process. Further, anions, for example, Cl^-, migrate into the pit and enhance the electrolyte conductivity of the medium in the pit. This leads to a speed-up of the corrosion process; pitting is autocatalytic (self-aggravating). By the nature of the process, pits usually penetrate downward; the retention of the corrosion products in the pit is aided by gravity. Material failure occurs when pits are deep enough to penetrate; a few deep pits are more damaging than a larger number of shallow ones.

Pitting attack is minimized under flow by avoiding stable formation of fixed anodic regions. Pitting attack, when halogens (e.g., chlorides) are present, can be minimized by alloying stainless steels with molybdenum. A high chromium content combined with molybdenum addition has a beneficial synergistic effect through formation of a protective film.

g. Compositional and Metallurgical Factors in Corrosion Behavior It was mentioned above that the nickel content of iron–chromium alloys determines susceptibility to chloride stress corrosion cracking. Also, the fact that the addition of molybdenum, in particular, to chromium-containing steels, imparts resistance to pitting attack. In general, the corrosion resistance of a given alloy will depend fundamentally on its chemical composition. In this regard, high purity, for example, as produced by electron beam melting, can have a significant beneficial effect on corrosion behavior. Of equal importance in determining corrosion behavior are metallurgical factors. The effect of heat and heat treatment in causing and preventing intergranular corrosion was noted earlier. Corrosion resistance often depends also on hardness, yield strength, and stress levels. Further, imperfections, surface impurities, inclusions (e.g., of mill scale) all play a role in practice and must be taken into account in the selection of metals for use in geothermal systems.

3. Tools for Study of Corrosion

A number of techniques and methods are available to investigate the corrosion behavior of metals and to understand and interpret such behavior. Some of the electrochemical techniques used for study of and interpretation of corrosion and measurement of corrosion rates are outlined below. This is followed by a short discussion of other related corrosion testing and diagnostic methods. For a more complete review of this material, the reader is referred to references cited.

a. Electrochemical Techniques (*1*) *Thermodynamics* The driving force behind corrosion and other electrochemical processes is the free energy change involved in the overall reaction, ΔG, which is related to the potential between anode and cathode, that is, the cell potential, $\mathscr{E}$, by the relation

$$\Delta G = -n\mathscr{F}\mathscr{E}$$

where n is the number of electrons transferred in the reaction and $\mathscr{F}$ the Faraday constant.

The cell potential will depend on the activities (effective concentrations) of reactants and products; it is related to the standard potential $(\mathscr{E}^0)$ for the case of unit activities by the Nernst equation:

$$\mathscr{E} = \mathscr{E}^0 + (\log_e 10)(RT/\eta\mathscr{F}) \log(\Pi\mathscr{A}_r/\Pi\mathscr{A}_p),$$

where R is the gas constant, T the absolute temperature, and $\Pi\mathscr{A}_r$ and $\Pi\mathscr{A}_p$ are, respectively, the products of the activities (with appropriate exponents) of reactants and products.

The corrosion behavior of a metal can often be better understood by use of potential–pH or Pourbaix diagrams. These are constructed based on the Nernst equation and the solubility equilibria of the various metal oxides and hydroxides involved. A simplified diagram of this sort for the iron–water system is shown in Fig. 8; the simplification being omission of all but one of

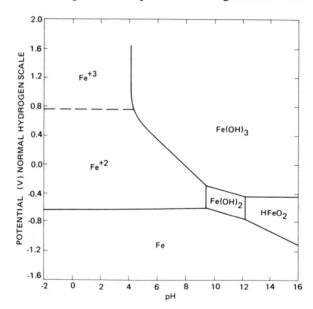

FIG. 8.   Simplified potential–pH diagram for the Fe–H$_2$O system. (From "Corrosion Engineering," by Fontana and Greene. Copyright © 1967 by McGraw-Hill Book Company. Used with permission.)

a family of lines for different species activities. In this diagram, the thermodynamically stable iron species at a given combination of potential and pH is shown. For example, where "Fe" is shown, iron is inert or immune and corrosion does not occur at the corresponding pH and potential values, while the region labeled "Fe^{2+}" is one under whose conditions the ferrous ion is the stable aqueous species. The region "Fe(OH)$_3$" corresponds to that of passivity where a protective film, a hydrous ferric oxide, is formed. Values of the potential at a surface and the medium pH can define the thermodynamically favored corrosion process (if any) to be expected.

(2) *Kinetics* At an electrode at equilibrium, the rate of oxidation and reduction are equal; this rate translated into a current equivalent is called the exchange current. For some processes, for example, hydrogen evolution, the magnitude of the exchange current density (current per unit area) i_0, is strongly dependent on the surface on which it is occurring.

When a current is flowing at an electrode, for example, when corrosion is occurring, the potential of an electrode, or between anode and cathode, departs from the equilibrium value. The departure, called polarization, η, can be attributed to the rate-determining (slow) step at an electrode (activation polarization, η_a), to depletion or accumulation of a reacting species at the electrode (concentration polarization, η_c), and to ohmic (IR) losses or polarization from both electronic and electrolytic conductor resistance. Total polarization is equal to the sum of each of these forms.

Excluding resistance, the effects of activation polarization usually are dominant at lower reaction rates while concentration effects become dominant at higher reaction rates where a limiting current, i_L, is obtained. This is shown in Fig. 9, where the total polarization η_t is shown as a function of log i, the current.

In metal dissolution processes, concentration polarization is not a factor; for $i > 50$ mA, the rate of the process as measured by the current is given by the Tafel equation:

$$\eta_a = \beta \log(i/i_0)$$

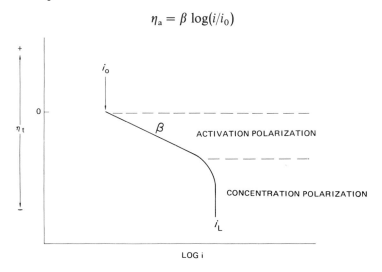

FIG. 9. Combined polarization curve–activation and concentration polarization. (From "Corrosion Engineering," by Fontana and Greene. Copyright © 1967 by McGraw-Hill Book Company. Used with permission.)

where β is called the Tafel constant and is the slope of the plot of η_a versus log i. In hydrogen evolution or oxygen reduction processes, the total polarization is given by

$$\eta_T = \eta_a + \eta_c$$

or

$$\eta_T = \beta \log(i/i_0) + (\log_e 10)(RT/n) \log(1 - i/i_L)$$

The mechanism of corrosion of a (single) metal is usually explained by the "mixed-potential" theory which postulates (1) that corrosion consists of individual anodic and cathodic processes which (2) occur at the same rate (same total electron transfer). The metal, a good electronic conductor, is essentially at a uniform potential which potential is fixed by balancing the rates of the anodic and cathodic processes, for example, metal dissolution and hydrogen evolution. These electrode processes are each polarized from their equilibrium potential values to the actual potential assumed by the metal. This potential is called the corrosion potential, E_{corr}; the current, corresponding to the rates of both anodic and cathodic processes is called the corrosion current, i_{corr}.

Many alloys and some metals (e.g., Ti, Cr, and Al) owe their corrosion resistance to the formation of a protective film which forms a barrier to electron and/or ionic transport. This phenomenon, called "passivity," can be produced with metals not usually exhibiting it, such as iron, by shifting their potential anodically. Beyond a certain potential, E_{pp}, the anodic process rate (current) drops to a very low value. Once formed, the film may continue to protect against corrosion until it is damaged or destroyed. If the potential is made sufficiently anodic, in some cases a transpassive region is entered and the corrosion rate once again increases. Protecting those metals which exhibit passivity against corrosion is sometimes done by application of an anodic potential (anodic protection).

(3) *Corrosion Rate Measurements*   The knowledge outlined above has formed the basis of two methods for measuring corrosion rates: Tafel extrapolation and linear polarization. Tafel extrapolation involves applying increasing current (usually cathodic) to the metal so that the linear (Tafel) curve may be determined. Extrapolating this curve linearly to the measured corrosion potential will give the corrosion current (value of current at the point of intersection).

It has been observed that within about 10 mV (anodic or cathodic) of the metal corrosion potential, the polarization produced (ΔE) is a linear function of the applied current (Δi_{ap}). The slope of the $\Delta E/\Delta i_{ap}$ curve is sometimes called the "polarization resistance"; it is related to the Tafel slopes of the anodic (β_a) and cathodic (β_c) reactions by

$$\Delta E/\Delta i_{ap} = \beta_a \beta_c/(\log_e 10)i_{corr}(\beta_a + \beta_c)$$

From linear or measured values of β_a, β_c, ΔE, and Δi_{ap}, i_{corr} may be calculated. If the values for β_a or β_c are not known, a close approximation may be made by assuming them equal and of value = 0.12 V.

These electrochemical methods are rapid, can measure low corrosion rates, and can often be applied to measurements on remote or fixed components. Instruments based on the linear polarization method are available commercially.

b. Mechanical and Related Corrosion Testing Methods   In evaluation of metals, alloys, and coatings for use in geothermal systems, both laboratory and on-site testing can be of value. In all cases, it is necessary to design the tests so that the results represent that which can be expected in service, or give a relative measure of resistance to attack. At the very least, the short-comings of any tests should be recognized and interpreted accordingly.

Considerations for selection of materials and specimens, surface preparation, exposure techniques, test periods, and post-test examination are discussed in several books on corrosion [23] and corrosion testing [26], to which the reader is referred. Test details and limitations should be noted.

A variety of standard tests have been formulated to evaluate resistance to specific forms of corrosion. For stainless steels, exposure to boiling 65% nitric acid for five 48-hr periods (Huey test; ASTM A262) is used to show susceptibility to intergranular corrosion. Resistance to stress corrosion cracking is measured using specimens under constant deformation (e.g., bolted U-bend samples) or constant load. The latter, providing a more certain knowledge of applied stress, also requires the more elaborate equipment. Exposure to boiling 42% $MgCl_2$ for up to 200 hr is a commonly used accelerated stress corrosion test for stainless steels.

In all testing procedures, it is necessary to take precautions to ensure conditions are as planned, and to interpret test results with caution. Mounting of several metals and alloys together on a test rack without ensuring that each is insulated from the others and from the rack will yield misleading results. If specimens are shorted together, the least noble will become anodic and show excessive attack while the others will be cathodically protected and appear relatively immune. Stress corrosion cracking of austenitic stainless steels in aqueous chloride media will usually occur at elevated temperatures. Scrupulous care to exclude oxygen from the system will often prevent such attack, whereas ordinary de-aeration procedures are inadequate and will lead to cracking.

For geothermal systems, testing will usually involve elevated temperatures. Many of the standard references discuss apparatus and equipment for such conditions [27]. A review of electrochemical measurements at temperatures over 100°C has been published [28].

D. Scale Deposition in Geothermal Systems

1. Introduction

This section presents a review of the nature and incidence of (noncorrosion) deposits from geothermal brine and steam, and of the basic chemistry involved in this deposition. Proposed methods for dealing with this scale formation are discussed in Section VI.

Except in rare incidences to be noted below, the scales formed in geothermal processing equipment contain numerous species. They form on process equipment or in combination with solids in brine or steam. Solids (sand, clay, etc.) carried along from the reservoir structure, corrosion products, and insoluble substances precipitating from the brine may be combined in such scales. Even quite soluble species may form part of the scale in steam-handling components.

The principal scale problems arise in handling the brine. These scales usually contain either carbonates (usually calcium carbonate) or silica as their principal component, the component which appears to determine whether or not the scale will form at all and which gives the scale its control and removal properties. These scales also contain sulfides, hydroxides, oxides, and sulfates of various cations. In some cases, silicate compounds may form (e.g., SiO_2 with FeO). Details on the actual composition of the scales encountered in geothermal processing are given below (Section IV). In the next section, the chemistry of the deposition of carbonates, silica, and other scale formers will be reviewed.

2. Chemistry of Scale Formation

a. Precipitation of Carbonates   Three key events occur in geothermal processes which lead to formation of scale deposits. These are: (1) reduction in temperature, (2) reduction in pressure leading to steam formation and escape of dissolved gases, and (3) concentration of species in the residual brines. The interactions of these factors may be illustrated by using, for example, the case of $CaCO_3$ (stable form, calcite). Figure 10 shows the solubility surface of calcite in the system $CaCO_3$–$NaCl$–CO_2–H_2O over the range shown ($NaCl = 1$ molal). The solubility decreases with increased temperature (at constant CO_2 pressures) and increases with increased CO_2 pressure (at constant temperature).

At any temperature, the solubility limit is determined by the solubility

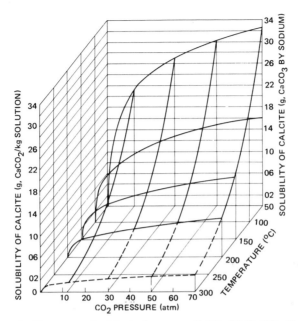

FIG. 10.   The solubility surface of calcite in the system: CaCO$_3$–NaCl–CO$_2$–H$_2$O between 50 and 300°C, between 0 and 70 atm, and at a NaCl concentration near one molal. (From "Geochemistry of Hydrothermal Ore Deposits," edited by H. L. Barnes. Copyright © 1967 by Holt, Rinehart and Winston, Inc. Reprinted by permission of Holt, Rinehart and Winston, CBS, Inc.)

product, K_{sp}, a function of temperature, where $K_{sp}(T) = \mathscr{A}_{Ca+2} \times \mathscr{A}_{CO_3} =$. The concentration and accordingly the activity of the carbonate ion is determined by the equilibrium

$$2\,HCO_3^- \rightleftarrows CO_2 + CO_3^{2-} + H_2O$$

($K_p = K_{A_2}/K_{A_1}$ where K_{A_i} refers to the carbonic acid ionization constants.)

Thus, as the geothermal brine is processed, the temperature decreases, increasing the solubility of the CaCO$_3$. On the other hand, the escape of the dissolved gases, principally CO$_2$, results in an increase in [CO$_3$ $^=$], while flashing to form steam results in a small amount of residual brine and a further increase in [CO$_3$ $^=$] and an increase in the [Ca^{+2}]; these tend to result in CaCO$_3$ deposition. The increase in brine salinity affects the activity coefficients and the formation of soluble complexes. These have relatively little effect on the behavior of CaCO$_3$. In many cases, CaCO$_3$ does tend to deposit, in particular where flashing occurs. Solubility products for a number of carbonates of concern in geothermal systems are given in Table 6 [29]. While all carbonates in general will follow a pattern similar to CaCO$_3$, inadequate data precludes accurate precipitation predictions at higher temperatures.

TABLE 6

Solubility Products for Carbonates

Substance	Logarithm of the solubility product $(\log K_{sp})$	Temperature (°C)	Substance	Logarithm of the solubility product $(\log K_{sp})$	Temperature (°C)
$BaCO_3$	-8.09	25	$MnCO_3$	-10.74	25
$CaCO_3$	$-7.91/-8.00/-8.09$	0/15/30	$PbCO_3$	-13.48	18
$CuCO_3$	Soluble complex formed	—	Ag_2CO_3	-11.09	25
			$SrCO_3$	$-9.96/-10.36$	25/40
$FeCO_3$	-10.46	25	$ZnCO_3$	-10.68	25
$MgCO_3$	$-3.51/-4.01/-4.68$	3.5/22/50			

b. Sulfide Precipitation The combination of reduced pressure (with loss of CO_2 and of H_2S, as well as with a consequent increase in pH), together with lower temperatures and higher brine concentrations, leads, in some cases, to precipitation of various cations as simple and mixed sulfides. These usually constitute a relatively minor component of the scale which will be mostly silica [30, 31].[2] The solubilities of some simple sulfides that may be anticipated are shown in Table 7 [29]. Accurate solubility data for simple and mixed sulfides at elevated temperatures and in brines are not available.

TABLE 7

Solubility Products for Sulfides

Substance	Logarithm of the solubility product (log K_{sp})	Temperature (°C)	Substance	Logarithm of the solubility product (log K_{sp})	Temperature (°C)
Cu_2S	− 48.58	25	MnS	− 14.96	25
CuS	− 37.49	25	Ag_2S	− 51.23	25
FeS	− 19.42	25	ZnS	− 26.13	25
PbS	− 28.17	25			

c. Silica Deposition An excellent review of the nature of dissolved silica in geothermal brines has recently appeared [32]. Other summary discussions are also available [33, 34] as well as other sources containing pertinent general information [35, 36]. Some investigators claim that silicates, such as magnesium silicate or ferrous silicate, rather than silica, are forming these scales. If so, deposition kinetics would differ but the thermodynamic (solubility) relations would be analogous to those for silica or other solids. Silica is encountered in nature in a variety of crystalline and amorphous forms. Quartz is the crystalline form stable at ambient temperature (up to 870°C); tridymite (stable between 870 and 1470°C); and cristobalite (stable over 1470°C) are other macrocrystalline forms found in nature.[3] Each of these forms undergoes structural transformations from a high temperature

---

[2] Evidence has been reported that suggests the silica deposits on metal sulfide particles already formed [30, 31]. Among other factors, this may help explain the widespread deposition of silica-type deposits on surface equipment with the Niland (Salton Sea, California) area brines as contrasted with the equally hot but lower mineral content brines at Cerro Prieto, Mexico. In the latter case, extensive (nearly pure) silica deposition occurs late in the process in reject brine lines.

[3] Chalcedony is another naturally occurring form of silica consisting of fibrous crystallites.

(β-form) to a low temperature (α-form) modification. These transformations, which occur at 573°C for quartz, 140°C for tridymite, and 240°C for cristobalite, are fast. The transformations between cristobalite, tridymite, and quartz are slow; this helps account for their occurrence in nature at ambient temperatures.

Various natural and manmade forms of amorphous silica are known (opal, silica gel); these generally contain significant quantities of water. The natural occurrence of these unstable crystalline and amorphous silica forms indicates that the crystallization of silica is a slow process. The solubilities of the various silica forms as functions of temperature are shown in Fig. 11.

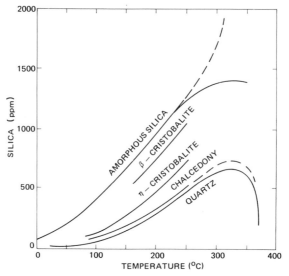

FIG. 11.   Solubilities of various silica phases along two-phase curve (water plus vapor) (from Fournier [34]).

Over the range of temperatures shown, quartz is the stable form, with chalcedony, α and β-cristobalites, and amorphous silica being progressively less stable and, accordingly, more soluble. The solubilities of all forms generally increase with temperature.

While rates of solution and deposition increase with rising temperatures, all other factors being equal, these rates are low under all geothermal processing conditions. At room temperature, time periods of months or even years may be required to reach solution equilibrium.

Silica dissolved in water is in the form of H_4SiO_4. At higher pH levels, approximately two units above neutral pH for a given temperature, this orthosilicic acid is dissociated [$K_1(298°K) = 10^{-9.8}$], thereby increasing its solubility. In forming solids from solution, the orthosilicic acid, a neutral molecule though often bearing a net charge, undergoes polymerization along with dehydration. This results in the formation of a colloidal sol. This is followed by aggregation leading to formation of a gel. This results in the deposition of a hydrated amorphous silica from water. Over the pH range in which silica is in the form of un-ionized H_4SiO_4, its solubility is essentially constant.

The solibility equilibrium of silica involves the reaction

$$SiO_2 + 2H_2O \rightleftarrows H_4SiO_4$$

The equilibrium constant K, is given by

$$K = \alpha_{H_4SiO_4}/\alpha_{SiO_2}\,\alpha_{H_2O}^2$$

The activity of silica, a solid phase, is a constant (= 1 for quartz). In dilute solutions or brines, α_{H_2O} is approximately 1. The solubility is little affected by salinity. However, in highly saline brines, α_{H_2O} is significantly less than 1 and the silica solubility is decreased accordingly.

Polymerization, gel formation, and silica deposition are rate processes. These rates will be functions of temperature, degree of supersaturation, and the details of the nucleation and accrual processes. A treatment of these topics is beyond the scope of the present discussion. The reader is referred to recent publications for this material [37].

Salinity, while not strongly affecting solubility, does affect the rate of silica gel formation (and deposition) in alkaline solutions. The pH also has a strong effect on the rate of the process as do certain catalytic ions. A graph showing the gel formation time as a function of pH is given in Fig. 12.

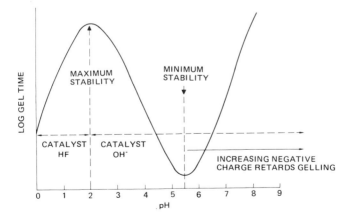

FIG. 12.   Sol stability: Relation between log (gel time) versus pH. (From "The Colloid Chemistry of Silica and Silicates," by R. Iler. Copyright © 1955 by Cornell University. Used by permission of Cornell University Press.)

Polymerized colloidal (sol) particles of silica are negatively charged in alkaline solution [33]. Mutual repulsion thereby prevents agglomeration. The particles can be coagulated by cations. At a pH between 5 and 7, the colloidal silica particles pass through the isoelectric point (no net charge) and may coagulate rapidly. Below this pH level, they take on a positive charge and the colloid is restabilized.

Hydrogen fluoride (up to a pH of about 2) and hydroxyl ion (down to this same pH) catalyze silica polymerization. With the effect of OH^- catalysis, it may be anticipated that the maximum polymerization rate would be at a pH somewhat over 5.5. Kitahara has found the maximum polymerization rate is maintained at a pH of 7.5 [38].

Dissolved silica can be effectively removed from solution by coprecipitation with MgO, $MgCO_3$, and Fe, Al, Mg, and Mn hydroxides, and Mg and Fe (III) sulfates. Colloidal silica can be coagulated by electrolytes of sea water and essentially quantitatively precipitated by aluminum and hydrous ferric oxide [32]. Solids accelerate the polymerization and precipit. colloidal silica from supersaturated solutions.

d. Other Scale Formers   Other compounds found in scales from seawater conversion and other water treatment processes are anticipated and are found in geothermal scales. These are the the various metal hydroxides or hydrous oxides, oxides, and sulfates.

Of the sulfates, calcium, strontium, and barium have been reported as scale components. In addition to the sulfate present in the brine down-hole, oxidation of the hydrogen sulfide in the fluid may produce additional sulfate ions.

Calcium sulfate has been encountered in seawater desalination. At ambient temperatures, the stable form is the dihydrate (gypsum), which has a limited solubility in water with a maximum at 38°C of 2.06 gm/liter. Above about 40°C, anhydrous calcium sulfate is the stable form, showing a decreasing solubility with temperature. Anhydrous $CaSO_4$ forms slowly, permitting the metastable hemihydrate ($CaSO_4 \cdot \frac{1}{2}H_2O$) to precipitate above 100°C. It has been observed that deposits of the hemihydrate are transformed in time to anhydrous $CaSO_4$ [39]. The solubilities of these different forms of $CaSO_4$ are strongly affected by the water's salt content.

The solubility products for several hydroxides and sulfates of potential interest in geothermal scaling are listed in Table 8.

IV. REPORTED GEOTHERMAL SCALING AND CORROSION

A short summary is given below of the types of scale and incidence of corrosion and erosion reported as found at the various locations in geothermal fluid production and disposal systems, and associated energy conver-

TABLE 8

Solubility Products for Hydroxides and Sulfates

Substance	Logarithm of the solubility product substance (log K_{sp})	Temperature (°C)	Substance	Logarithm of the solubility product substance (log K_{sp})	Temperature (°C)
$BaSO_4$	− 10.06	18	$Fe(OH)_2$	− 14.84	25
$BaSO_4$	− 9.96	25	$Fe(OH)_3$	− 36.35	25
$BaSO_4$	− 9.70	50	$Mg(OH)_2$	− 10.95	25
$BaSO_4$	− 9.58	100	$Mn(OH)_2$	− 12.72	25
$CaSO_4$	− 4.625	25	$Pb(OH)_2$	− 14.93	25
$PbSO_4$	− 7.80	25	$Pb(OH)_4$	− 65.49	25
$SrSO_4$	− 6.49	25	$AgOH$	− 7.73	25
Cu_2O	− 14.0	17	$Zn(OH)_2$	− 17.48	25
$Cu(OH)_2$	− 19.66	25			

sion components. This is done for a number of geothermal sites throughout the world where the nature of the fluids differ considerably, and where the scales found and corrosion encountered also differ accordingly.

A. Wells and Liners

Scale formation in the well is usually associated with steam flashing. This occurs where the pressure on the column of fluid equals the vapor pressure. These and related processes are discussed by White [4]. In the wells of the Niland area (Imperial Valley, California), a hard, black, glassy scale has been found in the upper well region [40, 41]. This is mostly silica and iron, with lead and other metal compounds reported. The actual compounds present have not always been identified. The iron is stated to be (nearly) stoichiometric with the silica. An iron (II) silicate may be present, but the scale is amorphous and exhibits no identifiable x-ray pattern. The lead and silver present have sometimes been identified as sulfides. It was observed that the downhole pipe was coated with crystalline magnetite. Analysis of pipe scale in the well showed relatively high Ca, As, Al, Cu, and Fe below 4000 ft, maximum amounts of Fe and Mo at about 2000 ft, and relatively large amounts of Ag, Si, Mg, Pb, Al, Cu, and Fe at 500 ft. Well scale samples were identified as containing Fe_3O_4 and iron arsenide.

At Cerro Prieto (Mexico), the well deposit at the flashing level is reported to be calcium carbonate. This results in a cone of calcite up the well. Silica deposits at adjacent levels are reported to overlap the calcite. The problem is dealt with by reaming out the scale every year or two [42].

At Heber, Schremp reported that with downhole pump operation there was minimal scale deposition [43]. The scale that did form was $FeCO_3$ with Fe_3O_4, PbS, and CuS.

Kuwada reported that in producing the low salinity brine (400°F downhole temperature) with high bicarbonate content at the Kizildere (Turkey) field, heavy scaling was encountered [44]. The well was blocked in six weeks with a dense calcite deposit.

Some scaling and corrosion of a downhole (black-iron pipe) heat exchanger at Klamath Falls (Oregon) was reported [12]. The fluid, with a 110°C maximum temperature, has 800 mg/liter dissolved solids, mostly Na^+ and $SO_4^=$. It was noted that the Langlier Saturation Index was followed [45].

Mild steel (e.g., J-55) oil well casings have been used for all geothermal wells reported to date. This is a natural result of the wholesale transfer of oil well technology to the geothermal area. Apparently satisfactory results with these casings have been obtained in most cases. A review of some of the mechanical and related considerations relative to well casings and cements has been published by Giovanaoni [19]. In this summary, he reports that some drill-pipe failure at Lardarello was attributed to corrosion fatigue.

In the Salton Sea KGRA, mild steel casings do not appear to give satisfactory life. The combination of high downhole temperatures (over 300°C), high acid gas content (CO_2, H_2S), high chloride and varied cation content combine to give a very high corrosion rate. The overall corrosion rate appears to be approximately proportional to depth (i.e., temperature), reaching levels of 3000 mpy at about 5000 ft.

Marsh reports that upon removal of a (mild steel) liner from a well, shut down for 6 yr after production, it was found to have undergone perforation " presumably while under the static conditions of a shut-in well " [40]:

These perforations were of two types. Above 277°C, the perforations were in the form of large pits with siderite (iron carbonate) scaling. At lower temperatures, there was a thick Fe_3O_4 scale and large chunks of metal had been removed, leaving a ' bloody stump.' Metallurgical examination of the area where this occurred showed that the metal had undergone high-temperature hydrogen attack. In other words, actual fission occurred as would be expected in high-temperature hydrogen attack in a refinery.

Although the casing had been surrounded by nonchloride containing water when the well was shut in, it was surmised that the outer casing had perforated as a result of a poor cementing job, thus allowing the brine from the formation to enter the annulus and thereby attack the production liner. This indicated that severe downhole corrosion could be expected, even though it was not observed in initial field testing due to the formation of the protective hard scale. This corrosion problem will have to be resolved before any successful commercial exploitation of Niland-type brines can take place.

Similar severe corrosion was noted by Berthelot at the Sinclair No. 4 well on a tube string, 1220 m long [46]. Only 336 m were recovered after a stay in the well.

A number of observers have reported that the hard glassy scale formed in the upper casing region protects against corrosion [40, 41]. This may be the case under some operating conditions. However, it was noted that severe attack with the formation of deep, wide pits occurred in the upper sections of a Salton Sea area well liner which had been in place about 2 yr and had operated only intermittently [47].

B. Production Lines and Surface Equipment

Scaling in surface components will usually resemble well deposits chemically, but their form and incidence will depend strongly on flashing, flow, and heat transfer parameters.

A very high rate of scale deposition was noted in the steam separator, valves, and surface piping of one of two nearby Salton Sea area wells. Up to 1-in.-thick crusts formed inside pipes and valves after seven to ten days of production. Relatively minor amounts of sulfides and oxides deposited in the steam lines. In contrast to this, relatively little scale was deposited in surface equipment in the other well.

In three months' operation of well IID No. 1 (Salton Sea area) 5-8 tons of dark scale precipitated in a 275-ft-long line [48]. The major scale component was silica with large amounts of Fe, Ag, and Cu.

In addition to the observation that a large amount of scale can form in a short time, it has been noted that

(1) scale is deposited from the flashing brine, even in high velocity flow zones, and changes in velocity (in the high range) have little effect;

(2) scale deposition in the line from well to separator can be reduced by maintaining as high a pressure as possible in this system.

Within limits, scaling may be localized by proper control of pressure and flow conditions. Berthelot reports that by operating the Sinclair No. 4 well "... choked back to the minimum point of stability necessary to maintain maximum wellhead temperature and pressure ... only an insignificant amount" of scale (1/32 or less) deposited in a five-month continuous flow period [46].

Marsh reported extensive deposits (1 in./week) of a soft scale (mostly silica with some iron oxide) in their separator and other surface equipment [40]. All scale control methods tested were ineffective or made things worse.

Bishop stated that they encountered extensive scaling in their steam heat exchanger [41]. This operation, at the Salton Sea, had an inefficient separator which gave steam with a 40,000–80,000 TDS content. The scale was hard and contained about one-third each of SiO_2, Fe, and Pb. PbS and $CaSiO_3$ were identified in the scale. Attempts to ream out the scale were halted when the reamer broke. Subsequently, with better separation and steam cleanup, they got steam with 20 TDS content and have had no scaling problem.

Bishop reported that their drum-type separator did scale heavily; about a wheelbarrow-full was produced in three months [41]. He estimates about one (24-hr) day downtime a month required for cleanout.

At the same site, Needham observed extensive scaling of corrosion test samples in brine cooled to 160°C [49]. The principal components were Si, Fe, and Pb.

Kuwada reported that at Kizildere (noted above), scaling occurred in the separator and heat exchanger [44]. It was a soft, powdery deposit identified as aragonite ($CaCO_3$). Its rate of deposition was related to the rate of well flow.

In tests using the brine from Mesa 6-1, Wahl reported both silica and carbonate type scaling [35, 50, 51]. With no flashing, $CaCO_3$ scale only was noted, and only at the exit line of the test unit. With maximum flashing, both $CaCO_3$ and SiO_3 (with Fe) scales were found at a probe in their test section. With an intermediate degree of flashing, $CaCO_3$ was deposited on the probe while SiO_2 (+ Fe) was deposited in the exit line.

A review on "Corrosion Characteristics and Geothermal Power Plant Protection" has been published by Allegrini and Bennenuti which essentially summarizes experience at Lardarello [52]. There, corrosion and scaling of steam lines have been encountered. Carbon steel (6–7 mm thick) gives reasonable life. 316 stainless steel was not satisfactory; pitting, stress corrosion, and intergranular attack were all observed. Alkaline washing of the steam is used to deal with these problems. At Lardarello, well-head equipment includes an allowance for the erosive properties of the fluid [52].

At Cerro Prieto, very rapid (2 hr to failure) erosion of elbow sections was observed. The problem was solved by use of sand "T"s [53]. Also, all copper alloys were removed from the plant after corrosion by H_2SO_4 (from S by bacterial action) and H_2S induced stress corrosion cracking were observed [54]. Failure of austenitic stainless steel plugs (possibly by stress corrosion cracking) in an otherwise carbon steel line was observed [53]. Carbon steel used in the shell and piping of the steam separators has shown an acceptably low parabolic corrosion rate.

At a New Zealand plant, several plastics (PVC, polyethylene) exhibited environmental stress cracking [55]. Also at the plant, use is made of Viton or synthetic fiber-reinforced rubber to resist the high temperature and erosive action of the fluid.

Corrosion of the well casing and surface pipes by moderately acid waters was reported as inhibiting the exploitation of the Tatun fields of North Taiwan [55].

For the Salton Sea area, Marsh reported corrosion rates on mild steel of 20 mpy in separated brine, but of 80 mpy in the steam [40]. The difference is attributed in part to the acidic noncondensible gases. Only alloys such as the Inconels held up well in the steam. (Titanium was not readily available at that time; it held up well also). It should be noted that air did enter their system.

Bishop reported that their surface equipment in the Salton Sea area (except for titanium heat exchangers and some valves) is of mild steel [41]. This has been satisfactory for 4–6 months of operation. There was some corrosion and sloughing but no loss of integrity. Some scaling formed which was thought to protect the metal. Instrument lines of type 304 SS developed cracks at bends.

At Kizildere, corrosion was minimized by exclusion of air [44]. The best materials for use were found to be the 300 series austenitic stainless steels. The 400 series stainless steels pitted and also were subject to sulfide stress corrosion cracking. This was also observed in New Zealand, where no chloride stress corrosion cracking of austenitic stainless steels was observed with a $[Cl^-] < 1500$ ppm.

At Heber (Imperial Valley, California), oxygen entry at pumps was found to accelerate corrosion and cause modular scaling [43]. Catalyzed sulfite was used successfully as an oxygen scavenger. Pitting of test coupons and in the gas phase region of two-phase separated brine lines was noted. While carbon steel exchanger tubes scaled, those of titanium showed minimal deposits.

Oxygen (through air entry) was also the principal cause of corrosion in the Reykjavik (Iceland) District geothermal heating system [56]. In this case, sodium sulfite was used as an oxygen scavenger. They also found that a scale of metal oxide and silica (or copper and zinc-coated iron) of or metal oxide, silica, and magnesia (on iron) provided protection against corrosion. The water is fairly noncorrosive, running 200–350 ppm TDS at 87°C with a pH ~ 9.

At the Geysers, the internals of surface pipe remain relatively free of corrosion [57]. Signs of water and dust erosion were found at turns and restrictions in the lines. The problem has been minimized by use of in-line separators. Low carbon steel (A53–Grade B) pipe is used, with cast steel slab gate valves and with stainless steel trim and stems on the steam flow valves.

C. Disposal Lines and Equipment

At Cerro Prieto, New Zealand, Italy [52], El Salvador [58], and elsewhere, silica deposition has been encountered in disposal lines and receiving ponds. The actual deposition process is seen to be relatively slow. At Cerro Prieto [42], acidification with HNO_3 to a pH ~ 1 prevents silica precipitation. In New Zealand [59], treating aged effluent brine with lime led to rapid precipitation of calcium silicate along with arsenic if the brine had been treated to preoxidize the arsenic to the $+5$ valence state. In El Salvador, ponding was used effectively to remove silica from the effluent brine.

Despite this lower temperature, effluent brines when exposed to air often are nearly as corrosive as the hotter ones from the well. This is particularly true of brines containing iron, which species is oxidized to Fe (III), precipitating $Fe(OH)_3$ and lowering the pH drastically. Also, any H_2S remaining in the brine may be oxidized to H_2SO_4.

D. Turbines

Turbines are the heart of geothermal power systems. Considerable care has been devoted to investigation of potential and actual scaling and corrosion problems of turbines, with fairly satisfactory results.

At Cerro Prieto [54], considerable preliminary testing was done in cooperation with Toshiba to select the proper materials and conditions for turbine construction and operation. Subsequent results have proved satisfactory. With proper steam separation, slight blade deposits (SiO_2, NaCl) only have been encountered [60]. Low alloy steel is used for the rotor. A Cr–Mo–V steel has proved better than a Ni–Cr–Mo–V steel which deteriorates in mechanical properties by intergranular corrosion. It was also found that the 12 Cr steel turbine blades suffered from a decreased fatigue endurance limit. This is a widely reported result[4] of operation with geothermal steam, apparently due to H_2S induced corrosion [61]. A modified blade with a 15 Cr–1.75 Mo steel strip and 12 Cr–low C (AISI 410) steel buckets is recommended [61].

In New Zealand, use of low strength steel turbine blades minimizes corrosion effects of H_2S. A 13% Cr steel is used there.

At Lardarello, turbine blades of 12–13.5 Cr steel with 0.5 max Ni and 0.15 max C have been used with a reported 2-yr life. A preferred composition would be 17.5 Cr–14.5 Ni–4.5 Mo with 0.05 C, and 0.30 max Si and Mn.

At the Geysers, conventional turbine, stator, rotor, and blading materials are used [57]. The 12% Cr SS blades have shown repeated fatigue failure. Also, turbine deposits apparently similar to the ones encountered at Matsukawa were found. These have included sodium borate of low enough solubility to remain during operation [62].

---

[4] Encountered at Wanrakei [55], Lardarello [52], Japan [61], and at the Geysers [57]. Notch sensitivity also increases in geothermal steam.

Toshiba has reported that they experienced precipitates and sedimentary materials on their turbines during the initial operation at Matsukawa. Various (soluble) sulfates together with rockdust, clay, and silica were found With a change in steam properties, no further deposits were observed.

E. Condensers

Some of the most serious geothermal corrosion problems reported have been encountered in condenser systems. The combination of mostly acidic noncondensible gases (CO_2, H_2S) plus air entry with oxidation of H_2S to H_2SO_4 is evidently responsible. At Lardarello, cast iron proved satisfactory. Superior results were obtained with AISI 316, but its use was considered economically unfeasible [52].

At Otake (Japan), NaOH was added to the cooling water to prevent corrosion. This led to sludge formation [63].

At the Geysers, aluminum alloy has been used for line hardware while these and austenitic stainless steel proved satisfactory for the condensate. Chemical additions—Fe (III)—to control H_2S emission has led to increased corrosion [57]. Severe localized corrosion in aluminum piping and some pitting of 316 SS pumps and condensers are reported. An extensive investigation is underway to obtain a basic understanding of the operative corrosion mechanisms.

Toshiba reported that 18 Cr–8 Ni (austenitic stainless) steel is used for the barometric condenser trays and steam jet gas ejectors at Matsukawa [64].

At Cerro Prieto, titanium alloys proved superior to the 18 Cr–8 Ni stainless steels in heat exchangers, while fiberglas reinforced epoxy resins proved satisfactory for handling condensate [61].

F. Laboratory and Field Testing

Considerable amounts of pertint test data relative to scaling, erosion, and corrosion problems to be encountered in geothermal systems has been obtained by on-site and laboratory testing. The laboratory studies in some cases preceded tests carried out in the field, while field tests have, at times, served to guide selection of material, design features, and modes of operation in actual production systems. At present, there is probably more laboratory and field data than actual operating data. This is especially true regarding the Salton Sea area where numerous investigations have generated much, mostly proprietary, data, and where there is not yet an operating geothermal system. While the data so obtained is valuable, conditions of tests generally differ from those of actual operation. Allowance must be made for this in interpretation of results.

In conjunction with the design and operation of the Matsukawa (Japan) generating station, extensive corrosion tests were carried out under different conditions [64]: moderate velocity steam; in condensate; and in accelerated steam. Subsequent experience in the operating plant essentially confirmed conclusions from tests. The geothermal steam was found to be more corrosive than conventional boiler steam; however, even carbon steel or cast iron can be used if an adequate corrosion allowance is made. Water droplet erosion is very severe, requiring removal by moisture separators and shielding of the tips of last-stage turbine blades by very hard metals (e.g., stellite). In the condensate, deoxidized copper is preferred to aluminum alloy, while epoxy-coated mild steel showed very good stability.

Two types of scale were encountered on the Matsukawa turbines; precipitates and sedimentary materials. The former, including Na_2SO_4, K_2SO_4, $FeSO_4$, and SiO_2, are soluble materials deposited as temperatures drop. The sediment represents material from the well carried along with the steam. It was concluded that an effective separator is needed to avoid such deposits as they had a great influence on turbine performance [65].

At Cerro Prieto, a similar extensive series of corrosion tests were carried out to select satisfactory component materials [61]. Specially prepared specimens for stress corrosion, erosion, tensile and fatigue testing, as well as atmospheric and general corrosion tests were used. Specimens were exposed to separated steam (as formed, aerated, and condensed both at high and low velocity), and to the atmosphere. It was concluded that corrosion is best prevented by excluding any entry of air into the steam, and by effective separation of droplets from the steam. Stainless steel with greater than 10% Cr is recommended for parts needing high corrosion resistance, while aluminum is satisfactory for less demanding conditions except in condensed steam. Epoxy-type coatings cured at high temperature showed resistance to corrosion and blistering in low and high velocity condensed steam.

The U.S. Bureau of Mines undertook a series of investigations to select suitable materials for use with Imperial Valley (California) geothermal brines. The results of a laboratory study with some commercial alloys using simulated Holtville (Mesa) and Niland (Salton Sea) brines, aerated and nonaerated, have been reported [66]. It was concluded that the Hastelloys, Inconel 625, E-Brite 26-1, Ti-1.7W, and Ti-10V, were the most corrosion resistant of the materials tested.

Subsequent to this study, the Bureau prepared and sent laboratory and corrosion test bed trailers to the Imperial Valley. Tests were made at the San Diego Gas and Electric Company Niland (Salton Sea) site. Considerable scaling was observed with the hot (to 320°C) brine produced. Major scale components were Si, Fe, and Pb. In specimens undergoing much corrosion (e.g., 4130 carbon steel), corrosion products were incorporated in the scale.

Testing was continued at the East Mesa region of the Imperial Valley (California) [49].

A series of laboratory and field tests on scaling were done by Wahl and coworkers [55]. Silica deposition from high salinity brines was found in laboratory tests to increase with increased Reynolds number, Ca^{2+} and Mg^{2+} content, and total CO_2, while such deposition decreased with added silica sand. Field tests on the low temperature, low salinity brines at the East Mesa (California) test site gave predominantly calcite scale, with lesser amounts of a silica plus iron scale. The deposition and location thereof were dependent on the degree of flashing allowed.

As part of their investigation of the total flow concept for geothermal power generation, workers at the Lawrence Livermore Laboratory are carrying out an extensive program to control scale deposition and to select suitable materials for such use [14].

In their study of silica-type deposition from Salton Sea area brines the significance of pH and sulfide precipitation was pointed out. Control of sulfides by oxidation [67] and pH adjustment to prevent silica polymerization [32] have been identified a promising scale control methods.

Erosion of components, especially nozzles and turbine blades, has been recognized as an important matter to be dealt with in a total flow system. Heavily anodized aluminum proved superior to the untreated alloy [68]. Field tests at the Sinclair No. 4 well (Salton Sea area) showed Teflon PFA superior in erosion resistance to TFE and other materials. In laboratory thermal stability tests using nonflowing brines at 300°C, a high degree of aromaticity combined with water-insensitive linkages gives stability. Also stable were high-melting fluorocarbons and highly crosslinked 1, 2-polybutadienes [69].

V. SELECTION OF CORROSION AND EROSION RESISTANT MATERIALS

A. Wells

The key factors in selection of well liner materials are temperatures, salinity, redox potential, pH, and the presence of H_2S, CO_2, and NH_3. Another practical factor is availability of the choice material in the form required.

Mild steel (e.g., J-55) oil well liners have been used for all test, development, and production geothermal wells throughout the world. Apparently, severe problems with this type of liner material have been encountered only in the Salton Sea KGRA. The results so obtained may serve as accelerated or long term indicators for regions of less severe corrosion and as guides if future problems arise.

The casing, cemented in place, serves to maintain the integrity of the bore hole and to permit flow only of fluids from a given depth (i.e., at which the casing ends or is perforated). The severe general attack of carbon steel well casings in their lower regions observed in the Salton Sea area may therefore be dealt with in some suitable cases by relying on the bore itself to transport the fluid to a level at which casing life is satisfactory.

The hard black silica-type scale reported for the upper regions of Salton Sea area geothermal wells, if of good integrity, may serve as a protective coating. Coated steel should also function well provided suitable coatings are found. One key requirement for metals and coatings here is resistance to the erosion by solids-laden brines. One difficulty in using coatings, either natural, forced-natural, or man-made is insuring their complete coverage of the casing. Incomplete coatings may serve to cause crevice corrosion in the vicinity of the break in coverage. Besides ceramic coatings of silica, alumina, and so forth, chromium-coated (plated and diffused into) steel has shown promise.

What is believed to be crevice corrosion has been observed in the upper regions of a Salton Sea area well liner and surface piping. The initiating cause appears to be formation (or existence) of deposits leading to severe attack below them.

Metals chosen as substitutes for J-55 or equivalent steel as well casings must be particularly able to resist pitting, crevice, and stress cracking corrosion in the fluid, as well as possible fatigue failure from the vibration attendant upon geothermal fluid production. The best of the usual austentic stainless steels (e.g., 316L) appear to have adequate resistance to general corrosion, but may not be adequately resistant to crevice and pitting attack. A number of more resistant steels, both austentic and ferritic, as well as other alloys such as Hastelloy C and the Inconels are either commercially available or are promised to be soon if demand indicates. Some of the steels included 18 Cr–1 Ti, 18 Cr–2 Mo, 18 Cr–8 Ni–2 Si, 18 Cr–28 Ni–2 Si, 18 Cr–18 Mn, 26 Cr–1 Mo (either electron beam melted with low interstitial content or more conventional), 29 Cr–4 Mo and 29 Cr–2 Mo–2 Ni, 20 Cr–24 Ni–6.5 Mo–1.5 Mn, and 20 Cr–6 Ni–8.25 Mn–2.5 Mo [25]. The high Cr content coupled with a relatively high Mo content imparts both pitting and crevice corrosion resistance to many of these steels. The ferritic steels, in general, are relatively resistant to stress corrosion cracking. Final choice of a suitable material will require testing under actual operating conditions.

More commonly in surface equipment, but also in wells, entry of air, especially likely during periods of shutdown, may cause corrosion problems which otherwise would not be present. Until or unless foolproof systems to prevent this are used, occurrence must be allowed for in materials selection.

B. Separators and " Demisters "

Separation of steam from brine produces two different fluids of different corrosivity. Entry of air is the single most aggravating corrosion factor. This, coupled with the presence of CO_2, H_2S, etc., can make the steam many times more corrosive than the brine (in which air does not readily dissolve). Where carbon steel is not satisfactory (even with a generous corrosion allowance), any of the steels mentioned for consideration in wells may prove suitable. As temperatures are lower than in wells, other alloys of lower Cr, Ni, Mo, etc., content should also be considered. Several Ti alloys, including Ti–38A, may also be considered. In all cases, testing under operating conditions is essential for proper selection.

C. Lines, Valves, and Disposal Lines

The importance of avoiding erosion has been stressed; with proper design, this should be possible. Carbon steel for lines should be satisfactory if adequate corrosion allowances are made. Air entry will require a much higher degree of corrosion resistance. This is the case mostly in disposal lines. For elevated temperatures, one of the recommended well casing metals may be required. At lower temperatures epoxy and plastic coatings may be used. Use of valve parts of corrosion resistant alloys with carbon steel bodies has been fairly common. This serves to protect the parts (cathodically), while sacrificing the bodies which of course can be made with an extra allowance for this purpose.

D. Turbines

The extensive tests done by Toshiba, and at Cerro Prieto [61] and the Geysers [57] in conjunction with use of Toshiba turbines, as well as by Kawasaki [70] have defined the problems quite well. Aside from problems with deposits, discussed earlier and dealt with in the following Section, the principle corrosion problem encountered is corrosion fatigue of the blades. There are, of course, other problems of erosion and corrosion.

Assuming a high, but practical, level of separation of entrained solids and droplets from the steam, the construction recommended for use at Cerro Prieto [60, 61] appears satisfactory. Provided air entry is avoided, carbon steel is satisfactory for the shell (core). Buckets are made of 12 Cr–low C (AISI 410) steel with strips of 15 Cr–1.75 Mo. A 1 Cr–1.25 Mo–0.25 V steel is used for the stator.

To this it should be added that stellite coatings have been required in some turbines where erosion by condensate droplets is a problem.

E. Condensers

The condensate has been found almost everywhere to be very corrosive, even where oxidation of H_2S has not been done purposely. Some air entry is likely the cause. In any case, the recommended best exchange and piping materials are, in increasing order of preference: cast iron, with adequate corrosion allowance; 316 stainless steel; titanium alloys. Only the latter two are satisfactory if an oxidant has entered the system.

VI. SCALE PREVENTION AND CONTROL

A. Projected Incidence of Scaling

1. Summary

Table 9 summarizes the type of scaling that can be anticipated on major components of the various cycles used for power generation. It should be noted that these are relatively pessimistic predictions; in actual cases some predicted scaling may not occur or may occur to only a limited extent. Also, the lack of actual experience is illustrated by the fact that no two-stage or three-stage system (test or otherwise) have ever been operated on high temperature/high salinity brines. In all systems, a gradation of scaling (type, quantity, hardness) may be anticipated from stage to stage, but in most cases there is no data to tell what will occur.

2. Discussion

a. Dry Steam Systems The deposits of concern in these systems (Geysers, Matsukawa, Lardarello) are mainly those on turbines from solids and droplets carried along with the steam. Proper removal of these prior to feeding the steam into the turbines will minimize deposits. Those which then form will generally be soluble.

b. High Temperature–Low Salinity Brines While containing concentrations of silica comparable to HT–HS brines, they differ in that (1) the rate of silica scale deposition is slower, and (2) there is little or no iron or metal sulfide content in the scale. As exemplified by the operation at Cerro Prieto, wells can be expected to get calcite and silica deposits in the flashing zone. Silica (white) will deposit in separators and brine lines and receiving ponds, especially downstream of the final flashing section.

c. High Temperature–High Salinity Brines In operation with these brines, as encountered in the Salton Sea (Niland) area, scaling in the well with a hard, glassy, black SiO_2 and iron-type scale can be anticipated. Its

TABLE 9

Projected Scaling[a] Sites and Types

					Component					
Fluid	Well	Down-hole pump	Valves and piping	Flash separation	Scrubber de-entrainer	HX	Turbine	Condenser	Brine lines	Pond
DS	N	—	—	—	N ED	—	ED[b]	N	—	N
HT/LS-F	S	—	S	S	S[c]	S[c]	ED[b]	N	S	S
HT/LS-NF	N	N	N	—	—	S	N	—	S	S
HT/HS-F	HBS	—	SBS/HBS	SBS/HBS	SBS[c]	—	ED[b]	N	SBS/HBS	SBS/HBS
HT/HS-NF	N	N	N	—	—	HBS	N	N	SBS/HBS	SBS/HBS
LT/LS-F	C	—	C	C	C[c]	C[c]	ED[b]	N	C[c]	C[c]
LT/LS-NF	N	N	N	—	—	C	N	N	C[c]	C[c]
LT/HS-F	C	—	C	C	C[c]	C[c]	ED[b]	N	C[c]	C[c]
LT/HS-NF	N	N	N	—	—	C	N	—	C[c]	C[c]

Fluid: F = Flashed, NF = Nonflashed

[a] Scale Types: HBS = Hard Black Silica-type
      SBS = Soft Black Silica-type
      S = White Silica-type
      C = Carbonate-type
      ED = Entrained Material Deposits
      N = None

[b] Deposits will not occur if steam scrubbed adequately or binary fluid used.

[c] Incidence and amount of scaling at downstream components depends on amount of prior deposition and effectiveness of scrubbers.

rate of deposition may be quite low ($< 1/2$ in./yr); it will depend in part on well flowrate, with greater deposition accompanying more nearly full flow.

In direct cycles with no heat exchanger, a softer silica-type scale can be expected to form in surface lines, valves, and separator(s). By proper design to minimize changes to the brine at sites other than the separator(s), scaling can be reduced in small components and intensified in the separator(s) from which it may be removed with least trouble. A fair amount of scaling can be tolerated in valves if the proper type with large tolerances and adjustable packing are used [48].

Slow deposition of silica can be expected in brine effluent lines and receiving tanks.

Care will be required to minimize carry-over of brine with the steam. With a TDS level of about 10 ppm in the steam, few scaling problems should occur in the steam portion of the system.

In cycles which have geothermal steam fed to one or more heat exchangers, clean-up of the steam should prevent serious problems with scaling in the heat exchangers. The scaling problem will be the same as in the direct cycles except that no scaling problem whatever should be encountered in the turbines.

It should be noted that in cycles with two or more stages of flash, the scaling problem will not be uniform. However, it is not possible without testing to determine where the regions of heavy scaling will be.

d. Low Temperature–Low Salinity Brines   In all the cycles concerned with this brine, little or no flashing occurs, and the well fluid is mainly a brine. Extrapolating the results reported by Chevron concerning their Heber operation, and the results obtained by several investigators at Mesa 6-1, a rate of scaling dependent on bicarbonate contents may be anticipated. It is likely to be predominantly carbonate or sulfate based, with lesser amounts of silica, iron, and so on. It should be possible to operate with relatively infrequent cleaning of lines, heat exchange tubing, and other components required.

e. Low Temperature–High Salinity Brines   These are not likely to be used as their low enthalpy and high corrosion potential make them the least attractive geothermal fluid. If they are ever exploited, their scaling tendencies will be like those of the other low temperature brines.

B. Methods for Scale Control

1. Introductory Remarks

These may be generally grouped as preventive methods and as control or removal methods. Of the preventive methods, some would function to prevent scale throughout the system while others are implemented at, and are applicable to, individual components. It should be noted that while all have some scientific or empirical basis, few have been tested, especially in geothermal systems, and none has proven to be fully satisfactory. Much more research and development are required to deal with geothermal scaling problems.

The methods worthy of note are listed in Table 10. Some discussion of each method is given below.

2. Discussion

a. Preventive Methods (*1*) *Localized Action* *Ultrasonic Vibration* may serve to agitate the brine interface of a component to prevent scale adhesion. It might have the opposite effect by providing nucleation sites and enhancing transport of brine components to the surface. A transducer (usually electrically driven) is required. The range of action is limited, and the power requirements will be significant. The method does merit testing.

Heat applied to a component will serve to raise brine temperature at the component-brine interface. This may decrease or eliminate supersaturation locally, thus preventing scaling. It is of limited applicability.

(*Plastic*) *Nonstick* surfaces may be coatings or may constitute the component material. They will probably not be wetted by the brine and may require a microscopically smooth surface. Erosion and other mechanisms for roughening the surface will ruin them. PFA Teflon has been reported [69] to show resistance to both erosion and scaling, but some scaling was observed and the testing was of short duration relative to equipment operating life time.

Oiled Nonstick Coatings will have the same characteristics and applicability as would other nonstick surfaces. Oiling may be accomplished by injection upstream of coated surfaces. The constantly renewed surface may enhance the likelihood that such a combination can be found which is effective in controlling scale. Successful operation of such a combination in preventing scaling is reported [71].

Injection of Water into the brine just upstream of and at the component surface to be kept free of scale should prevent scaling for a short distance. The water used may be recycled brine from which scale formers have been removed.

Use of a *liquid–liquid heat exchanger* implies a binary (fluid) cycle. In systems using a heat exchanger, scaling on that component is most deleterious to efficient operation. Preventing such scaling by doing away with the solid surface can make the difference between an operable system and one that is not. The requirements for the binary fluid to which the brine transfers heat are those in general of a binary fluid (high vapor density, large heat of vaporization, normal boiling point below maximum brine temperature, but above ambient temperatures, high but not excessive vapor pressure at maxi-

TABLE 10

Summary: Methods for Scale Control

Method	Scale applicable to[a]	Components applicable to	Requirements		Estimated cost		Likelihood of success
			Equipment	Materials	Cap.	Opr.	
1. *Preventive methods*							
a. Localized action: Prevent scale on individual components							
Ultrasonic vibration	C, S	Any; small preferred	Ultrasonic transducers and electric power	None	High	Mod.	Unknown
Heat	C, S	Any; small preferred	Heater and supply source	None	Mod.	Mod.	Unknown
Plastic; nonstick coating	C, S	Any but heat exchanger	None	Suitable coatings (none yet defined)	Mod.	Low	Unknown
Oiled nonstick coating	C, S	Any but heat exchanger	Oil injection separation subsystems	Suitable coatings (none yet defined)	Mod.	Low	Reported successful
Water injection	C, S	Any; separate injection required for each component	Water injection units (+ heaters if required)	Water or descaled brine	Mod.	Mod.	Unknown
Liquid–liquid heat exchanger	C, S	Heat exchanger	None	Suitable liquid or liquids	Low to high	Low to high	Reports of success; special fluids, equipment needed

b. General action: Prevent scale throughout system

Method	Type		Equipment required	Materials/reagents required			Status/comments
pH Adjustment	C, S	—	Injection and mixing unit; control unit	Chemicals (acid or base); acid resistant materials of construction throughout	Mod. to very high	High	Good
Seeding	C, S requires treatment for each	—	Injection, collection, preparation and recycle units	Seed materials	Mod.	Mod.	C—good S—unknown
Oil injection	C, S	—	Injection and recovery units	Suitable oil (none yet defined)	Mod. to high	Low to mod.	Unknown
Chelating agent	C	—	Injection system (possible: recovery system)	Chelating agent; recovery reagent (e.g., HCl)	Low to mod.	High to very high	Prob. good
Scale preventor or modifier	C, S	—	Injection system (possible: recovery system)	Suitable reagent(s) (none yet defined)	Low to mod.	Mod. to high	None tested so far work
Chemical precipitation (various; see text)	C, S	—	Injection system (possible: recovery system)	Suitable reagent(s) (none yet defined)	Low to mod.	Mod. to very high	Some application promising; others unknown
Chemical transformation	S	—	Injection system; control unit	Reagent (e.g., $SO_3^{=}$)	Mod. to high	High	Unknown
Electric potential	C—yes S—maybe	—	Insulated probes; DC power supply	None	High	Low	Commercial units available for carbonate scale. reported successful on sulfate scale. Unknown on silica scale

(continued)

TABLE 10 (Continued)

2. Control and removal methods

Method	Scale applicable to	Requirements		Estimated cost		Likelihood of success
		Equipment	Materials	Cap.	Opr.	
Scraping (mech. removal)	C, S on large smooth surfaces —	Scraper	None	Mod. to high	Mod. to high	C—good S—proven on soft; not on hard scales
Acid wash	C —	Injection, pumping, and collection system	Acid, inhibitor	Mod.	Mod. to high	Good
Chelating agent wash	C —	Injection, pumping, and collection system	Chelating agent; recovery agent (e.g., HCl)	Mod.	Mod. to very high	Good
Other chemical wash	S —	Injection, pumping, and collection system	Suitable chemicals (e.g., HF)	Mod.	Mod. to very high	Unknown to fair
Cavitation	C, S —	Pump, flow lines, nozzle	Water	Low to mod.	Low	Proven effective in laboratory

a C = Carbonate-type Scale; S = Silica-type Scale

mum brine temperature, chemical inertness) plus immiscibility with the brine. If losses to the brine are a problem, a third fluid may be used between the brine and the "binary" fluid. This fluid should be immiscible with both brine and binary fluid and preferably have the properties of high heat capacity, low vapor pressure, low viscosity, chemical inertness, and low cost. Such a system has been suggested by the Ben Holt Company, Inc.

(2) *General Action*   These methods, once implemented, are designed to prevent scaling throughout the system downstream of use.

Adjustment of pH has been suggested for prevention of both carbonate and silica-type scales. The mode of action is different in each case; indeed, both raising and lowering of pH is thought to offer benefits, though costs and problems in raising pH may be much greater.

Raising pH in a system tending to form carbonate-type scales will result in precipitation of the carbonate (e.g., $CaCO_3$) by the reactions:

$$CO_2 + 2OH^- \rightarrow CO_3^= + H_2O$$

and

$$HCO_3^- + OH^- \rightarrow CO_3^= + H_2O$$

followed by

$$Ca^{2+} + CO_3^= \rightarrow CaCO_3$$

Fe (II), Fe (III), Mn (II), Mg (II), and so forth, can also be expected to precipitate as $Fe(OH)_2$, $Fe(OH)_3$, $Mn(OH)_2$, and $Mg(OH)_2$, respectively.

$BaSO_4$ may also precipitate at the high pH if the solubility product is exceeded. By carrying this operation out prior to a solids removal operation, subsequent scaling in the system may be prevented. The quantities of caustic required will probably be great and the costs high (perhaps prohibitively so).

Raising the pH somewhat (to about 7.5–8.5) in a brine tending to deposit silica-type scales, may increase the rate of sol polymerization to form a gel deposit. At the same time, the precipitation of iron, manganese, aluminum, and other hydroxides and other sulfides may occur, which tends to speed the codeposition of silica. This may prevent (or at least reduce) subsequent silica deposition. The precipitate formed would probably have to be removed prior to further brine processing.

Lowering the pH to a value of 2–4 should prevent carbonate deposition. Dropping pH to 1 was found to prevent silica deposition at Cerro Prieto. It is reported that a somewhat higher pH (~ 4) may be low enough to prevent such deposition and require considerably less in cost of acid and acid-resistant materials of construction. In a total-flow system where brine temperatures are relatively low after passing through the conversion nozzle, such a low pH may be acceptable. In a cycle using hot brine the corrosion resistant material costs may be prohibitive.

Seeding with $CaCO_3$ and with $CaSO_4$ has been tested successfully in preventing scale in seawater desalination. Such seeding may, therefore, be expected to work in geothermal systems producing carbonate-type scale. The concept is based on providing already nucleated sites for crystal growth. The seed must be of the same form as the scale formed. For example, gypsum seed is not effective if calcium hemihydrate is the (meta-) stable form that deposits under operating conditions. With silica, growth is believed to occur by agglomeration of colloidal polymeric silica sol particles rather than deposits of silica molecules on a seed or nucleated particle. However, a large number of quality seed particles, of high surface area with surface defects to provide a growth mechanism path, may tend to seed growth and function to prevent scaling.

Injection of a suitable oil may act to prevent scaling of all types, while at the same time, it may reduce corrosion. The suitability of an oil will be determined by how well it meets the following requirements:

(1) wets the surface preferentially compared to the brine,
(2) provides a nonstick surface,
(3) is essentially immiscible with the brine,
(4) can be easily separated from the brine (i.e., does not emulsify, foam, etc.),
(5) is relatively nonvolatile at brine temperatures,
(6) is chemically stable and inert, and
(7) is nontoxic and inexpensive.

Preliminary testing by the author showed some promising results. The method has promise, but considerable development will likely be required to arrive at a suitable oil.

Chelating agents, such as EDTA, NTA, and so forth, act to form soluble complexes with scale-forming cations, such as Ca^{2+}, Fe^{2+}, etc. They should be effective in preventing carbonate-type scales. They are relatively expensive and recovery for reuse is not simple and not very efficient.

Scale preventive and modifying agents have been tested and proved of some value in preventing carbonate-type scales. Chemicals such as starches, tanins, and a wide variety of reagents have been suggested for this use. None so far has proved of any value in preventing silica-type scale. Some low molecular weight acrylic acid polymers have been successfully tested in preventing $CaSO_4$ scale. These or other related materials should be tested on silica scales, but there have been no tests reported to date.

Chemical Precipitation for scale prevention as it applies to pH adjustment has already been discussed. Addition of sulfides to HT–HS brines, together with some adjustment upward of pH, will precipitate Fe, Mn, Pb, Ag, and so on. This may serve to provide nuclei for silica deposition or codeposition, thus reducing silica scaling. It may function to remove these cations from the brine, leaving behind a metastable system from which silica deposition is slow, as at Cerro Prieto. In either case, scaling may be controlled thereby. No tests have been reported as yet. Addition of lime [Ca(OH)$_2$] has been effective in precipitating calcium silicate fairly rapidly from brine effluents. Tests on lime injection into brine to precipitate silica have been made but the system plugged up with heavy deposits. The method may warrant further study.

Chemical Transformation, in this case specifically sulfide oxidation may transfer a HS–HT brine to a LS–HT brine in its scaling properties. The resulting delayed and less severe deposition of silica would represent a vast improvement. Care in oxidation is required as corrosion problems may be intensified. If residual oxygen is present, they may become very severe.

Devices using *applied low voltage DC* have been marketed widely under the claim that they prevent and even remove scales from water systems. CaCO$_3$ scaling is usually the type involved, though claims are made for CaSO$_4$ scale as well. Anodic DC potential has been shown to nucleate CaSO$_4$ on platinum surfaces. Proper application of this potential may serve to nucleate scale on one surface (anode) and prevent it on another (cathode). Lawrence Livermore Laboratory tests with high DC potential showed no benefits. Tests by the author showed possible benefits on a cathodically polarized SS surface as far as silica scaling is concerned, but deleterious effects on the passivity of the surface since corrosion occurred.

Of the methods for prevention of scale, the use of pH adjustment, seeding (carbonates and sulfates), and local heating are the surest of success. Of the untested methods, oil injection, chemical precipitation (e.g., with S$^=$), chemical scale modifiers, and seeding (for silica) are judged most worthy of evaluation.

b. Control and Removal Methods *Mechanical removal* of carbonate-type scales is relatively easy to carry out. For heat exchange tubes and other regular surfaces, this can provide an effective method for insuring operation, provided the rate of buildup is low enough. Removal of silica-type scales of the hard, glassy type has not been done successfully. The softer silica-type deposits may be removed mechanically, e.g., as at SDG&E's Niland test facility.

Chemical removal of scale can be carried out. Carbonate-type scales may be removed using inhibited strong acids or weak acids, even in carbon steel equipment. Chelating agents (e.g., EDTA) have been used effectively on these types of scale. With silica-type scales, caustic or fluoride washes have been used with some success. Tests by the author showed that the scale is often modified rather than dissolved by such treatment. It may be that the scale will fall away from or be easily removed from surfaces after such treatment. A cavitating water jet has been shown to be both fast and effective in laboratory tests of silica scale removal [74].

If a general scale control method cannot be devised, or is not available, local control of scale formation at individual components will be required. Table 11 summarizes recommendations for such methods.

C. Well Reinjection

The preceding discussion has not concerned itself with problems of reinjection of cooled brine and/or condensate into a well. Such reinjection may

TABLE 11

Summary of Component Scale Control

Component	Recommended control method
1. *Silica scale* (HT–HS *Brine*)	
Well	(1) Allow to scale if slow-forming black silica scale
	(2) Ream out periodically if fast-forming silica or carbonate scale
Valves, orifices	(1) Replace periodically
	(2) Apply heat locally
Piping	Use oversize; clean or replace periodically
Flash evaporator	Clean periodically
Heat exchanger	(1) Avoid use of component
	(2) Clean frequently with reamer; chemicals
Turbine	Clean up steam
Brine discharge lines	Use oversized lines; clean or replace periodically
2. *Carbonate scale* (LT–LS *Brine*)	
Well	Ream out periodically
Valves, etc.	(1) Replace and clean periodically
	(2) Use local control method
Piping and heat exchanger	Clean periodically by reaming or chemical wash
Brine discharge lines	Use oversize lines; clean periodically

be required to prevent subsidence and possibly to recharge the underground reservoir. For this, the principal problems anticipated are (1) plugging of the well porosity with silica and metal (mainly iron) hydroxides, and (2) change in the producing structure through dissolution or chemical reaction, in particular, oxidation.

The limited experience to date has not shown these effects to be problems. However, no long-term reinjection experience is available. It is obvious that to insure trouble-free operation, the injected fluid must have:

(1) a redox potential and pH level near that of the production fluid,
(2) no entrained solids.

For this it may be necessary to process for solids removal and to deaerate or otherwise remove dissolved oxygen or other oxidants. Also, it may be necessary to add caustic or acid for pH adjustment, or modify processes being used for control of scale or corrosion so that the fluid is more suitable for reinjection.

REFERENCES

[1]   W. R. McSpadden *et al.*, *Proc. Conf. Res. Dev. Geothermal Energy Resources* NSF-RA-N-74-159, pp. 213–224 (September 23-25, 1974).

[2]   M. C. Smith, *Proc. Conf. Res. Dev. Geothermal Energy Resources* NSF-RA-N-74-159 pp. 207-212 (September 23-25, 1974).

[3]   A. Calamai *et al.*, *Proc. U.N. Symp. Dev. and Use of Geothermal Resources* pp. 305-313. U.S.E.R.D.A. (1976).

[4]   D. E. White, *Geothermics, Spec. Issue 2* **1**, 58–80 (1970).

[5]   J. Combs and L. J. P. Muffler, *in* " Geothermal Energy " (P. Kruger and C. Otte, eds.), pp. 95–128. Stanford Univ. Press, Stanford, California, 1973.

[6]   C. J. Banwell, *in* "Geothermal Energy" (H. C. H. Armstead, ed.), pp. 41–48. UNESCO, Paris, 1973.

[7]   J. Combs, *Proc. U.N. Symp. Dev. and Use of Geothermal Resources, 2nd* pp. LXXXI-LXXXVI. E.S.E.R.D.A. (1976).

[8]   T. Boldizsar, *Geothermics, Spec. Issue 2* **2**, 99. (1970).

[9]   I. M. Dvorov *et al.*, *Proc. U.N. Symp. Dev. and Use of Geothermal Resources, 2nd* pp. 2109–2116. U.S.E.R.D.A. (1976).

[10] H. C. H. Armstead, *Proc. U.N. Symp. Dev. and Use of Geothermal Resources, 2nd* pp. CXI-CXV. U.S.E.R.D.A. (1976).

[11] R. J. Shannon, *Proc. U.N. Symp. Dev. and Use of Geothermal Resources, 2nd* pp. 2165–2172. U.S.E.R.D.A. (1976).

[12] J. W. Lund, *Proc. U.N. Symp. Dev. and Use of Geothermal Resources, 2nd* pp. 2147–2154. U.S.E.R.D.A. (1976).

[13] W. Fernelius, *Proc. U.N. Symp. Dev. and Use of Geothermal Resources, 2nd* pp. 2201–2208. U.S.E.R.D.A. (1976).

[14] A. L. Austin, *Proc. U.N. Symp. Dev. and Use of Geothermal Resources, 2nd* pp. 1925-1936. U.S.E.R.D.A. (1976).

[15] R. A. McKay and R. S. Sprankle, *Proc. Conf. Res. Dev. Geothermal Energy Resources* NSF-RA-N-74-159, pp. 301–309 (September 23-25, 1974).

[16] J. H. Wolf, U. S. Patent 3,911,683 (October 14, 1975).
[17] B. Holt and J. Brugman, *Proc. Conf. Res. Dev. Geothermal Energy Resources* NSF-RA-N-74-159, pp. 292–300. (September 23–25, 1974).
[18] M. Nathenson, *Proc. U.N. Symp. Dev. Use of Geothermal Resources* pp. XCV-XCVI. U.S.E.R.D.A. (1976).
[19] A. Giovannoni, *Geothermics, Spec. Issue 2*, **1**, 81–90 (1970).
[20] F. Fabbri and A. Giovannoni, *Geothermics, Spec. Issue 2*, **2**, 742 (1970).
[21] V. Cigni *et al.*, *Proc. U.N. Symp. Dev. Use of Geothermal Resources* pp. 1471-1481 U.S.E.R.D.A. (1976).
[22] T. R. W. Final Rep. No. 26405-6001-RU-00, Vol. 2. Prepared for the National Science Foundation, Appendix F3 (December 31, 1974).
[23] M. G. Fontana and N. D. Greene, "Corrosion Engineering." McGraw-Hill, New York, 1967.
[24] F. H. Coley, Extended Abstracts, 142nd National Meeting, The Electrochemical Society, Fall, Miami Beach, Florida, Abstract No. 325, pp. 797–798 (1972).
[25] J. R. Maurer, *Proc. Workshop Mater. Probl. Associated with the Dev. of Geothermal Energy Syst. 2nd.* U.S. Bureau of Mines Grant No. PO 152088, pp. 105–119 (May 16–18, 1975).
[26] F. A. Champion, "Corrosion Testing Procedures," 2nd ed. Wiley, New York, 1965.
[27] F. L. LaQue and H. R. Copson, "Corrosion Resistance of Metals and Alloys," 2nd ed. Van Nostrand–Reinhold, Princeton New Jersey, 1963.
[28] D. deG. Jones and H. G. Masterson, *Adv. Corros. Sci. Technol.* **1**, 1–49 (1970).
[29] "Stability Constants." The Chemical Society London, 1958.
[30] B. J. Skinner *et al.*, *Econ. Geol.* **62**, 316–330 (1967).
[31] A. L. Austin, Lawrence Livermore Laboratory Rep. No. UCID 16721 (1975).
[32] L. B. Owen, Lawrence Livermore Laboratory Rep. No. UCRL-51866 (June 1975).
[33] R. K. Iler, "The Colloidal Chemistry of Silica and Silicates." Cornell Univ. Press, Ithaca, New York, 1955.
[34] R. O. Fournier, *Proc. Int. Symp. Hydrogeochem. Biogeochem. Tokyo, September 7-9* pp. 122–139. The Clarke Co., Washington, D.C., 1970.
[35] E. F. Wahl *et al.*, Silicate Scale Control in Geothermal Brines: Final Report, Garrett Research and Development Co., for the Office of Saline Water, Washington, D.C., Contract No. 14-30-3041 (September 1, 1974).
[36] F. G. R. Gimblett, "Inorganic Polymer Chemistry." Butterworths, London, 1963.
[37] A. G. Walton, "The Formation and Properties of Precipitates." Wiley (Interscience), New York, 1967.
[38] S. Kitahara, *Rev. Phys. Chem. Jpn.* **30**, 131–137 (1960).
[39] Personal observation of the author.
[40] G. A. Marsh, *in* Materials Problems Associated with the Development of Geothermal Energy Resources, USBM Grant No. PO 150296, pp. 10–11. Geothermal Resources Council, Davis, California (May 1975).
[41] H. K. Bishop, *in* Materials Problems Associated with the Development of Geothermal Energy Resources, USBM Grant No. PO 150296, pp. 14–15. Geothermal Resources Council, Davis, California (May 1975).
[42] M. Reed, *in* Materials Problems Associated with the Development of Geothermal Energy Resources, USBM Grant No. PO 150296, pp. 8–9. Geothermal Resources Council, Davis, California (May 1975).
[43] Fred Schremp, *in* Materials Problems Associated with the Development of Geothermal Energy Resources, USBM Grant No. PO 150296, pp. 16–17. Geothermal Resources Council, Davis, California (May 1975).

[44] J. Kuwada, op. cit, *in* Materials Problems Associated with the Development of Geothermal Energy Resources, USBM Grant No. PO 150296, pp. 18–19. Geothermal Resources Council, Davis, California (May 1975).

[45] W. F. Langelier, *J. Am. Water Works Assoc.* **28**, 150 (1936); **38**, 169 (1946).

[46] B. W. Berthelot, *in* Materials Problems associated with the Development of Geothermal Energy Resources, USBM Grant No. PO 150296, pp. 12–13. Geothermal Resources Council, Davis, California, May 1975.

[47] H. L. Recht *et al.*, Evaluation of Corrosion in a Geothermal Well Liner, Abstract No. 118. The Electrochemical Society, Fall Meeting, Las Vegas, Nevada, October 17–22, 1976.

[48] Private communication to the Author.

[49] Paul B. Needham Jr. *et al.*, *Proc. Workshop on Mater. Probl. Associated Dev. of Geothermal Energy Syst.*, *2nd* U.S. Bureau of Mines Grant No. PO 152088, pp. 45–62 (May 16–18, 1975).

[50] E. F. Wahl, *Proc. Workshop Mater. Probl. Associated Dev. of Geothermal Energy Systems*, *2nd* U.S. Bureau of Mines Grant No. PO 152088, pp. 15–43 (May 16–18, 1975).

[51] E. F. Wahl and I-Kuen Yen, *Proc. U.N. Symp. Dev. and Use of Geothermal Resources, 2nd.* pp. 1855–1864 U.S.E.R.D.A. (1976).

[52] G. Allegrini and G. Benvenuti, *Geothermics, Spec. Issue 2* **2**, Part I, 865 (1970).

[53] Personal Observation of the Author.

[54] E. Tolivia, *Geothermics, Spec. Issue 2* **2**, 1596–1601 (1970).

[55] H. C. H. Armstead, *Geothermics, Spec. Issue 2* **1**, 106–111 (1970).

[56] S. Hermannassom, *Geothermics, Spec. Issue 2* 1602–1612 (1970).

[57] F. J. Dodd *et al.*, *Proc. U.N. Symp. Dev. and Use of Geothermal Resources, 2nd* pp. 1959–1964 U.S.E.R.D.A. (1976).

[58] G. Cuéllar, *Proc. U.N. Symp. Dev. and Use of Geothermal Resources 2nd* pp. 1343–1348. U.S.E.R.D.A. (1976).

[59] H. P. Rothbaum and B. H. Anderton, *Proc. U.N. Symp. Dev. and Use of Geothermal Resources* pp. 1417–1426. U.S.E.R.D.A. (1976).

[60] A. Mañom, M., *Proc. Workshop on Mater. Probl. Associated with the Dev. of Geothermal Energy Syst.*, *2nd* U.S. Bureau of Mines Grant No. PO 152088, pp. 69–85 (May 16–18, 1975).

[61] E. Tolivia M. *et al.*, *Proc. U.N. Symp. Dev. and Use of Geothermal Resources*, pp. 1815–1820. U.S.E.R.D.A. (1976).

[62] M. Reed, Oral presentation at workshop, *Proc. Workshop on Mater. Probl. Associated with the Dev. of Geothermal Energy Syst.* U.S. Bureau of Mines Grant No. PO 152088 (May 16–18, 1975).

[63] H. Yasutake and M. Hirashima, *Proc. U.N. Symp. Dev. and Use of Geothermal Resources*, *2nd.* pp. 1871–1877. U.S.E.R.D.A. (1976).

[64] H. Nakanishi *et al.*, Toshiba Review (November 1970).

[65] H. Yoshida *et al.*, *Proc. Am. Power Conf.* **30**, 965–973 (1968).

[66] S. D. Cramer *et al.*, *in* Materials Problems Associated with the Development of Geothermal Energy Resources, USBM Grant No. PO 150296, pp. 20–28. Geothermal Resources Council, Davis, California, May 1975.

[67] D. D. Jackson and J. H. Hill, Lawrence Livermore Laboratory Rep. UCRL-51977 (January 19, 1976).

[68] A. L. Austin *et al.*, Lawrence Livermore Laboratory Rep. UCRL-51366 (1973).

[69] L. E. Lorensen *et al.*, *Proc. U.N. Symp. Dev. and Use of Geothermal Resources, 2nd* pp. 1725–1731. U.S.E.R.D.A. (1976).

[70] S. Okazalsi and H. Nakamura, On Corrosion Fatigue of Steels in Geothermal Steam. Kawasaki Heavy Industries, Ltd., Kobe, Japan (July 1975).

[71]   R. W. Erwin, U.S. Patent 3,891,496 (June 24, 1976).
[72]   A. P. Thiruvengadam, Cavitation Descaling Techniques for Geothermal Applications, pre-
       sented at *Conf. on Scale Management in Geothermal Energy Dev.* sponsored by the NSF
       and administered by the USERDA, Univ. of California at San Diego (August 2–4, 1976).

Chapter 6

Materials for Thermonuclear Fusion Reactors

DIETER M. GRUEN*

ARGONNE NATIONAL LABORATORY
ARGONNE, ILLINOIS

* Based on work performed under the auspices of the U.S. Energy Research and Development Administration.

325

I. INTRODUCTION

Nuclear energy obtained from the thermonuclear fusion of light nuclei is a goal to which an increasing world-wide effort is being committed. The demands on energy reserves and resources are continually increasing as ever more countries achieve modern industrial status. All projections agree that conventional means of energy production must be supplemented and indeed supplanted by new methods. Only the date at which the transition becomes imperative is subject to debate. The promise of fusion energy ultimately to provide a clean, cheap, dependable and potentially inexhaustible energy source provides potent motivation for a large, long range attempt at fulfillment.

If there were illusions at the start of the quest for controlled thermonuclear power that solutions would be easily found, the past two decades have dispelled them. Unwarranted optimism has been replaced by a realistic recognition of the immense scientific and technological challenges that arise in bringing about practical fusion energy. Broadly speaking, the problems can be put into two categories—those having to do with heating the fuel to thermonuclear temperatures at high enough particle densities and for sufficiently long confinement times to yield a net power return and those having to do with the actual construction of a power producing fusion reactor.

Most of the past and present fusion effort is devoted to the solution of the first set of problems. Indeed this must be the case at least until the scientific feasibility of the fusion power concept is demonstrated. However, as optimism for the eventual success of the project grew in the last few years, sparked by the outstanding performance of Tokamak-type devices, the need for a closer examination of the problems associated with the operation of an actual fusion reactor has come to be felt ever more strongly.

It has been appreciated for a long time that to obtain a net energy gain from fusion reactions, plasma must be contained at the ignition temperature for a minimum time, t, which depends on the plasma density, n. The requirement is given approximately by $nt > 10^{14}$ cm^{-3} sec (Lawson criterion). Since the power density is also a function of the plasma density, there is a wide range of possible reactor conditions, varying from low density state systems ($n \approx 10^{14}$, $t \approx 1$ sec) to explosive releases of energy at very high densities ($n \approx 10^{24}$, $t \approx 10^{-10}$ sec).

Using magnetic confinement, a wide range of geometries are possible involving either open ended or closed configurations. Both theory and experiments suggest that end losses from the open systems will make it difficult to fulfill the Lawson criterion without a high level of external recirculating power, so that the economics of a reactor based on such configurations have to take this factor into account. Closed geometries have theoretical containment times which scale with the physical size of the system and much of the work in the United States and abroad in plasma containment experiments focuses on several alternatives including stellarators, tokamaks and toroidal pinches.

In order to place the material requirements for fusion reactors in perspective, a brief discussion will be given of conceptual designs for both a Tokamak and a θ-pinch reactor. Conceptual design studies are currently being actively pursued in several National laboratories and universities for both the inertial (laser and relativistic electron beam heating) and magnetic confinement approaches to thermonuclear fusion. The selection of systems singled out here for discussion is meant to be purely illustrative.

A. Design Features of a Conceptual D–T Tokamak Reactor

The conceptual design [1, 2] of the Princeton Plasma Physics Laboratory (PPPL) Tokamak reactor is based on a "steady-state" machine in which the D–T plasma is brought up to "ignition" temperature by a pulse from a 3000-MJ ohmic-heating supply which supplies about 350 MW to a set of circular conductors parallel to the toroidal plane, above and below the vacuum chamber. The current induced in the plasma is 15 MA. At ignition the α-particles from the D–T reactor, which are trapped in the plasma, heat it by collisions with the D–T ions to such an extent that the burning can be maintained in a steady state as fuel is introduced in the form of 1-mm solid D–T pellets with argon cores. The plasma attains a temperature on axis of 48 keV at a density of 1.0×10^{14} cm^{-3}, corresponding to a magnetic field, β_θ(max), on axis of 0.14. Fuel "passes through" the plasma with a burn-up fraction of 2.9% per pass, corresponding to an average ion confinement of 2.8 sec. Thus a D–T feed rate of 34 kg/day (corresponding to a tritium inventory of 5.6 kg in the plasma and blanket) results in the production of 1.3 kg of He^4 "ash" per day.

In order to maintain a low level of He^4 in the plasma and a constant plasma density, the plasma must be continually removed at its outer periphery. This is accomplished by means of the diverter. Here the cross-sectional shape of the magnetic lines is not circular at the outer edge of the plasma. Instead the lines have a cusp at the inside of the vacuum chamber, so that plasma diffusing across to these magnetic lines is then led along them and up (down) to the diverters where the spent plasma is collected and its heat removed.

In this conceptual design the plasma has a major radius $R = 10.4$ m, minor radius $a = 3.2$ m, and a volume of 21100 m^3. The toroidal field β_θ is 60 kG at the plasma center and is furnished by 3.5×10^8 ampere-turns of

Nb$_3$Sn superconductor arranged in 50 D-shaped coils. A plasma q value of 2 is chosen.

The stainless steel first wall is cooled by a neutron-moderating and absorbing blanket, 0.80 m thick, which also protects the superconducting magnet coils. In order to avoid the high pumping pressure and losses associated with liquid lithium flow across the magnetic field, the blanket fluid (operating at 1000°C is chosen to be FLIBE (Li$_2$BeF$_4$) which is relatively nonconducting. The blanket is cooled by a flow of helium in 2.2-cm stainless steel tubes. The helium exists at 620°C and exchanges with steam which is the working fluid of a set of turbine generators. In the power balance 4790 MW are produced by the neutrons, blanket reactions, electromagnetic radiation and plasma in the diverter. Of the 2350 MW of electrical output from the thermal convertor (efficiency = 49%), 1920 MW provide useful output, with an 18% recirculating power fraction accounted for by helium pumping and other auxiliary uses. The net plant efficiency is 40%. The reactor core is enclosed in a concrete biological shield which in turn is enclosed in a shell to prevent the escape of tritium to the atmosphere.

B. Design Features of a Conceptual θ-Pinch Reactor

The θ-pinch Reactor [3, 4] is designed for repetitive pulsed operation. A shock-heating magnetic field is applied suddenly and sustained for a time sufficient for the equilibrated plasma to be "picked up" by the compression field. The rise time of the compression field is taken as 10 msec, and the compression field is held constant during the burn time of 100 msec. At the end of the burn the plasma contains approximately 10% helium ions. The magnetic field is then relaxed to some lower value which allows expansion of the plasma column radially to the vicinity of the wall and extinguishes the burn. Neutral gas flows between the wall and the plasma boundary, removing heat from the column and neutralizing the plasma. During the remainder of the cycle "off time," the plasma and hot gas are flushed out of the system and replaced by fresh plasma with negligible helium content.

The ratio of the cycle time τ_c to the burning time τ (inverse duty factor) determines the average thermal power loading P_W on the reactor first wall, which is the implosion-heating coil. The cycle time $\tau_c = 10$ sec is chosen to limit P_W to 3.5 MW/m^2 (350 W/cm^2), giving a reactor duty factor of 1%. During the off-time of the reactor cycle, heat from liquid lithium at the first wall and in the blanket reduces the material temperatures from their temperature excursions during the burning pulse to their ambient values.

An important feature of the pulsed reactor with a small duty factor is that during heat transfer the lithium need not be pumped across the magnetic field, and it does not incur electromagnetic pumping losses. Also, during heat transfer, the lithium flow is not along magnetic lines, since the field is absent. Thus the flow retains its turbulent character, and the field-free heat transfer coefficients apply.

The implosion heating coil with a number of radial transmission line feeds is surrounded by a Li–Be–C blanket which has three functions: to absorb all but a few percent of the 14-MeV neutron energy from the plasma, protecting the compression coil and providing heat for the electrical generating plant; to breed tritium by means of the $^7Li(n,n\alpha)T$ reaction for fast neutrons and the $^6Li(n,\alpha)T$ reaction for slow neutrons; and to cool the first wall.

Outside the inner blanket region is the multiturn compression coil which is energized by the slowly rising current from the secondary of the cryogenic magnetic energy store. The compression coil consists of the coiled-up parallel-sheet transmission lines which bring in the high voltage to the feed slots of the implosion-heating coil.

Next after the compression coil and its titanium coil backing comes the remainder of the neutron blanket for "mopping up" the last few percent of neutron energy and breeding the last few percent of tritium. Unlike the inner blanket, which would run at $\sim 800°C$ to provide high thermal efficiency of the generating plant, this portion of the blanket could run substantially cooler. Surrounding the outer blanket is a radiation shield, and beyond the shield the radially emerging transmission lines are brought around to make contact with the secondary coil current feeds and the high voltage implosion heating circuits.

The ceramic coating of the first wall and of the separate metal cells of the blanket may be polycrystalline oxide or glassy ceramic. The coating must hold off electric fields of the order of 6 kV/cm and retain its insulating properties at $\sim 800°C$ under the integrated effects of neutron radiation damage. However, the coating is not required to hold off high voltage during the intense radiation of the burning pulse, only during the shock-heating phase which precedes it when there is little radiation from the plasma.

The performance of ceramic coatings on the first wall of θ-pinch reactors exemplifies the stressfull conditions imposed by the fusion reactor regions on materials of construction.

Statements such as the following, taken from the U.S. National Academy COSMAT Summary Report (1974) on "Materials and Man's Needs" voice the concern felt by some about materials requirements for fusion reactors:

"Engineers and designers have grown steadily more confident that new materials somehow can be developed, or old ones modified, to meet unusual requirements. Such expectations in the main have been justified, but there are important exceptions. It is by no means certain, for example, that materials can be devised to withstand the intense heat and radiation that would be involved in a power plant based on thermonuclear fusion, although the fusion reaction itself is not primarily a materials problem".

In view of these sobering considerations, it is probable that only a well-conceived, interdisciplinary effort drawn from all fields of science and engineering can hope to cope with materials challenges of this magnitude. In this chapter, some of the major materials problem areas are discussed. Only the highlights can be touched on in this chapter as a considerable amount of work has already been done on materials relevant to fusion reactor operations.

II. EFFECTS OF RADIATION ON CTR MATERIALS

There are two main classes of radiation effects problems that must be considered: bulk and surface or near surface effects. The former result primarily from high energy neutron bombardment and the latter from photon and primarily charged particle irradiation. Although the magnitude of the problems due to each type of damage is not necessarily the same nor are the effects necessarily independent of each other it is easier to discuss the two subjects under separate headings.

A. Bulk Radiation Effects

The high flux of 14-MeV neutrons bombarding the first or vacuum wall of a fusion reactor can result in swelling due to transmutation reactions leading to He bubble formation and to displacement damage leading to void formation and precipitation of interstitials. Mechanical property changes lead to radiation hardening resulting in higher tensile and yield strength of metals, enhanced creep rates, increase in ductile-to-brittle transition and helium embrittlement.

The challenges facing materials development in coping with these effects are manifold since one cannot as yet predict the amount of swelling or define material compositions and operating temperatures to minimize swelling. In any event, the following considerations must be kept in mind:

(1) Swelling will be very dependent on the purity, composition and metallurgical state of the material.

(2) Synergistic effects of transmutation products, interstitial impurities, displaced atoms and accumulation of He and hydrogen isotopes on mechanical properties remain to be explored.

(3) The desired goal is the manipulation of the available parameters to develop an alloy, probably a very complex alloy, that will provide an optimum reduction of swelling and resist degradation of mechanical properties.

In the case of fusion reactors of the θ-pinch type, the first wall must be constructed of a very good insulator to withstand the several keV/cm electric field gradients that are developed during a certain phase of the pulse. Most likely, the walls would have composite structures consisting of insulating ceramic coatings applied to refractory metals. Neutron-induced radiation effects on insulators are less well understood than in the case of metals. Some of the areas which will require extensive work therefore will have to deal with neutron-induced atomic displacements and transmutation which can result in dimensional changes, electronically active point defects, and departures from stoichiometry. It remains to evaluate the detailed behavior of transmutation products in insulators with respect to their tendency to remain in solution, precipitate, escape via diffusion or migration, stabilize defect structures, change the electrical conductivity, and affect the thermal conductivity.

In all likelihood, new classes of insulators with desirable properties will have to be developed. Materials, such as yttria, which have intrinsic defect structures could conceivably be substances with an inherent resistance to swelling. The insulator should generally have good thermal and electrical properties at high temperatures, sustain irradiation without unacceptable degradation of electrical and mechanical properties, not undergo large dimensional changes as a result of irradiation and the buildup of transmutation products, and resist chemical attack at elevated temperatures by hydrogen isotopes. The insulator must also be chemically and mechanically compatible with the metal substrate.

1. Metals

The first wall will probably be constructed from an alloy based on a bcc refractory metal. The choice seems to be restricted to those alloy systems based on niobium, vanadium, and molybdenum. The discussion of damage will be separated into the two broad topics of component swelling and loss of ductility.

The swelling produced by irradiation will be important because the resulting changes in linear dimensions must be accommodated between the vacuum wall (which will be dimensionally unstable) and the connected vacuum, fuel handling, heat exchange, magnet and support systems. The effect of swelling will be magnified greatly if gradients in neutron flux and/or temperature are present over the first wall. Successful reactor operation will require a wall material that retains at least limited ability to deform without fracture after several years of bombardment by the intense neutron flux of the CTR environment.

The CTR flux is more damaging than an equivalent fission reactor flux, both in terms of the production of transmutation products by inelastic processes and in terms of the production of displacement damage by elastic scattering processes. The transmutation reactions, dominated by the highest energy 14-MeV neutrons, produce hydrogen, helium, and elements with atomic numbers near that of the target metal. Of these impurities produced in the wall material, helium is believed to be the most damaging. At the temperatures at which a CTR first wall will operate, both vacancies and interstitials will be mobile. Most of the defects produced will be lost by annihilation at sinks or by recombination, but the few that survive in stable defect aggregates produce technologically important property changes. The nature of the stable damage produced under expected CTR irradiation conditions will be discussed below. Detailed calculations of damage rates under neutron irradiation have been made and the results evaluated by comparing the damage effectiveness of 14-MeV neutrons compared to neutrons of lower energies [5, 6]. Such calculations, necessary to make a comparison of fission reactor data with expected results in CTR irradiation, show that in niobium a 14-MeV neutron is 2.5–4 times as damaging as a 1-MeV neutron.

The service requirements of the first wall will, to a large extent, determine both the type and concentration of radiation damage to be expected in the wall material. By far the most important of these parameters (aside from the neutron flux) is the operating temperature. The range of possible operating temperatures that has been suggested is 500–1200°C, with a more likely range for the eventual choice being 600–1000°C. The choice of reactor coolants, with helium, lithium, potassium, and molten salts having been suggested, will influence the choice of materials for the first wall but will have little or no effect on the radiation damage processes. On the other hand, tritium concentrations maintained in the CTR metal components could influence the radiation damage, but neither models nor data to establish possible effects are presently available. The following discussion is based on considerations given by Wiffen to radiation damage in CTR's [7].

a. Radiation Produced Swelling   Swelling in metals will occur both as a result of the transmutation and the displacement component of the CTR first wall radiation damage. Some information is available on swelling due to either of these processes separately, but none on the possible synergistic effects.

Transmutation reaction induced swelling is mainly a result of the precipitation in bubbles of the helium from α-producing reactions but it can also be due to the presence of the solid transmutation products themselves. The latter effect arises from the slightly different lattice parameter of the alloy compared to the material before transmutations occur.

Swelling produced by displacement damage in the approximate temperature range 0.25–0.5 T_m (where T_m is the melting temperature of the metal) results from the precipitation of vacancies to form voids, accompanied by the precipitation of interstitials in dislocation loops. The largest amount of swelling observed to date is 11% in stainless steel irradiated by 1.45×10^{23} neutrons/cm^2 (0.1 MeV) at 410°C. The maximum amount of swelling reported in a bcc metal is 4.8% volume increase in niobium at 585°C. The available data indicate that the swelling produced by void formation, acting alone, is at least an order of magnitude greater than the swelling produced by the formation of helium bubbles under equilibrium conditions.

b. Irradiation Effects on Mechanical Properties Among the changes produced in mechanical properties, the loss of ductility will have the most serious consequences. Very little data are available on mechanical properties of the bcc refractory metals irradiated at temperatures and fluences representative of CTR service. This section will consider briefly the effect of irradiation on strength properties, ductility and creep.

During irradiation at temperatures below about 0.5 T_m radiation damage in the form of defect clusters accumulates in the metal lattice. In general, damage builds up more rapidly the lower the temperature, and at a fixed temperature increases with increasing fluence. Defect clusters, whether dislocation loops or cavities, provide obstacles to the dislocation motion which is the atomic scale process responsible for plastic deformation. As a result, all measurements of strength properties show increases produced by neutron irradiation. If the strength properties are measured in a tensile test, the yield strength is increased, often to several times the pre-irradiation value. The ultimate tensile strength is also usually increased, but is not affected as much as the yield strength. In a creep test the strength increase produced by irradiation is usually reflected in a decreased minimum creep rate [8]. A more detailed discussion of the mechanism responsible for irradiation hardening is given in Moteff's paper [8].

All of the bcc refractory metals show a transition from low temperature brittle behavior to higher temperature ductile behavior at a temperature known as the ductile-to-brittle transition temperature (DBTT). The DBTT, dependent on the exact condition of the specimen being tested and on the method of test, is usually raised by any process that hardens the metal including neutron bombardment. It has been demonstrated in molybdenum that low temperature, low fluence neutron irradiation can produce DBTT shifts of up to +120°C [9]. Data for higher temperature irradiation to 3.0×10^{22} neutrons/cm^2 (> 0.1 MeV) show that fractures may be completely brittle, with zero elongation, to test temperatures at least as high as 550°C. While the available data on irradiated niobium and vanadium do not define a change in DBTT for these metals [9], some increase would be anticipated. The low values of the DBTT for the unirradiated metals suggests that the effect may be of less importance in these cases than it is in molybdenum.

Neutron bombardment causes a decreasing ductility with increasing fluence, accompanied by an increase in the strength property. Examples of ductility loss in tensile tests are known for irradiated Mo, Nb, and Nb–10% Zr [10], V [11] and austenitic stainless steels [12].

Helium produced in transmutation reactions also has an effect on reducing the ductility of many metals. In fcc metals the helium generally becomes detrimental to ductility at temperatures above about 0.5 T_m [13, 14]. In the only bcc metal (other than commercial steels) in which the effect of helium was studied, a concentration of 1×10^{-6} atom fraction helium introduced by α-implantation was found to reduce the tensile ductility of vanadium alloys at test temperatures above 800°C (about 0.5 T_m) but to have no effect below this temperature [15].

Helium embrittlement effects have not been seen in reactor-irradiated bcc metals because fluences have probably been too low to produce significant amounts of gas, and test temperatures have generally been below the 0.5 T_m level at which the effect might become important.

It is likely that metals stressed in the irradiation environment of the CTR will deform by creep at rates higher than would be predicted from results of out of reactor tests. The creep rate, set by the rate of motion of dislocations, will be enhanced by the very high concentrations of vacancies and interstitials in the metal and the high flux of these defects to the dislocation [16]. It is possible, too, that deformation in the irradiation field may enhance the amount of deformation that a metal can sustain before failure. This very important consideration, which might provide some relief from the severe limitations on ductility determined from postirradiation testing, has not been studied experimentally in any detail.

c. Scope of Developmental Work   One of the most crucial pieces of information that must be obtained before any large scale fusion reactors can be built is the synergistic effect of interstitial impurities, displaced atoms and high energy neutron reaction products such as He and H, on the *ductility* of potential CTR materials. This data is required over the temperature range from 500–1000°C for Nb, Mo, V, and alloys of the refractory metals with Ti, Cr, and Zr and stainless steel. Information is also required on mechanical and physical properties of Cu and Al during lower temperature (RT to 300°C) high fluence irradiations for pulsed systems. The effect of irradiation-induced defects on the creep properties and fatigue life of the proposed metals and alloys is vital to the design of reliable structural components. An almost equally important area is the effect of gases such as hydrogen and helium on the nucleation, growth and thermal stability of voids in metals. Measurements of the resistivity of Cu and Al after neutron fluences of 10^{22} neutrons/cm^2 and data on the stability of superconducting magnets are required. More information on the characteristics of graphite irradiated to more than 2–4 $\times$ 10^{21} neutrons/cm^2 is needed with particular emphasis on dimensional stability. A considerable number of experiments in the near term must be performed utilizing current fission reactor and accelerator facilities. In order to properly interpret these experiments, it is necessary to correctly assess the damage produced by 14-MeV and fission spectrum neutrons as well as by high energy heavy ions. Such an analysis can be accomplished theoretically and by carefully designed experiments. *In situ* experiments of irradiated insulating materials must be conducted at high temperatures in order to assess any electrical degradation problems that might arise.

2. Insulators

Important consequences of neutron radiation on the bulk properties of insulators will result from atomic displacements and the introduction of lattice impurities by neutron-induced transmutations. These phenomena can result in dimensional changes, the formation of electronically active point defects, and departures from stoichiometry. Whether any of these effects will lead to insulator failure must be determined experimentally for specific candidate materials. Many of the anticipated problems can be solved, alleviated, or circumvented by alterations in mechanical design, chemical composition, and microstructure of the first-wall insulator.

The major task of the first-wall insulator is to prevent electrical breakdown between the plasma and the blanket segments during the implosion heating stage. Once the plasma is compressed and ignited, radiation effects at the first wall could conceivably lead to a significant increase of electrical conductivity. *The insulator, however, is not required to sustain large voltage gradients after the implosion heating stage.* The design dielectric properties of the insulator must be regained, within about 10 sec after the plasma has cooled. Re-establishment of these properties after periodic intense irradiation represents a major, short-term materials problem for the θ-pinch reactor. Insulators which anneal rapidly could prove relatively insensitive to pulsed reactor radiation damage insofar as electrical properties are concerned. To accelerate the recovery of dielectric properties, manipulation of the defect structure so as to alter charge trapping and detrapping kinetics may be possible.

Neutron induced transmutations could eventually change the stoichiometry of the insulator. In the case of Al_2O_3, transmutation of aluminum and oxygen changes the oxygen/aluminum ratio slightly in the direction of hyperstoichiometry. These changes are accentuated by the build-up of other impurities such as MgO which may form at grain boundaries, form solid solutions or precipitate as spinel. Other impurities such as hydrogen, carbon, and helium are formed at even greater rates than magnesium. In general, the role of transmutational impurities on electrical properties will depend on whether the impurities remain in solution or precipitate. Some commercial refractory insulators (including alumina), however, already contain significant concentrations of impurities without serious reduction in electrical performance. This behavior is exhibited also by many glassy materials, which represent strong candidates for the first-wall insulator. Some impurities could exert beneficial effects on the electrical properties of the insulator. Impurities may behave similarly to displacement-generated point defects if they remain in solution. Transmutation products could cause localized field effects and premature dielectric breakdown if precipitated as well as lead to a degradation of structural properties.

The behavior of impurities in insulators can be expected to be vastly different from one insulator to another. For example, transmutation induced helium can escape rapidly from some glasses, but might precipitate as gas bubbles in ceramics. The use of gettering or precipitating agents in the insulator to control deleterious effects of transmutational impurities can be simulated by ion injection of hydrogen, helium, carbon, and magnesium or by using sputtering techniques to build-up insulator samples while simultaneously introducing impurities.

Neutron bombardment creates vacancies and interstitials, which as isolated point defects or as defect clusters can result in volume changes. The changes are usually small and may anneal at reactor operating temperatures. In anisotropic materials, defect agglomerates may orient in preferential crystalline directions and cause swelling in preferred directions. For example, it has been found [17] in fast fission reactor exposures of single-crystal Al_2O_3 at $1023°K$ that swelling along the c-axis is $\sim 1.8\%$ after a fluence of $\sim 5 \times 10^{25}$ neutrons/m² (> 0.1 MeV), whereas along the a-axis the swelling is only $\sim 0.2\%$. Such swelling in polycrystalline materials with randomly oriented grains has led to grain separation [18]. The problem possibly can be alleviated by orienting the grains with their maximum swelling directions perpendicular to the substance where some dilation can be tolerated. One possible method to obtain such an orientation would be by utilization of sputtering deposition techniques. Since cubic materials are not subject to anisotropic swelling, insulators with this structure may prove preferable to Al_2O_3.

It has also been found that the use of fine-grain material leads to a reduction of swelling in Al_2O_3. For this case, also, sputtering techniques could be helpful, since sputtered materials often have very small grains.

Recent evidence indicates that metals can be stabilized against swelling by the introduction of small precipitates. The precipitates reduce swelling apparently by inducing a more efficient recombination of vacancies and interstitials or by stabilizing the defect structure. Since the mechanisms of swelling in metals and ceramics are similar, this technique holds considerable promise in controlling swelling in insulators.

Important to note is that a small amount of first-wall swelling can be tolerated by designing the reactor to accommodate slight dimensional changes. Of course, the insulator and its metal substrate must undergo similar dimensional changes to prevent separation or fracture. Evidence exists that materials with an intrinsic defect structure exhibit an inherent resistance to swelling. Yttria represents an example of such a material [18, 20].

Heat transport in opaque insulators is almost entirely by lattice (phonon) conduction, and lattice damage caused by neutron irradiation of crystalline materials disrupts this conduction. Hence, another problem may be a decrease in thermal conductivity resulting from radiation damage. For example, it has been found [19] that the thermal conductivity of Al_2O_3 decreases by a factor of ~ 2 after exposure to a fission reactor fluence of 10^{26} neutrons/m^2 (> 0.1 MeV) at 973°K. The higher temperature at which the first-wall insulator is expected to operate may substantially reduce this effect by accelerating the annealing rate of damage.

The effect of photon bombardment on insulators is highly variable; some dielectrics (mainly alkali halides) have been found to suffer severe displacement damage from ionizing radiation, but the mechanism thought to be responsible (the Pooley–Hirsch effect) is not operative in most ceramic materials. Lattice damage from photons usually results only from the inefficient process of high energy photo-electron impacts with ions. Photon-induced displacements therefore could only be caused by gamma rays, and this contribution is calculated to be small relative to neutron-induced displacements. The creation of photo-electrons, however, will be much more plentiful than the ionization electrons generated by neutron bombardment and may contribute strongly to electronic charge transport. This induced conductivity is expected to be highly transient and to decay with a time constant which is much shorter than the reactor duty cycle.

B. Surface Effects of Radiation

During the operation of a thermonuclear fusion reactor, components such as the first wall, blanket shield, diverter, and beam dump of the injector region will be exposed to primary and secondary plasma radiation. The following discussion will deal with certain phenomena that may occur when hot surfaces are struck by energetic particles and photons.

The energetic particles and photons emanating from the plasma cause a variety of physical and chemical processes on wall surfaces. These processes include physical and chemical sputtering, secondary electron emission, x-ray emission, backscattering of particles and photons, release of absorbed and adsorbed gases, radiation blistering, photo-decomposition of surface compounds, particle entrapment, re-emission of trapped particles, and the like.

The totality of these phenomena can lead to serious wall erosion with subsequent mechanical failure and to plasma contamination leading to excessive power losses due to increased bremsstrahlung and recombination radiation [20–23]. In contradistinction to noble gas projectiles (e.g., He^+), D^+ and T^+ are chemically reactive species. Bombardment of metal and insulator surfaces with kilovolt D^+ and T^+ can lead to various "chemical" phenomena such as chemisorption, "trapping," "chemical" sputtering and hydride molecule formation.

The high temperatures of the fusion reactor walls not only may enhance the yields of some of the processes mentioned above (including gas permeation and diffusion) but in addition may cause vaporization of the target material and of the imbedded impurities, thermal desorption, thermal emission of electrons and ions, and whisker growth [24–27].

Plasma contaminants have an important effect on reactor power losses due to bremsstrahlung, and therefore impurity control is a vital aspect of the operation of fusion reactors making use of the magnetic confinement principle. For a hydrogen isotope plasma ($Z_1 = 1$), the ratio R of the power losses with and without the contaminant is $R = 1 + f(Z_2 + Z_2^2) + f^2 Z_2^3$, where f is the fractional concentration of the impurity and Z_2 is its atomic number. One notices that R increases rapidly with Z_2. On the assumption that the efficient operation of a fusion reactor limits the increase in the power loss that can be tolerated due to bremsstrahlung to 10% ($R = 1.1$), this expression can be solved for an upper limit of f. For fully ionized impurity atoms, some such limits are $f_{max} = 4.9 \times 10^{-3}$ for Be, 1.8×10^{-4} for V, 5.8×10^{-5} for Nb, and 5.5×10^{-5} for Mo. For any impurity whose atoms are not fully stripped, the values of f_{max} must be kept significantly smaller than those indicated.

It has been suggested that the structural integrity of the vacuum wall would be compromised if more than 20% of its thickness were lost. If a wall thickness of 1 cm is assumed and a wall lifetime of 20 yr is considered as desirable, then the maximum permissible annual thickness loss is $\Delta_{tot} = 0.1$ mm $= 100$ μm. Therefore it is necessary to sum over the individual thickness losses due to sputtering, thermal evaporation and other erosion processes. Particles with the same Z and A but different energy are treated as different particles since sputtering yields are energy dependent. One must consider separately the sputtering yields due to 14-MeV neutrons, 3.3-MeV alpha particles, 10-keV deuterium atoms, 20-keV deuterium atoms, and so forth.

Plasma contamination and wall erosion processes will be discussed individually in more detail in the remainder of this section.

1. Phenomena Induced by Particle and Photon Impact

When energetic neutral or charged particles impinge on and penetrate into solid targets, the resulting primary processes include momentum transfer between the projectile and the target atom, changes in the internal energy states of the projectile and/or the target atom, nuclear reactions, and so forth. The primary processes in turn can cause displacement of lattice atoms, lattice atom excitation and ionization, x-ray emission, and other secondary processes. The primary and secondary processes may lead to escape of the primary projectiles by backscattering [28, 29] or by re-emission after entrapment [30, 31] as well as to the emission of secondary particles. At the same time the structure, chemical composition, and thickness of the irradiated surface can be altered by chemical reactions due to compound formation, chemical sputtering, and embrittlement. The influence of each of these phenomena on plasma contamination and on wall erosion needs to be determined for the wall materials, fuels, and operation conditions envisioned for fusion reactors.

Blistering of surfaces as a result of energetic ion bombardment is a process which is under intensive investigation [32]. Radiation blistering has been observed for certain noble gas projectiles (e.g., for He^+ on Cu [33], He^+ on silicon [34], and A^+ on Cu [35], as well as for energetic D^+ [36] and H^+ [37] projectiles which have markedly higher solubility and larger diffusion coefficients in many metals.

As a result of current experimental work on blistering in candidate wall materials, the basic mechanisms underlying the blistering process are becoming better understood, and it may become possible to make reliable theoretical predictions of the amounts of gas released and the size, shape, and number of blisters formed in Nb, Nb alloys, V, or other materials considered for the vacuum wall under typical reactor operating conditions. Studies of the dependence of blistering on such important parameters as the type of projectiles, their energy flux, fluence and angle of incidence, the target material, structure and temperature all need to be made in order to reach an overall assessment of eroxion rates due to the blistering phenomenon.

The process of target particle emission under the impact of neutral or charged particles on the surface of a solid target is called sputtering. One often distinguishes between two cases, "physical" and "chemical" (or "reactive") sputtering, although in many experiments both physical and chemical sputtering occur simultaneously. Physical sputtering occurs when the kinetic energy of the impinging projectile is high enough to displace target atoms from their sites by momentum transfer in a collision cascade which imparts sufficient energy to surface atoms to overcome their surface binding energies and eject them from the target.

Chemical (or reactive) sputtering occurs whenever chemically reactive gas particles (e.g., primary beam projectiles or component particles of the residual gas) interact chemically with the surface of the solid resulting in the ejection of molecular sputtered products. The mechanism of chemical sputtering is complex and depends in detail on the parameters such as partial molar enthalpies of vaporization of the various sputtered species [38].

As an example of chemical sputtering the profound effect of deuterium loading on the sputtering rate of a Nb–Zr surface [39] may be mentioned. Its influence on trapping surface lifetimes needs to be determined.

An increase of sputtering rate with target loading is to be expected on the basis of the model used by Harrison *et al.* [40] for computer simulation of sputtering. As the amount of deuterium in the material increases, primary collisions are no longer all between deuterium ions and niobium but also between deuterium and deuterium. This greatly increases the scattering of the primary and secondary particles and thus would increase the "mole" sputtering mechanism. Furthermore, under saturation conditions which will prevail at trapping surfaces, sputtering mechanisms which have hitherto not been encountered can be expected to come into play. In particular, the possibility that in addition to metal atoms, metal deuteride and tritide molecules are ejected from strongly "hydrided" surfaces must be considered. Only an extremely limited number of measurements of sputtering yields for the various types and energies of projectiles such as D^+, T^+ and He^+ are available. This is especially true for the extreme operating conditions envisioned for fusion reactors e.g., for vacuum walls at temperatures of 600–1000°C bombarded by high fluxes of several species of energetic particles simultaneously. Synergistic effects due to simultaneous bombardment by several types of particles and photons could clearly be very important and need to be studied.

The preceding discussion has not taken account of wall erosion by energetic neutrons, though this process would occur for most fuel cycles and for most types of fusion reactors considered to date. As MeV neutrons penetrate a solid wall, they undergo elastic and inelastic scattering by the nuclei of the lattice atoms. Since the neutrons carry no charge, they impart momentum directly to the nuclei with which they collide. For a 14.1-MeV neutron, for example, the maximum energy transferable to the nucleus of a Nb lattice atom is $E_{max} = 593.6$ keV, and the mean energy E of a primary knock-on (a lattice atom displaced by a neutron) has been estimated to be ~ 150 keV [41, 42].

The particle release from cold-rolled niobium surfaces under 14-MeV-neutron impact to a total dose of 4.6×10^{15} neutrons/cm^2 was investigated under ultrahigh vacuum conditions and at ambient temperature [43].

The type and amount of material released and deposited on a substrate surface was determined independently by four analytical techniques. Surprisingly, there were two types of deposits; one in the form of large chunks; the other a more even layer covering the surface.

The mechanism for the emission of chunks is not clearly understood. A possible mechanism for the ejection of chunks could be related to the energy deposited by a 14-MeV neutron interacting with niobium lattice atoms in the near-surface region via elastic and inelastic collisions. This deposited energy can lead to localized thermal and ionization (electron) spikes which in turn may cause the generation of shock waves. The interference of such shock waves in a small volume in the near surface region in which the stored energy is very high (as a result of cold working) may set up stresses large enough to release energy by initiating submicroscopic cracks or by propagating already existing microcracks and cause the emission of chunks. However, one cannot exclude the possibility that other mechanisms may contribute to or dominate the chunk ejection process. The neutron sputtering yields appear to be strongly dependent on surface roughness and it is therefore difficult to estimate wall erosion rates due to this process.

The desorption of gases from solid surfaces under impact of sufficiently energetic electrons has been observed by several authors [44–46] usually at electron energies ranging from 15 to 500 eV. The desorbed species leaving the surface include neutral atoms, molecules, and positive and negative ions in their ground states or with varying amounts of excitation energy.

Desorption by electron impact can very well contribute significantly to plasma contamination. In an operating fusion reactor, the large fluxes of energetic photons and of particles (both ions and neutral atoms formed by charge exchange) cause copious emission of electrons from the walls. Some of these are turned back by the magnetic confinement field and cause further electron emission by electron impact. The energies of the electrons resulting from these diverse processes cover a wide range and their impacts on the walls will be sufficiently energetic to desorb gas species adsorbed on the surface.

Energetic photons (e.g., synchrotron radiation, bremsstrahlung, X rays, and γ rays from nuclear reactions) impinging on fusion reactor components can in principle cause photo-desorption of adsorbed or absorbed gases, photo-decomposition of surface compounds, photo-catalysis leading to a reaction between adsorbed molecules and the vacuum wall, and photo-electron emission. Sputtering by the conversion of high electronic excitation energy into displacement energy, rapid vaporization as a result of the high surface temperature and temperature gradients produced by photon absorption in layers near the wall surface, are other possible concomitants of photon impact.

These processes in turn can contribute to wall erosion. Contamination of the plasma with desorbed gases (1 at.% of fully stripped oxygen will cause a 77% increase in the power lost through bremsstrahlung) must be minimized by discharge cleaning prior to start-up, for example. The outgassing by synchrotron radiation has been studied in electron storage rings, in which the electrons producing the radiation are much more energetic than those envisioned in the operation of fusion reactors [47, 48].

Several studies of gas desorption from solids irradiated by visible or ultraviolet photons are also available [49–51]. However, the information is still too fragmentary to provide any reliable estimate of the rates of outgassing and erosion of materials considered for reactor components under the conditions of temperature and photon fluxes (especially X rays and γ rays).

When a fusion reactor is in operation, not only the vacuum wall but also such internal components as beam limiters and diverters are heated both by the energetic particle and photon flux from the plasma and by secondary radiations. For particle radiations, the thermal loadings depend on the mass, atomic number, and charge state; and for all radiations, the loadings depend on the energy. For example, low energy photons and particles have very short ranges in a solid so their energies heat the layers near the surface. Depositing all the energy in a shallow layer could result in high surface temperatures and temperature gradients.

Power deposition rates in solid reactor components therefore must be controlled to minimize thermal evaporation and desorption, thermal emission of electrons and ions, whisker growth, enhancement of the secondary particle emissions and acceleration of such processes as gas permeation and diffusion.

2. Surface Effects on Insulators

Most of the discussion so far has dealt with surface effects on metals. In the case of insulators, ion and neutral atom sputtering, neutron sputtering, blistering, electron and photon interactions need to be separately considered because of properties inherent in insulator materials. Although preliminary calculations indicate that the wall flux of energetic particles will be negligible, plasma instabilities could inadvertently cause non-negligible bombardment of the first wall of a θ-pinch reactor, thereby locally damaging the insulator. Sputtering by 14-MeV neutrons of Al_2O_3 single crystals has given no evidence of chunk emission [52].

Blistering of a θ-pinch reactor first-wall ceramic insulator, might give rise to gas bubbles which could move to the metal/oxide interface. The resulting debonding could have serious consequences in the long term, if the decreased thermal conduction through the first wall results in overheating of the insulator.

Sputtering of insulators needs to be examined much more closely than heretofore. Although sputtering rates of oxides appear to be comparable to those for some metals, special chemical sputtering effects resulting in non-stoichiometric surface layers and surface compound formation have been shown to occur in H^+ and D^+ bombardments of aluminum oxide [38].

III. FIRST-WALL MATERIALS AND DIVERTER TRAPPING SURFACES

One of the principal scientific and technical problems associated with building and operating a fusion reactor is the choice of materials for a vacuum wall. The choice of wall materials will be dependent on various physical properties of the material.

In the 1000 MW power range, a low-β toroidal reactor will have a vacuum vessel of the order of 10–20 ft minor diameter and 30–60 ft major diameter. The vacuum wall must be constructed of a suitable material—probably a refractory metal—in the form of plate probably much less than but up to 1 in. thick. The wall will be surrounded by a breeding blanket and coolant, neutron shielding and superconducting magnet coils. The whole structure may weigh over 1000 tons and will have to be supported in a manner which does not interfere with the main functions of the reactor. The vacuum vessel itself will be under nonuniform external pressure and will have a variety of penetrations for pumps, divertors, fuel injectors and diagnostic probes. It must, therefore, be as strong as possible but the walls must be thin to minimize thermal stresses. Ingenuity will be required in the design of the vessel to produce a strong yet thin structure, which in heating to 1000°C will expand linearly more than one foot.

Thermal stresses in the vacuum wall are generated by differential thermal expansion which depends on the mode and rate of energy deposition at the wall. The 14-MeV (for the D–T reaction) neutrons will only deposit a small fraction of their energy in the walls with most given to the blanket. The energy deposition due to ions or neutral atoms will depend on their flux to the wall which in turn depends on the presence and efficiency of divertors as well as a number of plasma parameters. There will be incident electrons and photons from bremsstrahlung and cyclotron radiation. Gamma rays will be generated within the wall by (n, γ) reactions and will also be backscattered from similar reactions in the blanket. The nature of the thermal stress pattern will depend on whether the reactor is operated in a steady state or a pulsed mode; in the latter case, fatigue stresses must not be allowed to exceed the yield stress.

A. Metal First Wall

There are various alternatives in designing a thermonuclear reactor that will directly affect the choice of materials used in the vacuum wall; for example, the mode of energy conversion and the thermal efficiency required of the wall. With direct energy conversion it would appear that the wall temperature would be lower than with a conventional dual-cycle heat exchanger. If a 50% thermal efficiency is required, the wall would be kept at a temperature of about 1000°C; if 40% efficiency is tolerable, the wall temperature could be approximately 700°C and the use of stainless steels could be considered.

For optimum performance of the reactor system, thermal conductivity in the wall must be maximized to allow the best heat transfer from the radiant thermal loading on the interior of the chamber through the vacuum wall to the lithium blanket.

Vapor pressure is an important quantity to consider in selection of wall materials, since the lower the pressure, the lower the plasma impurity level.

Elasticity is also an important consideration for the structural integrity of the wall. This quantity should be maximized as it determines the stresses and thermal loadings the wall can sustain before reaching the elastic limit. On the other hand, creep, the stress loading deformation after a period of time should be minimized. The wall material must satisfy the criteria discussed above as well as a number of others.

TABLE 1

Comparison of Relevant Properties for Prospective Vacuum Wall Materials

Property	Niobium	Molybdenum	Vanadium	Iron and nickel base alloys
Absorption of 1 MeV neutrons	Mod.–high	Mod.–high	Low	Fairly low
Decay heat	High	Mod.	Low	Mod.
Thermal conductivity	High	High	High	Low
Linear coefficient of thermal expansion	$\sim 7 \times 10^{-6}/°C$	$\sim 5 \times 10^{-6}/°C$	$\sim 7 \times 10^{-6}/°C$	$\sim 18 \times 10^{-6}/°C$
Approximate melting point	2400°C	2600°C	1900°C	1500°C
0.3–0.5 melting point	500–1100°C	600–1200°C	350–800°C	300–650°C
Corrosion resistance (to Li)	Good	Good	Good (?)	Med.–poor
Sputtering ratio,[a] relative	Low	Larger	Low	Largest
Tritium diffusion rate	High	Low	Very high	Low
Fabrication cost	High	Med.	High	Low

[a] Measured in ultrahigh vacuum laboratory experiments; information not available for expected fusion reactor conditions.

In an attempt to arrive at a choice of wall material, all the various factors must be taken into consideration and inevitably the final choice will be a compromise one. A comparison of only some of the relevant properties for prospective vacuum wall material are given in Table 1.

Vanadium and molybdenum appear to be good choices except for the possible problem of swelling, with vanadium having an advantage in decay heat and molybdenum in cost. For first generation reactors, iron or nickel base alloys may be suitable although they will also operate in their swelling range.

The questions of fabrication and design for heat removal can only be addressed properly after a detailed design study of the vacuum vessel has been completed. It is already clear that all the metals listed can be fabricated into many different shapes, joined together and inspected. However, no vessels as large or as intricate as a fusion reactor vacuum vessel have been made and development will be required for methods of forming and joining plate of the required thickness. For coolant supply and removal, the size, shape, and availability of particular flow channels will need to be evaluated.

B. Insulator First Wall

The operation of θ-pinch reactors and certain design features of such reactors have already been discussed. The requirement of an insulating first wall capable of holding off electric fields of the order of 6 kV/cm at high temperatures while undergoing intense neutron, photon and ion bombardment was pointed out. In this section, the material measurements imposed on such a wall will be discussed in more detail.

Generally, the insulator should (1) have good thermal and electrical properties at high temperatures, (2) sustain irradiation without unacceptable degradation of electrical and mechanical properties, (3) not undergo large dimensional changes as a result of irradiation and the buildup of transmutation products, and (4) resist chemical attack at elevated temperatures by hydrogen isotopes. The insulator must also be chemically and mechanically compatible with the metal substrate. In the following paragraphs, the relationship between the first-wall operating conditions and the anticipated behavior of insulator materials is discussed.

The dielectric property of the insulator must be sufficient to withstand the large implosion heating fields for ~ 0.1 μsec prior to the thermonuclear burn once every 10 sec.

Electrical resistivity and dielectric strength of insulators normally decrease at elevated temperatures, whether charge transport is by electronic or ionic motion. A typical design requires an electrical resistivity $\gtrsim 10^6 \, \Omega$ cm and a dielectric strength ≥ 100 kV/cm at operating temperatures. The better insulators can meet the resistivity requirement, but their dielectric strength is often less than 100 kV/cm at operating temperatures. Fortunately, the voltage in the pulsed reactor is applied across the first-wall insulator for less than 1 μsec, and dielectric strength at elevated temperatures for many materials is considerably higher under pulsed conditions [53].

Stress represents a second, major short-term problem for the first-wall insulator. Both thermal and dielectric shock loading of the insulator must occur without fracture, spallation, or debonding of the insulator from its metal substrate.

Fatigue failure of either the insulator or the metal/insulator bond may represent a major long-term materials problem. The phenomena of crack nucleation and growth as well as creep with local thinning, may be ameliorated in the presence of a periodic, intense radiation field. The effects of the implosion process on microcracks and porosity in the insulator, however, are not known.

The chemical compatibility of an oxide insulator with (1) the reducing environment of the plasma chamber, (2) the reducing action of hydrogen generated within the insulator, and (3) the reducing environment of the metal backing needs to be evaluated [54]. However, if very hot, protonic hydrogen impinges on the ceramic insulator, significant reduction of Al_2O_3, for example, may occur [55]. This situation is avoided in theory by the use of a neutral gas layer situated between the plasma and the first-wall insulator, although the effectiveness of this neutral gas layer is yet to be tested by experiment and more realistic calculation.

Although the compatibility of Al_2O_3 and Nb at elevated temperatures is good, DeVan [56] reports that alumina bonded to niobium can undergo chemical reduction at 1350–1470°K if lithium is in contact with the uncoated niobium surface; oxygen from the alumina is apparently transported through the niobium and into the lithium while BeO under similar conditions did not exhibit reduction. This area must await further analysis and experiment before being put into perspective.

C. Diverter Trapping Surfaces

In toroidal fusion reactors of the Tokamak type, special provisions for plasma extraction have been considered. A class of such devices, called diverters, have been proposed which direct plasma particles out of the reactor along magnetic field lines whose direction is determined by magnetic field coils of special design [57]. During confinement, plasma ions drifting toward the reactor wall are skimmed off by the diverters and magnetically guided into relatively narrow channels which are extensions of the plasma containment vessel.

Various schemes for attenuating the energetic charged particles in the diverter region have been mentioned, including direct conversion, glancing collisions with surfaces, collisions with cold gas, reactive (chemical) and nonreactive trapping. Chemical trapping occurs when energetic particles, and in particular, H^+, D^+ and T^+, interact with certain metals to form hydrides, deuterides and tritides. The particles are captured by the metal and quickly come to thermal equilibrium. In the case of He^+ particles, no chemical trapping can occur, but a certain fraction of incident particles will be retained by implantation in the metal.

1. The Mechanism of Chemical Trapping of Deuterium and Tritium in Metals

Experimental measurements on deuterium trapping in solid metal targets [58–64] have shown that this may be a promising technique for "pumping" a useful fraction of the ion flux in the diverter.

The total particle flux from a ~ 6 GW(t) reactor will be $\sim 6 \times 10^{23}$ ions/sec. To maintain a sufficiently large mean free path in the diverter region for the incoming ions, (say 10^{-5} Torr pressure) would require a pumping speed of $\sim 10^8$ liter/sec at thermal energies. Handling the required throughput of D, T and He by conventional pumps would pose formidable problems. Chemical trapping could reduce pumping speed requirements by a factor of about 20 assuming 100% trapping efficiency, 5% burnup and zero trapping of He. A great incentive therefore exists for investigating trapping mechanisms and efficiencies.

The trapped D and T could be recovered by heating the trapping surface outside the reactor proper to temperatures at which decomposition pressures of 5×10^{-2} Torr are produced. Vapor booster pumps handling $\sim 10^3$ t liter/sec at these pressures could then be employed.

Studies of trapping in solid targets have shown that the trapping efficiency of hydrogen isotope ions in solids depends on ion energy, bombardment time and target temperature. Trapping also has been found to be most efficient in those metals which form stable metal hydrides. Some results obtained by McCracken *et al.* [60] are shown in Fig. 1, where trapping efficiency after a fixed arbitrary bombardment time (total dose of 5×10^{18} ions/cm² of 18-keV D^+) is plotted against temperature range for each of four metals. It is to be noted that the trapping is efficient over only a certain temperature range.

These results can be understood on the basis of the following considerations [60]. The 18-keV D^+ ions have a mean range of $\sim 2 \times 10^{-5}$ cm [66]. After slowing to thermal energies, deuterium diffuses through the lattice without being able to escape from the surface over a certain temperature range because the activation energies for diffusion are much lower [67–69] than the heats of formation of the metal deuterides [70–72]. The decreasing

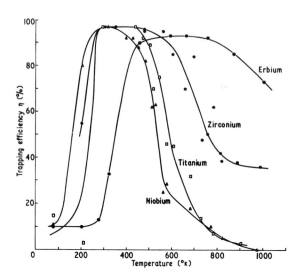

FIG. 1. Trapping efficiency of Nb, Ti, Zr and Er for 18 keV D^+ ions as a function of temperature. Total ion dose $= 5 \times 10^{18}$ ions/cm^2.

trapping efficiencies at higher temperatures are due to the increasing probability of the diffusing D atoms to overcome the potential barrier for escape from the surface. The expected correlation between the heats of formation of the (stoichiometric) metal deuterides and the temperature at which the trapping efficiency has decreased to some arbitrary value is observed [60].

At temperatures in the liquid nitrogen range and below, the diffusion coefficient of deuterium in metals falls to very low values and therefore the deuterium concentration rises rapidly in a layer whose depth equals the projected range of the incident ions. It has been found experimentally for all the metals studied to date [60] that at low temperatures, the trapping efficiency decreases quite abruptly at a dose equivalent to 10^{18} ions/cm^2. If distributed evenly throughout the solid to a depth equivalent to the initial projected range of the ions in the metal, the M/D ratio would be of the order of unity. It appears that as the surface layers are converted to a bulk hydride phase, no further uptake of deuterium occurs.

At higher temperatures, the trapping efficiency remains high at doses larger than 10^{18} ions/cm^2 because the diffusion coefficient is large enough to prevent the buildup of saturation layers near and on the surface.

2. Characteristics of a Trapping Surface in a Thermonuclear Reactor

The preceding discussion has made clear that trapping is inefficient both at low temperatures (after a dose of $\sim 10^{18}$ ions/cm^2) and at high temperatures. The low temperature limit is determined by the low diffusion rate of the gas in the solid leading to saturation of the surface layers of metal. The high temperature limit is determined by the decomposition pressure of deuterium in the metal deuteride. However, at intermediate temperatures, trapping efficiencies greater than 90% may be obtained. For practical application in a thermonuclear reactor, it is important to determine whether the total fluence which can be trapped with high efficiency, fulfills the requirements placed on a trapping surface.

The 6×10^{22} particles leaving the 6 GW(t) reactor per second carry 150 MW of power. With approximately 5×10^5 cm^2 of trapping surface (some 10% of the reactor vacuum wall area), the power loading becomes 300 W/cm^2. The maximum ion current density which has to be considered is $\sim 1 \times 10^{17}$ ions/cm^2/sec or 15 mA/cm^2. Such ion current densities incidentally are experimentally accessible with ion sources of the duoplasmatron type.

In a recent study [64], a tensimetric technique was developed for measuring 15-keV D^+ trapping in Ti metal. The formation of the TiD$_2$ phase in the surface and near surface regions was monitored by decoupling the θ–2θ scan of an x-ray diffractometer as well as by scanning electron microscope studies. Micron-sized "reaction zones" appear to serve as nucleation centers for the TiD$_2$ phase. Insight into the mechanism of chemical trapping was obtained by correlating the results from the three types of measurements.

IV. BLANKET MATERIALS

A. Introduction

A thermonuclear reactor that fuses deuterons with tritons yields 17.6 MeV per fusion event mostly in the form of very energetic neutrons. Such a device, as already briefly mentioned in the discussion of Tokamak and θ-pinch conceptual designs, requires a blanket system capable of performing at least two functions. These functions, achieved by a more or less complex composite of several materials, are (1) absorption of the energy carried by the energetic fusion reaction products and transfer of heat generated in the blanket to the power-producing portion of the reactor, and (2) generation of tritium, in a manner such as to enable its ready recovery, to replace that consumed in the fusion reaction. The first of these functions can be satisfied by a suitable heat transfer fluid, and the second by the presence of a sufficient amount of lithium, from which tritium can be effectively bred. The two functions are separable in principle. It is possible, to visualize a blanket in which tritium is bred in stationary lithium metal, lithium alloys, or lithium salts, with the necessary cooling accomplished by other fluids such as molten sodium, pressurized helium or molten salts. Nonetheless, there are probably considerable advantages to having the blanket function both in a breeding and cooling capacity.

Such a blanket must meet several criteria. The blanket materials must have adequate neutronics to satisfy the tritium breeding requirement; have good heat transfer and hydrodynamic behavior in the presence of large magnetic fields; be noncorrosive toward metals of construction in the blanket region, the pumps, and the power generation equipment; possess a relatively low vapor pressure; be compatible with auxiliary blanket subsystems such as neutron moderators (graphites) or neutron multipliers. Finally, the blanket materials should be of a nature such as to permit recovery of the tritium for return to the fusion cycle.

B. Liquid Lithium Metal

To explore the potential of liquid lithium metal as the major component of a breeding blanket, a number of properties such as natural abundance, thermodynamic and transport behavior, characterization, analysis, and control of species in lithium, and corrosion of materials including electronic insulators need to be discussed. The importance of individual properties in affecting the overall design considerations of an operating reactor and areas where information is lacking and additional work needed are pointed out.

Reviews of the properties of liquid lithium [73] and of its use as a blanket in fusion reactors [74, 75] have appeared.

1. Natural Abundance of Lithium Relative to Potential CTR Needs

The fuel cycle for the D–T concept consumes deuterium and lithium, and therefore it is important to consider the natural availability of these substances in terms of a fusion economy. Reserves of deuterium are virtually inexhaustible, since heavy water constitutes 0.016% of the worlds oceans. Deuterium would therefore provide energy which could meet the world's power requirements for the indefinite future. A fuel cycle dependent on tritium however, is limited by the natural abundance of lithium. Lithium utilization studies [74] have shown that the amount of energy per gram of natural lithium can be maximized to a value of ~ 25 MWh(t)/gm of natural lithium. Assuming a 50% thermal-to-electrical conversion efficiency, one can calculate the amount of electrical energy which the quantity of available lithium can generate. Assuming present day electrical energy production rates in the United States, burnup of only 10% of the currently known lithium resources means that the reserves would last about 2000 yr. Assuming that methods can eventually be developed for recovery of lithium from the ocean in an economically competitive manner, the quantity of energy corresponding to the total world's supply of lithium would then be expected to meet electricity demands for millions of years [74]. At the current market price for high purity lithium, it would cost 15 million dollars to provide the 10^6-kg inventory, necessary for a 6 GW(t) fusion reactor, a small fraction of the total cost of such a plant. These estimates indicate that the supply and cost of lithium are not likely to be a near-term problem for fusion reactors unless competing demands develop.

2. Materials Compatibility and Chemistry

It is beyond the scope of this chapter to give a review of the thermodynamic and transport properties of lithium. Nonetheless, they are of fundamental importance for an understanding of lithium as a blanket material, and the interested reader is referred to the appropriate literature [75]. However, the compatibility of liquid lithium with materials of construction needs to be discussed.

At elevated temperatures, the compatibility depends to a significant extent on the concentrations of various impurities in the lithium and in the metal or ceramic material of construction. The magnitude of the effect of a given concentration of a given impurity on the corrosion rate depends upon the thermodynamics and phase relationships for the impurity–lithium, the impurity–metal, or impurity–ceramic system, and upon the ternary interactions. For this reason, it is important to have phase diagram and thermodynamic data for the various impurities in lithium and in the metals and ceramics of interest. The following discussion is based on considerations given to this subject by Cairns *et al.* [75].

a. The Li–LiH, Li–LiD, and Li–LiT Systems Data on phase equilibria in the Li–LiH, Li–LiD, and Li–LiT systems are quite limited [76]. For the Li–LiH system, there are several studies dealing with the dissociation pressure of hydrogen as a function of temperature and composition [77–82]. The earlier data however lacked consistency, particularly in the region which was dilute in hydrogen where serious disagreements exist. Because of these inconsistencies it was not possible to make reliable estimates of pressure–composition relationships in the region of dilute solutions of hydrogen in lithium. More recently [81, 82], pressure–composition isotherms in the Li–LiH systems at temperatures in the range 700–900°C have been measured very carefully. These results represent the most consistent set of isotherms yet obtained for the Li–LiH system, and have permitted the estimation of Sieverts' law constants, K_S, for dilute solutions of hydrogen and deuterium in lithium.

Heumann and Salmon [80] have also measured pressure–composition isotherms in the Li–LiT systems, which show a significant isotope effect on the partial pressure at a given mole fraction.

Swain *et al.* [83] have derived the equation

$$Q_H/Q_T = \alpha(Q_H/Q_D)^{1.442}$$

to describe the relationship between the isotope effects of tritium and deuterium on the thermodynamic quantity, Q, as measured for hydrogen. The
value of α has been shown to lie in the range 0.907–1.157 for a variety of
properties over a wide temperature range. Using this equation with $\alpha = 1.0$
and assuming $P_D/P_H \approx 2.0$ ($Q = P =$ partial pressure), one estimates that
$P_T/P_H \approx 2.7$ for H and T in Li solutions having the same mole fraction of H
or T at the same temperature. Using these values for the P_D/P_H and P_T/P_H
ratios and newer results for the Li–LiH system, one can estimate Sieverts
constants for the dilute hydrogen isotope regions of the Li–LiD and Li–LiT
systems [75]. Further data are needed to refine the accuracy of these values.

b. Solubility of Nonmetals in Liquid Lithium   The solubilities of oxygen
and nitrogen in lithium were determined in the range 250–450°C [84]. The
heat of solution for oxygen was found to be 9.4 kcal.

Despite the uncertainty about the reliability of some of the nitrogen
measurements and the wide scatter exhibited by the oxygen results, the data
have proven useful for they provide an explanation for the fact that oxygen
in lithium can be reduced to levels of 100 ppm by cold trapping, but that
nitrogen cannot [84, 85]. The solubility of nitrogen at temperatures just
above the melting point of lithium is too high for cold trapping to be
effective. Above 600°C, nitrogen is thought to exist as the nitride. Below
600°C, the chemistry of nitrogen and perhaps of oxygen may be more
complex.

The lithium–carbon system has been studied using thermal and x-ray
diffraction methods [86, 87].

The partial phase diagram derived from these findings [88] exhibits the
compound, Li_2C_2 which exists in four polymorphic modifications, and a
eutectic between Li and Li_2C_2 at 1650°C and a concentration of 1 atom per
carbon. However, another study of this system demonstrated that Li_2C_2, in
samples that were annealed at various temperatures between 350° and 700°C
and then quenched, exists in only one crystalline form. This finding is not
consistent with the earlier x-ray results [86] which have been thrown into
question [89]. Above 600°C, carbon is thought to exist as the carbide.

It is of the utmost importance to develop methods for the detection and
determination of hydrogen, tritium, oxygen, carbon and nitrogen impurities
in lithium. Such methods have been developed for other alkali metals and
some may have potential use with lithium.

For example, a method that is finding applicability in sodium systems for
the determination of carbon at activities as low as 10^{-3} is the carbon diffu-

sion meter [90]. This meter which may also be useful for the determination of carbon activity in Li systems consists of a 10 mil wall iron probe through which a decarburizing gas (Ar–5% H_2–0.5% H_2O) is passed. Carbon diffuses through the probe and reacts with the decarburizing gas to form carbon monoxide. The carbon monoxide is subsequently converted to methane and determined by means of a flame ionization detector. Since pure iron has been reported to exhibit good resistance to corrosion by lithium, the use of iron as a probe material should be investigated.

c. Corrosion of Selected Materials in Lithium Of special interest to CTR's as structural materials are the refractory metals and their alloys. This interest has arisen principally from the fact that some of these metals possess favorable neutronic properties, low solubility and high corrosion resistance to lithium.

Studies of the corrosion of metals by liquid lithium generally have been performed under more poorly defined conditions than the investigations of solubilities. For this reason, most of the corrosion results can be used only as guidelines to the behavior of metals and alloys in liquid lithium.

Cowles and Pasternak [73] have reviewed the literature up to 1968 on the corrosion of various materials by lithium. It can be seen from their review that the corrosion rates of nonrefractory metals and alloys are generally too high at temperatures approaching 1000°C to be of significant interest for controlled thermonuclear fusion reactors. Impurities such as oxygen, both in the metal and in the lithium, can have a profound effect on the corrosion and penetration rates. However, the concentrations of impurities were generally not known or controlled in many of the earlier corrosion studies. Nonetheless, selected corrosion studies on refractory metals and their alloys, and some ceramic materials are discussed below.

The metals which are most resistant to attack by lithium at temperatures in the vicinity of 1000°C are those of groups IVB, VB, VIB, and VIIB of the periodic chart. Those which have received the most attention are Ti, Zr, Hf, V, Nb, Ta, Mo, W, and Re, and their alloys.

The problem of lithium penetration due to oxygen contamination seems to be greatly ameliorated by the use of oxygen " getter " alloying agents such as Ti, Zr, and Hf, to form alloys such as Nb–1% Zr, Nb–28% Ta–10.5% W–0.9% Zr (FS-85) and Ta–8%W–2%Hf (T-111). The corrosion resistance of a number of these alloys is tabulated in Cairns *et al.* [75]. In order to maximize the benefit obtained by the use of oxygen getters, it is necessary to employ a heat treatment that encourages the reaction between O and Zr or Hf before the refractory alloy is exposed to lithium. It has been found that an annealing treatment at 1200–1300°C for 1 hr is appropriate for Nb–1%Zr, Nb–28%Ta–10.5% W–0.9% Zr, and Ta–8%W–2%Hf [92]. This treatment results in the combination of sufficient oxygen with the getter to reduce the O concentration below the threshold level, resulting in an alloy with good corrosion resistance to lithium.

More recently, Sessions and DeVan [92] investigated the effect of heat treatment and test temperature on the compatibility of several advanced refractory alloys with lithium. Their results, showed that (1) attack by lithium at 500°C was less severe than anticipated based on current corrosion models, (2) at higher oxygen levels (> 2000 ppm) corrosion at 500°C was both transcrystalline and intergranular, while attack at 1000°C was predominantly intergranular, (3) in general, heat treatment at 1300°C enhanced resistance of the materials tested to corrosion by lithium, and (4) hydrogen pickup during decontamination can be a significant problem with advanced niobium and tantalum-based alloys when rapid, hydrogen-producing reactions occur.

The corrosion resistance of the Group VIB metals Cr, Mo, and W in lithium has not been as thoroughly investigated as that of Nb, Ta, and their alloys. In general, alloys rich in Mo and W show good corrosion resistance to lithium, even under flow conditions, at temperatures well above 1000°C, for at least 1000 hr.

Of the group VIIB metals, only Re has received attention in connection with resistance to lithium attack. Rhenium is very expensive, and probably is not practical as a material of construction, except under unusual circumstances. It does show good corrosion resistance at 1370°C for at least 1000 hr. In addition, rhenium has been used as an alloying agent together with molybdenum and tungsten.

In some CTR designs it has been proposed that the lithium coolant be electrically isolated from the metallic channel walls by an electronically insulating layer, in order to minimize the pumping power requirements [93]. The number of known materials which might serve in this capacity and might exhibit a significant lifetime in the high temperature flowing lithium is very small. This is partly because the corrosion resistance of insulating ceramics in lithium has not been studied thoroughly, and partly because lithium is extremely aggressive toward ceramics.

A number of insulating ceramics have been tested [94] in lithium at 375°C. Although the temperature was lower in these experiments than the temperatures of interest for CTR use, the relative corrosion rates of the ceramics may provide some useful guidelines. Purity and method of preparation were found to be important parameters, in the case of BeO for example, in influencing resistance to lithium.

Similar effects can be expected for other ceramics, and have already been noted for BN. The most promising ceramics for resistance to lithium at 375°C are hot-pressed, high-purity BeO, $Y_3Al_5O_{12}$, ThO_2, $MgO \cdot Al_2O_3$, AlN, and Y_2O_3.

As can be seen from the paucity of verified results on the corrosion of metals, alloys, and electronically insulating ceramics by lithium under well-known purity conditions, there remains a great deal of work to be done before the most suitable materials of construction for CTR application can be identified. The starting point has been indicated by past work—refractory metal alloys, perhaps with getters such as Ti, Zr, or Hf, and stable oxide, nitride, or double-oxide ceramics. Carbides probably will not provide the desired electronically insulating properties.

C. Molten Salts

This section attempts to assess the potential of molten salts as blanket fluids in controlled thermonuclear reactors, using molten Li_2BeF_4, probably the most studied and best understood of the possible candidates, for this assessment. The following discussion is based on considerations of this subject by Grimes and Cantor [95].

The problems which must be surmounted by a molten salt CTR blanket system depend upon the detailed design of the reactor such as power level, size, temperature, temperature differential, and coolant passage patterns within the blanket region. To illustrate the nature and general magnitude of the problems, a 2.2 GW(t) toroidal CTR, blanketed and cooled with molten Li_2BeF_4, is chosen.

Although the radiation density in the blanket near the plasma confinement (first) wall may be larger than those proposed for the MSBR (molten salt breeder reactor), it is not likely to approach the maximum density at which molten salts have been tested [96]. On the other hand, the effects of strong magnetic fields on the behavior of molten salts needs to be examined in some detail.

1. Effects of Strong Magnetic Fields

For a liquid metal, pumping the fluid across magnetic field lines can lead to large pumping power losses due to magnetically induced turbulence [93]. In the case of molten salts whose conductivities are lower by a factor of 10^3 compared to liquid metals, pumping power losses are considered to be negligible but there can still be pronounced effects of the magnetic field on the fluid dynamics and the chemical stability of the liquids.

a. Effects on Chemical Stability From electromagnetic theory it is known that the electric field induced in a conducting fluid crossing a magnetic field is given by the cross product of fluid velocity and magnetic field:

$$\mathbf{E} = \mathbf{V} \times \mathbf{B}$$

Molten Li_2BeF_4 (or any conducting fluid) flowing at 10 m/sec in a pipe of 5 cm diam with its axis aligned perpendicular to the lines of force of an

80 kG (8 V sec/m^2) magnetic field will experience at right angles to both the magnetic field and the flow direction, an induced potential difference of 4 V between the salt and the pipe wall. Although LiF and BeF$_2$ are both very stable compounds [97], 4 V are equivalent to destabilization by 92 kcal/mole which would make for considerable corrosivity to the metallic tube walls.

Homeyer [98], who seems to have been the first to consider such electrolytic corrosion in a CTR blanket system, noted that the problem could be alleviated by (1) reducing fluid velocities perpendicular to the magnetic field, and (2) using a series of parallel pipes to reduce the pipe dimension where flow across magnetic field lines is necessary.

b. Effects on Fluid Dynamics   To avoid induced electric fields as discussed above, flow of the blanket fluid within the torus will probably have to be aligned with the magnetic lines of force. However, in that case the magnetic field will exert a force opposed to eddies within the fluid and will tend to damp turbulent flow. Heat loadings in the blanket structure of a CTR, and especially at the first (plasma confining) wall, will be very large, and molten Li$_2$BeF$_4$ must develop turbulent flow if it is to cool this wall effectively. It is important, therefore, to assess this damping effect of the field on turbulent flow in the salt. Extrapolations of data on magnetic damping with liquid mercury to molten salts, indicate that molten Li$_2$BeF$_4$ can be made to flow turbulently within the blanket system. Clearly however, actual experimental data need to be obtained on molten salt systems flowing in magnetic fields [95].

2. Production of Tritium

It is necessary to use the neutrons produced in fusion to breed the tritium required to fuel a D–T reactor.

The only practical neutron reactions which will yield tritium sufficient for the needs of a D–T fusion reactor are ^{6}Li(n,α)T and ^{7}Li(n,αn$'$)T. The latter is a high energy reaction with a threshold at 2.5 MeV. Neutron capture cross sections of ^{6}Li become significant only at energies of 0.5 MeV or less. The ^{7}Li reaction is particularly favorable in principle since the product neutron (n$'$) can react with ^{6}Li to yield a second tritium atom. Because the blanket contains elements which scatter and absorb high energy neutrons, the production of tritium from ^{7}Li is not very efficient. When the lithium in the blanket is in natural isotopic abundance (92.58% ^{7}Li, 7.42% ^{6}Li), the greater fraction of tritium is in fact produced from ^{6}Li.

Several studies have been reported concerning the breeding of tritium in fusion reactor blankets [99–102]. In cases where the lithium is in natural isotopic abundance, tritium breeding ratios greater than unity are calculated for lithium metal blankets while the breeding ratios for lithium salts are distinctly less favorable.

Three options, whether alone or in combination, can be considered for upgrading tritium production when the breeding ratio is marginal as appears to be the case for Li_2BeF_4. The options are to design a blanket including a region of metallic lithium; to increase the Be content of the blanket by adding a region of Be or Be_2C thus enhancing neutron multiplication and providing more ^6Li; to enrich (modestly) the blanket material in ^6Li.

The second option was briefly treated by Bell [102] who showed that if a blanket region (40 cm thick adjoining a 1 cm first wall of molybdenum) were changed from Li_2BeF_4 to an equal thickness of Be and Li_2BeF_4, the tritium breeding ratio would increase from 0.95 to 1.50.

The third option has also received some attention. Impink [100] reported that small increases in ^6Li enrichment of the Li_2BeF_4 blanket led to modest gains in breeding ratio.

In light of present knowledge of the pertinent cross sections, it appears that the breeding capability of Li_2BeF_4 is marginal. This material would, therefore, probably need to be augmented by one of the methods indicated above, or by other means. It is clear that better cross-section data are needed so that this point can be decided.

3. Recovery of Tritium

Approximately 270 gm of tritium are consumed per day by fusion in a 2.2 GW(t) D–T reactor and 300 gm (about 50 moles of T_2 or 100 moles of TF) must be produced and recovered. The mean residence time of the fluid in the blanket is 5 min per cycle, in which time 0.174 moles of T_2, or 0.348 moles TF is produced. If the fluid entering the blanket contains no tritium species, the fluid emerging from the blanket will contain about 1.74×10^{-7} moles of T_2 (or alternatively about 3.48×10^{-7} moles TF) per liter assuming complete homogeneity. The problems of recovery and management of the tritium depend significantly on whether the material exists as T_2 or as TF. These two situations, and the extent to which the mode of tritium behavior can be controlled, are briefly described in the following.

Solubility of H_2 in molten Li_2BeF_4 has been shown to increase linearly with pressure of H_2; at 1000°K, the solubility is approximately 7×10^{-5} moles H_2 per liter of salt per atmosphere of H_2 [103]. No studies of tritium solubility have been reported. If the bred tritium occurs as T_2, and if the solubility behavior of T_2 and H_2 are similar, the emerging blanket fluid carries T_2 (generated during its pass through the blanket) equivalent to a saturating pressure of about 2.5×10^{-3} atm. Equilibration of the emerging salt with a relatively small volume of inert gas will result in stripping of a very large fraction of the dissolved T_2 from the salt.

The solubility of HF in molten Li_2BeF_4 also depends linearly on pressure of the solute gas, with a Henry's law constant of 10^{-2} moles HF per liter of salt per atmosphere HF at 1000°K [104]. The TF produced during each cycle of coolant through the blanket region corresponds to about 3.5×10^{-7} moles TF per liter of Li_2BeF_4. Because the saturation pressure of TF (3.5×10^{-5} atmospheres) is about 100 times lower than T_2, TF is more difficult to strip from the salt than T_2. However, TF does not diffuse through the metal walls, and if its chemical reaction with the walls can be held to a low level, the TF concentration can be allowed to increase so that the rate of processing the blanket fluid can be correspondingly reduced.

4. Chemical Transmutations

Several types of chemical transmutations can occur in molten Li_2BeF_4 when serving as the blanket fluid in a fusion reactor, and their deleterious effects must be held to a minimum.

Transmutation of lithium is, of course, essential to the production of tritium. The overall reactions can be represented as

$$^7LiF + n \rightarrow {}^4_2He + TF + n'$$

and

$$^6LiF + n \rightarrow {}^4_2He + TF.$$

These reactions are not inherently oxidizing or reducing, but the generated TF can oxidize reactive structural metals to form metal fluorides which dissolve in the melt. On the other hand, the transmutation of beryllium leads to corrosion of metal systems since the disappearance of Be^{2+} is equivalent to release of fluorine. The two reactions may be represented as

$$BeF_2 + n \rightarrow 2n + 2{}^4_2He + 2F \quad (\text{or } F_2)$$

and

$$BeF_2 + n \rightarrow {}^4He + {}^6_2He + 2F$$

followed by

$$^6_2He \xrightarrow{\text{0.8 sec half-life}} {}^6_3Li$$

and

$$^6_3Li + F \rightarrow {}^6LiF.$$

In a 2.2 GW(t) reactor, the two reactions yield, respectively, the equivalent of 500 gm and 70 gm of fluorine per day. A redox buffer capable of reducing $F°$ to F^- must therefore be provided. It is also necessary, if Ni, Mo, or W are constituents of the container system, that the redox buffer be consistent with maintenance of the tritium as TF. The couple $Ce^{3+} = Ce^{4+}$ may possibly serve this function. For example, with the concentration of cerium in the melt set at 10^{-4} mole fraction, the blanket contains 6×10^3 mole of $Ce^{3+} + Ce^{4+}$, and the Ce^{3+}/Ce^{4+} ratio would require chemical adjustment on a cycle time of many days. If, on the other hand, the container metal is Nb (or some other metal which reduces TF in dilute solution) the redox couple must be chosen so as to be considerably more reducing. The couple must deal with the F_2 generated by transmutation of beryllium but it must also be capable of reducing the 100 moles per day of TF produced by transmutation of LiF. Such a buffer system would require adjustment on a cycle of a few days.

In addition to the above reactions, transmutation of fluorine occurs due to capture of neutrons of energy above about 3 MeV. This reaction may be represented by

$$^{19}_9F + n \rightarrow {}^{16}_7N^- + {}^4_2He.$$

The nitrogen isotope decays, with a 7.3 sec half-life, to an oxygen isotope

$$^{16}N^- \rightarrow {}^{16}O^- + \beta^-,$$

with resulting conversion, most probably, of O^- to $O^=$. The absolute quantity of ^{16}N formed by this reaction is relatively uncertain; it is estimated to be, within a factor of three, 120 gm/day. The very short half-life of this isotope guarantees that all the ^{16}N decays within the CTR blanket. Although the concentration of ^{16}N, in whatever chemical form, cannot exceed 1.1 parts in 10^{11}, some fraction will undoubtedly react with the CTR containment metal. Furthermore, the decay of this isotope will lead to formation of metal oxides with several undesirable consequences. If all the $^{16}N^-$ decayed within the blanket salt, the oxide concentration would increase about 60 ppb per day. Since 10–50 ppm of oxide is almost certainly tolerable, a process for removal of oxide on a cycle time of several months to several years should suffice.

Finally, it should be noted that the transmutation reactions shown all generate He, giving a daily production of helium of about 125 gm at. or nearly 100 standard cubic feet. Because helium is relatively insoluble in molten Li_2BeF_4 [105], its production rate per pass of blanket salt corresponds to a saturation pressure of 2.6×10^{-3} atm. If no sparging were attempted, the helium pressure would reach 1 atm in about 30 hr.

5. Compatibility of Li_2BeF_4 with Various Materials

LiF and BeF_2 are much more stable than the metal fluorides of the structural metals. Consequently, corrosion due to chemical reactions with the structural components should prove minimal. Indeed, experience with the Molten Salt Reactor Experiment [106] has shown negligible corrosion by liquid Li_2BeF_4 on a nickel based alloy (Hastelloy N).

Melts such as Li_2BeF_4 are chemically inert toward, and do not wet, graphite [107]. However, the possibility that such salts will transfer graphite and carburize metals such as Mo or Nb cannot be discounted.

In any system involving heat exchangers and hot flowing liquids, there exists a real and finite probability that leaks will occur.

The reaction of steam with Li_2BeF_4 yields HF and BeO,

$$H_2O(g) + BeF_2(l) \rightarrow BeO(c) + 2HF(g)$$

Although this reaction is only somewhat exothermic, both H_2O and HF are likely to corrode the metal in contact with the salt and corrosion-product fluorides will dissolve or be otherwise carried by the salt. Since BeO is only very slightly soluble (125 ppm at 500°C) in Li_2BeF_4 [108], leakage of steam would soon lead to the precipitation of BeO in the salt circuit.

In general, although inadvertent mixing of Li_2BeF_4 (or most other molten salts) with other CTR fluids would prove troublesome such mixing would probably not lead to violent or explosive reactions.

6. Comparison of Other Molten Salts with Li_2BeF_4

Obtaining breeding ratios greater than unity with molten salts alone may pose a real difficulty. Molten LiCl or Li_2CO_3 in reasonable but not optimized blanket configurations give breeding ratios that are unsatisfactory. Breeding ratios have also been calculated for $LiNO_2$ [100] and $LiNO_3$ [101] with results which tend to be quite unfavorable. Impink [100] for example, obtained the value 0.82 for $LiNO_2$ in the same configuration which gave 1.15 for Li_2BeF_4. No calculations appear to have been made for Li_2SO_4, but the high cross sections for S(n,α) and S(n,p) reactions almost certainly will reduce the breeding ratio below that for Li_2CO_3. Moreover, $LiNO_2$ and $LiNO_3$ lack the thermal stability required of truly high temperature coolants, and Li_2CO_3 and Li_2SO_4 tend to oxidize many CTR structural materials.

The phase diagram for LiF–BeF_2 [109] shows that a melting temperature as low as 363°C is available in this system. Unfortunately, the viscosity of the melt increases with BeF_2 concentration, and mixtures with > 40 mole% BeF_2 have viscosities greater than 50 cP at 450°C [10]. The optimum salt mixture of low melting temperature and acceptable viscosity which still has a reasonably good tritium breeding ratio is at 33 mole% BeF_2, corresponding to the Li_2BeF_4. Decreasing the BeF_2 concentration below 33 mole% may have a modest beneficial effect upon breeding ratio which might, possibly, offset disadvantages due to the increased liquidus temperature.

The best blanket coolant salt in which to breed tritium would appear to be LiF, but its melting point of 848°C limits its usefulness only for cooling a blanket that operates above this temperature. If the blanket coolant must also transfer heat to the steam system, Li_2BeF_4, or some modest variant of this composition, appears to be the best choice.

7. General Comparison of Molten Salts with Lithium in Fusion Reactors

Lithium metal is clearly superior to any molten salt for tritium breeding purposes. Certain approaches to fusion, of which laser fusion is the best example, can certainly breed sufficient tritium using molten salts alone. However, the indication that tritium breeding is marginal for the salts in some (if not most) reactor designs represents the greatest drawback to their use. Fluoride salts are also inferior to lithium in that the salts are more intense gamma sources and cause increased gamma heating of the vacuum wall [99], primarily because of fluorine's relatively high cross section for inelastic neutron scattering.

Tritium recovery should prove much more simple if molten salts are used; this is particularly true if the tritium can be maintained as TF to minimize diffusion through metallic walls.

Several of the physical properties of lithium (thermal conductivity, specific heat, viscosity and melting point, for example) are superior to those of the molten salt. However, since the magnetic field will prevent turbulent flow in the blanket for lithium, but not for the salt, it may be that the molten salt is a better heat transfer medium in reactors based on the magnetic confinement scheme.

Lithium is compatible with relatively few structural metals. Moreover, lithium reacts with graphite, and this material must certainly be clad if it is to serve in a lithium-cooled blanket assembly. Molten Li_2BeF_4 is compatible with graphite and with a wide variety of structural metals. Corrosion by the salt is possible, through interaction with strong magnetic fields, but such corrosion can apparently be avoided by careful design. Transmutations within the salt could lead to corrosive reactions. However, adequate means can probably be found to deal with this problem.

Reactions of salts with steam or with air produce accelerated corrosion but, unlike similar reactions of lithium, lead to no inherently hazardous conditions.

It is clearly not possible at this stage of the technology to predict with confidence how the problems inherent in use of molten salts (or of lithium) will be solved. It is entirely possible that both lithium and molten salts will be useful. In any event, and regardless of the ultimate choice, it is clear that much research and development is required to solve the problems associated with the CTR uses of these materials.

V. SUPERCONDUCTING MATERIALS

Superconducting magnets are being considered for use in producing the confining fields in several fusion reactor conceptual designs. It has been estimated that the cost of these magnets will represent a considerable fraction of the capital cost of a fusion power plant [111–116]. The performance of these magnets will have an impact on the plasma, as well as the power density of the plasma. Therefore, materials research that may improve the performance limits of the large superconducting magnets may also improve the performance of the fusion power reactor as a whole, as well as possibly reduce the capital costs of the power plant. If it can be determined how material properties of the various components in a superconducting magnet affect its performance, then it may be possible to identify areas where materials research would have the largest potential impact on superconducting magnet performance. Also the potential effects of radiation damage on the transport properties of the superconductor and stabilizer materials may be assessed.

The application of superconducting magnet technology to thermonuclear fusion reactors of the θ-pinch type will require the development of inductive storage capacity. Large magnets capable of extremely rapid charging and discharging rates as well as rotating machinery with superconducting field coils for the rapid and reversible transfer of extremely large quantities of stored energy are needed.

Another need is in the area of power transmission from large power production centers. Serious proposals are under consideration for the development of superconducting power transmission lines for the highly efficient transmission of large blocks of power over large distances and through connecting regional and national grids.

Currently available superconducting materials are just barely adequate for projected requirements even in systems cooled by liquid helium. Certainly, the discovery of new materials with only slightly better properties would be very valuable. It is also true that materials with the highest transition temperatures are brittle and satisfactory techniques for the fabrication of these materials have not been developed. Thus, materials development must include a continuing effort to improve fabrication techniques as well as a continuing search for new materials with enhanced superconducting properties.

A. Theoretical and Empirical Correlations

It was not until the high critical field capability of Nb_3Sn was discovered by Kunzler *et al.* in 1961 [117] that earlier hopes were revived for practical, high field, superconducting magnets.

During the past decade, many laboratories have been engaged in the search for improved superconducting materials. This effort has resulted in the discovery of hundreds of superconducting compounds. However, despite many optimistic predictions of high temperature superconductivity, the highest known transition temperature has increased only from 18.3°K for Nb_3Sn to ~ 20.3°K for Nb_3Ga [118] and ~ 20.5–21.0°K for $Nb_3(Al,Ge)$ [119, 120]. From the viewpoint of magnet technology, the most significant developments have been the previously mentioned discovery of the high field capability of Nb_3Sn and the discovery of ductile alloys of Nb–Zr and Nb–Ti with transition temperatures of 10.8–9.8°K and critical fields (at 4.2°K) up to 120 kG [121–123].

Potentially, the compound $Nb_3(Al,Ge)$ is the best material known at this time for use in high field magnets. In addition to the high T_c, critical fields > 400 kG have been measured at 4.2°K [124] and ~ 200 kG at 14°K [125]. However, fabrication techniques for this material have not been developed. The problems involved, together with possibilities for fabrication, will be discussed later.

To make a first-principles calculation of the superconducting properties of any metal requires a knowledge of electron and phonon spectra together with details of electron–phonon interaction matrix elements and the Coulomb interaction between electrons. Thus, it is not surprising that only limited success has been realized in attempts to calculate effective interactions even for simple metals.

In the absence of sufficient knowledge of normal state properties for the application of theory in the search for better superconductors, many attempts have been made to correlate observed T_c values with simple, readily measured normal state parameters. For example, when T_c is related to the average number of valence electrons per atom in an alloy or compound, T_c maxima are found for electron/atom values near 5 and near 7 [126]. This concept has been extended by including molar volumes. Thus, the valence electron density is related to T_c [127].

A large fraction of the known high temperature superconductors have been found in A_3B compounds in the A-15 (β-W) crystal structure with Nb or V in the A positions and nontransition elements of group 3 and 4 in the B positions. The structure has attracted the attention of theorists and has been cited as evidence for the value of empirical rules. Unfortunately, attempts to explain widely different T_c values either by theoretical arguments or by empirical rules have been confused by effects due to nonstoichiometry in the samples studied. The great significance of stoichiometry in the A-15 structure has not been widely recognized until quite recently.

It is interesting to speculate whether stoichiometric, ordered A-15 structures containing Nb and nontransition elements (if they could be prepared) might all exhibit similar high T_c values. Such speculation would be consistent with the conclusion [128] that A_3B compounds in the A-15 structure should all have similar band structures. Perhaps, therefore, the differences in T_c that appear to exist, even for the ordered structures, result from differences in phonon spectra. This could explain the different T_c values found for the corresponding V compounds. It would be very valuable to be able to correlate different T_c values with measurable normal state properties. Unfortunately, very few studies have been performed on stoichiometric, ordered structures. Consequently, attempts to develop empirical correlations that are valid for a variety of crystal types will be useful only after additional studies have been made on well-characterized materials.

From the viewpoint of a search for new, high temperature superconductors, the possible relationship between crystallographic instability and high temperature superconductivity implies that new high temperature superconductors may be very difficult to find because they are masked by more stable phases. Consequently, it has been suggested that one might stabilize desired structures by the use of additives to form ternary or even quaternary phases [129, 130]. Unfortunately, there are very few useful criteria to use in the attempt to prepare a new metastable phase. For the immediate future at least, experiment and empirical correlations will still be required. Indeed, new correlations must be developed.

B. Fabrication Development

Fabrication techniques have not been developed for the effective utilization of the highest T_c and critical field materials in magnet fabrication. From the viewpoint of proposed designs for fusion reactors, techniques for the fabrication of such materials in contrast to current technology based on NbTi alloys would have a major impact on the development of CTR magnets for steady state as well as pulsed reactors. For example: (1) the current density under high field (at 4°K) could be increased by at least a factor of 3, reducing the volume of superconductor required (and also the cost) correspondingly; (2) the raising of current density would relieve structural problems and costs; (3) at chosen magnetic fields, the operating temperature could be higher, thus reducing refrigeration costs, or reducing insulator (and blanket) thickness requirements; (4) the working magnetic fields could actually be significantly higher. This could be important to plasma properties and the demonstration of scientific feasibility.

For a typical ductile superconductor like NbTi which can be coextruded and coreduced, the upper useful field is ~ 85 kG. Above this field the compounds Nb_3Sn and V_3Ga could be used. These are commercially available only in the form of thin ribbons at the present time. Although the Nb_3Sn and V_3Ga tapes have been used in small high field magnets, they have some disadvantages which may render them unsuitable for the large fusion reactor magnets. They are brittle, fragile, not easily joined in lossless joints, exhibit unstable performance in perpendicular fields, and are not readily stabilized electrically or thermally.

Multifilament Nb_3Sn and V_3Ga metallurgically bonded in a copper matrix have been developed. In the case of V_3Ga, complete reaction of the vanadium core was achieved, but in the Nb_3Sn wires the niobium core remains with a coating of Nb_3Sn. In both materials an outer jacket of copper of low resistivity was obtained. The potentially very important binary compound, Nb_3Al (with an upper critical field of 295 kG and a critical temperature of $18.7°K$) as well as the more complicated pseudobinary $Nb_3(Al_xGe_{1-x})$ (with an upper critical field of 410 kG and a critical temperature of $20.7°K$) are not developed even in ribbon form. For the most part, all the experiments have been on small-sized samples, but a long length (2000 ft) of V_3Ga has been successfully prepared and wound into a small solenoid.

There are at least four NbTi superconducting magnets with bores of 1 m or larger in operation: the 1.0 m, 41 kG, 10.5 MJ BIM magnet at Saclay [131]; the 2.44 m, 28 kG, 64 MJ bubble chamber magnet at Brookhaven [132]; the 4.8 m, 18.5 kG, 80 MJ bubble chamber magnet at Argonne; and the 4.8 m, 30 kG, 400 MJ bubble chamber magnet at the National Accelerator Laboratory [133]. In addition to these solenoids, the baseball (minimum B design) magnet at Livermore [134] has approximately a 1.2 m bore and has operated with a maximum field at the windings of 55 kG and stored energy of 10 MJ. All of these magnets use the cryogenic stability design concept (i.e., superconductor embedded in a large amount of copper and liquid helium in intimate contact with all the windings). No large adiabatically stabilized NbTi magnet or large high field Nb_3Sn or V_3Ga magnets have been built. The IMP quadrupole magnet (minimum B design; 30 cm bore) at ORNL is perhaps the most complex Nb_3Sn magnet, and it has been operated with a maximum field at the winding of about 60 kG [135].

Less work has been done on pulsed systems. Small magnets mostly for synchrotron programs ($E_s < 100$ kJ), have been developed which can be pulsed at 100 kG/sec up to 50 kG without undergoing a quench [136]. The largest energy storage magnet appears to be the 600 kJ, 76 cm bore coil built at the Laboratory de Marcoussis of Compagnie Generale d'Electricite [137].

C. Physical and Material Constraints on Magnet Performance

In order to assess the full impact of an improvement in a given property of a superconductor on magnet design, it is necessary to study optional coil design that satisfy all the constraints imposed by a specified set of material properties. Such a study has been made for a toroidal magnet system of a Tokamak reactor concept [138]. The coil was assumed to be a circular coil with a rectangular cross section made up of a superconductor, a stabilizer material, a structural material, and a coolant within internal coolant passages. The coolant passages separate subcoils which make up an individual coil.

A computer analysis was used to maximize the magnetic field by finding that particular configuration of superconductor, stabilizer, structural material, and coolant that maximizes the total current in the coil and thus maximizes the magnetic field in the plasma region while satisfying *all* the inequality constraints imposed on the total system.

Four basic material combinations were studied. Two superconductors, Nb_3Sn and NbTi, and two stabilizers, copper and aluminum, were considered in four designs. Material properties of the superconductor taken as parameters were

(1) critical current density,
(2) critical field (varied from 0.5 to 1.5 times the reported value),
(3) percent deflection limit, and
(4) the parameter τ, $J_c(\partial J_c/\partial T) - 1$.

Material properties of the stabilizer materials taken as parameters were

(1) the electrical and thermal conductivity,
(2) the yield stress limit, and
(3) the modulus of elasticity (Young's modulus).

In addition, the modulus of elasticity of the structural material was varied over two orders of magnitude to represent ways of improving the coils apparent stiffness by improvements in their mechanical support. Furthermore, the heat transfer coefficient between coolant and stabilizer was allowed to vary in the calculations to represent changes in this quantity induced by pressurizing the helium cryostat.

The results of the computer analysis showed that subject to the various systems constraints, Nb–22% Ti is superior to Nb_3Sn as a superconductor

[138], primarily because of the larger percent deflection specified for Nb–22% Ti. Nb_3Sn was limited to 0.04%, while Nb–22% Ti was limited to 0.1%. These two values assume the coil will be wrapped in compression.

Increasing the stiffness of the coil resulted in largest improvements in the field particularly when the superconductor is Nb_3Sn. Nb–22% Ti soon became limited by its critical field limit.

It is possible, using existing materials, to build the magnets envisioned for CTRs, although it will be a difficult and expensive task. Thus, efforts should continue in the search for better superconducting materials and the development of known materials to commercial availability. The desired new materials should exhibit intrinsic stability and would remain superconducting at higher fields, current densities, and (especially) temperatures, to ease the stability problems in the large magnet design. They should be economical, ductile, and easy to fabricate.

Of equal or greater importance is development of superconducting materials which are now available. Compounds such as Nb_3Sn, V_3Ga, V_3Si, or $Nb_3(Al_xGe_{1-x})$ are known to be capable of operation at high magnetic fields, but they are also very difficult to fabricate and expensive. Development of these materials to exhibit better fabricability and better methods of fabrication such as plasma spraying, diffusion bonding, powder metallurgy, electron beam welding, electroplating, sputtering, and other advanced fabrication techniques, may significantly reduce the cost of magnet fabrication with these advanced materials.

Closely coupled to the fabrication of the superconducting materials is the fabrication of the normal conductor and the reinforcing structure. The structure accounts for a major portion of the magnet cost, even with present day costs of superconducting magnet materials. It has been postulated [139] that the cost of the superconducting material itself may be expected to reduce by factors of ~ 3–9 by the time large amounts of material are required for fabrication of CTRs. If this occurs, the cost of the supporting structure then becomes an overwhelming part of the magnet costs. Backing material currently used in large magnets has been stainless steel. Although the allowable yield or creep stress of this material has been improved, elasticity is a more crucial structural property since the superconducting compounds are generally strain limited. Thus, if fabrication methods cannot be found to eliminate the strain limitation of the superconductors, it will be necessary to develop a backing structure to give very low strain even at high stresses. This is the most critical task to be accomplished for reduction of the costs of superconducting magnets.

Work should continue on the development of cryoresistive magnets, that is, magnets made with winding materials with very low resistance such as high purity aluminum or sodium. This work should be considered as a backup to the superconducting magnet program. The structural problems are equally severe in these cases. If it is possible to operate these coils with only a small portion of the CTR power then they might become economic compared to either superconducting or normal room temperature magnets.

Superconducting materials are discussed in more detail in Chapter 10.

VI. TRITIUM PERMEATION BARRIERS

Prototypical fusion power plants will require tritium inventories in excess of 10^7 Ci and it is likely that the permissible level for tritium release to the environment from such plants will be limited to a few parts per million of the total inventory per day. Should future standards for radioisotope emissions become more stringent than they currently are, tritium releases in excess of 10–20 Ci/day may eventually be construed as an objectionable radiologic hazard to the environment. In addition to the environmental considerations, there are other potentially significant consequences of the tritium in fusion reactors and they too must be assessed. These include hold-up of tritium in the structure, accelerated corrosion and loss of mechanical integrity due to hydriding, as well as assorted maintenance problems.

The principal objective with regard to control and handling of tritium is to minimize the total inventory and loss rate of tritium in every sector of the plant. Both economics and safety, therefore make it mandatory to have reliable information on the permeation of tritium through the various components of the reactor.

The rate of permeation of tritium through the first wall and through the tubing in heat exchangers will be unavoidably high because: (1) the temperatures will be high (600–1000°C), thereby increasing the permeability of any material; (2) two of the candidate wall materials (Nb and V) have extremely high permeabilities, and a third candidate (Mo) has a moderate permeability.

Since all materials and coatings have finite permeabilities and leakages, there will be a finite rate of transport of tritium to the steam loop, to the evacuated region between the blanket and the magnet, and to the immediate surroundings. The problem is to ensure that this rate is below the level determined by safety considerations and by economics.

In this section, the permeability of hydrogen and its isotopes through material walls be discussed as well as the possibilities for developing materials to provide a contiguous system of barriers (both metal and ceramic) which are relatively impervious to the hydrogen isotopes in any chemical form, and which could help to solve the problems associated with tritium containment.

The important properties of the hydrogen isotopes that will determine their distribution and leakage characteristics in currently conceived fusion reactors are their solubilities in the materials with which they come in contact, their diffusion rates through the working fluids and structural components, and the stable chemical forms in which they exist. Before discussing the more conventional thermodynamic and kinetic aspects of hydrogen permeability, it is necessary to consider the possible influence of fusion reactor conditions on this property.

A. Permeation of Tritium through the First Wall of a Fusion Reactor

The concentration of tritium in the plasma will depend upon the net permeation through the first wall, and this rate may change in both magnitude and direction during the transient conditions associated with bringing the plasma up to the desired steady state (or periodically pulsed) operating conditions. For example, the direction of net permeation may be from the plasma to the blanket initially when the concentration of tritium in the blanket is very low; however, the direction may reverse at a later time when the concentration in the blanket increases as a result of tritium production by neutron–lithium reactions. At present, it is not possible to make a reliable estimate of this permeation rate because of the following complications:

(1) The highly nonequilibrium conditions existing at the plasma side of the first wall prevent direct application of the conventional permeation equations. Specifically, the tritium particles escaping from the plasma will have sufficient energy (up to ~ 20 keV) to be implanted (i.e., to penetrate the surface of the wall) in the lattice, thereby circumventing the steps of adsorption, dissociation, and absorption (penetration) that are necessary in the conventional case where the particles impinging on the surface are molecules. Even if a theoretical treatment were available for calculating the permeation rate resulting from implanted tritium, the flux, energy distribution and directional distribution of the impinging tritium particles are not well defined.

(2) Accurate data are not available for the solubility, diffusivity, and permeability of tritium in the candidate materials for the first wall. Since isotopic effects can be estimated, available data for hydrogen should provide a reasonably accurate basis for extrapolations to tritium. However, there is very little data available on the diffusivity and permeability of hydrogen in niobium and vanadium, and there is evidence that the solubility and diffusivity of hydrogen in these metals are quite sensitive to the common impurities, O, N, and C. It is conceivable that the penetration through the first wall may change with time as the impurity content is altered by degassing or by transfer of impurities to or from the lithium.

(3) The properties of the wall will change with time as the result of bombardment by energetic particles (n, D, T, He) and by plasma radiation. Radiation damage leads to dislocations, pores, and blisters as already discussed which may influence the solubility, diffusivity, and permeability by introducing trapping sites (e.g., sites where the bond between tritium and the metal is stronger than for the usual interstitial site) or cavities where tritium comes out of solution to form either gaseous molecular tritium or a molecular species involving impurities such as C or O. The properties of the wall will also change because of transmutations and subsequent decay processes, thereby causing the composition of the alloy to vary with time. There is some evidence that the solubility, diffusivity, and permeability of hydrogen in alloys generally are influenced substantially by the composition of the alloy.

(4) Calculation of the permeation of tritium from the wall into the lithium blanket requires knowledge of (a) the concentration of tritium in the lithium layer adjacent to the wall, and (b) the concentration and temperature dependence of the chemical potential of tritium in lithium. The first point poses a problem because the concentration will depend upon the diffusivity of tritium in lithium, the motion of the lithium resulting from convection and magnetohydrodynamic effects, and the spatial distribution of the production of tritium in the blanket by neutron–lithium reactions. The second point will benefit from current experimental studies of the thermodynamic properties of the hydrogen–lithium system [76].

B. Permeation of Hydrogen Isotopes

Webb [140] and Stickney [141] have reviewed the principles of the permeation process for hydrogen isotopes. The expression which relates permeation to the operating parameters is given by

$$\Phi = (A/X)(P_H^{1/2} - P_L^{1/2})B \exp(-Q_p/RT)$$

where Φ is the flux of hydrogen through the interface, A and X the area and thickness of the interface, P_H and P_L the hydrogen pressures on each side of the interface $(P_H > P_L)$, B and Q_p parameters related to the material under consideration, and R and T the gas constant and absolute temperature. Values of B and Q_p for the permeation of protium through several materials considered for use in various CTR reactor designs are listed in Table 2. In most cases only limited data are available for deuterium and tritium permeation through these materials. However, one can estimate the deuterium and tritium permeation rates by using the values of B and Q_p for protium and multiplying the result by the square root of the mass of the hydrogen isotope of interest.

TABLE 2

Parameters for the Permeability, Φ, of Protium[a] Through the Reference Design Construction Materials

| Material | $\Phi \sim B \exp(-Q_p/RT)$ | | Reference |
	B (g cm/cm^2-day-Torr$^{1/2}$)	Q_p (cal/mole)	
PE-16	0.65×10^{-2}	16,100	b
Incoloy 800	0.65×10^{-2}	16,100	b
Niobium	1.6×10^{-2}	5,200	142
Copper	0.86×10^{-2}	18,562	18V
Aluminum	2.0	30.932	19V

[a] Values of Φ for D_2 and T_2 were obtained by multiplying the values for H_2 by $\sqrt{2}$ and $\sqrt{3}$, respectively.

[b] The values of Φ for hydrogen in PE-16 and Incoloy 800 are not expected to be significantly different from those for 304-SS.

C. Permeation Barriers

Using permeation ratios obtained experimentally as estimated in the manner described in Section B above, calculations on tritium leakage rates from a fusion reactor have been performed by several workers [142–144]. In particular, there seems to be general agreement that the heat transfer system represents the major and potentially most hazardous tritium loss route. Since consideration of permeation-resistant coatings arises from the need for reducing the permeation of tritium through the walls of heat exchangers primary interest centers on coatings that are suitable for iron or nickel base alloys at 500–600°C and exposure to either potassium or steam.

Oxide coatings on metals have been studied to some extent as permeation barriers. Results on hydrogen diffusion in oxides are available on a limited number of materials [145–150]. Little or nothing is published on other large classes of compounds, such as nitrides, carbides, and halides. Of the papers available on the oxides, all but one restrict their attention to diffusion coefficients of hydrogen in the host lattice, with no attention paid to either the temperature or pressure dependence of solubility on overall permeability.

From measurements of the changes of the spectroscopic and electronic properties of the solids, it has been established that the hydrogen dissolves in the atomic form, forming hydrogen bonds with the oxygen. Solubility should therefore be dependent on the square root of pressure, just as it is for metals.

The one case where solubility has been carefully measured (ZnO) shows the square root dependence.

The calculated permeability shows ZnO to be less permeable than austenitic stainless steel [151] but more permeable than tungsten at very high temperature [152]. It appears that ZnO is far from the most effective barrier material. A better choice would be alumina, and beryllia seems to be even better.

The solubility of hydrogen in alumina is probably less than that in ZnO, based on the known solubility mode in these two materials combined with knowledge of the oxygen–metal diatomic bond strengths. Assuming a somewhat lower value of solubility of H_2 in Al_2O_3 than in ZnO and reasonable diffusion coefficients, the permeability is 2.5×10^{-7} times less than that of stainless steel at $1000°K$. If a 1-mm-thick wall of stainless steel were used, the permeation rate could then be reduced by 10^3 by applying a 10-μin. layer of alumina [144].

Steigerwald [153] tested a number of metallic oxide, and glass coatings on types 303 and 304 stainless steel and on Haynes Alloy No. 25 (a high temperature cobalt base alloy). The metallic coatings included Ag, Cu, Cr, Al, W and Si, as well as the combinations W–Si, Mo–Si, Zr–Si, and V–Si. W coatings were applied to the stainless steels and Haynes 25 by two different methods, vapor dtposition in vacuum and chemical vapor deposition using tungsten hexachloride, but in all cases the coatings cracked and spalled when annealed at 1000°C. As an attempt to remedy this problem by improving the bonding between the W coating and the substrate, a flash nickel plating was applied to the surface before deposition of W. The results were unsatisfactory, however, because the coating developed porosity and cracks during annealing. The same difficulty was encountered with Cr coatings, which agrees with the results of earlier tests by Flint [154]. Cu and Ag reduced the permeability of stainless steel but the reduction was insufficient because the permeability of Cu and Ag are not extremely low.

Although Steigerwald's results indicate that aluminum is not a satisfactory coating material for type 304 stainless steel, the results of Rudd and Vetrano [155] demonstrate that permeation through type 430 stainless steel (sample thicknesses ranging from 1.5 to 3.1 mm) may be reduced by approximately 100-fold at 650°C by calorizing (aluminizing) the surface. A calorized coating appears to consist mainly of an aluminum–iron intermetallic compound with a surface film of aluminum oxide. Furthermore, Flint [154] has reported that the permeability of type 347 stainless steel at 860°C is reduced approximately 100-fold by calorizing plus oxidation. The discrepancy may arise from the effect of temperature on the structural and chemical properties of the calorized surface layer; for example, Steigerwald annealed his coatings at $\sim$ 900°C, whereas Rudd and Vetrano did not heat their coatings above 760°C. It is also possible that the discrepancy may be related to differences in the calorizing process or the substrate material.

In the hope of forming a stable intermetallic compound having low permeability, Steigerwald tested coatings of various metal silicides on Haynes 25. W–Si proved to be the best coating, and, on the basis of extrapolation of data taken at slightly higher temperatures, it appears to reduce the permeability of a 1-mm-thick sample by approximately 100-fold at 600°C.

Steigerwald evaluated five glass coatings on type 304 stainless steel and on Haynes 25, and concluded that Solaramic (proprietary composition and application process) possessed the best overall properties on the basis of long-term stability, ease of application, and permeability. Since Solaramic reduces the permeability of a 1-mm-thick sample of type 304 stainless steel only by slightly more than one order of magnitude at 600°C, it does not appear to be suitable for the potassium–steam heat exchanger. Steigerwald's results indicate by extrapolation that the permeability of a 1-mm-thick sample at 600°C may be reduced by more than 200-fold by other glass coatings, such as A. O. Smith #3308 and Nucerite SC-30. Reductions of this order of magnitude have also been reported by others [154, 155] for glass coatings on types 304 and 347 stainless steel. It should be emphasized, however, that the results of these studies clearly illustrate that the use of glass coatings is severely limited by their limited ability to withstand thermal cycling.

There have been a number of investigations of the reduced rate of permeation of hydrogen through samples having oxide coatings formed by intentional oxidation of the surfaces [153, 154, 156]. A thorough study of the influence of oxidation processes on the permeability of type 347 stainless steel [154], found that the permeation rate at 630°C was not affected by changing from dry hydrogen to wet hydrogen (i.e., an $H_2 + H_2O$ mixture formed by bubbling H_2 through water at room temperature). However, heating the sample to $\sim 1000°C$ in wet hydrogen for 3 hr produced a green oxide coating that greatly reduced the permeation rate at that temperature and below. For example, upon cooling to $\sim 725°C$ the measured rate was more than 100-fold lower than the rate for an unoxidized sample at the same temperature. Furthermore, the effectiveness of the coating at $\sim 725°C$ was stable even when the wet hydrogen was replaced by dry hydrogen; at $\sim 1000°C$, however, the coating was stable when wet hydrogen was continuously supplied, but dry hydrogen quickly reduced the oxide to the degree that the permeation rate approached that of an unoxidized sample. The effectiveness of the coating was degraded by thermal cycling to lower temperatures; for example, the permeability at $\sim 725°C$ increased six-fold after the coated sample was cooled to room temperature and then reheated to the same level. The degradation of the coating may be the result of partial disintegration of the oxide by: (1) thermal stresses produced by differences in the coefficients of thermal expansion of the oxide and the metal; and/or (2) stresses generated at the oxide–metal interface by the rapid evolution of hydrogen from the metal during temperature and/or pressure transients. As evidence in support of the second point, it was found that the permeability of an oxide-coated sample increased 15-fold after a temporary reversal in the direction of hydrogen flow [154]. (Disintegration of the coating is expected to be greater during reversed flow because the hydrogen concentration at the oxide–metal interface will be higher when the flow direction is from metal to oxide rather than from oxide to metal.) It appears that the stability of the coating is improved if the sample is first exposed to wet hydrogen and then to steam, both exposures being at $\sim 1000°C$. Results for types 304, 316, and 410 stainless steel were similar to those for 347, whereas the results for type 321 indicate that its permeability is not influenced significantly by exposure to wet hydrogen at $\sim 1000°C$. Since the composition of type 321 stainless steel is similar to the other types tested except for the addition of a small amount (0.6%) of titanium, the unique behavior of type 321 may be related to the segregation of titanium to the surface where it is oxidized to form a discontinuous or porous coating that does not have low permeability.

Other promising approaches to the development of permeation barriers are metal composite materials or laminate compounds for example of ceramvar and copper. Questions of bond formation, mechanical integrity and temperature stability of these materials are among those requiring further investigation [157, 158].

REFERENCES

[1] R. G. Mills (ed.), "A Fusion Power Plant." Princeton Plasma Physics Laboratory, 1974 (in preparation).

[2] G. Lewin and F. H. Tenney, *Proc. Int. Vac. Congr., 6th; Jpn. J. Appl. Phys. Suppl. 2* Pt. 1, 221 (1974).

[3] S. C. Burnett, W. R. Ellis, and F. L. Ribe, *Texas Symp. Technol. Contr. Thermonucl. Fusion Exp. and the Eng. Aspects of Fusion Reactors* Paper 1, Session II (November 1972).

[4] F. L. Ribe, *Proc. Int. Conf. Nucl. Solutions to World Energy Probl., Washington, D.C.* p. 226 (November 13–17, 1972).

[5] M. T. Robinson, *in* "Nuclear Fusion Reactors," pp. 364–378. British Nuclear Energy Society, London, 1970.

[6] D. G. Martin, Radiation Damage Effects in the Containment Vessel of a Thermonuclear Reactor, CLM-R-103 (March 1970).

[7] F. W. Wiffen, *Proc. Int. Working Sessions on Fusion Reactor Technol.* CONF-710624, ORNL (July 1971).

[8] J. Moteff, Irradiation embrittlement in the BCC metals, *Proc. Int. Working Sessions on Fusion Reactor Technol.* CONF-710624, p. 175, ORNL (July 1971).

[9]   M. Kangilaski, Radiation Effects Design Handbook, Sect. 7, Structural Alloys, NASA-CR-1873 (October 1971).

[10] W. J. Stapp and A. R. Begany, Seventh Annual Report, AEC Fuels and Materials Program, GEMP-1004, pp. 137–145 (March 29, 1968).

[11] H. Bohm, W. Dienst, H. Hauck, and H. J. Laue, *in* "The Effects of Radiation on Structural Materials," ASTM-STP-426, pp. 95–106. American Society for Testing Materials, Philadelphia, Pennsylvania, 1967.

[12] E. E. Bloom and J. O. Stiegler, *in* "Irradiation Effects on Structural Alloys for Nuclear Reactor Applications," ASTM-STP-484, pp. 451–467. American Society for Testing Materials, Philadelphia, Pennsylvania, 1971.

[13] D. R. Harries, *J. Brit. Nucl. Energy Soc.* **5**, 74–87 (1966).

[14] G. J. C. Carpenter and R. B. Nicholson, *in* "Radiation Damage in Reactor Materials," Vol. II, pp. 383–400. International Atomic Energy Agency, Vienna, 1969.

[15] K. Ehrlich and H. Bohm, *in* "Radiation Damage in Reactor Materials," Vol. II, pp. 349–355. International Atomic Energy Agency, Vienna, 1969.

[16] S. D. Harkness, A means of studying radiation-controlled creep in refractory metals for fusion reactor design, *Proc. Int. Working Sessions on Fusion Reactor Technol.* CONF-710624 p. 183 ORNL (July 1971).

[17] W. A. Ranken, SEPO Quarterly Progress Report for Period Ending October 31, 1972, Los Alamos Scientific Laboratory Rep. LA-5113-PR (1972).

[18] J. W. Patten, E. D. McClanahan, and J. W. Johnston, Room temperature recrystallization in thick bias-sputtered copper deposits, *J. Appl. Phys.* **42**, 4371 (1971).

[19] W. H. Reichelt, W. A. Ranken, C. V. Weaver, A. W. Blackstock, A. J. Patrick, and M. C. Chaney, Radiation induced damage to ceramics in the EBR-II reactor, presented at the *Thermionic Convers. Specialists Conf., Miami, Florida* (October 26–29, 1970).

[20] L. Spitzer *et al.*, Princeton Univ. Rep. NYO-6047 (1954).

[21] R. Carruthers, P. Davengport, and J. Mitchell, Culham Lab. Rep. CLM-R85 (1967).

[22] D. Rose, Oak Ridge National Lab. Rep. ORNL-TM-2204 (1968).

[23] M. Kaminsky, *Proc. Int. Summer Inst. Surface Sci.* Univ. of Wisconsin, Milwaukee, Wisconsin (August 1975).

[24] M. Kaminsky, "Atomic and Ionic Impact Phenomena on Metal Surfaces." Springer-Verlag, Berlin and New York, 1965.

[25] G. Carter and J. Colligon, "Ion Bombardment of Solids." American Elsevier, New York, 1968.

[26] M. Kaminsky, *J. Vac. Sci. Technol.* **8**, 14 (1971).

[27] M. Kaminsky, *in* "Recent Developments in Mass Spectrometry" (K. Ogata and T. Hayakaw, eds.), p. 1167ff. University Park Press, Baltimore, Maryland, 1970.

[28] R. Behrisch, *Can. J. Phys.* **46**, 527 (1968).

[29] G. M. McCracken and N. J. Freeman, *J. Phys. Ser. B* **2**, 661 (1969).

[30] K. Erents and G. M. McCracken, *Radiat. Effects* **3**, 123 (1970).

[31] K. Erents and G. M. McCracken, *Brit. J. Appl. Phys.* **2**, 1397 (1969).

[32] M. Kaminsky, *IEEE Trans. Nucl. Sci.* **NS18**, No. 4, 208 (1971).

[33] R. Barnes and D. Mazey, *Proc. Roy. Soc. London* **275**, 47 (1963).

[34] W. Primak, *J. Appl. Phys.* **35**, 1342 (1964).

[35] R. S. Nelson, *Phil. Mag.* **9**, 343 (1964).

[36] M. Kaminsky, *Adv. Mass Spectrom.* **3**, 69 (1964).

[37] R. Behrisch and W. Heiland, *Proc. Symp. Fusion Technol. 6th, Aachen.*

[38] D. M. Gruen, P. A. Finn, and D. L. Page, *Nucl. Technol.* **29**, 309 (1976).

[39] O. Yonts, *BNES Nucl. Fusion Reactor Conf., Culham, England* p. 424 (1969).

[40] D. E. Harrison, N. S. Levy, J. P. Johnson, and H. M. Effran, *J. Appl. Phys.* **39**, 3742 (1968).

[41] M. Robinson, *Proc. Nucl. Fusion Reactors Conf., Culham Lab.* p. 364ff (September 1969).

[42] B. Myers, *Proc. Nucl. Fusion Reactors Conf., Culham Lab.* p. 379ff (September 1969).

[43] M. Kaminsky, J. H. Peavey, and S. K. Das, *Phys. Rev. Lett.* **32**, 589 (1974).

[44] D. Menzel and R. Gomer, *J. Chem. Phys.* **41**, 3311, 3329 (1964).

[45] D. Lichtman and T. K. Kirst, *Phys. Lett.* **20**, 7 (1966).

[46] D. R. Sandstrom, J. K. Leck, and E. E. Donaldson, *J. Chem. Phys.* **48**, 5683 (1968).

[47] G. Fischer and R. Mack, *J. Vac. Sci. Technol.* **2**, 123 (1965).

[48] M. Bernardini and L. Malter, *J. Vac. Sci. Technol.* **2**, 130 (1965).

[49] J. Haber and F. Stone, *Trans. Faraday Soc.* **59**, 192 (1963).

[50] W. J. Lange, *J. Vac. Sci. Technol.* **2**, 74 (1965).

[51] R. O. Adams and E. E. Donaldson, *J. Chem. Phys.* **42**, 770 (1965).

[52] F. W. Clinard, Jr. and J. M. Bunch, Private communication (January 1974).

[53] D. B. Watson and W. Heyes, *J. Phys. Chem. Solids* **31**, 2531 (1970).

[54] D. Stull and H. Prophet, JANAF Thermochemical Tables, 2nd ed., NBS Rep. NSRDS-NBS-37 (1971).

[55] J. W. Tester, R. C. Feber, and C. C. Herrick, Heat Transfer and Chemical Stability Calculations for Controlled Thermonuclear Reactors (CTR), USAEC Rep. LA-5328-MS (July 1973).

[56] J. H. DeVan and R. L. Klueh, *Trans. Am. Nucl. Soc.* **17**, 149 (1973).

[57] G. Lewin and F. H. Tenney, *Proc. Int. Vac. Congr., 6th* (1974); *Jpn. J. Appl. Phys. Suppl.* 2 Pt. 1, 221 (1974).

[58] G. M. McCracken, J. H. C. Maple, and H. H. H. Watson, *Rev. Sci. Instrum.* **37**, 860 (1966).

[59] G. M. McCracken and J. H. C. Maple, *Brit. J. Appl. Phys.* **18**, 919 (1967).

[60] G. M. McCracken, D. K. Jefferies, and P. Goldsmith, *Proc. Int. Vac. Congr., 4th* Part I, p. 149 (1968).

[61] G. M. McCracken, and D. K. Jefferies, *Symp. Fusion Techn., 5th, Oxford* Paper No. 149 (1968).

[62] G. M. McCracken, and S. K. Erents, *BNES Nucl. Fusion Reactor Conf., Culham, England* Paper No. 4.2, p. 353 (1969).

[63] D. M. Gruen, " The Chemistry of Fusion Technology," p. 215. Plenum Press, New York, 1972.

[64] I. Sheft, A. Reis, D. M. Gruen, and S. W. Peterson, *J. Nucl. Mater.* **59**, 1 (1976).

[65] O. C. Yonts and R. A. Strehlow, *J. Appl. Phys.* **33**, 2903 (1962).

[66] H. E. Schiøtt, *I. Dan. Vidensk. Selsk. Mat.-Fys. Medd.* **35**, No. 9 (1968).

[67] A. Sawatsky, *J. Nucl. Mater.* **9**, 364 (1963).

[68] D. Zamir and R. M. Cotts, *Phys. Rev.* **134**, A666 (1964).

[69] R. P. Marshall, *Trans. Met. Soc. AIME* **233**, 1449 (1965).

[70] W. M. Albrecht, W. D. Goode, and M. W. Mallett, Battelle Memorial Inst. Rep., No. B.M.I. 1332 (1959).

[71] J. R. Morton and D. S. Stark, *Trans. Faraday Soc.* **56**, 354 (1960).

[72] W. M. Mueller, J. P. Blackledge, and G. G. Libowitz (eds.), " Metal Hydrides." Academic Press, New York, 1968.

[73] J. O. Cowles and A. D. Pasternak, Lithium Properties Related to Use as a Nuclear Reactor Coolant, USAEC Rep. UCRL-50647 (April 1969).

[74] J. D. Lee, *in* " The Chemistry of Fusion Technology" (D. M. Gruen, ed.), Chapter 2. Plenum Press, New York, 1972.

[75] E. J. Cairns, F. A. Cafasso, and V. A. Moroni, *in* " The Chemistry of Fusion Technology" (D. M. Gruen, ed.), Chapter 3. Plenum Press, New York, 1972.

[76] E. Veleckis, *in* Physical Inorganic Chemistry Semiannual Report, July-December 1971, USAEC Rep. ANL-7878, p. 11 (March 1972).

[77] L. L. Hill, Unpublished Ph.D. Thesis, Univ. of Chicago, Chicago, Illinois (1938).
[78] C. E. Messer, A Survey Report on Lithium Hydride, USAEC Report NYO-9470 (October 1960).
[79] M. R. J. Perlow, Unpublished Ph.D. Thesis, Univ. of Chicago, Chicago, Illinois (1941).
[80] F. K. Heumann and O. N. Salmon, The Lithium Hydride, Deuteride and Tritide Systems, USAEC Rep. KAPL-1667 (December 1956).
[81] E. Veleckis and E. Van Deventer, ANL-7923, Argonne National Lab., Argonne, Illinois (1972).
[82] C. E. Messer, E. B. Damon, P. C. Maybury, J. Mellon, and R. A. Seales, *J. Phys. Chem.* **62**, 220 (1958).
[83] C. G. Swain, E. C. Stivers, J. F. Reuwer, Jr., and L. J. Schaad, *J. Am. Chem. Soc.* **80**, 5885 (1958).
[84] E. E. Hoffman, Solubility of Nitrogen and Oxygen in Lithium and Methods of Lithium Purification, in ASTM Special Tech. Publ. No. 272, p. 195 (1960).
[85] E. E. Hoffman, Corrosion of Materials in Lithium at Elevated Temperatures, USAEC Rep. ORNL-2674 (March 1959).
[86] P. I. Fedorov and M. T. Su, *Hua Hsueh Hsueh Pao (J. Chinese Chem. Soc.)* **23**, 30 (1957).
[87] D. R. Secrist, A Study of the Lithium-Boron-Carbon System, USAEC Rep. KAPL-2182 (1962).
[88] R. P. Eliott, "Constitution of Binary Alloys," 1st Suppl., p. 219. McGraw-Hill, New York, 1965.
[89] D. R. Secrist, A Study of the Lithium-Boron-Carbon System, USAEC Rep. KAPL-2182 (1962).
[90] J. M. McKee, W. H. Caplinger, and M. Kolodney, *Nucl. Appl.* **5**, 236 (1968).
[91] J. H. Devan, A. P. Litman, J. R. DiStefano, and C. E. Sessions, Lithium and potassium corrosion studies with refractory metals, in "Alkali Metal Coolants," p. 675. IAEA, Vienna, 1967.
[92] C. E. Sessions and J. H. DeVan, Effects of Oxygen, Heat Treatment and Test Temperature on the Compatibility of Several Advanced Refractory Alloys with Lithium, USAEC Rep. ORNL-4430 (April 1971).
[93] M. A. Hoffman and G. A. Carlson, Calculation Techniques for Estimating the Pressure Losses for Conducting Fluid Flows in Magnetic Fields, USAEC Rep. UCRL-51010 (1971).
[94] M. L. Kyle and J. R. Pavlik, in Development of High Energy Batteries for Electric Vehicles, Progress Report for the Period July 1970-June 1971, USAEC Rep. ANL-7888, p. 59 (1971).
[95] W. R. Grimes and S. Cantor, in "The Chemistry of Fusion Technology" (D. M. Gruen, ed.), Chapter 4, Plenum Press, New York, 1972.
[96] W. R. Grimes, Materials Problems in Molten Salt Reactors, in "Materials and Fuels for High Temperature Nuclear Energy Applications" (M. T. Simnad and L. R. Zumwalt, eds.). MIT Press, Cambridge, Massachusetts, 1969.
[97] JANAF Thermochemical Tables, 2nd ed., Nat. Std. Ref. Data Ser., NSRDS-NBS 37, U.S. Nat. Bur. of Std. (June 1971).
[98] W. G. Homeyer, Thermal and Chemical Aspects of the Thermonuclear Blanket Problem, Tech. Rep. 435, M.I.T. Res. Lab. of Electron. (1965).
[99] D. Steiner, The nuclear performance of fusion reactor blankets, *Nucl. Appl. Tech.* **9**, 83 (1970).
[100] A. J. Impink, Jr., Neutron Economy in Fusion Reactor Blanket Assemblies, Tech. Rep. 434, M.I.T. Res. Lab. of Electron. (1965).
[101] W. B. Myers, M. W. Wells, and E. H. Canfield, Tritium Regeneration in a D-T Thermonuclear Reactor Blanket, UCID-4480. Lawrence Livermore Radiation Lab. (1962).

[102] G. I. Ball, Neutron Blanket Calculations for Thermonuclear Reactors, LA-3385-MS. Los Alamos Scientific Lab. (1965).

[103] A. P. Malinouskas and D. M. Richardson, MSR Program Semiannual Rep. ORNL-4782, Oak Ridge National Laboratory (February 28, 1972).

[104] P. E. Field and J. H. Shaffer, *J. Phys. Chem.* **71**, 3218 (1967).

[105] G. M. Watson, R. B. Evans, III, W. R. Grimes, and N. V. Smith, Solubility of noble gases in molten fluorides, *J. Chem. Eng. Data* **7**, 285 (1962).

[106] R. E. Thoma, Chemical Aspects of MSRE Operations, ORNL-4658, pp. 68–71. Oak Ridge National Laboratory (December 1971).

[107] W. R. Grimes, *Nucl. Appl. Tech.* **8**, 137 (1970).

[108] B. F. Hitch and C. F. Baes, Jr., Reactor Chemistry Division Annual Progress Rep., ORNL-4076, p. 20, Oak Ridge National Laboratory (December 31, 1966).

[109] K. A. Romberger, J. Braunstein, and R. E. Thoma, New electrochemical measurements of the liquidus in the LiF-BeF$_2$ system. Congruency of Li$_2$BeF$_4$, *J. Phys. Chem.* **76**, 1154 (1972).

[110] S. Cantor, W. T. Ward, and C. T. Moynihan, *J. Chem. Phys.* **50**, 2874 (1969).

[111] R. Carruthers, P. A. Davenport, and J. T. D. Mitchell, The Economic Generation of Power from Thermonuclear Fusion, Culham Laboratory, Rep. CLM-R-85 (October 1967).

[112] A. P. Fraas, Conceptual design of a fusion power plant to meet the total energy requirements of an urban complex, *BNES Nucl. Fusion Reactor Conf.*, *Culham Lab.* CR 70-45 (1969).

[113] I. N. Goloven, U. N. Dnestrovsky, and D. P. Kostomarov, Tokamak as a possible fusion reactor—Comparison with other CTR devices, *BNES Nucl. Fusion Reactor Conf. Culham Lab.*, CR 70-47 (1969).

[114] A. Gibson, R. Hancox, and R. J. Bickerton, On the economic feasibility of stellarator and tokamak fusion reactors, *Conf. Plasma Phys. and Controlled Nucl. Fusion Res.*, *4th Madison, Wisconsin* CN-28/K-4 (1971).

[115] J. R. Powell, Design and Economics of Large DC Fusion Magnets, Brookhaven National Laboratory Rep., BNL 16748 (1972).

[116] W. C. Young and R. W. Boom, Materials and Cost Analysis of Constant-Tension Magnet Windings for Tokamak Reactors, FDM 27, Presented at *Int. Magn. Conf.*, *4th, Brookhaven* (1972).

[117] J. E. Kunzler, C. Buehler, F. S. L. Hsu, and J. H. Wernick, *Phys. Rev. Lett.* **6**, 89 (1961).

[118] G. W. Webb, *Am. Inst. Phys. Proc.* No. 4; " Superconductivity in d- and f-band Metals " (D. H. Douglas, ed.), American Institute of Physics, New York, 1972; G. W. Webb, L. J. Vieland, R. E. Miller, and A. Wicklund, *Solid State Commun.* **9**, 1769 (1971).

[119] B. T. Matthias *et al.*, *Science* **156**, 645 (1967).

[120] G. Arrhenius *et al.*, *Proc. Nat. Acad. Sci. U.S.* **61**, 621 (1968).

[121] W. Desorbo, *Phys. Rev. A* **140(3)**, 914 (1965).

[122] J. Sutton and C. Baker, *Phys. Lett.* **21**, 601 (1966).

[123] S. J. Williamson, *Phys. Lett.* **23**, 629 (1966).

[124] S. Foner, E. J. McNiff, Jr., B. T. Matthias, T. H. Geballe, R. H. Willens, and E. Corenzwit, *Phys. Lett.* **31A**, 349 (1970).

[125] S. Foner, E. J. McNiff, Jr., B. T. Matthias, and E. Corenzwit, *Proc. Int. Conf. Low Temp. Phys.*, *2nd* **II**, 1925 (1968).

[126] B. T. Matthias, *Phys. Rev.* **97**, No. 1, 74 (1955).

[127] B. W. Roberts, *in* " Intermetallic Compounds " (J. H. Westbrook, ed.), Chapter 29. Wiley, New York, 1967.

[128] L. F. Mattheiss, *Phys. Rev.* **138**, A112 (1965).

[129] B. T. Matthias, E. Corenzwit, A. S. Cooper, and L. D. Longinotti, *Proc. Nat. Acad. Sci. U.S.* **68**, 56 (1971).

[130] B. T. Matthias, *AIP Conf. Proc.* (D. H. Douglas, ed.), No. 4, p. 367. American Institute of Physics, New York, 1972.

[131] A. Berruyer *et al.*, *Adv. Cryogen. Eng.* **15**, 158 (1970).

[132] A. G. Prodell, *Proc. Int. Conf. Bubble Chamber Tech.*, *Argonne* (June 1970).

[133] J. Purcell and H. Desportes, Private communication.

[134] C. D. Henning, R. L. Nelson, M. O. Calderon, A. K. Chargin, and A. R. Harvey, *Adv. Cryogen. Eng.* **14**, 98 (1969).

[135] K. R. Efferson *et al.*, *IEEE Trans. Nucl. Sci.* **NS-18**, 265 (1971).

[136] W. B. Sampson, R. B. Britton, P. F. Dahl, A. D. McInturff, G. H. Morgan, and K. E. Robins, *Part. Accel.* **1**, 173 (1970).

[137] Company Data Sheet (July 1971).

[138] D. W. DeMichele and J. B. Darby, Jr., MSD/CTR/TM-13, Argonne National Lab., Argonne, Illinois (   ).

[139] J. R. Powell, BNL-16580, Brookhaven National Lab., Upton, New York (1972).

[140] R. W. Webb, Permeation of Hydrogen Through Metals, USAEC Rep. NAA-Sr-10462 (1965).

[141] R. E. Stickney, *in* "The Chemistry of Fusion Technology" (D. M. Gruen, ed.), Chapter 7. Plenum Press, New York, 1972.

[142]   . Johnson,   . Steiner, and   . Fraas, *Nucl. Safety* **13**, 353 (1972).

[143] V. A. Maroni, CEN/CTR/TM-9, Argonne National Lab., Argonne, Illinois (1974).

[144] R. G. Hickman, *Proc. Conf. Tech. Controlled Thermonucl. Fusion Exp., Austin, Texos* CONF-721111, p. 105 (1972).

[145] V. J. Wheeler, *J. Nucl. Mater.* **40**, 189 (1971).

[146] D. G. Thomas and J. J. Lander, *J. Chem. Phys.* **25**, 1136 (1956).

[147] G. J. Hill, *Brit. J. Appl. Phys.* **1**, 1151 (1968).

[148] K. Hauffe and D. Hoeffgen, *Ber. Bunsenges. Phys. Chem.* **74**, 537 (1970).

[149] E. W. Roberts and J. P. Roberts, *Bull. Soc. Fr. Ceram.* **77**, 3 (1967).

[150] K. T. Scott and L. L. Wassell, *Proc. Brit. Ceram. Soc.* **7**, 375 (1967).

[151] J. C. Collins and J. C. Turnbull, *Vacuum* **11**, 114 (1961).

[152] R. Frauenfelder, *J. Vac. Sci. Technol.* **6**, 388 (1969).

[153] E. A. Steigerwald, TRW Electromechanical Division Rep. E-5623, NASA CR-54004, Cleveland, Ohio (November 1963).

[154] P. S. Flint, General Electric Knolls Atomic Power Laboratory Rep. KAPL-659 (1951).

[155] R. W. Rudd and J. B. Vetrano, Atomics International Rep. NAA-SR-6109 (1961).

[156] R. Gibson, P. M. S. Jones, and J. A. Evans, United Kingdom Atomic Energy Authority Rep. AWRE-O-47/65 (1965).

[157] W. G. Perkins and D. R. Begeul, SC-DC-714493, Sandia Lab., Albuquerque, New Mexico (March 1972).

[158] V. A. Maroni, ANL-7948, Argonne National Lab., Argonne, Illinois (June 1972).

Chapter 7

Development of Fuel Cells—A Materials Problem

H. BEHRET, H. BINDER, AND G. SANDSTEDE

BATTELLE-INSTITUT E.V.
FRANKFURT/MAIN
GERMANY

Fuel cells are electric energy generating devices which can, in principle, be used both in small power units and in power stations. Originally they were developed because of their high efficiency. This is still a favorable feature, but there are—in most cases—other properties which make them attractive, for example, their neutrality to environment, silent operation, quick start, and so on—properties which would compensate for the high capital expenditure likely to be incurred during the introductory period before costs of manufacture and materials come to match those of conventional energy generators. In addition to their operation as self-contained electricity generators, hydrogen fuel cells combined with water electrolysis can constitute a power storage system for load leveling.

This review is devoted to materials and present and future materials problems that have to be solved. For information on fuel cells in general and their historical development the reader is referred to the comprehensive books and reviews on this topic [1–26].

The short description of fuel cells together with their classification will enable the reader to discern the critical points regarding their final development and judge the relative significance of the various materials problems. By considering fundamental aspects of electrochemical energy generation, it is possible to compare the materials problems with the potential merits of the various new energy converters.

I. BACKGROUND OF ELECTROCHEMICAL POWER GENERATION

A. Electrochemical Reactions

In a fuel cell electricity is generated by electrochemical combustion. Direct chemical combustion, that is, the reaction between fuel and oxygen which includes the direct electron transfer from the fuel to oxygen molecules, is prevented by having them separated—by an electrolyte. The electrolyte allows charge transfer via ions (ionic conductor) but not via free electrons. The overall process is split into two electrochemical reactions taking place simultaneously at separate electrodes: at the anode the take up of electrons from the fuel (also called anodic oxidation or electrooxidation), and at the cathode, the take up of electrons by, e.g., the oxygen (also called cathodic reduction or electroreduction). Thus, taking the hydrogen–oxygen cell with acid electrolyte as an example, hydrogen is deelectronated and transferred to hydrogen ions [see Fig. 1, Eq. (1)]:

$$2H_2 \rightarrow 4H^+ + 4e^- \tag{1}$$

Driven by an electric field the hydrogen ions migrate through the electrolyte

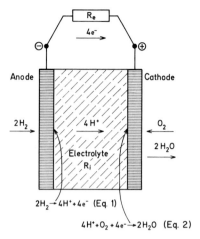

FIG. 1.   Hydrogen–oxygen fuel cell.

solution to the other electrode, the cathode, where oxygen is electronated and at the same time reacts with the hydrogen ions to form water

$$4H^+ + O_2 + 4e^- \rightarrow 2H_2O \qquad (2)$$

as reaction product [see Fig. 1, Eq. (2)]: The addition of both reactions gives the well-known combustion reaction

$$2H_2 + O_2 \rightarrow 2H_2O \qquad (3)$$

where the reaction product is formed in a direct process.

While the hydrogen ions migrate through the acid electrolyte with an electrical resistance R_i (resistance of the electrolyte, not of the total cell), the electrons simultaneously flow from the anode to the cathode through the load with an electrical resistance R_e; the water molecules formed at the cathode have to be removed (e.g., by evaporation).

The voltage of the cell can be derived from the free enthalpy of the combustion reaction (3). It is called the reversible cell voltage or electromotive force:

$$U^0 = -\Delta G/nF \qquad (4)$$

where

$$\Delta G = \Delta H - T\Delta S \qquad (5)$$

and ΔG is the free energy per mole of product, ΔH the enthalpy (negative value of the heat of combustion per mole of product), T the absolute temperature, F the Faraday constant (96.62 kJ/mole V), and n the number of moles of electrons per mole of product (for H_2O, $n = 2$).

The cell voltage can be considered as the difference between the two electrode potentials

$$U^0 = E_c^0 - E_a^0 \tag{6}$$

Both the cathode potential E_c^0 and the anode potential E_a^0 are measured with respect to the same reference potential, usually the potential of the normal hydrogen electrode (NHE),[1] which is arbitrarily set to zero. For the oxygen electrode and for the hydrogen electrode the pH dependence is

$$E_c^0 = E_{c,\,pH=0}^0 - (2.3\ RT/F) \cdot pH \tag{7}$$

$$E_a^0 = E_{a,\,pH=0}^0 - (2.3\ RT/F) \cdot pH \tag{8}$$

$$2.3\ RT/F = 59\ mV \qquad \text{at } 25°C$$

As $U^0 = 1229$ mV for reaction (3) and the reversible hydrogen electrode potential $E_a^0 = 0$ mV at pH = 0, the reversible oxygen electrode potential is $E_c^0 = 1229$ mV at pH = 0. At pH = 14 ($\hat{=}$ 1 N KOH solution) the reversible potentials would be $E_a^0 = -828$ mV and $E_c^0 = 401$ mV.

Now, in fuel cell research it is convenient to put the potential of the reversible hydrogen electrode potential equal to zero at any pH value. Thus, the pH term cancels in Eqs. (7) and (8). The potential of each electrode is measured with respect to the hydrogen electrode (HE) in the same solution.

Figure 2 shows the dependence of the cell voltage and the electrode potentials on the current density.[2] As the load increases (decreasing resistance R_e, see Fig. 1) the current increases and so does the internal voltage drop IR_i in the electrolyte of the cell and the voltage drop across the electrode–electrolyte interface (polarization). Consequently, the O_2 potential drops and the H_2 potential rises to give a lower voltage of the working cell:

$$U = E_c - E_a - IR_i \tag{9}$$

Hence, by definition a fuel cell is a galvanic cell consisting of two invariant electrodes with an invariant electrolyte in between and generating current as long as fuel and oxidant (preferably oxygen) are continuously supplied and the reaction products are continuously removed (cf. Fig. 1).

The main materials problem lies in the necessity of "invariance," whereas the supply of fuel and oxidant belongs to the sphere of engineering.

---

[1] $P_{H_2} \approx 1$ bar, $c_{H^+} \approx 1$ mole/1 $\hat{=}$ pH = 0.

[2] The alkaline cell developed at the Battelle-Institut in Frankfurt/Main (non-noble metal containing; hydrogen–air) has been used as an example. With minor modifications the result is applicable to all kinds of cells.

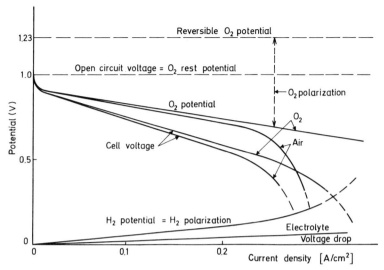

FIG. 2. Performance of a fuel cell with alkaline electrolyte and hydrophobic electrodes (Battelle).

B. Efficiency

The electrochemical combustion is an isothermal reaction. Its efficiency therefore is not dependent on the Carnot cycle between heat source and sink. The degree of efficiency is theoretically given by

$$\eta_{\text{isotherm}} = \Delta G/\Delta H = 1 - T\,\Delta S/\Delta H \tag{10}$$

The data for various fuels are summarized in Table 1. The theoretical efficiency of hydrogen combustion is relatively low because the entropy of the reaction is negative.

The high theoretical efficiency of carbon monoxide, hydrazine, formic acid and formaldehyde cannot be utilized because it is difficult, if not impossible, to obtain anode potentials of less than 0 mV. Therefore the voltage efficiency (see below) is lowered.

As shown in Fig. 2, the operating voltage is smaller than the reversible voltage. Therefore the voltage efficiency is below 100%:

$$\eta_{\text{volt}} = W_{\text{elec}}/W_{\text{max}} = U/U^0 \tag{11}$$

The electric work IUt of the cell has to be divided by the highest possible work (at U^0). We now can define the electrical efficiency as follows:

$$\eta_{\text{elec}} = \eta_{\text{isotherm}}\eta_{\text{volt}} \tag{12}$$

TABLE 1

Reversible Cell Voltages $U°$, Reversible Anode Potentials $E_a°$ versus HE, and Theoretical Degrees of Efficiency $\eta_{isotherm}$ (Assuming the Reaction Product Water to be Liquid) for the Electrochemical Combustion of Various Fuels at 25°C.

Fuel	$U°$ (mV)	$E_a°$ (mV)	$\eta_{isotherm}$ (%)
C	1022	207	100
CO	1333	−104	91
H_2	1229	0	83
NH_3	1172	57	89
N_2H_4	1561	−332	99
HCOOH	1425	−196	87
HCHO	1301	−72	97
CH_3OH	1198	31	97
CH_4	1059	170	92
C_3H_8	1087	142	95

This quantity gives us both the actual useful electric energy and the heat developed by the cell:

$$Q = (U° - U)nF - T\Delta S = -\Delta H(1 - \eta_{elec}) \tag{13}$$

In this case Q is obtained per mole of water formed. We may now subtract the heat of vaporization, if the water is removed from the cell in the gaseous state. Instead of relating the heat to the product, one can relate it to the electric work of the cell:

$$Q_{cell} = W_{elec}[(1 - \eta_{elec})/\eta_{elec}] \tag{14}$$

Losses other than the voltage losses may occur:

(1) The fuel may not be consumed completely;
(2) the reaction may lead to products other than CO_2 and H_2O (e.g., with methanol).

Therefore, in order to take full account of all the effects involved, further degrees of efficiency (e.g., Faraday efficiency) have to be added as multipliers. However, by taking proper measures these values can reach almost 100%.

C. Electrocatalysis

Naturally, the electrodes must be electronically conductive. Unfortunately, this is not the only necessary condition. They must also be able to activate the molecules before the reaction so that the electron transfer from or to the electrode can take place. And this process must occur at potentials as close to the reversible potentials as possible. Only then will it be possible to obtain high current densities at high efficiency. Therefore, a catalyst has to be incorporated in the electrode. As we are dealing with electrochemical reactions, the catalyst is called an electrocatalyst. The first step in the reaction sequence is the adsorption, that is, the formation of a more or less strong bond between the reactant and the surface of the electrocatalyst A.

Hydrogen is dissociatively adsorbed (chemisorbed) even in the first step:

$$2A + H_2 \rightarrow 2A\text{-}H \tag{15a}$$

$$A\text{-}H \rightarrow A + H^+ + e^- \tag{15b}$$

Reaction (15b) has to occur twice for reaction (15a) to take place.

In the case of oxygen, the reaction is much more complex because four electrons have to be taken up by the oxygen molecule. Adsorption involves, say, an $A\text{-}O_2$ complex in the first step. In one of the intermediate steps, the adsorption complex $A\text{-}O_2H^-$ is produced, which can decompose to form the hydrogen peroxide anion O_2H^- which is dissolved in the electrolyte.

With regard to adsorption, one has to take into account that each adsorption step is a replacement reaction, because the catalyst surface in the electrolyte solution is normally covered with water molecules—at low temperatures completely, at elevated temperatures or/and at high electrolyte concentrations partially. At higher temperatures, especially in the solid oxide electrolyte cell (see below), oxygen is dissociatively chemisorbed:

$$2A + O_2 \rightarrow 2A\text{-}O \tag{16}$$

Without going into details of electrode kinetics it should be noted that a potential gradient is formed across the electrode–electrolyte interface by accumulation of negative ions in the electrochemical double layer adjacent to the electrode if the electrode (cathode) is positively charged by the presence of the oxidant, or by accumulation of positive ions if the electrode (anode) is negatively charged by the presence of the fuel, for example, according to reaction (15).

The basic equation of electrode kinetics is written

$$i = i_o A \left(a_{rd} e^{\eta/b+} - a_{ox} e^{-\eta/b-} \right) \tag{17}$$

which holds for the reaction

$$Rd \rightleftharpoons Ox + e^- \tag{18}$$

where a_{rd} is the activity of the reductant[3] at the electrode surface, for example, the coverage of the electrode by H atoms; a_{ox} the activity of the oxidant[3]

---

[3] In the equation for the cathode the reductant Rd is H_2O and the oxidant Ox is O_2.

formed from the reductant, for example, the H^+ concentration at the electrode surface and/or fraction of surface not covered by H atoms (electrochemical reaction orders of the activities have been neglected); η the overpotential; A the internal surface area of the electrode per cm^2 of geometric electrode area; i the current density, that is the current per geometric electrode area; i_o the exchange current density (A/cm^2); and b_+ and b_- the Tafel factors.

The quantity i_o is obtained when the overpotential is zero and the activities put equal to one. Thus, at the reversible potential there is an electron transfer back and forth across the electrode–electrolyte interface in such a way that no change in charge and matter results. The higher i_o, the faster is the reaction (the higher is the current), as can be seen from Eq. (17). If the overpotential exceeds a few hundred millivolts, the second term can be neglected and we arrive at the Tafel equation

$$\eta = b_+ \ln i + b_+ \ln a_{rd} A i_o \qquad (19)$$

The exchange current density and the Tafel factor b_+ can be determined from the semilogarithmic plot of the overpotential versus the current density:

$$b_+ = RT/\alpha_+ F \qquad (20)$$

α_+ is the so-called transfer coefficient from which inferences can be made as to the reaction mechanism.

For methanol at platinum electrodes $b_+ = 60$–70 mV, (see Fig. 3). The value for i_o is very small—10^{-14} A/cm² at 25°C, 10^{-11} A/cm² at

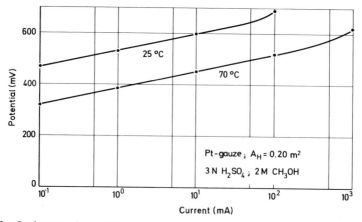

FIG. 3. Stationary galvanostatic current–voltage curves for methanol oxidation on platinum gauze. From Binder et al. [27].

70°C—because Pt is a rather poor electrocatalyst for methanol. The exchange current density is, incidentally, just the quantity where electrocatalysis comes in. For an anodic reaction it is given by

$$i_o = nFk_+^{\,0}\,\exp(E_a^{\,0}/b_+)\,\exp(-\Delta G_a^{\ddagger}/RT) \tag{21}$$

and for a cathodic reaction we may write

$$i_o = nFk_-^{\,0}\,\exp(E_c^{\,0}/b_-)\,\exp(-\Delta G_c^{\ddagger}/RT) \tag{22}$$

We will not consider the unimportant influence of the reversible electrode potentials E^0 or the pre-exponential factors k^0, but go a little further into the activation energy $\Delta G^{\ddagger}$. This term is exactly the same as in ordinary heterogeneous catalysis. The energy of the transition state, that is, the complex between the reacting molecule and the electrode surface, must be as low as possible. The electrocatalyst has to be selected according to its ability to provide a low activation energy. As usual, $\Delta G^{\ddagger}$ is derived from the Arrhenius plot, that is, the temperature dependence of the reaction rate (current density). Figure 4 shows an example of how the activation energy and hence the catalytic activity of platinum can be improved by modifying its surface properties. An adsorbed layer of sulfur increases the activity for carbon monoxide electrooxidation.

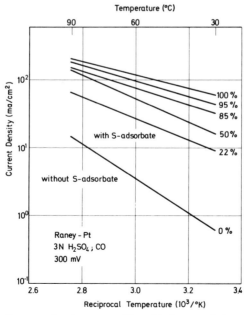

FIG. 4. Arrhenius diagram of carbon monoxide oxidation on platinum with and without sulfur adsorbate layers of different coverage. From Binder *et al.* [27, p. 70].

Further examples of surface modification will be described in Section IV. Only one point should be noted here: From Eq. (17) it is obvious that a high surface area is necessary for an active electrocatalyst. This can best be achieved by depositing the catalyst material on a large surface area support, for example, active carbon.

The requirements to be met by an electrocatalyst may be summarized as follows:

(1) Adsorption (weak) of the reactant;
(2) low activation energy resulting in high exchange current density;
(3) large surface area—small particle size;
(4) electronic conductivity;
(5) corrosion resistance (see below);
(6) small amounts in order to be cheap and available in sufficient quantity.

Most of these requirements involve extreme and partially contradictory materials problems.

D. Corrosion

Corrosion problems have already been touched upon in the foregoing discussions of chemical stability. Corrosion, in its general sense, is the effect of an aggressive environment on a material. In most cases, especially in the corrosion of metals, it is an (electrochemical) oxidation reaction which plays the important part.

In the context of chemical stability, the swelling and dissolution of plastics in an inorganic or organic solvent is likewise considered in this section.

Most metals undergo severe corrosion if in contact with an electrolyte solution, notably with an acid solution. From a thermodynamic point of view all those metals whose standard potential is more negative than that of the standard hydrogen electrode, dissolve in strong acids (pH = 0) with evolution of hydrogen gas, provided that the acid does not contain oxygen and/or does not have an oxidizing ion such as $NO_3{}^-$.

All metals having a standard potential more positive than the normal hydrogen electrode (NHE) are not susceptible to corrosive attack; they are "immune." All the others could, in principle, undergo corrosion, but in some cases a dense passivating layer is formed on the metal surface, thus protecting the underlying metal from further attack. They are then in a passive state. Considering other cases, for example, lead, the overpotential for the accompanying hydrogen evolution is so high that the metal does not dissolve. Hence, in addition to thermodynamic effects kinetic processes must not be neglected. (With respect to materials problems, potential–pH diagrams are of interest; see the Atlas by Pourbaix [28]).

If fuel cells are discussed, it is necessary to consider that the potential of the anode (fuel electrode) is somewhat more positive than the hydrogen electrode in the same solution while that of the cathode (oxygen or air electrode) is even far more positive; the materials of the cathode work under the same condition as in concentrated nitric acid $(E_0 = +0.95 \text{ V})$. Sometimes a passive state can be reached (at $E > E_p$, cf. Fig. 5); sometimes the material is already in the transpassive state (cf. Fig. 5). For more details the reader is referred to corrosion textbooks, such as those by Evans [29] and Kaesche [30].

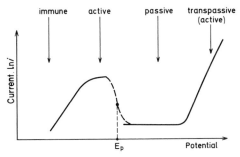

FIG. 5.   Corrosion; current potential curve.

Regarding passivity, it is obvious that some oxides (forming the passive layer) must be relatively resistant to electrolytic attack. Yet in fuel cell electrodes it is conductivity that counts, so that a variety of metals—and metal oxides—can only be used under certain conditions. For instance, tantalum protected by a dense oxide layer does not permit the transfer of electrons from the electrolyte into the metal provided it is possible to apply only small voltages, as in a fuel cell. Among the oxides, chromium containing oxides of the spinel type are highly resistant to the attack of acids even at the potential of the oxygen electrode, but all of these compounds are poor electronic conductors. Tungsten bronzes, such as $Na_{0.3}WO_3$ are good conductors and resistant to acids. In alkaline solution, some conductive spinels, such as $NiCo_2O_4$, are stable at the potential of the oxygen electrode. Nevertheless, the range of materials as candidates for application in a fuel cell is drastically limited by severe oxidation at the cathode.

Among plastics, certain esters have to be excluded because they are hydrolyzed under fuel cell conditions. Others such as polyethylene and polypropylene undergo swelling in the presence of hydrocarbons. The most stable material so far employed in fuel cells is polytetrafluorethylene

(PTFE). In using PTFE as a binder for the active layer, its highly hydrophobic character is a disadvantage. New materials might well be superior here. For this reason there is a permanent search for new materials even in the field of plastics.

II. BASIC TYPES OF FUEL CELLS

Fuel and electrolyte and hence electrode and catalyst are the most important determinants for the type of fuel cell chosen. The operating temperature of the cell is a variable which has to be fixed according to the electrolyte used.

The electrolyte has to meet the following requirements:

(1) High ionic conductivity;
(2) no electronic conductivity;
(3) chemical stability;
(4) presence of a hydrogen- or an oxygen-containing ion.

These requirements lead to cells with acid (H_2SO_4, H_3PO_4), alkaline (KOH), fused carbonate and solid oxide (both at high temperature) electrolytes. For neutral electrolytes including biochemical cells, see Section IVC.

A. Gaseous Fuel

The basic types of fuel cells according to the electrolytes used are shown in Fig. 6. The electrolyte of the high temperature solid-oxide fuel cell contains a modified ZrO_2 ceramic material. At a temperature as high as, say, 1000°C this material is an excellent conductor for oxygen ions. The only disadvantage is the high temperature needed to achieve a reasonable

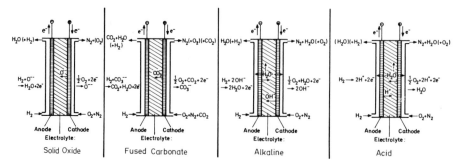

FIG. 6. Fuel cells for gaseous fuels with various electrolytes which determine the electrode reactions shown. From Sandstede [23, p. XIII].

conductivity. Other oxides investigated so far have either a lower conductivity or partial electronic conductivity which leads to an internal short circuit and hence to a loss in efficiency apart from the extra heat evolved.

The electrode reactions are simple. Oxygen from the air, which passes through the cathode chamber, is either dissociatively adsorbed or even absorbed (dissolved), if the cathode consists of silver, and the atoms take up electrons at the electrode–electrolyte interface to give oxygen ions, O^{2-}. These migrate through the electrolyte to the anode where they react with hydrogen leaving the electrons "in" the anode. Carbon monoxide as a fuel is also easily oxidized but gives rise to a somewhat higher polarization.

The high temperature fused carbonate cell contains a melt of a mixture of alkali carbonates as electrolyte. The migrating ion by which the electricity is transported is the carbonate ion $CO_3{}^{2-}$. Therefore, carbon dioxide has to be added to the oxygen (air) so that this ion can be formed at the cathode (see Fig. 6). At the anode the carbon dioxide is set free again and can be recycled to the cathode. Instead of hydrogen, carbon monoxide can also be oxidized and used as a fuel.

In the low and medium temperature (about 200°C) alkaline fuel cell, potassium hydroxide solution is used as electrolyte. For the medium temperature cell, the electrolyte may be considered as fused potassium hydroxide containing about 10% water. In the alkaline cell, the current transporting ion is the hydroxide ion which is formed at the cathode by electronation of oxygen under reaction with water molecules (see Fig. 6). At the anode hydrogen reacts with the hydroxide ions leaving the electron "in" the anode and forming water—in twice the amount used at the cathode—as a reaction product. Therefore water has to be transported—not necessarily through the electrolyte—from the anode to the cathode.

If air is used, carbon dioxide has to be removed because it is dissolved in the electrolyte by reaction with the hydroxide ions:

$$CO_2 + 2OH^- \rightarrow CO_3{}^{2-} + H_2O \tag{23}$$

Thus the hydroxide solution is converted to a carbonate solution which is much less conductive and results in much higher electrode polarizations.

The last of the four basic types to be mentioned is the cell with acid electrolyte, for example, sulfuric acid solution (about 20% H_2SO_4) below 100°C or phosphoric acid (H_3PO_4) at about 150°C with a water content of, say, 10%. The electrochemical reactions (see Fig. 6) have already been described in Section IA. In contrast with the cell with alkaline electrolyte, the reaction product water is here formed at the cathode.

In addition to hydrogen, also carbonaceous fuels, for example, hydrocarbons, may be used. These fuels can be anodically oxidized completely to carbon dioxide and water with a remarkable current density, but only with platinum in an intolerably large amount as a catalyst. Hydrocarbons still have to be reformed before being fed to the anode (see Section III,A).

B. Dissolved Fuel

There are fuels which can be dissolved in the electrolyte satisfactorily. Therefore, the cells can be fueled by dissolving the fuel in the circulating electrolyte. Anodes in fuel cells, which are fed with dissolvable fuel, need not have a gas side and can, therefore, be completely immersed in the electrolyte. They are called immersion electrodes. Such cells are shown in Fig. 7. For

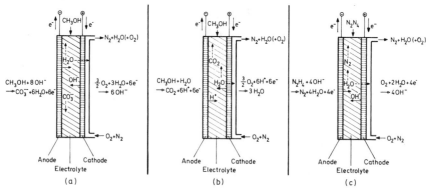

FIG. 7. Fuel cells with immersion electrodes for electrolyte dissolvable fuel: (a) Electrolyte consumable; (b), (c) electrolyte invariant.

example, methanol is used as a fuel which reacts at the anode with water to give carbon dioxide, hydrogen ions and electrons, if an acid electrolyte, for example, sulfuric acid solution, is used. The hydrogen ions migrate to the cathode just as in a fuel cell fed with gaseous fuel. Formic acid and formaldehyde have also been used as fuels.

For the hydrazine cells (right-hand side of Fig. 7) an alkaline electrolyte can be used. It remains unaffected because carbon dioxide is not formed with this fuel. The reaction products are nitrogen and water. Hydrazine would be the ideal fuel if it were not so expensive and toxic. Compared with hydrazine, ammonia which also gives nitrogen and water as the reaction products, is a cheap fuel.

C. Cells with Consumable Electrolyte

If an alkaline solution is used as electrolyte for a methanol cell, the cell will stop functioning when most of the hydroxide ions are converted to carbonate by the carbon dioxide formed during the reaction of methanol at the anode (see Fig. 7). In this case the electrolyte either has to be wasted or to be renewed by electrodialysis. Instead of methanol, formic acid or rather potassium formate can be used [25]:

$$HCOO^- + 3OH^- \rightarrow CO_3^{2-} + 2H_2O + 2e^- \tag{24}$$

These cells operate as refillable primary cells and are therefore more economical than, for example, Leclanché cells.

Another interesting organic fuel in this connection is ethylene glycol, which is oxidized to potassium oxalate [31, 32]. Refillable cells can also use a consumable metallic electrode, for example, the zinc–air cell equipped with replaceable anodes. The zinc itself acts as an anode because it is electrically conductive and does not need a catalyst:

$$Zn + 4OH^- \rightarrow Zn(OH)_4^{2-} + 2e^- \tag{25}$$

As can be seen from Eq. (25), in this case the alkaline electrolyte is consumed as well.

D. Regenerative Fuel Cells

Under this heading numerous special types of fuel cells can be summarized, all of which have in common that the fuel is regenerated in one way or other. The following four systems have not been realized in practice:

(1) Redox cells (chemically regenerative cells);
(2) electrochemical heat converters (thermally regenerative cells);
(3) radiolytically regenerative cells;
(4) photochemically regenerative cells.

Electrically regenerative cells constitute a kind of secondary battery. They include:

(1) Metal–air cells. Zinc or iron anodes are coupled with an air cathode where oxygen is developed during recharging. A special zinc–air cell uses zinc powder as a fuel and the resulting zincate solution is replaced by fresh electrolyte and the zinc processed at the "filling station."

(2) Redox cells (electrically regenerative cells). The systems Ti–Fe salt solution or Cr salt solution are rechargeable and are currently being discussed for electricity storage [33].

(3) Water electrolysis cells–fuel cells combination [3, 34]. In this system hydrogen and oxygen are formed which could be stored and used for power production in an $H_2–O_2$ fuel cell. There are even so-called valve electrodes which permit the electrolysis to be carried out in the fuel cell [3].

III. BREAKDOWN OF FUEL CELL UNITS AND SYSTEMS

A. Components and Cells

A variety of flat single cells are usually combined to form a fuel cell stack. A whole fuel cell system consists of a number of stacks and additional devices for the fuel and oxidant supply (see Fig. 8). If pure hydrogen is used as the fuel, it can be fed direct into the cell stacks. If, however, hydrocarbons, for example, natural gas, are taken as fuel, they have to be processed (see

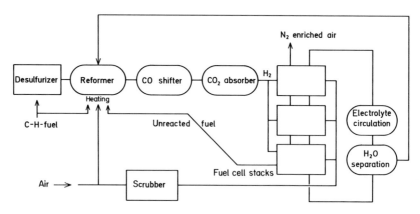

FIG. 8.   Fuel cell system.

Fig. 8). After desulfurization (this step depends on the quality of the reforming catalyst) the fuel is reformed at, say, 700°C in the reformer supplied with water vapor from the fuel cell. In the case of alkanes we have

$$C_nH_{2n+2} + (2n - m)H_2O \rightarrow (n - m)CO_2 + mCO + (3n - m + 1)H_2 \qquad (26)$$

Further hydrogen can be produced using a shifter where the carbon monoxide is converted:

$$CO + H_2O \rightarrow CO_2 + H_2 \qquad (27)$$

This operation is not absolutely necessary if the carbon monoxide can be tolerated in the fuel gas. For methanation and other potential processing methods, for example, partial oxidation, see Section VA.

If fuel cells with alkaline electrolyte are used, the carbon dioxide has to be removed from the fuel gas (and from the air) because it neutralizes the alkali by carbonate formation [Eq. (23)]. However, this is not necessary in cells with acid or solid electrolytes, and carbon dioxide can be fed into the anode chambers together with the hydrogen. It will leave the cell together with unreacted fuel. An air scrubber removing carbon dioxide is not needed in these cells.

In low temperature fuel cells, the electrolyte is usually circulated by pumping. It may be necessary to remove liquid water from the electrolyte stream. This is not the case if medium temperature cells (e.g., with H_3PO_4) are used where electrolyte circulation can be avoided. Thus the diagram in Fig. 9 contains the maximum number of system components. A possibly necessary cooling system, other than by air or electrolyte circulation, is not included.

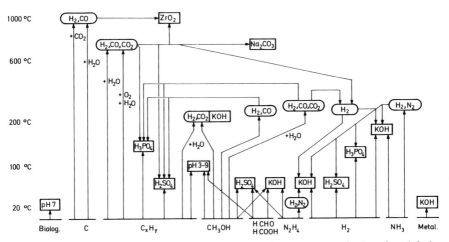

FIG. 9. Types of fuel cells and converters classified by temperature (ordinate) and fuel (abscissa). The fuels are represented by the formula of the electrolyte; the reformers, crackers and purifiers are represented by the gases produced.

Various types of reformers, crackers and fuel cells are summarized in Fig. 9. The diagram not only gives a survey of the systems available but also shows which reformer or cracker can be connected with which type of fuel cell. In addition to the biocell we find the high temperature fuel cells on the left-hand side of Fig. 9. The one with solid electrolyte may be operated either with gasified carbon or reformed hydrocarbons. There is no heat loss in the reformer if it is operated at the same temperature as the fuel cell next to it. As hydrocarbons can be decomposed to carbon in a fused carbonate cell, it has to be equipped with a reformer. The operating temperature of about 600–700°C gives optimum efficiency for a cell connected to a reformer. At higher temperatures the isothermal (thermodynamic) efficiency decreases. In connection with low temperature cells, on the other hand, an energy loss is encountered in the reformer owing to the endothermal reaction and the dissipation of heat involved.

Nevertheless, most of the research effort is concentrated on low temperature cells, because their prospects are brightest from a technical point of view. Theoretically, the phosphoric acid cell is very promising because it supplies the heat required for the vaporization of water and rejects the carbon dioxide. The sulfuric acid cell operating at temperatures below 100°C does not supply the heat necessary for vaporization. Its prospects are all the more favorable because highly active nonplatinum metal catalysts which also convert carbon monoxide have been found.

An attempt to use an internal reforming cell of the type shown in the center of Fig. 9 failed because of the unsatisfactory lifetime of the palladium membrane.

B. Electrode and Battery Construction

The most important types of electrodes are shown in Fig. 10:

(1) Electrodes with a conductive cover layer of small pore size (usually sintered nickel), also called double porosity electrodes (Bacon-type, in Vielstich [7]);

(2) electrodes with a nonconductive cover layer of small pore size (e.g., asbestos), with a liquid electrolyte in between, called supported electrode (Siemens-type, in Sturm [19]);

(3) an electrode–electrolyte vehicle combination, for example, with asbestos diaphragms as electrolyte vehicle (Allis-Chalmers-type, in Bockris and Srinivasan [20]); an alternative version is the electrode–ion exchange membrane–electrolyte combination (General Electric type, in Liebhafsky *et al.* [17]);

(4) hydrophobic electrodes, which operate at normal pressure (Kordesch-type, in Berger [18]) [35]; today sintered PTFE electrodes are

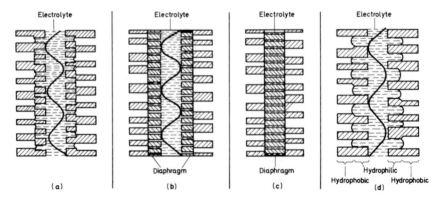

FIG. 10. Basic types of fuel cells according to the electrode structure.

widely used; the catalyst and electrically conductive particles are incorporated in the structure at the electrolyte side of the electrode.

In addition, there are also other types, such as flow-through electrodes (Varta-type, in Döhren and Euler [3]) or immersion electrodes [2]. The electrolyte can also be mixed with nonconducting particles, for example, aluminum oxide so that a slurry is formed. In the case of a liquid electrolyte, a plastic grid is needed to keep the two electrodes apart.

In a technical fuel cell stack, several single cells (or double cells, sometimes equipped with bipolar electrodes) must be combined into a battery. The filter press principle is normally applied making use of end plates of high mechanical strength (see Fig. 11). In this case the single

FIG. 11.   Filter press type fuel cell (WC–carbon, sulfuric acid), laboratory scale; end plates of high mechanical strength. From Bohm [36].

components—electrodes, spacers and cell walls—need not be mechanically stable per se. Mechanical stability is achieved in the completed stack. Materials problems arise from the electrode frames which form the cell and especially the gaskets in between. In KOH cells creeping of the electrolyte is a severe problem.

The inlet and outlet provided for the electrolyte must consist of chemically and mechanically resistant tubes and channels. In most cases plastics will do the job. However, a compromise still has to be made between price and durability. PE, for example, is only stable up to, say, 100°C, whereas PTFE, which is stable at temperatures up to about 250°C, even in alkaline solution, is rather expensive.

IV. ELECTRODE MATERIALS AND ELECTROCATALYSTS

A. Electrodes in Alkaline Electrolyte

1. General Considerations

One of the main features of the fuel cell with an alkaline electrolyte is that potassium hydroxide is used rather than sodium hydroxide. This is due to the reduced performance of oxygen electrodes in NaOH solution—at least where normal operating conditions are concerned [17]. In addition, KOH solutions have a higher specific conductivity at given concentration and temperature (about 50% higher in 5 M solutions at normal temperature).

Another feature is the resistance to alkaline solutions; this applies not only to the electrically conductive electrode material, but also to the current collector which is necessary to operate the cell with the currents available at electrode current densities of several hundred milliamps. Preferred materials include silver wire, silver-coated wires of other metals (copper), and graphitized and graphite-coated materials. Nickel may also be used as anode material. Nickel wire can even be employed as current collector in the cathode if it is protected by a thin layer of PTFE [37].

Electrode materials may consist of one or more materials which form the electrode structure, and the electrocatalyst is incorporated in this support. The electrode structure is composed of electronically conductive materials; the electrocatalyst may, in principle, consist of a nonconductive material in which electron tunneling to the electrode structure is possible. Electrodes for fuel cells with alkaline electrolyte are in the most advanced stage of development since materials problems are less severe than in acid electrolyte.

2. Hydrogen Electrodes

Within the group of hydrogen electrodes the most prominent example is the electrode used in the hydrogen–oxygen fuel cell for the Apollo project. Materials problems are rather critical since the cell type used in this concept requires a highly concentrated KOH solution (about 75%) and a comparatively high operating temperature (above 200°C). The problems were solved, at least for an operating time exceeding that needed for the space missions, by using sintered Ni powder as electrode and catalyst materials [38].

The development of sintered metal electrodes has led to mechanically stable electrodes in which the catalyst is incorporated. For fuel cells operating at temperatures below 100°C, where corrosion problems are less severe, a more active catalyst material than normal porous nickel is required. Hence, Raney nickel has been introduced [1, 2] in fuel cell electrodes using double skeleton catalyst electrode (DSK) and successfully applied in industry [39, 40]. In this case, the active catalyst is formed after sintering of the nickel electrode matrix by leaching the aluminum out of the Raney nickel–aluminum alloy in KOH solution (activation). By this treatment the catalyst cannot age during sintering. Figure 12 shows a micrograph of a DSK electrode [40]. The electrode structure simultaneously constitutes the current collector. The electrode is characterized by a system of medium-sized pores; an electrode under gas pressure may even show many pores filled completely with electrolyte.

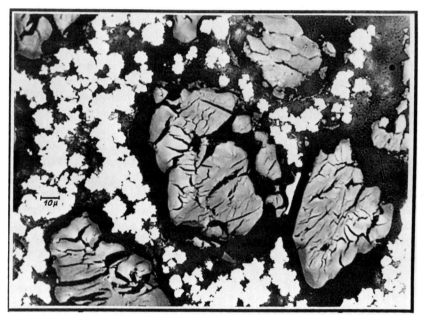

FIG. 12. Micrograph of a DSK electrode. White: carbonyl nickel support; Grey: Raney Ni-catalyst; Black: pores. From Holthusen *et al.* [40].

If PTFE is used as the structure-forming material of the electrode, Raney nickel, activated prior to the formation of the electrode by pressing, can be applied since at the sintering temperature of PTFE ($\sim 370°C$) no severe aging is observed at the catalyst [41, 42]. Sintering of the electrode can be avoided by using the concept of the "supported electrode" ("gestützte Elektrode") [43] where the Raney nickel catalyst mixed with a few percent of finely divided PTFE (PTFE suspension) is fixed between an asbestos sheet and a nickel wire mesh serving simultaneously as support and current collector (cf. also Section IV,B and Fig. 16). If PTFE is used not only for bonding but also for the (highly hydrophobic) backing layer, the electrode can be used at normal pressure [42].

In order to achieve high power densities and long term stability at an operating temperature below 100°C, nickel catalysts less active than Raney nickel have been improved by adding small amounts of platinum (Allis–Chalmers, in Bockris and Srinivasan [20]). Another development uses active carbon as a support catalyzed by traces of palladium or platinum [37, 44]. In any case, rather pure hydrogen has to be used as feed gas since traces of poison (e.g., H_2S) would drastically reduce the catalyst activity.

3. Hydrazine Electrodes

When selecting electrode and catalyst materials for hydrazine electrodes it is necessary to note that hydrazine undergoes undesired side reactions instead of being oxidized to give water and nitrogen as reaction products. In contact with nearly all metals which could serve as anode catalyst, a non-electrochemical decomposition is observed giving nitrogen and hydrogen. Even during the electrochemical oxidation at, say, Raney nickel catalysts, ammonia may be formed as a harmful byproduct. Therefore, the Raney nickel must be deactivated in part by the addition of mercury to obtain an amalgamated nickel surface [45]. Other catalysts used are nickel boride and platinum metals supported by nickel or carbon (cf., e.g., [6, 7]).

A new method of preparing "surface-catalyzed anodes" [46] for hydrazine fuel has been described, where a porous layer of Mond Ni powder and Ni powder is codeposited on an electroformed Ni sheet substrate; this deposit is catalyzed by nickel boride precipitated chemically onto the inner surface of the porous Ni layer.

4. Ammonia Electrodes

Ammonia does not undergo catalyzed decomposition under fuel cell conditions. Nevertheless, a more active catalyst than Raney nickel is required. Direct ammonia electrodes have been operated at temperatures between 25°C and 140°C in aqueous KOH; PTFE-bonded electrodes contain either unsupported platinum black or graphite-supported platinum [47, 48].

Electrocatalysts containing iridium and platinum iridium have shown encouraging levels of performance when used as ammonia anodes. These materials were much more active than platinum alone. Pt–Ir alloys can be used supported on boron carbide and graphite [48, 49].

5. Methanol and Formate Electrodes

For methanol and formate electrodes the same structure-forming materials can be used as in hydrogen electrodes, but the catalysts must be superior to those used in the case of hydrogen. Since formate is produced as an intermediate in the oxidation of methanol, the catalyst must be active for the oxidation of formate as well. Hence, palladium–platinum alloys provide a good compromise [50, 51]. Even lead may be used as a component in a platinum alloy [52]. A high activity has also been achieved by using a Raney alloy of palladium and silver [53].

6. Oxygen Electrodes

Semiconducting nickel oxide, NiO, has been used as a catalyst for the oxygen reduction in alkaline electrolyte at a temperature of 200°C [54]. It has to be doped with lithium oxide in order to get a conductivity sufficient for the large current densities that can be achieved in this temperature range. This catalyst has also been used successfully in the Apollo missions at temperatures up to 250°C.

At temperatures below 150°C lithium-doped nickel oxide is sufficiently active only if it is prepared by spraying and freeze-drying [55]; this treatment gives a very uniform distribution of very small sized lithium oxide particles in the nickel oxide.

The same method can be used for the preparation of a nickel cobalt spinel, $NiCo_2O_4$ [56], so that this catalyst is active even at normal temperature, whereas conventional methods lead to a material which is active only at elevated temperatures [57].

In general, silver and carbon, with and without additions of metal, are the preferred catalyst materials for cells operated at temperatures between the ambient temperature and 100°C. Raney silver electrodes have been prepared [2, 3] in the form of a DSK-type electrode containing at least 20% of supporting skeleton material. When preparing Raney silver the starting compound should not be an alloy of silver with aluminum only, since these alloys are too ductile for grinding. Preferably, calcium should be added as a third component [41]. On the other hand, the Raney alloy powder could be produced by spraying [58].

Silver exhibits a tendency to aging even at room temperature so that the most active sites are lost in favor of lattice planes with low indices. Nevertheless, rather active electrodes can be prepared by heating an intimate mixture of PTFE and silver carbonate up to 370°C [42, 59, 60]. These electrodes have been loaded with current densities of more than 100 mA/cm^2 over a period of several thousands of hours without serious degradation.

In order to reduce the amount of silver, carbon was used as a catalyst support [37]. It has been found that a mixture of specific types of carbon with no addition of metals used in rather thin PTFE-bonded layers gave very active electrodes which have been operated at more than 200 mA/cm^2

[41, 42]. In 6.5 N KOH at a temperature of 50°C and a current density of 100 mA/cm^2, a potential of 760 mV was measured (versus hydrogen in the same solution). When air was used instead of pure oxygen, the potential was lower by only 30 mV. These electrodes were operated for more than 3000 hr without noticeable degradation [41]. A silver catalyst doped with oxides of Bi, Ni and Ti was developed [61] and used in a 7-kW fuel cell battery [62].

B. Electrodes in Acid Electrolyte

1. General Considerations

Acid electrolytes in a fuel cell are not affected by carbon dioxide either from the air, or as a pollutant in the fuel or possibly as a reaction product of carbonaceous fuels. Hence, the electrolyte does not have to be regenerated and air does not have to be purified. These are the main reasons why the development of fuel cells with acid electrolyte is so attractive.

On the other hand, corrosion problems with the materials used in sulfuric or phosphoric acid are far more serious than in alkaline electrolyte. Only a few materials are available which resist the attack of these acids under strong oxidizing conditions at the oxygen electrode. (For general aspects of carbon electrodes in acidic fuel cells, see [63].)

Phosphoric acid is the electrolyte preferred in fuel cell development in the United States, whereas sulfuric acid is mostly used in European countries. Phosphoric acid can be used at temperatures above 100°C so that heat and water balances are achieved more readily than in sulfuric acid. On the other hand, corrosion problems are very severe, since concentrated phosphoric acid (about 85%) does attack nearly all metals and metal oxides. This is one reason why sulfuric acid is used in fuel cells despite the heat and water balance problems encountered. Furthermore, in sulfuric acid higher current densities can be achieved at a particular catalyst.

In any case PTFE has to be used as electrode binder—except for electrodes to be operated at ambient temperature where PE is suited as well.

Apart from the catalyst, the current collector presents one of the most severe materials problems. Only carbon, gold, and platinum or tantalum are resistant to the attack of hot sulfuric or even phosphoric acid. Stainless steel is sometimes used as cathode material. As platinum is a less favorable conductor (compared with silver or copper) the *IR* drop is relatively large; in addition, the price of platinum as well as of gold is prohibitively high. As far as cost is concerned, an optimum conductivity–price ratio would be achieved with aluminum or copper, but these are not stable in acids.

2. Hydrogen Electrodes

In phosphoric acid the only practical catalysts are platinum or platinum metals alloys. The amounts of platinum required are below 1 mg/cm^2

provided that the catalyst is deposited as a thin layer on a carbon or graphite support [64–66]. At temperatures above, say, 130°C platinum (if alloyed with rhodium, in particular) tolerates substantial amounts of carbon monoxide in the feed hydrogen, whereas at lower temperatures, and in sulfuric acid, for example, special measures have to be taken to avoid poisoning of the platinum catalyst by carbon monoxide. In this case it is necessary either to remove CO completely, or to cover the platinum catalyst with a monolayer of sulfur or selenium [67, 68], or to add molybdenum or tungsten oxides to the platinum catalyst [69].

On the other hand, materials other than platinum can be used as catalyst in sulfuric acid. Good results have been achieved by using tungsten carbide, WC [70–74] or, recently, cobalt phosphide, CoP_3 [75]. The activity of tung-

FIG. 13.   Ground section of an Ag–WC electrode. 200 mg catalyst/cm^2, grain size 40–63 μm; magnification 100. Layers from top to bottom: asbestos/catalyst/carbon tissue. From Mund *et al.* [77].

sten carbide can be increased by special anodic activation in the presence of reducing agents [76] or by adding silver [77]. Figure 13 shows a view of a supported Ag–WC electrode for sulfuric acid (cf. Section IVA2). It should be noted that it is not really the stoichiometric compound WC which shows optimum activity but rather an oxygen- (and hydrogen- ?) containing species on the surface of the carbide. On the other hand, tungsten oxides alone do not exhibit any activity. None or relatively low catalytic activities have been observed with a series of other carbides and nitrides [78, 79].

When platinum is used for the oxidation of hydrogen in phosphoric acid at temperatures of 150°C and higher, where carbon monoxide is tolerated, another problem arises: the reduction in surface area of the finely divided catalyst particles during operation. Although the platinum is not dissolved in the acid, the rearrangement of the most active surface sites in the presence of the electrolyte leads to the deactivation of the catalyst [66]. This problem has not yet been overcome, not even where the platinum is supported by very stable substances such as boron carbide or carbon.

3. Hydrocarbon Electrodes

In acid electrolyte even paraffins, such as methane, ethane, and propane, can be oxidized electrochemically. Higher species of the paraffin family, such as decane, are likewise oxidized to give carbon dioxide (Esso, in Bockris and Srinivasan [20, p. 563]). Yet only platinum will successfully catalyze the oxidation of these hydrocarbons, whether in sulfuric acid at 100°C, or in phosphoric acid at 150°C, or even in a mixture of CsF and HF at 150°C. Unfortunately, rather large amounts of platinum are required for the oxidation of hydrocarbons, and the most serious problem is how to extend the surface area of platinum. High current densities have been achieved using the Raney method (catalyst surface area about 40 m^2/gm), but the amount of platinum remained high and current densities were largest during the oxidation of ethane and propane [80]. The deposition of platinum on the anode by the use of boron carbide, B_4C, as support has reached a value of 6 mg/cm^2. Thus, in a PTFE-bonded electrode a specific current of 15 mA/mg has been achieved with propane in H_3PO_4 at 150°C and a potential of 300 mV [81].

Nevertheless, even if a finer distribution were possible, platinum alone will not provide a successful direct hydrocarbon fuel cell catalyst, since during operation the catalyst slowly deteriorates by being covered with a carbonaceous residue formed as intermediate. At this stage there is hardly any chance of finding an appropriate support or an alloying component.

4. Methanol Electrodes

The oxidation of methanol, which is the most attractive liquid fuel next to the hydrocarbons, does not give rise to such a catalyst-blocking residue.

However, when pure platinum is used current densities are rather low. Performance can be improved substantially by using a platinum–ruthenium alloy as catalyst. This finely divided alloy is produced either by precipitating it from a mixed solution of complex chlorides of platinum and ruthenium and adding a solution of sodium borohydride [82–84], or by making use of the Raney method [85]. In both cases a real intermetallic compound is obtained at ambient temperatures. Maximum catalytic activity is achieved if the fcc phase of platinum contains approximately 40% ruthenium.

At this alloy catalyst the oxidation of formaldehyde takes place at a rate lower than that of methanol, so that formaldehyde, an intermediate of the electrochemical oxidation, is released from the electrolyte into the atmosphere unless special precautions are taken. During the oxidation of ethanol on a platinum–ruthenium catalyst large amounts of brownish-black polymers are formed [23, 41].

A suitable catalyst for the oxidation of formaldehyde is tungsten carbide (WC) [71, 74, 86]. This reaction gives formic acid which is oxidized at a low rate. The oxidation of formic acid was most successful in the presence of Raney platinum partially covered with sulfur or selenium [68, 87], whereas in the presence of pure platinum self-poisoning occurs, which suggests that a chemisorption product or an intermediate product is firmly attached to and thus blocking the active platinum sites [87].

A good catalyst for the oxidation of both, methanol and formic acid, is platinum supported by molybdenum sulfide [88]. Formic acid itself is oxidized using the metal chelate Co tetraazaannulene as electrocatalyst [89], whereas methanol is oxidized by Pt–Sn catalysts [90].

5. Oxygen Electrodes

For the oxygen electrode, platinum on a carbon support is still required if current densities of more than, say, 100 mA/cm² are to be achieved. Such a current density can also be obtained by using platinum-free activated carbon which, however, leads to a much lower potential. The best results have been achieved after activating the carbon in an ammonia atmosphere at temperatures as high as 1000°C [41, 91]. Similar results have been obtained by heating polyacenequinones in ammonia; the substances thus obtained showed only a relatively low decrease in activity over a period of 1 yr [41].

Since nitrogen is incorporated during this activation process, it seemed plausible to heat a compound already containing nitrogen in the molecule, such as polyacrylonitrile, in order to get a satisfactory catalyst [92].

In search for a real substitute for platinum in sulfuric acid, several metal chelates were found to activate the reduction of oxygen. These include Fe phthalocyanine [93–102], Co tetraazaannulene [74, 103–107] or Co tetraphenylporphyrins [23, 74, 107, 108]. All these substances—cf. [74, 89, 104]—are nonconductors and require a conducting material as support. It became apparent that carbon would do the job, but the activity of the catalyst strongly depended on the carbon material used. Unfortunately, all these metal chelates are resistant to the attack of sulfuric acid under oxidizing conditions only for a period of several hundreds of hours. Polymeric phthalocyanines are claimed to have a higher stability [109].

Another interesting group of catalysts are the thiospinels of transition metals [74, 110]. They are good conductors and many of them are surprisingly stable in sulfuric acid. These catalysts, however, are destroyed (dissolved) within a matter of several weeks when used in oxygen electrodes (cf. Section ID).

C. Neutral Electrolytes

At first glance, in a neutral electrolyte corrosion problems would seem to be less severe than in strong alkaline or acid solutions. Unfortunately, at a pH value of about 7 the conductivity of the electrolyte is rather poor compared with that of strong acids or alkalies. Furthermore, current densities attainable at this pH value are more than one magnitude below those in acid or alkaline solutions.

Attention has been paid to the so-called "equilibrium electrolyte," where the electrolyte consists of a buffering solution containing carbonate and bicarbonate. In such an electrolyte carbonaceous fuels, for example, formic acid [111], can be oxidized and the resultant carbon dioxide is released as in an acid electrolyte, but current densities are low. And what is more, the nickel catalyst is attacked by the electrolyte, whereas it is stable in strong alkalies. For these reasons this concept has already been abandoned.

Fuel cells with neutral electrolyte have attracted new interest for application as a power source in the human body, for example, in pacemakers which require much smaller current densities than a technical cell. Several research groups have worked in this field [112–133].

In most concepts the glucose contained in the blood is the fuel and platinum the catalyst. The problem here is the fact that with platinum the reaction stops at an intermediate point so that less energy is available than calculated for carbon dioxide as final product and, furthermore, the platinum is poisoned by the intermediate product which remains chemisorbed at the catalyst surface. Further research and development is required to overcome this difficulty, especially if the fuel cell is intended to be used to power an artificial heart.

Another problem is the catalyst for the oxygen electrode. If platinum is used, glucose must be prevented from reacting at this electrode because it would cause a substantial potential drop. Another choice is to use as cathode catalyst a material which does not activate the oxidation of glucose, for instance, Fe phthalocyanine. On the other hand, oxygen is not allowed to reach the anode. One of the published concepts consists of two cathodes with the anode in between [112]. The whole cell is surrounded by a membrane material (organic silicon polymer) which allows penetration of glucose and oxygen from the blood. Oxygen is reduced at the outer carbon cathode, and the glucose can thus diffuse in an oxygen-free liquid to the platinum-containing inner anode.

The most serious problem in the field of implantable fuel cells is the compatibility of the material which is in direct contact with the blood, for example, the housing of the cell or the membranes through which oxygen and glucose can diffuse to the electrodes whereas larger molecules are retained.

Some work on fuels other than glucose has been done, such as de-ammoniation in neutral solution [134] and oxidation of amino acids [135]. However, no general new material aspects evolved from this work.

D. Fused Carbonate and Solid Oxide Electrolytes

Cells with fused carbonates or solid oxide electrolyte are attractive as these high temperature systems involve a high reaction rate. According to Eqs. (17), (21), and (22), the current density increases exponentially as the temperature rises and is high even if the activation energy for the particular reaction is high. Unfortunately, practical systems are not yet available because the high temperature gives rise to (1) high corrosion rates, and (2) unsatisfactory bonding of the various solid parts due to thermal expansion.

Further research work would be required especially for the fused carbonate cell, but hardly any publications have appeared in recent years. The operating temperature lies in the ideal range between $600°$ and $700°C$ (cf. IIIA1). A mixture of Li_2CO_3, Na_2CO_3 and K_2CO_3 is used as electrolyte. The melting point of the eutectic mixture is $390°C$. The melt is mixed with a solid powder in order to get a paste. At first MgO was used as solid powder. Much better results were later obtained by using a mixture of solid aluminates such as $LiAlO_2$, $NaAlO_2$, and $KAlO_2$ [136], which ensured firm contact between the anode and the electrolyte during operation. Carbon dioxide is evolved and the aluminate seems to act as a CO_2 buffer:

$$2 LiAlO_2(s) + CO_2 \rightleftharpoons Li_2CO_3(l) + Al_2O_3(s) \tag{28}$$

A current density of 200 mA/cm^2 was reached with the aluminate mixture. The electrodes were made of nickel and silver. Instead of silver CuO or NiO can be used as cathode materials. The corrosion problems still remain to be solved.

After the solid electrolyte fuel cells had been empirically investigated in the early part of this century, work was taken up again by several groups between 1961 and 1964 [137–145]. Suitable electrolytes are $ZrO_2 +$

15 mole % CaO with a resistivity of about 50 Ω cm at 1000°C and ZrO_2 + 10 mole % Y_2O_3 which has a resistivity smaller by a factor of five but which is more expensive. Ion conduction proceeds via oxygen vacancies in the lattice. An aging effect—a decrease in conductivity with time [146]—is observed during operation, which cannot yet be completely avoided by proper sintering. The original conductivity can, however, be regained by heating the electrolyte to a temperature higher than the operating temperature of the cell.

At first sight silver would appear to be the best cathode material because it is highly conductive and dissolves oxygen. These advantages are, however, offset by its low melting point [147]. Doped indium oxide and mixed oxides have been investigated by several groups [148, 149]. Recently $LaNiO_3$, a perovskite-type compound, has been found to be a very active cathode material [150]. It has a resistivity of 0.01 Ω cm at 1000°C. Suitable anode materials are cermets of nickel or cobalt with zirconium oxides. The three layers—anode, electrolyte, cathode—have to be sintered together in order to achieve a satisfactory bond. With such cells a specific power of 400 mW/cm^2 has been reached [150]. Similarly high values are difficult to achieve in low temperature fuel cells and would permit the construction of batteries with a power per weight ratio of below 10 kW/kg.

In principle disk-type cells can be used [139, 142]. Figure 14 shows such a cell combined with a reformer (containing Ni gauze as catalyst) operating as an integrated system at the same temperature. With this system it was demonstrated that hydrocarbons can be converted completely. An interesting development is the thin film concept which was designed in order to connect a large number of cells in series to get a higher voltage. It has the additional advantage that the expensive electrolyte is very thin. The multiple cell structure is supported by a porous alumina tube (see Fig. 15) onto which the various layers are sprayed using suitable masks. The electric connection between two cells is achieved by applying an interconnection material which

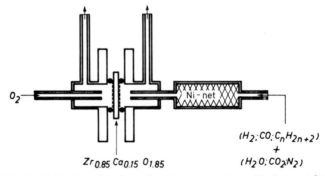

FIG. 14.   Solid electrolyte fuel cell with converter. From Binder *et al.* [138].

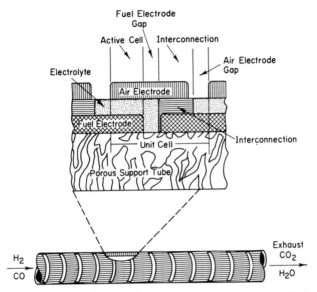

FIG. 15.   Cross section through the wall of a thin-film high temperature fuel cell battery; vertical scale exaggerated. From Sverdrup *et al.* [151].

must be electronically rather than ionically conductive and stable in both oxidizing and reducing atmospheres. In the arrangement shown manganese-doped cobalt chromite was used [152]. Its resistivity is 6 Ω cm at an oxygen partial pressure of 0.2 bar (air) and 50 Ω cm at an oxygen partial pressure of 10^{-14} bar (fuel gas mixture). The problem here is that the interconnection material is not allowed to react with electrode or electrolyte. On the other hand, it must adhere firmly to electrode or electrolyte and must separate the cells in a completely gas-tight manner.

It should be pointed out that further problems consist in finding suitable materials for the casing of the cell stacks, for the reformer and for the heat exchanger of the air. The cells operate very satisfactorily on air but heat exchange problems arise due to the specific heat of the accompanying nitrogen. The conversion of coal at 1000°C has also been demonstrated. Further work on these still promising cells is in progress [146, 151–154].

V. MATERIALS ASPECTS OF FUEL CELL SYSTEMS

A. General Considerations

Despite all the problems encountered in fuel cell research and development a variety of fuel cells have reached a technological stage. In these cases the fuel cell has to be integrated in a complete system of fuel supply and control, water and heat balance, where mainly engineering problems have to be overcome. Nevertheless, materials problems are involved as well, for example, in pumps, housings, current leads, and so forth. Some of these problems are outlined in the following sections where complete systems are introduced. One major problem is the supply of hydrogen.

B. Hydrogen Generation

Hydrogen–air fuel cells, which are in the most advanced state of development, have one disadvantage: the hydrogen gas on which they rely cannot easily be stored and transported. Its energy density is low compared with that of most liquid fuels, notably of the hydrocarbons. Even in liquid form—cryogenically stored—hydrogen has a comparatively low energy density. On the other hand, hydrogen is always a secondary fuel which has to be generated from fossil fuels or by electrolysis of water. It therefore appeared obvious to use a hydrogen fuel cell which provides high power densities by attaching a generating device which produces the required hydrogen *in situ*. Only a few examples will be cited here: methanol can be split into H_2 and CO at a relatively low temperature (220°C) in the presence of a nickel—copper catalyst [76]; steam-reforming of methanol or alkanes gives hydrogen containing CO and CO_2 impurities which often have to be removed [155, 155a]; ammonia can be split into hydrogen and nitrogen [156, 157]; alkanes may be converted into a hydrogen-rich feed gas by partial oxidation [158, 159] at temperatures above 1000°C, a process which does not require additional water or a catalyst. Thus, catalyst poisoning will not occur if a sulfur-rich oil is converted, and desulfurization prior to the reforming step is not necessary. On the other hand, the electrode catalyst has to tolerate the sulfur.

The hydrogen generator based on one of these principles works in an integrated power generating system in accordance with the requirements of the heat and mass balance. In addition, in case of an alkaline fuel cell system the CO has to be converted into CO_2 in a shift reactor, residual traces have to be converted in a special methanation reactor into harmless methane. Finally CO_2 has to be removed in an absorber.

Very recently major interest has again been focused on fuel cells that can be fed with high purity hydrogen either generated electrolytically or purified prior to use in large combined steam reforming and purification plants. In the new concept of a "hydrogen economy," hydrogen rather than electricity is to be transported over great distances. The primary energy for hydrogen generation would be supplied by electricity (electrolysis) or nuclear heat, solar or geothermal power (thermal water splitting) (cf. [160, 161] and Chapter 8).

For a survey on fuel cells with auxiliary systems, which include hydrogen generation, the reader is referred to the literature (e.g., [22, 34]).

C. Alkaline Electrolyte Battery

1. Hydrogen Fuel Cell Battery

Pure hydrogen was used as fuel for the alkaline fuel cell system of the Apollo manned spaceflights manufactured by Pratt and Whitney. The main solid material of the Apollo system [38]—in addition to electrodes materials and leads—is the PTFE isolation and matrix material for a working temperature of 200–260°C. Gaseous product water diffusing into the H_2 compartments is condensed and separated by a centrifugation blower operating independent of gravitational forces. The reaction gases used are cryogenically stored and the system is purged only every 7 hr because of the low amount of impurities present. The problem of heat exchange is solved by a separate cooling cycle and radiation cooling. The whole fuel cell battery is surrounded by an N_2 atmosphere serving as a pressure buffer, a measure to prevent an H_2 explosion, an additional heat transfer medium, and for pressurizing the electrolyte in each of the cells.

The space shuttle concept calls for a reusable vehicle; hence, the fuel cell system should be available for additional missions: major changes, compared with the Apollo concept, are increased life requirement at equivalent reliability and multiple restart capability. NASA has commissioned parallel technology assessment of cell systems developed at General Electric and Pratt and Whitney [162–164]. The Pratt and Whitney approach uses a base electrolyte matrix cell technology.

Other materials problems in hydrogen fuel cell systems have been solved, for example, by ÅSEÅ, EPS/Chloride, Allis-Chalmers, Institut Français de Pétrole (and others in France [24, 165]), VARTA (e.g., [166]) and others.

Figure 16a shows an exploded view of the cell materials of a Siemens system. Details are described, including the use of polysulfone as hydrophilic matrix material [62, 167–169].

The VARTA system (3.5 kW, see Fig. 16b) powering a fork lift truck can be fueled by stored hydrogen or by hydrogen produced in an integrated ammonia splitting apparatus. In this case an additional regulation system is required for ammonia splitting besides the control of temperature, electrolyte concentration, product gas concentration, pumps and air compressor [166]. The cut face of the system is shown in Fig. 16c. Meanwhile (1976) units of higher power output including 15 kW units have been built. A hydrogen-fuel-cell-powered car has been developed which is driven by Kordesch (Union Carbide) [171, 172]. The system is supported by lead acid secondary batteries, so that the power is generated by a hybrid system, the secondary batteries serving for peak power demand.

Moreover, fuel cell systems are used or have been proposed for boats, trucks, especially for buoys and unattended beacons [173–175].

(a)

(b)

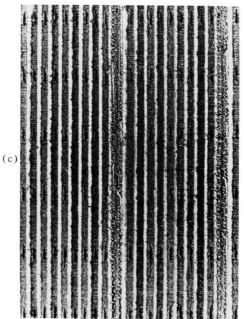

(c)

FIG. 16. (a) Exploded view of a cell arrangement with supported electrodes. From front to back: waved Ni foil, contact net, Raney Ni anode, asbestos layer, electrolyte spacer, asbestos layer, Ag cathode, contact net, waved Ni foil. From Sturm [167].   (b)   View of a 3.5-kW alkaline electrolyte system. From Sprengel [166] and Krämer *et al.* [170].   (c)   Cut face of the system in part (a) White: asbestos. Dark–black: electrodes. Grey (e.g., middle of the picture): electrolyte. From Krämer *et al.* [170].

2. Hydrazine Fuel Cell Battery

Two construction materials and principles were used for the hydrazine-O_2 fuel cell system developed and operated by Siemens [176]:

(1) Plastic frames were machined from Plexiglas or injection molded from polystyrene. The system was sealed with O rings.

(2) Instead of normal frames and O rings, asbestos sheets, reinforced along the edges with asbestos packing material were used in a filter-press-type system.

The injection of the hydrazine fuel into the circulating electrolyte is controlled by a new hydrazine sensor or the battery voltage. A catalytic H_2O_2 decomposer is used for the oxygen supply. A plate-type heat exchanger cools both the O_2 of the H_2O_2 decomposer and the waste gas from the battery in order to extract the water vapor.

Shell developed a hydrazine system [177] to power a car, driven by a secondary battery-fuel cell hybrid. The hydrazine fuel is supplied to the stack as a dilute 1–2% solution in the caustic potash electrolyte. The concentration is maintained by a fuel pump which is itself controlled by an electrochemical fuel concentration monitor.

When undertaking fuel cell materials research and development the usual procedure is to start with the catalyst—reputedly the thorniest problem—then to move on to the electrode, from there to the unit cell, thence to the battery, and finally to the cell system with its auxiliaries. The following alternative approach has likewise been pursued by Alsthom/Exxon: Starting with a system concept a battery structure was devised, which in turn led to a special type of electrode and a specific electrolyte composition, which finally led to a new approach to the catalyst problem [178]. According to the battery concept, a special type of single cell was developed which involved, for example, the simultaneous injection of gas and electrolyte (fuel and electrolyte) into the cells. The single cells consist of electrodes of wafer structure separated by membranes. Electric connection is in series, each element is only 0.5 mm (or less) thick. The fuel used is hydrazine. Automatic decarbonization—if necessary—of the electrolyte is effected by membranes [178].

A series of other hydrazine fuel cell systems has been under investigation as well (see especially [7, 20]).

3. Methanol Fuel Cell Battery

Varta developed direct methanol fuel cell systems. The low power system [179] worked with circulating electrolyte at 40°C; methanol is oxidized in 8 N KOH. In a modified high power methanol system [180], to control the vapor pressure of methanol, the operating temperature is likewise limited to 40–60°C. An electrolyte–fuel cascade—starting with 8 N KOH—is used for the battery systems, each cascade step having its own circulating system. The first and the last cascade step is provided with a sensor element to monitor the depletion of the KOH–CH$_3$OH mixture. The active material of this sensing element is an ion exchange membrane [181].

Two systems, the Bosch system [182] and the Battelle system [183, 184], for the direct oxidation of methanol show the performance of Raney-type precious metal alloys as catalyst materials in 6 M KOH–3 M CH$_3$OH electrolyte. The Bosch self-regulating system includes automatic fuel supply. The systems are fed either with oxygen from the air (cf. Section IVA) or with stored O$_2$.

Other methanol systems, for example, by I.F.P. [24] and Allis–Chalmers (cf. [7]) are known.

4. Special Considerations

Some materials problems of auxiliary systems and fuel cell system parts are known from chemical technology, but other problems had to be solved especially for the fuel cell technology. This may be illustrated by the example of a magnet pump consisting of stainless steel; the magnet is made of a special Ni–Co–Al alloy, which is resistant to KOH. Alkali resistant plastic housings and graphite pump bearings had to be constructed as well [185].

During continuous operation the water content can be controlled by a special water transporting material: in a method proposed by Allis–Chalmers (cf. Vielstich [7]) for H$_2$–O$_2$, cell water is withdrawn from the cells using an asbestos membrane wetted with KOH and combined with a porous nickel plate. At a given temperature the KOH concentration in the two membranes and the rate of removal of water from the cell becomes directly dependent on the pressure outside the porous nickel plate.

Another special feature is the use of additional materials as corrosion resistant supports, such as the separator papers used for the "supported electrode." Such separators have also been developed for acid fuel cells: hydrophilic asbestos paper—consisting of 35% asbestos, 50% cation exchange material and 15% synthetic rubber additive [186]—was specially developed for fuel cell use. Another separator matrix consists of a (sulfonated) organic polymer to which Ta$_2$O$_5$ is added to improve the wetting properties [187].

D. Sulfuric Acid Electrolyte Battery

H$_2$–O$_2$ and H$_2$–air fuel cell systems are the most advanced systems for acid electrolytes at present, though laboratory cells oxidizing formic acid on sulfur- or selenium-covered platinum, formaldehyde (at a relatively low current density, however) on tungsten carbide and methanol on Raney platinum–ruthenium have been investigated [23, 41]. Even the methanol–air sulfuric acid fuel cell system of AEG-Telefunken [188] is an indirect system which consists of an integrated methanol cracker and an H_2–air fuel cell battery (cf. Section VB). The methanol is cracked catalytically at 300–330°C without addition of other substances such as water. The resultant gas contains H_2 and CO; the hydrogen is oxidized anodically in the fuel cell stacks, while the carbon monoxide is burned to heat the cracking reactor after having passed the battery.

The H_2 fuel cell itself contains a WC anode and a carbon–Pt cathode [188]. This carbon cathode contains a few milligrams of Pt per square centimeter of surface area. The same system shows reduced performance when activated carbon without the platinum additive is used [189, 190]. Figure 17

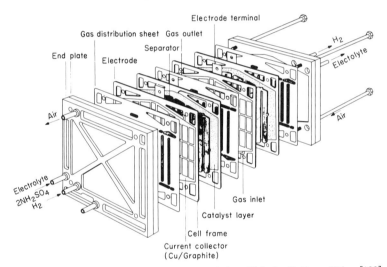

FIG. 17.   Materials of a AEG-Telefunken sulfuric acid fuel cell. From Böhm [190].

shows an exploded view of the AEG-Telefunken fuel cell assembly; the whole system is shown in Fig. 11. Small battery systems without noble metals showed outstanding operating times [189].

The greatest difficulties have been encountered in finding suitable contact materials for fuel cell systems. With the exception of noble metals and tantalum, all metals corrode in dilute acids under fuel cell conditions. Two concepts have been developed for the solution of this problem: electrodes with external contact (see Figs. 11 and 17) and bipolar electrodes. For both designs graphite is used as the conductive material. The electrodes with external contact consist of a perforated steel foil coated on both sides with a layer of graphite; in the bipolar electrodes the oxygen chamber of one cell is separated from the hydrogen chamber of the next one by a flexible graphite foil 0.2–0.3 mm thick [189].

A laboratory model based on the same principle of cracking methanol and using hydrogen in WC–carbon fuel cells was developed at Battelle [191].

Sulfuric acid electrolyte systems were also used by Shell [13], American Cyanamid Co., and S.E.R.A.I. [22], among others.

In General Electric's system for the Apollo mission the ion exchange membrane is in principle an acid electrolyte. In General Electric's system for the space shuttle program, the solid polymer electrolyte functions as a sulfonic acid electrolyte fuel cell where the acid groups $(-SO_3^-)$ are chemically linked to the polymer chain where they are immobilized. The—sulfonated—polymer has been developed by DuPont and is a fluorocarbon analogous to PTFE [192]. The capability of the system of 2000 hr of maintenance-free life and the potential of 5000–10,000 hr of useful life with invariant performance together with the start–stop procedures have been described [162, 193].

E. Phosphoric Acid Electrolyte Battery

Highly concentrated phosphoric acid in a matrix between the electrodes is used at about 150°C as electrolyte for a fuel cell system tested by UTC–Pratt and Whitney [64] for commercial and especially residential electricity supply. The whole system is supplied by natural gas and includes a reformer system for hydrogen production. The efficiency of the system, including a dc–ac converter, can be calculated at 42%. The units, that is, the single integrated systems of this project named TARGET are automatically controlled 12.5 kW aggregates. Figure 18 shows a scheme of a typical cell assembly of the system [193].

A 26 MW power station is United Technology Corporation's (formerly Pratt and Whitney's) second proprietary project of fuel cell systems with phosphoric acid electrolyte [65]. The subunits are under investigation. A stack of 376 phosphoric acid cells which is prepared for pilot test is shown in Fig. 19. To achieve the desired 26 MW level will require assembling and interconnecting about 100 such stacks [194]. The fuel is hydrogen produced by conversion of natural gas, gasification of carbon or other future hydrogen producing reactions. Construction materials in principle are the same as with the system described above. Both systems use oxygen from air.

A third, commercially available, system supplied by Engelhard Industries Mining and Chemical Company, is based on H_3PO_4 at about 150°C using precious metals as electrode catalysts. The main difference is the production of hydrogen fuel from ammonia [195, 196].

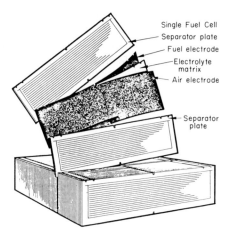

FIG. 18.   Typical phosphoric acid fuel cell assembly, Pratt and Whitney. From Crowe [193].

FIG. 19.   Development stack of 376 phosphoric acid fuel cells in preparation for pilot test. From [194].

VI. OUTLOOK OF RESEARCH IN FUEL CELL MATERIALS

Improvement of the corrosion stability of materials under fuel cell conditions and the search for new electrocatalysts are the two main problems of future materials research. Precious metals should be replaced by cheaper materials or their concentration at least be drastically reduced.

Only a short period has elapsed between the introduction of the word "electrocatalysis" in 1963 [197] and today's scientific research trend in electrocatalytic materials [198]. Electrocatalysis is far from being understood. Practical men still wait for more stimulation from theorists. Stability problems—chemical and electrochemical—associated with electrode and electrocatalyst are the most serious among many other materials problems. The study of electrocatalysis should include not only electron transfer but also ion transfer between catalyst and electrolyte which in many cases is responsible for the deactivation of the catalyst. Research on ion transfer in other areas may help to solve this problem [199–201].

An important recent development is the use of carbon in various forms as a structural component in fuel cell systems. Carbon may replace many expensive materials or help to reduce the required amount.

A necessary condition for future fuel cell systems is to minimize the auxiliary equipment. This, of course, calls for materials design, electrodes and cells that do not require complex mechanical and electrical attachments.

REFERENCES

[1]   E. Justi, M. Pilkuhn, W. Scheibe, and A. Winsel, "Hochbelastbare Wasserstoff-Diffusions-Elektrode." Franz Steiner Verlag, Wiesbaden, 1959.
[2]   E. Justi and A. Winsel, "Kalte Verbrennung." Franz Steiner Verlag, Wiesbaden, 1962.
[3]   H. H. v. Döhren and K. J. Euler, "Brennstoffelemente," 6th ed., Varta Fachbuch Band 6. VDI Verlag, Düsseldorf, 1971.
[4]   G. J. Young (ed.), "Fuel Cells," Vols. I and II, Van Nostrand–Reinhold, Princeton, New Jersey, 1960, 1963.
[5]   W. Mitchell, Jr. (ed.), "Fuel Cells." Academic Press, New York, 1963.
[6]   W. Vielstich, "Brennstoffelemente." Verlag Chemie, Weinheim, 1965.
[7]   W. Vielstich, "Fuel Cells." Wiley, New York, 1970.
[8]   G. Sandstede, Galvanische brennstoffzellen, in "Dechema-Monographien 49." Verlag Chemie, Weinheim and Bergstr., 1964.
[9]   G. Sandstede, Elektrochemische brennstoffzellen, in "Fortschritte der chemischen Forschung 8." Springer-Verlag, Berlin and New York, 1967.
[10] "Fuel Cell Systems," Vols. I and II. American Chemical Society, Washington, D.C., 1965 and 1969.
[11] B. S. Baker (ed.), "Hydrocarbon Fuel Cell Technology." Academic Press, New York, 1965.
[12] "Les Piles à Combustible." Inst. Français du Pétrole, Ed. Technip, Paris, 1965.
[13] K. R. Williams (ed.), "An Introduction to Fuel Cells." Elsevier, Amsterdam, 1966.

[14] M. Barak, "Fuel Cells—Present Position and Outstanding Problems," Vol. 6, Advanced Energy Conversion. Pergamon, Oxford, 1966.
[15] A. B. Hart and G. J. Womack, "Fuel Cells: Theory and Application." Chapman and Hall, London, 1967.
[16] J. Euler, "Aus der Frühzeit der galvanischen Brennstoffelemente." Varta AG, Frankfurt, 1966.
[17] H. A. Liebhafsky and E. J. Cairns, "Fuel Cells and Fuel Batteries." Wiley, New York, 1968.
[18] C. Berger (ed.), "Handbook of Fuel Cell Technology." Prentice Hall, Englewood Cliffs, New Jersey, 1968.
[19] F. v. Sturm, "Elektrochemische Stromerzeugung." Verlag Chemie, Weinheim and Bergstr., 1969.
[20] J. O'M. Bockris and S. Srinivasan, "Fuel Cells: Their Electrochemistry." McGraw-Hill, New York, 1969.
[21] M. W. Breiter, "Electrochemical Processes in Fuel Cells." Springer-Verlag, Berlin and New York, 1969.
[22] A. Winsel, *in* "Ullmann's Encyklopädie der technischen Chemie," 3rd ed., p. 703. Urban and Schwarzenberg, München, 1970.
[23] G. Sandstede (ed.), "From Electrocatalysis to Fuel Cells." Univ. of Washington Press, Seattle, Washington, 1972.
[24] Y. Bréelle, O. Bloch, P. Degobert, and M. Prigent, "Principes, Technologie Applications des Piles à Combustible." Editions Technip, Paris, 1972.
[25] W. Vielstich, *Int. Congr. Pure and Appl. Chem., 24th, Hamburg, 1973* Vol. 5, Applied Electrochemistry, p. 79. Butterworths, London, 1974.
[26] F. v. Sturm, *Int. Congr. Pure and Appl. Chem., 24th, Hamburg, 1973* Vol. 5, Applied Electrochemistry, p. 49. Butterworths, London, 1974.
[27] H. Binder, A. Köhling, and G. Sandstede, *in* "From Electrocatalysis to Fuel Cells" (G. Sandstede, ed.), Univ. of Washington Press, Seattle, Washington, 1972.
[28] M. Pourbaix, "Atlas d'Equilibres Electrochimiques." Gauthier-Villars, Paris, 1963.
[29] U. R. Evans, "The Corrosion and Oxidation of Metals: Scientific Principles and Practical Application." Arnold, London, 1960.
[30] H. Kaesche, "Die Korrosion der Metalle." Springer-Verlag, Berlin and New York, 1966.
[31] H. Ewe, E. Justi, and M. Pesditschek, *Energy Convers.* **15**, 9 (1975).
[32] F. v. Sturm, *Chem. Ing.-Tech.* **48**, 91 (1976).
[33] K.-D. Beccu, *Chem. Ing. Tech.* **46**, 95 (1974).
[34] J. M. Burger, P. A. Lewis, R. J. Isler, F. J. Salzano, and J. M. King, Jr., *Intersoc. Energy Convers. Eng. Conf. 9th*, San Francisco, California (1974).
[35] K. V. Kordesch, *Proc. Power Sources Conf., 21st* p. 14. PSC Publ. Committee, Red Bank, New York, 1967.
[36] H. Böhm (AEG), Private communication.
[37] K. V. Kordesch, *in* "Handbook of Fuel Technology" (C. Berger, ed.), p. 359. Prentice Hall, Englewood Cliffs, New Jersey, 1968.
[38] *Chem., Ing. Tech.* **41**, A1085 (1969).
[39] *Int. Symp. Fuel Cells, 3rd, Bruxelles* Presses Acad. Eur., Bruxelles, 1969.
[40] H. Holthusen, R. Wendtland, and A. Winsel (Varta), *in Int. Symp. Fuel Cells, 3rd, Bruxelles* p. 154. Presses Acad. Eur., Bruxelles, 1969.
[41] Battelle-Institut, Frankfurt/M., unpublished results.
[42] H. Binder, A. Köhling, W. H. Kuhn, W. Lindner, and G. Sandstede, *in* "From Electrocatalysis to Fuel Cells" (G. Sandstede, ed.), p. 131. Univ. of Washington Press, Seattle, Washington, 1972.

[43] H. B. Gutbier (Siemens), Reprint of "Forschung und Technik", Neue Zürcher Zeitung, No. 490, 22.10.73.

[44] A. R. Duckworth, M. I. Gillibrand, and F. Twentyman (EPS), *Nature (London)* **226**, 846 (1970).

[45] M. Jung (Varta), in "Fuel Cells," p. 112. Wiley, New York, 1970; cf. also *ISE Meeting, 23rd, Stockholm* Abstracts of papers, p. 441 (1972).

[46] S. G. Meibuhr (GM), *Electrochim. Acta* **12**, 1059 (1967); *J. Electrochem. Soc.* **121**, 1245 (1974).

[47] F. L. Simons, E. J. Cairns, and D. J. Surd (GE), *J. Electrochem. Soc.* **116**, 556 (1969).

[48] E. L. Simons, D. W. McKee, and E. J. Cairns (GE), in *Int. Symp. Fuel Cells, 3rd, Bruxelles* p. Presses Acad. Eur., Bruxelles, 1969.

[49] D. W. McKee, A. J. Scarpellino, I. F. Danzig, and M. S. Pak (GE), *J. Electrochem. Soc.* **116**, 562 (1969).

[50] H. Schmidt and W. Vielstich, *Z. Anal. Chem.* **224**, 84 (1967).

[51] P. G. Grimes and H. M. Spengler, in "Hydrocarbon Fuel Cell Technology" (B. S. Baker, ed.), p. 121. Academic Press, New York, 1965.

[52] E. Schwarzer and W. Vielstich, *Chem. Ing. Tech.* **45**, 201 (1973).

[53] H. Binder, A. Köhling, W. Kuhn, W. Lindner, and G. Sandstede, *Chem. Ing. Tech.* **40**, 171 (1968).

[54] F. T. Bacon, *Ind. Eng. Chem.* **52**, 301 (1960); cf. Vielstich [7, p. 130].

[55] H. L. Bevan, A. C. C. Tseung, *Electrochim. Acta* **19**, 201 (1974).

[56] W. J. King, A. C. C. Tseung, *Electrochim. Acta* **19**, 485, 493 (1974).

[57] H. G. Oswin, *Int. Symp. Fuel Cells, 2nd, Brussels* p. 321 (1967).

[58] K. Mund, *Siemens Forsch. EntwicklungsBer.* **5**, 209 (1976).

[59] Battelle-Institut, Frankfurt/M., unpublished results; cf. Binder *et al.* [53].

[60] K. Brill, *Int. Symp. Fuel Cells, 3rd, Bruxelles* p. 39. Presses Acad. Eur., Bruxelles, 1969; cf. Dietz and Jahnke [182, p. 16].

[61] K. Höhne, *Siemens Forsch. Entwicklungsber.* **1**, 31 (1974).

[62] H. Grüne, H. B. Gutbier, and K. Strasser, *Power Sources Symp.*, Brighton (1976).

[63] K. V. Kordesch and R. F. Scarr, *Proc. Intersoc. Energy Convers. Eng. Conf.* p. 12, San Diego, California p. 12 (1972).

[64] N. R. Iammartino, *Chem. Eng.* **81**, 62 (1974).

[65] *Adv. Battery Technol.* **10** (No. 2), 1 (1974).

[66] P. Stonehart, K. Kinoshita, and J. A. S. Bett, in *Proc. Symp. Electrocatal.*, San Francisco, California (M. Breiter, ed.), p. 275. The Electrochemistry Society, Princeton, New Jersey, 1974.

[67] H. Binder, A. Köhling, and G. Sandstede, in "From Electrocatalysis to Fuel Cells" (G. Sandstede, ed.), p. 59. Univ. of Washington Press, Seattle, Washington, 1972; cf. Binder *et al.* [87].

[68] H. Binder, A. Köhling, and G. Sandstede, *Angew. Chem.* **79**, 903 (1967); *Angew. Chem. Int. Ed. Engl.* **6**, 884 (1967); cf. *Nature (London)* **214**, 268 (1967).

[69] L. W. Niedrach and I. B. Weinstock, *Electrochem. Technol.* **3**, 270 (1965).

[70] H. Böhm and F. A. Pohl, *Wiss. Ber. AEG-Telefunken* **41**, 46 (1968).

[71] H. Binder, A. Köhling, and G. Sandstede, Preprints of papers, *Am. Chem. Soc. Meeting N.Y.* **13**, No. 3, 99 (1969).

[72] G. Louis and H. Böhm, *Int. Symp. Fuel Cells, 4th, Antwerpen* (1972).

[73] H. Binder, A. Köhling, W. Kuhn, W. Lindner, and G. Sandstede, *Nature (London)* **224**, 1299 (1969).

[74] H. Behret, H. Binder, and G. Sandstede, in *Proc. Symp. Electrocatal. San Francisco, California* (M. Breiter, ed.), p. 319. The Electrochemistry Society, Princeton, New Jersey, 1974.

[75] K. Mund, G. Richter, R. Schulte, and F. v. Sturm, *Ber. Bunsenges.* **77**, 839 (1973).
[76] K. v. Benda, H. Binder, A. Köhling, and G. Sandstede, *in* "From Electrocatalysis to Fuel Cells" (G. Sandstede, ed.), p. 87. Univ. of Washington Press, Seattle, Washington, 1972.
[77] K. Mund, G. Richter, and F. v. Sturm (Siemens), *Collect. Czech. Chem. Commun.* **36**, 439 (1971).
[78] D. Baresel, W. Gellert, W. Sarholz, and G. Schulz-Ekloff, *IUPAC Congr. 24th, Hamburg* Abstracts of Papers, p. 496 (1973).
[79] J. Heidemeyer, D. Baresel, W. Gellert, and P. Scharner, *Collect. Czech. Chem. Commun.* **36**, 944 (1971).
[80] H. Binder, A. Köhling, H. Krupp, K. Richter, and G. Sandstede, *J. Electrochem. Soc.* **112**, 355 (1965).
[81] W. T. Grubb and E. J. Cairns, *in* "Fuel Cells and Fuel Batteries," p. 458. Wiley, New York, 1968.
[82] H. C. Brown and C. A. Brown, *J. Am. Chem. Soc.* **84**, 1493 (1962).
[83] D. W. McKee and F. J. Norton, *J. Phys. Chem.* **68**, 481 (1964).
[84] O. A. Petry, B. I. Podlovchenko, A. N. Frumkin, and Hira Lal, *J. Electroanal. Chem.* **10**, 253 (1965).
[85] H. Binder, A. Köhling, and G. Sandstede, *in* "Hydrocarbon Fuel Cell Technology" (B. S. Baker, ed.), p. 91. Academic Press, New York, 1965.
[86] H. Binder, A. Köhling, W. Kuhn, and G. Sandstede, *Angew. Chem.* **81**, 748 (1969); *Angew. Chem. Int. Ed. Engl.* **8**, 757 (1969).
[87] H. Binder, A. Köhling, and G. Sandstede, *Adv. Energy Convers.* **7**, 121 (1967).
[88] H. Binder, A. Köhling, and G. Sandstede, *Energy Convers.* **11**, 17 (1971).
[89] H. Jahnke, M. Schönborn, and G. Zimmermann, *Topics Current Chem.* **61**, 133 (1976).
[90] M. R. Andrew, J. S. Drury, B. D. McNicol, C. Pinnington, and R. T. Short, *J. Appl. Electrochem.* **6**, 99 (1976); B. D. McNicol, A. G. Chapman, and R. T. Short, *ibid.* **6**, 221.
[91] H. Böhm, *Wiss. Ber. AEG-Telefunken* **43**, 241 (1970).
[92] G. Richter and G. Luft (Siemens), Bunsentagung, Hamburg, 1972; *Int. Symp. Fuel Cells, 4th, Antwerpen* (1972).
[93] R. Jasinski, *Nature (London)* **201**, 1212 (1964); *J. Electrochem. Soc.* **112**, 526 (1965).
[94] H. Jahnke, Bunsentagung, Augsburg (May 1968); Abstract, *Ber. Bunsenges. Phys. Chem.* **72**, 1053 (1968).
[95] H. Jahnke and M. Schönborn, "Comptes Rendus, Troisièmes Journées International d'Etude des Piles à Combustible," p. 60. Presses Académiques Européennes, Bruxelles, 1969.
[96] M. Schönborn, H. Jahnke, and G. Zimmermann, *Int. Symp. Fuel Cells, Antwerpen* (October 1972).
[97] H. Jahnke, M. Schönborn, and G. Zimmermann, *Bosch Tech. Ber.* **4**, 2 (1973).
[98] A. Kozawa, V. E. Zilionis, and A. J. Brodd, *J. Electrochem. Soc.* **118**, 1705 (1971).
[99] L. Y. Johanssen, J. Mrha, and R. Larsson, *El. Acta* **18**, 255 (1973).
[100] R. Larsson and J. Mrha, *El. Acta* **18**, 391 (1973).
[101] A. J. Appleby and M. Savy, *El Acta* **21**, 567 (1976).
[102] H. Kropf and F. Steinbach (eds.), "Katalyse an Phthalocyaninen." Thieme Verl., Stuttgart, 1973.
[103] F. Beck, *Symp. Appl. Electrochem., Vienna* (February 1971).
[104] F. Beck, "Elektroorganische Chemie." Verlag Chemie, Weinheim, 1974.
[105] F. Beck, W. Dammert, J. Heiss, H. Hiller, and R. Polster, *Z. Naturforsch.* **28a**, 1009 (1973).
[106] F. Beck, *Ber. Bunsenges.* **77**, 353 (1973).
[107] H. Alt, H. Binder, and G. Sandstede, *J. Electroanal. Chem. Interfacial Electrochem.* **31**, 19 (1971).

[108] H. Alt, H. Binder, and G. Sandstede, *J. Catal.* **28**, 8 (1973).
[109] H. Meier, W. Albrecht, U. Tschirwitz, and E. Zimmerhackl, *Ber. Bunsenges.* **77**, 843 (1973).
[110] H. Behret, H. Binder, and G. Sandstede, *Electrochim. Acta* **20**, 111 (1975).
[111] G. Grüneberg, *in* "Fuel Cells." Wiley, New York, 1970.
[112] F. v. Sturm (Siemens), Bunsentagung, Erlangen (1973).
[113] J. R. Rao and G. Richter, *Naturwissenschaften* **61**, 200 (1974).
[114] J. R. Rao, G. Richter, F. v. Sturm, E. Weidlich, and M. Wenzel, *Biomed. Eng.* **9**, 98 (1974).
[115] J. R. Rao, G. J. Richter, F. v. Sturm, and E. Weidlich, *Bioelectrochem. Bioenerget.* **3**, 139 (1976).
[116] U. Gebhardt, J. R. Rao, and G. J. Richter, *J. Appl. Electrochem.* **6**, 127 (1976).
[117] R. F. Drake *et al.* (Monsanto), *Annu. Summary Rep. Implantable Fuel Cell for an Artif. Heart, 2nd* Contract No. PH 43-66-976 (July 1968).
[118] R. F. Drake, *Proc. Artif. Heart Program Conf.* p. 869. U.S. Dept. Health, Education and Welfare (1969).
[119] R. F. Drake, B. K. Kusserow, S. Messinger, and A. Matsuda, *Trans. Am. Soc. Artif. Int. Organs* **16**, 199 (1970).
[120] S. Messinger and R. F. Drake, *Proc. Intersoc. Energy Conf.* p. 361 (1969).
[121] M. L. B. Rao and R. F. Drake, *J. Electrochem. Soc.* **116**, 334 (1969).
[122] M. Beltzer and J. S. Batzold (Esso), *Electrochim. Acta* **16**, 1775 (1973).
[123] J. Batzold and M. Beltzer, *Proc. Artif. Heart Program Conf.* p. 817 (1969).
[124] M. Beltzer and J. S. Batzold, *Proc. Int. Energy Conf., 4th* p. 361 (1969).
[125] A. J. Appleby and C. van Drunnen, *J. Electrochem. Soc.* **118**, 95 (1973).
[126] A. J. Appleby, D. J. C. Ng, and H. Weinstein, *J. Appl. Electrochem.* **1**, 79 (1970).
[127] J. Giner and G. Holleck (Tyco), *in* "From Electrocatalysis to Fuel Cells" (G. Sandstede, ed.), p. 283. Univ. of Washington Press, Seattle, Washington, 1972.
[128] J. Giner, P. Malachesky, *Proc. Artif. Org.* p. 839 (1969).
[129] J. Giner, Bunsentagung, Erlangen (1973).
[130] A. Kozawa, V. E. Zilionis, and R. J. Brodd (Union Carbide), *J. Electrochem. Soc.* **117**, 1470, 1474 (1970).
[131] A. Kozawa, V. Zilionis, R. Brodd, and R. Powers, *Proc. Artif. Org.* p. 849 (1969).
[132] J. Fishman and J. Henry (Leesona Moos), *Proc. Artif. Org.* p. 825 (1969).
[133] S. Y. Yao, M. Michuda, F. Markley, and S. K. Wolfson, Jr. (M. Reese Hosp.), *in* "From Electrocatalysis to Fuel Cells" (G. Sandstede, ed.), p. 291. Univ. of Washington Press, Seattle, Washington, 1972.
[134] S. J. Yao, S. K. Wolfson, *Trans. Am. Soc. Artif. Int. Organs,* **18**, 60 (1972).
[135] H. Jahnke, G. Zimmermann, and H. Metzger (Bosch), *Ber. Bunsenges.* **75**, 1140 (1971).
[136] G. H. J. Broers and H. J. H. van Ballegoy, *in Int. Symp. Fuel Cells, 3rd, Bruxelles* p. 77. Presses Acad. Eur., Bruxelles, 1969.
[137] J. Weissbart and R. Ruka, *J. Electrochem. Soc.* **109**, 723 (1962).
[138] H. Binder, A. Köhling, H. Krupp, K. Richter, and G. Sandstede, *Electrochim. Acta* **8**, 781 (1963).
[139] H. Tannenberger, H. Schachner, and P. Kovacs, Comptes Rendues Journées—Int. d'Etude des Piles à Combustible, Revue Energie Primaire III, p. 19. Publication trimestrielle, Editée par L'K. I. Lv., Bruxelles, 1965.
[140] H. H. Moebius and B. Rohland, Comptes Rendues Journées—Int. d'Etude des Piles à Combustible, Revue Energie Primaire III, p. 27. Publication trimestrielle, Editée par L'K. I. Lv., Bruxelles, 1965.
[141] E. B. Schultz, K. S. Vorres, L. G. Marianowski, and H. R. Linden, *in* "Fuel Cells" (G. J. Young, ed.), Vol. II, p. 24. Van Nostrand-Reinhold, Princeton, New Jersey, 1963.

[142] H. J. Böhme, and F. J. Rohr, *in Int. Symp. Fuel Cells, 3rd, Bruxelles* p. 120. Presses Acad. Eur., Bruxelles, 1969.
[143] T. Takahashi, H. Iwahara, and Y. Suzuki, *in Int. Symp. Fuel Cells, 3rd, Bruxelles* p. 113. Presses Acad. Eur., Bruxelles, 1969.
[144] G. Robert, M. Forestier, and T. Deportes, *in Int. Symp. Fuel Cells, 3rd, Bruxelles* p. 97. Presses Acad. Eur., Bruxelles, 1969.
[145] T. L. Markin, *in* " Power Sources " (D. L. Collins, ed.), Vol. 4, p. 583. Onil Press, Newcastle upon Tyne, 1973.
[146] W. Baukal, *in* " From Electrocatalysis to Fuel Cells " (G. Sandstede, ed.), p. 247. Univ. of Washington Press, Seattle, Washington, 1972.
[147] C. S. Tedmon Jr., H. S. Spacil, and S. P. Mitoff, *J. Electrochem. Soc.* **116**, 1170 (1969).
[148] A. Isenberg, W. Pabst, and G. Sandstede, DOS 1571991 = FP 1549424 (1966) and DOS 1571995 = FP 1549425 (1971).
[149] A. Isenberg, Extended Abstracts, Electrochemistry Society Atlantic City, New Jersey, p. 43 (1970).
[150] H. H. Eysel, H. Kleinschmager, and A. Reich, Electrochemistry Society Meeting, Extended Abstracts (May 1972).
[151] E. F. Sverdrup, C. J. Warde, and A. D. Glasser, *in* " From Electrocatalysis to Fuel Cells " (G. Sandstede, ed.), p. 255. Univ. of Washington Press, Seattle, Washington, 1972.
[152] C. C. Sun, E. W. Hawk, and E. F. Sverdrup, *J. Electrochem. Soc.* **119**, 1433 (1972).
[153] H. Tannenberger, *in* " From Electrocatalysis to Fuel Cells " (G. Sandstede, ed.), p. 235. Univ. of Washington Press, Seattle, Washington, 1972.
[154] W. Fischer, H. Kleinschmager, F. J. Rohr, R. Steiner, and H. H. Eysel, *Chem. Ing. Tech.* **43**, 1223 (1971); **44**, 726 (1972).
[155] H.-J. Henkel, C. Koch, H. Mentschel, H. Stamm, and E. v. Szabo, (Siemens), *in Int. Symp. Fuel Cells, 3rd, Bruxelles* p. 273. Presses Acad. Eur., Bruxelles, 1969.
[155a] O. Bloch, C. Dezael, and M. Prigent, *in Int. Symp. Fuel Cells, 3rd, Bruxelles* p. 286. Presses Acad. Eur., Bruxelles, 1969.
[156] H. Laig-Hörstebrock (Varta), *IUPAC Congr., 24th, Hamburg* (1973); *Chem. Ing. Tech.* **46**, 118 (1974).
[157] M. F. Collins, R. Michalek, and W. Brink, *Intersoc. Energy Convers. Eng. Conf., 7th* p. 32 (1972).
[158] G. R. Frysinger and I. Trachtenberg, *in* " Hydrocarbon Fuel Cell Technology " (B. S. Baker, ed.), p. 12 and 256. Academic Press, New York, 1965.
[159] H.-J. Henkel (Siemens), *Erdöl Kohle* **27**, 251 (1974).
[160] J. O'M. Bockris, *Science* **176**, 1323 (1972).
[161] H. Behret, *Tech. Zukunft* **8**, 35 (1974).
[162] J. R. Huff (ed.), Seventh Status Report on Fuel Cells. U.S. Army Mobility Equipment Res. and Dev. Center, Fort Belvoir, Virginia (1972).
[163] General Electric, Fuel Cell Technology Program, Contract No. NAS-9-11033, cited in Huff [162].
[164] United Aircraft, PAW Div. Fuel Cell Technology Program Contract No. NAS-9-11034, cited in Huff [162].
[165] G. Feuillade, *Tech. Mod.* **55** (1963); **56** (1964).
[166] B. Sprengel, *IUPAC Congr. 24th, Hamburg* (1973); *Chem.-Ing. Tech.* **46**, 118, 967 (1974).
[167] F. v. Sturm, Private communication.
[168] E. W. H. Justi, *Phys. Blätter 29*, 21, 71 (1973); *VOB Kraftwerkstech.* **52**, 363 (1972).
[169] H. Ewe and E. W. Justi, *Chem.-Z.* **97**, 469 (1973).
[170] J. Krämer, A. Winsel, and B. Sprengel, Private communication.
[171] K. V. Kordesch, *Ber. Bunsenges. Phys. Chem.* **77**, 751 (1973).

[172]   K. V. Kordesch. *J. Electrochem. Soc.* **118**, 812 (1971).
[173]   M. I. Gillibrand and J. Gray (EPS), 2. Journées Int. d'Etude des Piles à Combustible, Brussels (1967).
[174]   M. I. Gillibrand and J. Gray, *Int. Oceanol. Conf., Brighton* (1969).
[175]   H. Binder, "INTEROCEAN 17," Vol. 2, p. 197. VDI-Verlag, Düsseldorf, 1971.
[176]   H. Hohlmüller, H. Cnobloch, and F. v. Sturm, *Power Sources* **3**, 373 (1971).
[177]   N. H. Andrew, W. J. Pressler, J. K. Johnson, H. T. Short, and K. H. Williams, *Automatic Eng. Congr., Detroit* (1972).
[178]   B. Warszawski, *Rev. Gén. Elec.* **78**, 531 (1969).
[179]   A. Winsel and G. Wolf, *in* "Aktuelle Batterieforschung." Varta AG, Frankfurt, 1966.
[180]   J. Gugenberger and D. Spahrbier, "Power Sources," p. 489. Pergamon, Oxford, 1972.
[181]   A. Winsel, *Chem. Ing. Tech.* **44**, 163 (1972).
[182]   H. Dietz, H. Jahnke, and H. Reber, *BOSCH Techn. Ber.* **3**, 3 (1971).
[183]   H. Binder, A. Köhling, W. H. Kuhn, W. Lindner, and G. Sandstede, *in* " From Electrocatalysis to Fuel Cells" (G. Sandstede, ed.), p. 131. Univ. of Washington Press, Seattle, Washington, 1972.
[184]   H. Binder, A. Köhling, W. H. Kuhn, W. Lindner and G. Sandstede, *Chem. Ing. Tech.* **40**, 171 (1968).
[185]   F. Kozdan (Siemens), *Siemens Z.* **44**, 392 (1970).
[186]   W. Naschwitz and H. Cnobloch (Siemens), *Gummi Asbest Kunststoffe* **25**, 1038 (1972).
[187]   H. J. Barger (Fort Belvoir), Private communication.
[188]   H. Böhm and K. Maass, *Proc. Intersoc. Energy Convers. Eng. Conf.*, 9th p. 836 (1974).
[189]   L. Baudendistel, H. Böhm, J. Heffler, G. Louis, and F. A. Pohl, *Proc. Intersoc. Energy Convers. Eng. Conf., 7th San Diego, California* (1972).
[190]   H. Böhm, Private communication.
[191]   K. v. Benda, H. Binder, A. Köhling, and G. Sandstede, *in* " From Catalysis to Fuel Cells" (G. Sandstede, ed.), p. 99. Univ. of Washington Press, Seattle, Washington, 1972; cf. Battelle-Institut [41].
[192]   L. E. Chapman, *Proc. Intersoc. Energy Convers. Eng. Conf., 7th, San Diego* (1972).
[193]   B. J. Crowe, "Fuel Cells: A Survey." NASA, Washington, D.C., 1973.
[194]   Industrial Research (August 1976).
[195]   O. J. Adlhart (Engelhard), *in* " From Electrocatalysis to Fuel Cells" (G. Sandstede, ed.), p. 181. Univ. of Washington Press, Seattle, Washington, 1972.
[196]   O. J. Adlhart, *Proc. Intersoc. Energy Convers. Eng. Conf., 7th, San Diego* (1972).
[197]   W. T. Grubb, *Nature (London)* **198**, 883 (1963).
[198]   M. W. Breiter (ed.), *Proc. Symp. Electrocatal., San Francisco*. Electrochem. Society, Princeton, New Jersey, 1974.
[199]   G. Feuillade, *Electrochim. Acta* **14**, 317 (1969).
[200]   G. Feuillade and R. Jacon, *Electrochim. Acta* **14**, 1297 (1969).
[201]   M. S. Whittingham, *in* "Fast Ion Transport in Solids" (W. van Gool, ed.), p. 429. North-Holland Publ., Amsterdam and American Elsevier, New York, 1973.

Chapter 8

The Role of Materials Science in the Development of Hydrogen Energy Systems

G. G. LIBOWITZ

CORPORATE RESEARCH CENTER
ALLIED CHEMICAL CORPORATION
MORRISTOWN, NEW JERSEY

I. INTRODUCTION

A hydrogen energy system involves using hydrogen for the storage, transmission, and/or utilization of energy. In the overall concept employing hydrogen energy systems (which has been termed the "hydrogen economy"), hydrogen may be viewed as an energy carrier, or as a secondary

427

source of energy. It is generated by some primary source (e.g., nuclear, solar, etc.) and permits the energy to be easily stored, transported, or utilized in a more convenient form.

A. Advantages of Hydrogen

The question arises as to why hydrogen is chosen as a possible energy carrier. The advantages of using hydrogen are as follows:

(1) Hydrogen is a very versatile fuel; it can be burned directly, it can be oxidized catalytically at lower temperatures, or it can be used to generate electricity via fuel cells.

(2) Using hydrogen as a fuel has no deleterious effects on the environment since its combustion product is merely water vapor.

(3) Hydrogen may be easily stored, either as a liquid, or as discussed in more detail in Section IIIB, as a solid metal hydride.

(4) Hydrogen may be conveniently transported as a gas or as a liquid. It has been suggested that the existing natural gas pipeline system could be used to transport hydrogen gas [1]. Over large distances, the cost should be considerably less than transmission of electrical energy [2].

(5) A source of hydrogen, water, is abundant and usually readily available.

B. Hydrogen Energy Systems

The best way to describe the hydrogen economy is to give some examples of possible hydrogen energy systems.

1. Off-Peak Power Storage

The demand for electrical power does not remain constant but varies considerably diurnally, as well as seasonally. For example, peaks in demand usually occur during the day or evening hours and also during winter or the hot summer months. If there were some way to store electricity, power plants could run at their normal operating capacity in the middle of the night and during spring and autumn, and the excess electricity stored would be used during periods of peak demand. Hydrogen offers this opportunity. Excess electricity available during low demand periods would be used to electrolyze water and the hydrogen thus formed would be stored. During periods of peak demand the stored hydrogen could be fed into fuel cells to generate excess electricity, or a conventional electricity generating plant could use the hydrogen as a boiler fuel or as a gas turbine fuel. This concept is also referred to as load leveling.

2. Vehicular Fuel

It has been shown that an automobile with a normal internal combustion engine can be run on hydrogen instead of gasoline with only minor adjustments in the carburetor [3]. In addition to the major advantage of minimized pollution, the hydrogen-fueled engine appears to run with greater efficiency, and there is less engine wear because unburned hydrocarbon particles from gasoline are no longer present.

Because of its light weight, liquid hydrogen has been suggested as a possible aircraft fuel.

3. Energy Carrier for Nuclear Energy

Nuclear power plants provide energy in the form of electricity. However, at present, only about 15% of United States energy use is electrical; the remainder is in the form of storable and combustible fuels, essentially oil and natural gas. As the availability of nuclear power increases and replaces fossil fuels, hydrogen can provide a means of converting nuclear energy to combustible fuels.

4. Intermittent Energy Sources

Some of the more advanced sources of energy under development such as solar and wind energy are intermittent sources. In order to efficiently utilize such energy sources, there must be a way of providing energy when the sun does not shine or the wind does not blow; hydrogen can serve that function during such periods.

C. Materials Science and the Hydrogen Economy

Before a hydrogen economy can be achieved, many technological problems must be solved. More economic ways of generating hydrogen should be developed; improved methods of transporting and storing hydrogen are necessary; and more efficient means for utilizing hydrogen are required. The bulk of this chapter deals with cases where materials science can be of value in solving problems related to each of these areas. No attempt is made to cover all possible applications of materials science to hydrogen energy systems, but rather to illustrate the types of problems which can be solved.

II. GENERATION OF HYDROGEN

A. Hydrogen From Coal

Although coal is a desirable fuel because it can be conveniently shipped and stored, there are still some advantages to converting coal to hydrogen for energy applications. Most importantly, hydrogen is a cleaner burning fuel. Also, for some applications, for example, as an automobile fuel, hydrogen is a more convenient form of energy.

One method of producing hydrogen from coal is to react the coal with steam at elevated temperatures to form synthesis gas, $CO + CO_2 + H_2$. The hydrogen yield can then be increased by the water shift reaction

$$CO + H_2O \rightleftarrows CO_2 + H_2$$

The free energy of this reaction becomes less negative with increasing temperature. Therefore, it is desirable to run the reaction at as low a temperature as possible. In order to maintain a reasonable reaction rate, catalysts must be used, preferably below 300°C. Unfortunately, such catalysts are poisoned by the presence of sulfur from the coal. With the increased use of high sulfur coals, one major materials problem is to find new catalysts which can operate in the presence of sulfur and sulfur-containing compounds. Since the subject of catalysis is thoroughly covered in the chapter by J. H. Sinfelt, the development of new catalysts will not be discussed here except to point out that various sulfides such as sulfo-spinels and layer-structured transition metal sulfides are being investigated as possible sulfur resistant catalysts.

B. Water Electrolysis

The generation of hydrogen by the electrolysis of water is a process normally used only when inexpensive hydroelectric power can be provided. However, this process could become more important with the growing availability of electricity from nuclear energy. In addition, the development of newer sources of energy (particularly intermittent sources) such as solar, wind, and ocean thermal gradients, which require a means of converting electricity into a storable form of energy, has increased the interest in using electrolysis to produce hydrogen.

Some of the disadvantages of electrolytic cells with aqueous electrolytes include corrosion and changes in electrolyte concentration due to evaporation of water. Therefore, the possibility of using solid electrolytes which are ionic conductors has been explored [4, 5]. The properties of one such material, zirconia doped with calcia or yttria to introduce vacancies into the lattice, is discussed by Whittingham in Chap. 9. Other materials being investigated include appropriately doped thorium oxide [6] and cerium oxide [6, 7].

Another advantage of ceramic-type solid oxide electrolytes is that they can operate at relatively high temperatures (up to 1000°C in the case of ZrO_2) leading to increased efficiencies. The voltage E required to dissociate water can be obtained from the expression

$$-2FE = \Delta G$$

where ΔG is the free energy of formation of water and F Faraday's constant. Therefore,

$$E = (1/2F)(-\Delta H + T\Delta S) \tag{1}$$

Using -68.3 kcal/mole and -39.1 cal/deg mole, respectively, for the enthalpy, ΔH, and entropy, ΔS, of formation of water, Eq. (1) becomes

$$E(V) = 1.48 - 8.5 \times 10^{-4} T°K \tag{2}$$

At room temperature, the voltage required to electrolyze water is 1.23 V, but as can be seen from Eq. (2), as the temperature is raised, the required voltage decreases by 0.85 mV per degree. At 1000°C (1273°K), the required voltage would be only about 0.9 V (taking the enthalpy and entropy of vaporization into consideration).

It should be emphasized that the migrating ion in solid electrolytes for water electrolysis must be either oxygen or hydrogen. All of the ceramic-type materials are oxide ion conductors. The only hydrogen ion conductor under consideration at present is a solid polymer of perfluorinated sulfonic acid whose manner of operation is illustrated schematically in Fig. 1 [8]. The cell

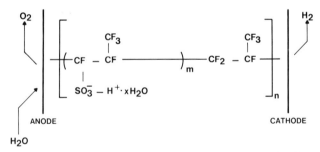

FIG. 1. Perfluorinated sulfonic acid polymer electrolyte for electrolysis of water (adapted from Nuttall *et al.* [8]).

consists of the polymer membrane with thin layers of platinum black acting as electrodes. Water is introduced at the anode and is electrolytically decomposed to form oxygen gas which is evolved, electrons which move through the external circuit, and H^+ ions which migrate through the polymer electrolyte as hydrated ions passing from one sulfonic acid group to the next. The hydrogen finally is evolved as H_2 gas at the cathode. Since the sulfonic acid groups are fixed in the polymer, the concentration of electrolyte remains constant.

Recently, a possible new solid electrolyte with a relatively high proton conductivity has been reported [9], hydrogen uranyl phosphate,

$HUO_2PO_4 \cdot 4H_2O$. The hydrogen tungsten bronzes H_xWO_3, where $x = 0$–0.6 have high proton conductivities [10], but unfortunately, they also have high electronic conductivities.

C. Thermochemical Water Splitting

It is possible to thermally decompose water by direct application of heat, but temperatures in excess of 2500°C would be required [11]; such high temperatures are not available from the usual energy sources. However, water can be indirectly thermally decomposed at lower temperatures by using a series of chemical reactions in which all the reactants (except water) are regenerated. Ideally, a two-step process, such as the one illustrated by Eqs. (3) and (4) should be used:

$$H_2O + (1/n)X \rightarrow (1/n)XO_n + nH_2 \tag{3}$$

$$(1/n)XO_n \rightarrow (1/n)X + \tfrac{1}{2}O_2 \tag{4}$$

where X represents a simple compound or element which could react with water in the manner shown. Unfortunately, there is no known substance, X, which would make Eqs. (3) and (4) thermodynamically favorable. For example, a metal, such as Mg, which would react readily with water, forms oxides that are much too stable to be easily thermally dissociated. Similarly, no thermodynamically feasible reactions such as Eqs. (5) and (6) are known:

$$H_2O + (2/n)X \rightarrow (2/n)XH_n + \tfrac{1}{2}O_2 \tag{5}$$

$$(2/n)XH_n \rightarrow (2/n)X + H_2 \tag{6}$$

Many multistage systems have been proposed [12, 13]. An example of one system suggested by Wentorf and Hanneman [12] is illustrated by Eqs. (7)–(11):

$$2Cu(s) + 2HCl(aq.) \xrightarrow{100°C} 2CuCl(s) + H_2(g) \tag{7}$$

$$4CuCl(s) \xrightarrow{30-100°C} 2CuCl_2(s) + 2Cu(s) \tag{8}$$

$$2CuCl_2(s) \xrightarrow{500-600°C} 2CuCl(s) + Cl_2(g) \tag{9}$$

$$Cl_2(g) + Mg(OH)_2(aq.) \xrightarrow{80°C} MgCl_2(aq.) + H_2O(l) + \tfrac{1}{2}O_2(g) \tag{10}$$

$$MgCl_2(s) + 2H_2O(g) \xrightarrow{350°C} Mg(OH)_2(s) + 2HCl(g) \tag{11}$$

The summation of Eqs. (7)–(11) is merely

$$H_2O \rightarrow H_2 + \tfrac{1}{2}O_2 \qquad (12)$$

The result is the thermal dissociation of water using no temperature higher than 600°C. Therefore, lower grade heat, such as that available from nuclear reactors, may be used to thermochemically split water.

One major problem, which is common to almost all thermochemical processes which have been proposed, is materials compatibility of containers. Note that two of the substances in the above process are HCl and Cl_2, both highly corrosive materials, particularly at elevated temperatures. Halogen acids or corrosive halogen compounds are present in most of the thermochemical cycles under consideration. In cases where no halogen is involved, other corrosive materials are present such as sulfuric acid, as illustrated in the cycle shown in Eqs. (13)–(16) [14]:

$$3SO_2 + 2H_2O \rightarrow 2H_2SO_4 + S \qquad (13)$$

$$2H_2SO_4 \rightarrow 2H_2O + 2SO_2 + O_2 \qquad (14)$$

$$3S + 2H_2O \rightarrow 2H_2S + SO_2 \qquad (15)$$

$$2H_2S \rightarrow 2H_2 + 2S \qquad (16)$$

Coen-Porisini and Imarisio [15] have been carrying out extensive testing on the compatibility of various materials, including alloys and ceramics, in corrosive environments related to thermochemical hydrogen production. In halogen compound atmospheres, molybdenum, tantalum, or zirconium alloys, and in some cases Inconel 625 (a Ni based alloy containing Cr and small amounts of Mo and Fe) and IN 691 (a Ni based alloy containing Cr and Co as major constituents) appear to be promising containment materials. However, much materials work must be done under various temperature and recycling conditions before thermochemical water splitting becomes feasible from a materials standpoint. It may be necessary to develop new alloys and ceramics for these particular applications.

D. Photoelectrolysis

Although the energy required to decompose water is 2.46 eV (which corresponds to 500 nm-wavelength light), direct photolysis of water does not occur in sunlight because water does not absorb radiation until well into the uv portion of the spectrum (190 nm) where solar irradiance is weak.

In 1972, Fujishima and Honda [16] proposed that water could be electrochemically photolyzed with sunlight by using semiconductor electrodes immersed in an electrolyte. The concept, now referred to as photoelectrolysis, is illustrated in Fig. 2, which is a schematic representation of the energy levels of an n-type and a p-type semiconductor immersed in an electrolyte relative to the redox levels in the electrolyte. The band bending which occurs at the semiconductor–electrolyte interface is due to the difference in work functions between semiconductor and electrolyte. That is, in order to equilibrate the Fermi levels, some electrons will be transferred from the n-type semiconductor to the electrolyte, which results in the electrolyte becoming slightly negative relative to the semiconductor. Therefore, electrons in the semiconductor will be repelled from the surface, creating a depletion region and band bending. The same situation (in reverse) holds for holes in

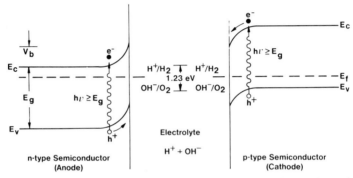

FIG. 2.   Schematic representation of the energy levels in a p–n junction analogue photoelectrolysis cell. E_c and E_v represent the bottom of the conduction band, and the top of the valence bond, respectively, in each semiconductor. E_g is the band gap, V_b the amount of band bending, E_f the Fermi level of the system, and H^+/H_2 and OH^-/O_2 are the redox levels in the electrolyte.

the p-type semiconductor. In a sense, the p–n photoelectrolysis cell may be viewed as a p–n junction interposed by an electrolyte at the junction (p–n junction analogue).

If the semiconductors are irradiated with light whose wavelength is such that $hv > E_g$, electron–hole pairs will be formed in each semiconductor electrode. Excess electrons will flow from the p-type semiconductor (cathode) into the semiconductor–electrolyte interface (as shown in Fig. 2) to reduce the H^+ ions in the electrolyte according to the reaction (in acidic electrolyte)

$$2H^+ + 2e^- \rightarrow H_2 \tag{17}$$

Similarly, holes, h^+, from the n-type semiconductor electrode (anode) will oxidize water as follows:

$$H_2O + 2h^+ \rightarrow 2H^+ + \tfrac{1}{2}O_2 \tag{18}$$

The two electrodes are, of course, connected through an external circuit to permit current flow.

If the electrolyte is alkaline, then the reactions corresponding to Eqs. (17) and (18) are

$$2H_2O + 2e^- \rightarrow H_2 + 2OH^- \tag{19}$$

and

$$2OH^- + 2h^+ \rightarrow \tfrac{1}{2}O_2 + H_2O \tag{20}$$

The sum of Eqs. (17) and (18), or of Eqs. (19) and (20), corresponds to the decomposition of water [Eq. (12)].

It should be pointed out that photoelectrolysis is fundamentally different from using a photovoltaic cell to electrolyze water. In the latter case, the electrons (and holes) which electrolyze the water are the majority carriers in the semiconductor, while in photoelectrolysis it is the minority carriers which are directly utilized in the reduction and oxidation of water, as seen from Fig. 2. However, as is true for the case of photovoltaic solar cells (see Chap. 4), the development of this process is mainly a materials problem; discovery of new semiconductors with properties appropriate for photoelectrolytic applications. Possible advantages of photoelectrolysis over electrolysis using photovoltaic cells for generating hydrogen include: (1) greater efficiency, since the former is a one-step process (the efficiency of the latter process is a product of the efficiencies of two processes, solar conversion and normal electrolysis); (2) less expensive polycrystalline thin film semiconductors may be used; (3) lower energy radiation is required.

Photoelectrolysis was first demonstrated [17–19] using a Schottky barrier analogue cell, illustrated in Fig. 3, which consists of one semiconductor electrode and a platinum metal counter electrode; the semiconductor anode was rutile, TiO_2. In this type of cell, the band gap must be greater than the

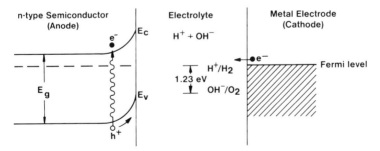

FIG. 3. Schematic representation of the energy levels in a Schottky-barrier analogue photoelectrolysis cell. E_c, E_v, and E_g represent the bottom of the conduction band, top of the valence band, and band gap energy, respectively, of the semiconductor.

1.23 eV required to electrolyze water in order for the excited electrons to have sufficient energy. However, as can be seen in Fig. 4, if the band gap is too large (> 4 eV) essentially none of the solar spectrum will be absorbed. TiO_2, which has a band gap of 3 eV only absorbs about 8–10% of the solar spectrum. Therefore, for maximum efficiency, the band gap of the semiconductor electrode in a Schottky barrier-type cell should be higher than 1.3 eV (additional energy is needed to overcome irreversible losses in the cell) and less than 2.5 eV in order to absorb a sufficient portion of the solar spectrum.

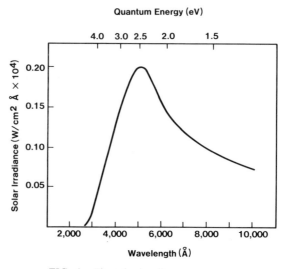

FIG. 4.   The solar irradiance curve.

In a p–n junction type cell, the total energy ideally available for photoelectrolysis is the sum of the band gaps of the n-type and p-type semiconductors (if two different materials are used) [20]. Therefore, the band gap of each semiconductor may be less than 1 eV, which means a larger percentage of the solar spectrum could be absorbed with correspondingly greater efficiencies of operation.

From this discussion, it can be seen that for photoelectrolysis, the band gap of a semiconductor is an important property to be taken into consideration. However, there are other requirements which must be met before a semiconductor can be utilized as a photoelectrolysis electrode. The semiconductor must be electrochemically stable. This is particularly true for n-type semiconductors, which tend to become oxidized when acting as an anode. For example, CdS and n–GaP which have band gaps in the appropriate range for a Schottky barrier type cell (2.25 and 2.4 eV, respectively) will undergo the following reactions and degrade when used as an anode [21, 22]:

$$CdS + 2h^+ \xrightarrow{h\nu} Cd^{2+} + S$$

$$GaP + 3H_2O + 6h^+ \xrightarrow{h\nu} Ga^{3+} + 3H^+ + H_3PO_3$$

The positions of the energy levels in the semiconductors relative to the redox levels in the electrolyte are also important. For example, the bottom of the conduction band, E_c, in the p-type semiconductor must be at a higher energy than the H^+/H_2 redox level (see Fig. 2) so that the photoexcited electrons do not have to overcome an energy barrier in order to reduce the H^+ ions [Eq. (17)]. Similarly, the top of the valence band, E_v, in the n-type semiconductor should be below the OH^-/O_2 redox level (since holes flow up). In a Schottky barrier type cell, a bias voltage can be used [23] to overcome the mismatch of energy levels, but this would decrease the efficiency of the cell to the point where its operation would no longer be economically feasible [24]. It should be pointed out that E_c in the p-type semiconductor should not be too far above the H^+/H_2 redox level (or E_v in the n-type semiconductor too far below the OH^-/O_2 level) because this energy difference is lost in the cell as overvoltage [25] and is not available for the dissociation of water, thus leading to a decrease in cell efficiency.

With respect to energy levels in the semiconductors, a further requirement is that the relative positions of the flat band potentials (the positions of the original Fermi level in the semiconductor before it equilibrates with the electrolyte) of the two semiconductor electrodes in the p–n junction type cell should not differ by too much because this would result in a large amount of band bending and a corresponding loss of efficiency. As can be seen in Fig. 2, the energy of the excited electron in the depletion region at the cathode–electrolyte interface is less than the energy of the electron at E_c in the bulk of the semiconductor. This difference in energy increases with the amount of band bending. The same is true for the difference, V_b, between E_v and the energy at the anode–electrolyte interface.

It is apparent that the requirements of semiconductors for photoelectrolysis are rather stringent, and that there is need for much further research to find appropriate materials.

III. TRANSMISSION AND STORAGE OF HYDROGEN

A. Hydrogen Embrittlement

The problem of hydrogen embrittlement is important in both the transmission and storage of hydrogen. If existing natural gas pipelines are used to ship hydrogen, as suggested in Section IA4, the possibility of hydrogen embrittlement must be considered. Hydrogen embrittlement may also be an important factor in choosing container materials for long term storage of hydrogen, particularly in the forms of compressed gas or liquid. There are three general types of hydrogen embrittlement of metals [26]: (1) hydrogen reaction, (2) internal, and (3) hydrogen environment embrittlement. However, the nature of hydrogen embrittlement is, by no means understood, and it is very likely that the basic mechanisms leading to each type of embrittlement may be strongly related.

1. Hydrogen Reaction Embrittlement

This type of embrittlement is due to the reaction of hydrogen with an impurity or with the metal itself to form an additional internal phase which creates stress within the metal. For example, in the case of carbon steels, hydrogen may react with the carbon to produce methane gas which forms high pressure bubbles along grain boundaries causing fissures in the metal. This kind of embrittlement can frequently be diminished by the addition of alloying elements, such as molybdenum, which form stable carbides and thus inhibit the formation of methane by reducing the thermodynamic activity of the carbon.

For the case of hydride forming metals (Group III, IV, and V transition metals including lanthanides and actinides), or for alloys containing hydride forming metals, the formation of the hydride will cause internal stresses because the volume of the hydride phase may be 10–25% larger than the corresponding metal. Also, the hydrides themselves are very brittle. The conditions under which hydride formation occurs is different for each metal, but the tendency to form the hydride phase always increases at lower temperatures and higher hydrogen pressures and concentrations.

2. Internal Hydrogen Embrittlement

Internal hydrogen embrittlement is caused by hydrogen which has been previously dissolved in the metal. It is usually introduced as a result of the reaction of the metal with hydrogen or a hydrogen-containing species (e.g., H_2O) during plating, melting, casting, pickling, and welding operations, or by corrosion. It is generally believed that embrittlement is associated with hydrogen concentrating at embrittlement locations such as the tips of cracks in the metal. Therefore, the diffusivity of hydrogen must be high, and it is possible that dislocation movement may be a mechanism for hydrogen diffusion [27]. However, the precise mechanisms by which hydrogen contributes to brittle crack growth or to the fracture process has not been established [28].

3. Hydrogen Environment Embrittlement

The third type of hydrogen embrittlement occurs only in the presence of hydrogen, and the process is frequently reversible in the sense that embrittlement disappears when hydrogen is removed. It is a temperature dependent process, with maximum embrittlement occurring at about room temperature. However, small amounts of oxygen impurity in the hydrogen gas will usually inhibit embrittlement; this is also sometimes true for SO_2 and CO_2 impurities.

As in the case of internal hydrogen embrittlement, the exact fracture mechanisms in hydrogen environment embrittlement are not known. It is believed that in many instances the mechanisms are the same for both types of embrittlement. Birnbaum [29] states that the source of the hydrogen does not determine the basic mechanism, but only the kinetics of the process. In the case of environment embrittlement, the rate controlling step appears to be the rate of transfer of hydrogen to the surface of the metal.

Embrittlement mechanisms which have been proposed include: (1) high pressure bubble formation at crack tips due to high supersaturation concentrations of hydrogen in the metal; (2) solute effects on plasticity—for example, hydrogen in solution increasing the resistance to plastic flow; (3) solid solution decohesion—that is, the weakening of metal–metal bonds due to the hydrogen solute; (4) change in surface energy due to adsorption of hydrogen, whereby plastic deformation is suppressed and tendency towards cleavage enhanced; (5) nucleation and precipitation of nonequilibrium hydrides at crack tips. However, despite the vast number of investigations on hydrogen embrittlement as a function of various conditions and types of metals (see, for example, Bernstein and Thompson [30] and Thompson and Bernstein [31]), it is apparent that there is a need for much further work before an understanding of the fundamental nature of embrittlement can be obtained.

B. Metal Hydrides for Hydrogen Storage

Hydrogen may be stored as a gas, as a liquid, or in the form of solid metal hydrides. The first two methods do not involve materials problems other than the choice of container materials as discussed in the previous section. However, with respect to volume, they are less efficient than the third method. Although the volume storage efficiency of liquid hydrogen is better than for gaseous hydrogen, there is still a need for heavy walled storage containers which are thermally insulated. Also, when storing for long periods of time, loss of hydrogen due to evaporation is not negligible.

1. *Advantages of Metal Hydride Storage*

As mentioned above, hydrogen can be stored as a metal hydride more efficiently than in liquid hydrogen. This is illustrated in Table 1, which shows the number density of hydrogen atoms in some representative binary metal hydrides as compared to liquid and solid hydrogen. Each hydride contains more hydrogen atoms per unit volume than does liquid hydrogen, and in most cases even more than that contained in solid hydrogen. In titanium and vanadium dihydrides the hydrogen content is more than twice that in liquid hydrogen.

TABLE 1

Hydrogen Densities in Some Hydrogen-Containing Compounds

Compound	No. of H atoms $\times 10^{-22}/cm^3$, (n_H)
Liquid hydrogen (20°K)	4.2
Solid hydrogen (4.2°K)	5.3
Water	6.7
LiH	5.9
MgH_2	6.6
YH_2	5.7
TiH_2	9.2
$VH_{0.8}$	5.1
VH_2	10.4
$PdH_{0.8}$	4.7
GdH_2	5.4
GdH_3	6.4
UH_3	8.2

As seen in Table 1, water also has a relatively high hydrogen density. This points up another major advantage of metal hydrides, the ease of reversibility of the formation reaction. As was discussed in the Section on the production of hydrogen, the recovery of hydrogen from water is not a simple process and requires the expenditure of a large amount of energy. Although the formation of a metal hydride, MH_n,

$$M(s) + (n/2)H_2(g) \rightleftarrows MH_n(s) \tag{21}$$

is normally a spontaneous exothermic reaction, the reaction can usually be easily reversed, and the hydrogen recovered by the application of a moderate amount of heat.

Metal hydrides provide an unusually safe method of storing hydrogen because the hydrides are generally quite stable below their dissociation temperature, that is, the temperature at which the equilibrium hydrogen pressure becomes 1 atm in the reverse reaction of Eq. (21). Even if the hydride is at or near its dissociation temperature, the loss of hydrogen will be slight if a leak develops in the system because the self-cooling effect of the endothermic reaction will bring the temperature down, well below the dissociation temperature. Lundin and Lynch [32] evaluated iron–titanium hydride in terms of explosibility and other safety factors and concluded that the hydride is a relatively safe material to handle.

2. Properties of Hydrides as a Storage Medium

In using metal hydrides to store hydrogen, they must meet certain requirements which are listed in Table 2. To store hydrogen efficiently, the hydrogen content of the hydride must be high, that is, the hydrogen densities

TABLE 2

Desired Properties of a Hydrogen Storage Material

High hydrogen retentive capacity
Low temperature of dissociation ($\leq 100°C$)
High rates of hydrogen uptake and discharge
Low heats of formation
Low cost of alloy
Light weight
Stable towards oxygen and moisture

such as those listed in Table 1 should preferably exceed 6×10^{22} H atoms/cm^3. A high hydrogen content is usually associated with a large value for the hydrogen-to-metal atomic ratio, but it is also a function of structure. For example, the highest H/M ratio known is 3.75 for thorium hydride, Th_4H_{15}. However, its n_H value (defined as H/atoms/cm$^3 \times 10^{-22}$) is only 7.9, which is exceeded by TiH_2, UH_3 and VH_2 as seen in Table 1. For some applications (e.g., H storage in automobiles) the number of H atoms per unit weight may be more important than n_H. MgH_2, which is about 50% less volume efficient than TiH_2 with respect to H storage, is more efficient with respect to weight, 4.6 and 2.5 $\times 10^{22}$ H atoms/gm for MgH_2 and TiH_2, respectively.

For most applications, in order to easily recover the hydrogen, the dissociation temperature should not be too high, preferably less than 100°C. In some cases it is desirable that the dissociation temperature be no higher than room temperature, which means that the hydride is stored under a positive hydrogen pressure. Several hydrides have dissociation pressures greater than one atm at room temperature; for example, VH_2 has a dissociation pressure of 1.5 atm at 21°C [33].

High rates of hydride formation and dissociation are important; that is, the times necessary to recover the hydrogen and to recharge the hydride should not be excessively long. There have been relatively few kinetic studies of hydrides, but many hydrides such as magnesium hydride, which are desirable in other respects, have low rates of hydride formation.

The enthalpy of formation of the metal hydride should be low in order to minimize the energy required to recover the hydrogen [reverse of Eq. (21)]. Enthalpies of formation of the binary metal hydrides range from under 10 kcal/mole for the Group V hydrides and PdH to over 50 kcal/mole for some of the rare earth hydrides [34]. Values under 10 kcal/mole would be desirable for storage applications. Another disadvantage of high enthalpies of formation is that this heat must be dissipated on formation of the hydride.

Light weight hydrides are important if the application involves a portable or mobile system such as hydrogen-powered vehicles. The use of heavy or high cost metals can be partly compensated for by using hydrides with high hydrogen retentive capacities, thus requiring the use of less material.

Finally, it is desirable that the metal and its corresponding hydride be chemically stable with respect to interaction with air and water vapor. Although most uses will involve closed systems, it would be preferable if it were not necessary to replace the hydride in the event of a leak.

Although all of the binary metal hydrides exhibit some of the properties listed in Table 2, none meet all, or even most, of these requirements. Therefore, in order to utilize hydrides in future hydrogen energy systems, it is necessary to explore new hydrides based on alloys or intermetallic compounds. Before, discussing the development of such new materials, however, some of the fundamental characteristics of binary metal hydrides will be reviewed.

3. Nature of Metal Hydrides

Figure 5 shows all the binary metal hydrides which can be formed by direct reaction of metal and hydrogen gas according to Eq. (21). The compounds in parentheses indicate that these hydrides have never been prepared because of the unavailability of sufficient metal, but they should have the formulas shown. It has been reported [35] that nickel hydride will form with difficulty according to Eq. (21) at pressures exceeding 8000 atm; but the hydride is unstable at normal working pressures and temperatures.

The hydrides shown in Fig. 5 have sometimes been referred to as solid solutions or interstitial compounds. However, in almost every case, the metal undergoes a definite structural change on formation of the hydride phase [34, p. 45]. The hydrides of the Groups I and II metals resemble the corresponding alkali and alkaline earth halides and are essentially ionic compounds. However, the nature of the bonding in Groups III–VIII transi-

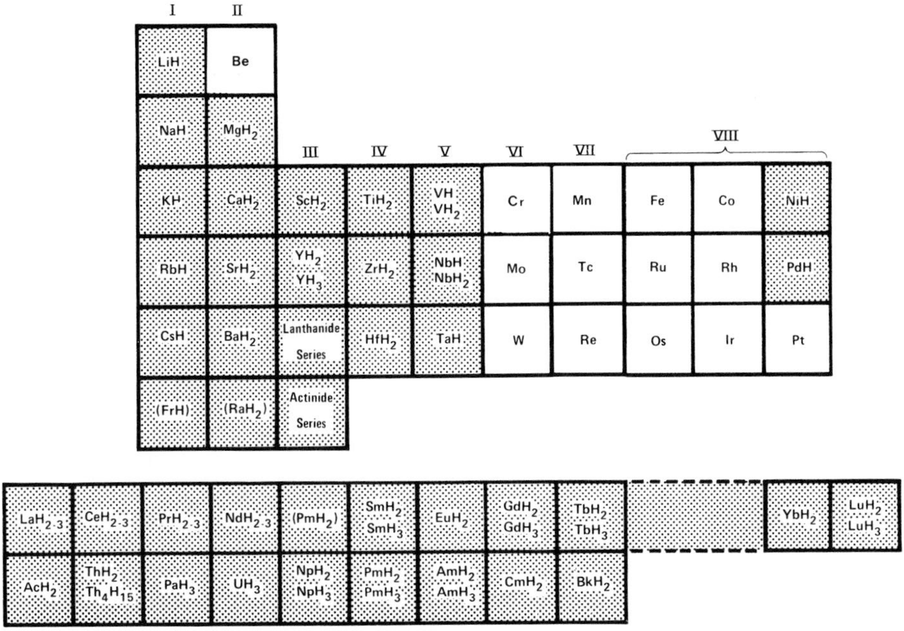

I	II	III	IV	V	VI	VII	VIII		
LiH	Be								
NaH	MgH$_2$								
KH	CaH$_2$	ScH$_2$	TiH$_2$	VH / VH$_2$	Cr	Mn	Fe	Co	NiH
RbH	SrH$_2$	YH$_2$ / YH$_3$	ZrH$_2$	NbH / NbH$_2$	Mo	Tc	Ru	Rh	PdH
CsH	BaH$_2$	Lanthanide Series	HfH$_2$	TaH	W	Re	Os	Ir	Pt
(FrH)	(RaH$_2$)	Actinide Series							

LaH$_{2.3}$	CeH$_{2.3}$	PrH$_{2.3}$	NdH$_{2.3}$	(PmH$_2$)	SmH$_2$ / SmH$_3$	EuH$_2$	GdH$_2$ / GdH$_3$	TbH$_2$ / TbH$_3$		YbH$_2$	LuH$_2$ / LuH$_3$
AcH$_2$	ThH$_2$ / Th$_4$H$_{15}$	PaH$_3$	UH$_3$	NpH$_2$ / NpH$_3$	PmH$_2$ / PmH$_3$	AmH$_2$ / AmH$_3$	CmH$_2$	BkH$_2$			

FIG. 5.   Binary metal hydrides formed by direct reaction of hydrogen and metal. Shaded areas indicate which metals in the periodic table form hydrides in this manner.

tion metal hydrides, including the lanthanides and actinides, have been the subject of controversy for about 20 yr [34, 36, 37]. Because of their metallic conductivities, these hydrides have been viewed as alloys of hydrogen and metal whereby the electrons from the hydrogen occupy the d-band of the transition metal such that the hydrogen atoms exist as screened protons in the metal sublattice. Therefore, this model has been referred to as the protonic model. An alternate model, first proposed by Libowitz and Gibb [38], the so-called ionic model, assumes that hydrogen takes electrons from the metal to form hydride anions, H$^-$, and metal cations, thus forming an essentially ionic compound. Experimental evidence is available to support each of these models [37]. However, recent energy band investigations by Switendeck [39] and others [40] appear to have resolved the controversy. When hydrogen is added to the metal to form a hydride, mixing and hybridization occurs between the $1s$ orbitals of the hydrogen and the energy band states of the metal to form a modified band, lowered in energy. In some hydrides, new low-lying states, associated with the hydrogen atoms, are also formed. In the former case, if the modified band is below the Fermi level, it will be already filled in the metal, and the added electrons from the hydrogen atoms will occupy the empty metallic states above the Fermi level. This resembles the protonic model since the hydrogen appears to donate its electrons to the metal. On the other hand, in cases where new low-lying states are formed, the additional electrons will preferentially occupy those states and the hydride will appear to be ionic. Thus, it can be seen that the experimental behavior of metal hydrides will be determined by the nature and position of energy states in the electronic structure of the hydride. Switendick's computations indicate that the hydrides of the Group V metals and palladium should behave as if the protonic model applied, while the rare earth hydrides should appear to fit the ionic model; and indeed, the results of various experimental studies appear to bear this out [37].

The thermodynamic properties of metal hydrides are usually obtained from pressure–composition–temperature measurements which yield isotherms such as those illustrated in Fig. 6. Consider the isotherm for T_1: As

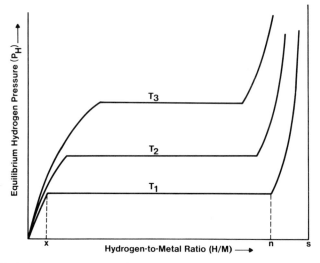

FIG. 6. Typical pressure–composition isotherms for a metal–hydrogen system; $T_3 > T_2 > T_1$. x represents the solubility limit of hydrogen in metal, n is the lower limit of the nonstoichiometric hydride, and s is the stoichiometric composition.

hydrogen dissolves in the metal, the equilibrium hydrogen pressure increases until the solubility limit is reached at $H/M = x$. At this point the nonstoichiometric hydride phase MH_n is formed and the pressure becomes constant because according to the phase rule

$$F = C - P + 2$$

an increase in the number of phases, P, by one, will decrease the number of degrees of freedom, F, by the same value. As hydrogen is added in the range *x* to *n*, metal is converted to hydride, and the two phases, hydrogen-saturated metal plus nonstoichiometric hydride, coexist. The two-phase plateau pressure, therefore, represents the dissociation pressure of the hydride at the particular temperature. It is an indication of the stability of the hydride; that is, stability varies inversely with plateau pressure.

At composition *n*, the metal has been completely converted to nonstoichiometric hydride, and with further addition of hydrogen, the hydrogen pressure again increases as the composition approaches the stoichiometric value, *s*. Note that the solubility limit, *x*, and the homogeneity range, (*s* − *n*), of the nonstoichiometric hydride both increase with temperature.

A plot of logarithm of plateau pressure, P_H, versus reciprocal temperature for a series of isotherms will yield a straight line whose slope yields the enthalpy of formation, ΔH_f, and whose intercept is a function of entropy of formation, ΔS_f, of the hydride according to the Van't Hoff equation

$$\partial \ln K_p / \partial T = \Delta H_f / RT^2 \tag{22}$$

where K_p is the equilibrium constant of Eq. (21). The integrated form of Eq. (22) can then be written [34, pp. 53–55]

$$\ln P_H = (2/nR)[(\Delta H_f/T) - \Delta S_f] \tag{23}$$

One problem which must be considered when using metal hydrides as a storage medium is the hysteresis effect. The plateau pressures in Fig. 6 are frequently different depending upon whether the isotherms are obtained on absorption or desorption. In those systems where hysteresis does occur, the desorption pressure is always lower than the absorption pressure. There have been several explanations for this phenomenon [34, pp. 83–86], but it is still not fully understood.

4. Intermetallic-Compound Hydrides

One approach to finding new hydrides which meet the requirements listed in Table 2, is to investigate the behavior of intermetallic compounds with hydrogen. At present, intermetallic-compound hydrides which have received the most attention with respect to hydrogen storage are iron–titanium hydride [41] and a group of hydrides of general formula RM_5H_n discovered at Philips–Eindhoven [42], where R is a rare earth metal and M is either nickel or cobalt. FeTiH is of interest because of its relatively low cost and high dissociation pressure (about 3.5 atm at room temperature) (i.e., low dissociation temperature). However, with respect to its hydrogen retentive capacity, this hydride is not too favorable, since n_H is only 3.4. The higher hydride of iron–titanium, $FeTiH_2$ has an n_H value of 6.0 and a room temperature dissociation pressure of about 7 atm, but because of sloping plateau pressures, the use of this hydride is more difficult to control. The iron–titanium hydrides, although attacked by oxygen, are easily regenerated since a stable oxide does not appear to form [43]. Another disadvantage of these hydrides is their weight—only 1.1×10^{22} H atom/gm for $FeTiH_2$.

The RM_5H_n hydrides exhibit extremely rapid rates of hydrogen uptake and discharge. Their hydrogen retentive capacities are only moderately good; the most investigated of these hydrides, $LaNi_5H_{6.7}$, has an n_H value of 6.1, which is probably the highest n_H value exhibited by this group. The room temperature dissociation pressures range from under 1 atm to very high values; for example, over 60 atm for $SmNi_5$ hydride; for $LaNi_5H_5$, P_H is 25 atm. Most of the RM_5 hydrides are relatively stable with respect to air and moisture [44]. Major disadvantages of this group of hydrides are their high cost (at present prices of the metals involved) and high weights; the number of hydrogen atoms per gram in $LaNi_5H_{6.7}$ is even less than in $FeTiH_2$—0.9×10^{22}. The enthalpies of formation of the iron–titanium and RM_5 hydrides are in the range 6–8 kcal/mole H_2, which is acceptable for hydrogen storage.

Obviously, the ideal metal hydride for hydrogen storage has not yet been found, and there is need for further work in the synthesis of new intermetallic compounds for this application. In general, the properties of intermetallic-compound hydrides bear little or no resemblance to the constituent metal hydrides. Therefore, a knowledge of the properties of the constituent metal hydrides will generally not be a reliable guide to predicting new intermetallic-compound hydrides. For example, the values of n in the RM_5H_n hydrides exceed the values which would be predicted on the basis of the rare earth hydrides, which have a maximum hydrogen-to-metal ratio of three; Ni and Co do not form hydrides under normal conditions.

The unpredictability of the properties of intermetallic-compound hydrides is illustrated in Table 3, which compares the properties of $ZrNiH_3$, the first intermetallic-compound hydride studied in any detail [45], with ZrH_2. Although the maximum hydrogen-to-metal ratio in zirconium

TABLE 3

Comparison of $ZrNiH_3$ with ZrH_2

	ZrH_2	$ZrNiH_3$
Structure	Tetragonal (distorted fluorite)	Orthorhombic
Dissociation pressure at 250°C	3×10^{-9} Torr	200 Torr
Zr–H distance	2.09Å	1.96Å
Closest H–H distance	2.22Å	2.04Å

hydride is two, there are three hydrogen atoms per zirconium atom in the intermetallic compound. Also, the crystal structures are different, and although the intermetallic-compound hydride is less stable (dissociation pressure is higher by a factor of 10^{11}), the Zr–H and H–H distances are smaller in that compound [46], which usually indicates stronger bonding.

Since there is an extremely large number of possible intermetallic compounds and an infinite number of compositional variations, it would be desirable to have some way of predicting which of these compounds will form hydrides, and of approximating their properties. Miedema and co-workers [47, 48] have proposed a "Rule of Reversed Stability" which, to some degree relates the stability of an intermetallic-compound hydride to its constituent metal hydrides. Stated in terms of enthalpies, ΔH, the rule can be expressed as follows:

$$\Delta H(\mathrm{MN}_a\mathrm{H}_{b+c}) = \Delta H(\mathrm{MH}_b) + \Delta H(\mathrm{N}_a\mathrm{H}_c) - \Delta H(\mathrm{MN}_a) \qquad (24)$$

where MN_a is the intermetallic compound, $\mathrm{MN}_a\mathrm{H}_{b+c}$ its hydride, and MH_b and $\mathrm{N}_a\mathrm{H}_c$ are the constituent hydrides. Equation (24) applies for cases where M is a strong hydride former, and $a \geq 1$. In any series of hydrides (e.g., rare earths), $\Delta H(\mathrm{MH}_b)$ will not vary much, and for weak hydride formers, $\Delta H(\mathrm{N}_a\mathrm{H}_c)$ will not have a very large value. Therefore, in any given series of intermetallic compounds, the greater the stability of the intermetallic compound, the less stable its hydride will be relative to other compounds in the series. This Rule of Reversed Stability has had some degree of success in predicting the properties of new intermetallic compounds but it has limited applicability because (1) it can only be used within a particular compositional series, (2) the thermodynamic stabilities of most intermetallic compounds are unknown and in most cases can only be approximately calculated [49], and (3) in some cases, the choice of constituent hydrides is arbitrary.

A better understanding of the relationships between the electronic structure of an intermetallic compound and its behavior with hydrogen would be of value in seeking new intermetallic-compound hydrides. Bond structure calculations, such as those of Switendick, discussed in Section IIIB3 have been valuable in explaining some properties, particularly of binary metal hydrides, but such calculations are complex and require some simplifying assumptions, and consequently, as yet, they cannot be used to predict. The importance of electronic structure is emphasized in some recent work of Maeland [50] on metallic glasses (sometimes called amorphous alloys). The intermetallic compound CuTi forms a hydride CuTiH [51] wherein the hydrogen atoms are situated in positions inside Ti tetrahedra in the lattice, such that the Ti–H distance is the same as in TiH_2 [52]. This would indicate that the stability of the hydride is due to the favorable structural characteristics of the intermetallic compound. However, a study of the corresponding metallic glass of CuTi showed that it formed an amorphous "hydride" of the same composition with comparable stability. Since the electronic structure of metallic glass alloys do not differ radically from the corresponding crystalline materials [53], it appears that the electronic structure may be more important than the crystal structure in determining the amount of hydrogen absorbed by an alloy.

5. Modification of Known Hydrides

Another approach towards developing hydrides having the required characteristics for hydrogen storage is to modify the properties of known hydrides by appropriate alloying. The thermodynamic stability can be changed in this manner. For example, Reilly and Johnson [54] found that the substitution of 10% nickel for iron in FeTi increased the stability (decreased the dissociation pressure) of iron–titanium hydride as illustrated in Fig. 7.

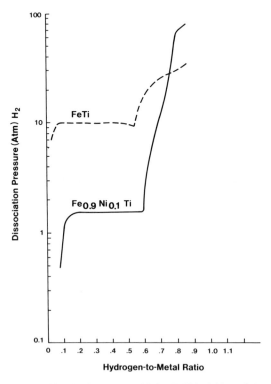

FIG. 7.   Pressure–composition isotherms at 50°C for FeTi hydride and $Fe_{0.9}Ni_{0.1}Ti$ hydride (adapted from Reilly and Johnson [54]).

In general, the dissociation pressure of a metal hydride, and therefore, its stability, is a function of the enthalpy and entropy of formation as shown in Eq. (23). The enthalpy of formation is an indication of the bonding strength in a compound. Since the formation of a metal hydride [Eq. (21)] is an exothermic reaction, ΔH_f is negative, and the stronger the M–H bond in the metal hydride, the more negative the value of ΔH_f. Consequently, in order to decrease the stability (increase the dissociation pressure) of a hydride, alloying elements which weaken the bonding should be introduced into the hydride to make ΔH_f less negative. The converse is true if increased stability is desired. The relationship between bond strengths and dissociation pressures is illustrated in Fig. 8, which shows a series of isotherms for nonstoichiometric lanthanum pentanickel hydride as a function of Ni/La ratio, a [55].

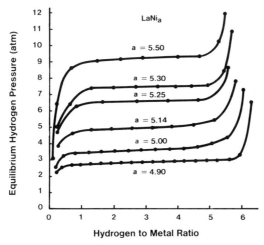

FIG. 8.   Pressure–composition isotherms at 40°C for $LaNi_aH_n$, as a function of Ni/La ratio, a (adapted from Buschow and van Mal [55]).

The hydride of stoichiometric $LaNi_5$ has a dissociation pressure of about 3.5 atm at 40°C, as shown. As Ni is substituted for La in the intermetallic compound (increasing a), the dissociation pressure increases (stability decreases). Since La forms a strong hydride and Ni forms a normally unstable hydride, the La–H bond would be stronger than the Ni–H bond. Consequently, the overall effect of Ni substitution would be to weaken the bonding, making ΔH_f less negative, and increasing the dissociation pressure according to Eq. (23). The reverse effect is obtained if La is substituted for Ni ($a < 5$) as shown.

The Rule of Reversed Stability can be employed to choose proper alloying elements for changing the stabilities of hydrides [47]. This is illustrated in

Fig. 9 which shows the effects of substituting another metal for 20% of the nickel in LaNi$_5$. As seen from the enthalpy values, ΔH_{IM}, for the intermetallic compounds listed, LaPd$_5$ is more stable than LaNi$_5$. Therefore, if Pd is substituted for some of the Ni in LaNi$_5$, the resulting hydride should be less stable than LaNi$_5$ hydride, according to the Rule of Reverse Stability. This is indeed the case as can be seen from the increased dissociation pressure of the hydride of LaNi$_4$Pd in Fig. 9. Conversely, the substitution of metals which form less stable intermetallic compounds with La (less negative or more positive values of ΔH_{IM}) such as Co and Cr, will yield more stable hydrides (lower dissociation pressures) as shown.

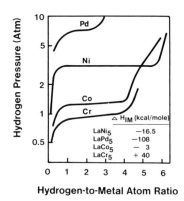

FIG. 9.　Effect of alloying elements on the dissociation pressure of lanthanum pentanickel hydride at 40°C (adapted from van Mal *et al.* [47]).

According to Eq. (23), the dissociation pressure, or stability of a hydride also can be varied by changing the entropy of formation. For instance, if it is desired to decrease the stability (increase the dissociation pressure) of an alloy hydride, the entropy of formation must be decreased. According to Eq. (21) this may be accomplished by decreasing the entropy of the product hydride by lowering the vibrational entropy. A lower vibrational entropy means a higher vibrational frequency which is usually associated with stronger bonding. However, as discussed above, this will also tend to lead to a more stable hydride in terms of a more negative value of ΔH_f.

Although the discussion in this section was devoted to methods of changing the thermodynamic stabilities of hydrides, it should be emphasized that it also would be desirable to find alloying elements which could improve the hydrogen retentive capacity, chemical stability, and kinetics of known hydrides. For example, Douglass [56] has been investigating the effect of alloying elements on the hydrogen discharge rate of magnesium alloys.

It is clear that there is need for much further research in materials science before a storage hydride will be found which ideally meets all the requirements listed in Table 2.

IV. UTILIZATION OF HYDROGEN

As mentioned in Section IA, hydrogen may be catalytically oxidized to provide flame-free heat at lower temperatures than normally obtained by direct hydrogen combustion. The advantages of this method of using hydrogen include safety and greatly diminished formation of oxides of nitrogen. Therefore, catalytic oxidation would be most desirable for home heating and in certain appliances such as space heaters and camp food warmers. However, one problem is the limited life and cost of the noble metal catalysts such as Rh, Pd, and Pt now being used. Consequently, there is a need for new, more efficient, low cost catalysts for this application.

Other energy related applications of hydrogen include its use in coal liquefaction and gasification and in the synthesis of methanol. The development of catalysts for these applications is adequately covered in the chapter by Sinfelt.

The design of hydrogen fuel cells to efficiently convert stored hydrogen to electricity involves some major materials problems with respect to electrodes, electrocatalysts, and electrolytes. These problems are fully discussed in Chap. 7 on fuel cells.

A possible new use of hydrogen which should not be strictly categorized under the hydrogen economy, since in this application hydrogen is not used as a fuel, is its utilization for thermal energy storage. There are two preferred existing methods of storing thermal energy: (1) sensible heat storage with water or rocks which have high heat capacities and are capable of retaining heat for long periods of time, or (2) the use of phase change materials, such as low melting eutectic salts [57], which use the enthalpy of melting (and fusion) to store heat. Recently, a method for using metal hydrides for thermal storage has been proposed [58] which employs the enthalpy of reaction of Eq. (21). The heat to be stored (e.g., solar) dissociates the hydride [reverse of Eq. (21)] and the hydrogen gas is separated from the metal. When it becomes necessary to recover the stored heat, the hydrogen gas is recombined with the metal and the heat of the exothermic reaction is used as needed. The major advantage of this method over the other two mentioned above is that the heat can be stored indefinitely, because it is not recovered until the metal and hydrogen are allowed to recombine. There have been proposed modifications of this concept whereby it is used for cooling [59, 60] and for driving a thermal expansion engine [60].

The requirements of the metal hydrides for this application are the same as those listed in Table 2 for hydrogen fuel storage except for one major difference. Since the enthalpy of formation of the hydride formation reaction is most important in this application, it should be as high as possible.

V. CONCLUSION

As stated in Section IC, this chapter does not deal with all the possible materials problems which could arise in the development of hydrogen energy systems for a future hydrogen economy. Nevertheless, it can be seen from the cases discussed that materials requirements include new catalysts, polymers, ceramics, semiconductors, structural materials, and both hydrogen-resistant and hydride-forming alloys. It is clear that there is a need for much further materials science research involving the application of the fundamentals of chemistry, physics, and metallurgy, before hydrogen energy systems can be effectively utilized to solve some of our energy problems.

REFERENCES

[1]　D. P. Gregory, *Sci. Am.* **228**(1), 13 (1973).
[2]　W. E. Winsche, K. C. Hoffman, and F. J. Salzano, *Science* **180**, 1325 (1973).
[3]　For example see R. J. Schoeppel, *Chem. Tech.* 476 (1972); R. E. Billings, *in* "Effect of Hydrogen on Behavior of Materials" (A. W. Thompson and I. M. Bernstein, eds.), p. 18. The Metallurgical Society of AIME, New York, 1976.
[4]　H. S. Spacil and C. S. Tedmon, *J. Electrochem. Soc.* **116**, 1618, 1627 (1969).
[5]　W. W. Aker, D. H. Brown, H. S. Spacil, and D. W. White, U.S. Patent No. 3,616,334, Oct. 26, 1971.
[6]　H. Obayashi and T. Kudo, *in* "Solid State Chemistry of Energy Conversion and Storage" (J. B. Goodenough and M. S. Whittingham, eds.) p. 316. American Chemical Society, Washington, D.C., 1977.
[7]　H. L. Tuller and A. S. Nowick, *J. Electrochem. Soc.* **122**, 255 (1975).
[8]　L. J. Nuttall, A. P. Fickett, and W. A. Titterington, *in Proc. Hydrogen Econ. Miami Energy Conf.* (T. N. Veziroglu, ed.), p. S9-33. Univ. of Miami, Coral Gables, Florida, 1974.
[9]　M. G. Shilton and A. T. Howe, *Mater. Res. Bull.* **12**, 701 (1977).
[10]　P. G. Dickens, D. J. Murphy, and T. K. Halstead, *J. Solid State Chem.* **6**, 370 (1973).
[11]　S. Ihara, *in Proc. World Hydrogen Energy Conf., 1st* (T. N. Veziroglu, ed.), Vol. II, p. 5B-55. Univ. of Miami, Coral Gables, Florida, 1976.
[12]　R. H. Wentorf and R. E. Hanneman, *Science* **185**, 311 (1974).
[13]　See a series of papers in, *Proc. Hydrogen Econ. Miami Energy Conf.* (T. N. Veziroglu, ed.), Vol. I, pp. 5A-1–5A-95, 6A-1–6A-86, 7A-1–7A-99, 8A-1–8A-114, 9A-1–9A-68. Univ. of Miami, Coral Gables, Florida, 1974.
[14]　M. G. Bowman, *Proc. World Hydrogen Energy Conf., 1st* (T. N. Veziroglu, ed.), Vol. II, p. 5A-27. Univ. of Miami, Coral Gables, Florida, 1976.
[15]　F. Coen-Porisini and G. Imarisio, *Proc. World Hydrogen Energy Conf., 1st* (T. N. Veziroglu, ed.), Vol. II, p. 7A-3. Univ. of Miami, Coral Gables, Florida, 1976.
[16]　A. Fujishima and K. Honda, *Nature (London)* **238**, 37 (1972).

[17] A. J. Nozik, *Nature (London)* **257**, 383 (1975).
[18] M. S. Wrighton, D. S. Ginley, P. T. Wolczanski, A. B. Ellis, D. L. More, and A. Linz, *Proc. Nat. Acad. Sci. U.S.* **72**, 1518 (1975).
[19] J. G. Mavroides, D. I. Tchernev, J. A. Kafalas, and D. F. Kolesar, *Mater.'Res. Bull.* **10**, 1023 (1975).
[20] A. J. Nozik, *Appl. Phys. Lett.* **29**, 150 (1976).
[21] R. Williams, *J. Chem. Phys.* **32**, 1505 (1960).
[22] A. J. Nozik, *Proc. World Hydrogen Energy Conf., 1st* (T. N. Veziroglu, ed.), Vol. II, p. 5B-31. Univ. of Miami, Coral Gables, Florida, 1976.
[23] T. Ohnishi, Y. Nakato, and H. Tsubomura, *Ber. Bunsenges. Phys. Chem.* **79**, 523 (1975).
[24] This conclusion is a result of calculations by R. W. Armbrust and A. J. Nozik of Allied Chemical Corp., Private communication.
[25] A. J. Nozik, *in* " Semiconductor Liquid-Junction Solar Cells " (A. Heller, ed.), p. 272. The Electrochemical Society, Princeton, New Jersey, 1977.
[26] W. T. Chandler and R. J. Walter, *Proc. Hydrogen Economy Miami Energy Conf.* (T. N. Veziroglu, ed.), p. S6-15. Univ. of Miami, Coral Gables, Florida, 1974.
[27] J. K. Tien, *in* " Effect of Hydrogen on Behavior of Materials " (A. W. Thompson and I. B. Bernstein, eds.), p. 309. The Metallurgical Society of the AIME, New York, 1976.
[28] I. M. Bernstein, R. Garber, and G. M. Pressouyre, *in* " Effect of Hydrogen on Behavior of Materials " (A. W. Thompson and I. M. Bernstein, eds.), p. 37. The Metallurgical Society of AIME, New York, 1976.
[29] H. K. Birnbaum, paper presented at the Fall Meeting of the Metallurgical Society of the AIME, Chicago, Illinois (October 1977).
[30] I. M. Bernstein and A. W. Thompson (eds.), " Hydrogen in Metals." American Society of Metals, Metals Park, Ohio, 1974.
[31] A. W. Thompson and I. M. Bernstein (eds.), " Effect of Hydrogen on Behavior of Materials." The Metallurgical Society of AIME, New York, 1976.
[32] C. E. Lundin and F. E. Lynch, *Proc. Intersoc. Energy Convers. Eng. Conf., 10th* p. 1386. American Institute of Chemical Engineers, New York, 1975.
[33] J. J. Reilly and R. H. Wiswall, Jr., *Inorg. Chem.* **9**, 1678 (1970).
[34] G. G. Libowitz, " The Solid State Chemistry of Binary Metal Hydrides." Benjamin, New York, 1965.
[35] B. Baranowski, K. Bochenska, and S. Majchrzak, *Rocz. Chem.* **41**, 2071 (1967).
[36] T. R. P. Gibb, Jr., *Prog. Inorg. Chem.* **3**, 315 (1962).
[37] G. G. Libowitz, *in* " MTP International Review of Science " (L. E. J. Roberts, ed.), Vol. 10, Solid State Chemistry, p. 79. Butterworths, London, 1972.
[38] G. G. Libowitz and T. R. P. Gibb, *J. Phys. Chem.* **60**, 510 (1956).
[39] A. C. Switendick, *Solid State Commun.* **8**, 1463 (1970); *Int. J. Quant. Chem.* **5**, 459 (1971); *Proc. Hydrogen Econ. Miami Energy Conf.* (T. N. Veziroglu, ed.), p. S6-1. Univ. of Miami, Coral Gables, Florida, 1974.
[40] J. H. Weaver, J. A. Knapp, D. E. Eastman, D. T. Peterson, and C. B. Satterthwaite, *Phys. Rev. Lett.* **39**, 693 (1977); J. H. Weaver and D. T. Peterson, *Phys. Lett.* **62**, 433 (1977); J. H. Weaver, R. Rosei, and D. T. Peterson, *Solid State Commun.* **25**, 201 (1978).
[41] J. J. Reilly and R. H. Wiswall, *Inorg. Chem.* **13**, 218 (1974).
[42] J. H. N. van Vucht, F. A. Kuijpers, and H. C. A. M. Bruning, *Philips Res. Rep.* **25**, 133 (1970).
[43] G. D. Sandrock, *Proc. Intersoc. Energy Convers. Eng. Conf., 11th* Vol. I, p. 967. American Institute Chemical Engineers, New York, 1976.
[44] R. H. Wiswall and J. J. Reilly, *Proc. Intersoc. Energy Convers. Eng. Conf., 7th* p. 1342. American Institute Chemical Engineers, New York, 1972.

[45] G. G. Libowitz, H. F. Hayes, and T. R. P. Gibb, Jr., *J. Phys. Chem.* **62**, 76 (1958).

[46] S. W. Peterson, V. N. Sodana, and W. L. Korst, *J. Phys. (Paris)* **25**, 451 (1964).

[47] H. H. van Mal, K. H. J. Buschow, and A. R. Miedema, *J. Less-Common Metals* **35**, 65 (1974).

[48] K. H. J. Buschow, H. H. van Mal, and A. R. Miedema, *J. Less-Common Metals* **42**, 163 (1975).

[49] A. R. Miedema, *J. Less-Common Metals* **32**, 117 (1973); **41**, 238 (1975); **46**, 67 (1976).

[50] A. J. Maeland, *Int. J. Hydrogen Energy* (1978) (to be published).

[51] A. J. Maeland, *Adv. Chem. Ser.* (1978) (to be published).

[52] A. Santoro, A. J. Maeland, and J. J. Rush, *Acta Cryst, Sect B* (1978) (to be published).

[53] D. S. Boudreaux, Allied Chemical Corp., Private communication.

[54] J. J. Reilly and J. R. Johnson, *Proc. World Hydrogen Energy Conf., 1st* (T. N. Veziroglu, ed.), Vol. II, p. 8B-3. Univ. of Miami, Coral Gables, Florida, 1976.

[55] K. H. J. Buschow and H. H. van Mal, *J. Less-Common Metals* **29**, 203 (1972).

[56] D. L. Douglass, *Metall. Trans.* **6A**, 2179 (1975); see also paper to be published in *Int. J. Hydrogen Energy* (1978).

[57] M. Telkes, *in* "Critical Materials Problems in Energy Production" (C. Stein, ed.), p. 440. Academic Press, New York, 1976.

[58] G. G. Libowitz, *Proc. Intersoc. Energy Convers. Eng. Conf., 9th* p. 322. American Institute Chemical Engineers, New York, 1974.

[59] G. G. Libowitz and Z. Blank, *in* "The Solid State Chemistry of Energy Conversion and Storage" (J. B. Goodenough and M. S. Whittingham, eds.), p. 271. American Chemical Society, Washington, D.C., 1977.

[60] D. M. Gruen, R. L. McBeth, M. Mendelsohn, J. M. Nixon, F. Schreiner, and I. Sheft, *Proc. Intersoc. Energy Convers. Eng. Conf., 11th* p. 681. American Institute Chemical Engineers, New York, 1976.

Chapter 9

Material Aspects of the New Batteries

M. S. WHITTINGHAM

CORPORATE RESEARCH LABORATORIES
EXXON RESEARCH AND ENGINEERING COMPANY
LINDEN, NEW JERSEY

I. INTRODUCTION

In a society relying progressively more on electrical power it is necessary to have a convenient way of storing this energy. Batteries are the predominant means used today; in these the electrical energy is readily mobile and is stored as chemical energy. However, the largest batteries made contain only a few kilowatt-hours of energy and at a relatively low density; thus 1 ft³ of a lead acid battery and of petroleum store 10^4 and 10^5 Btu, respectively. It is therefore necessary to increase this density considerably, at the same time maintaining, or preferably increasing, the power density and reducing the cost. The purpose of this chapter is to describe the key problems which are all materials related, the materials themselves and where future research effort might best be expended. Fuel cells, which are like all primary batteries generators of electricity and not strictly a storage medium, are described in Chapter 7 and will not be discussed here in any detail.

In order to gain a perspective of where batteries can fit into the energy picture it is useful to look at a typical electricity load curve. Such a curve is shown in Fig. 1 [1]. Key points to note are the peak demand each midday and the dip at night; there is also a generally lower demand at weekends.

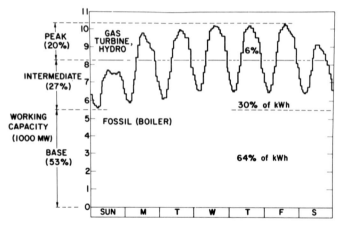

FIG. 1.   Typical electricity load curve (after Federal Power Commission [1]).

The broad base load is fed by fossil fuels and to an ever increasing extent by nuclear sources. Most of these plants are new, capital intensive, and require running 24 hr a day (although a small cutback in output is technically feasible, it is not economically desirable). Historically, the intermediate load has been supplied by the older fossil fuel stations, which are predominantly coal fired. The peak load is now increasingly provided by the burning of oil in turbines and to a minor extent by hydroelectric power. Although this peak load regime provides only 6% of the electricity, it makes up 20% of the generating capacity, and hence is expensive.

The problem that the utilities are now facing is that all modern baseload generating equipment has a high capital cost and is also designed to operate around the clock. It is thus not suitable for downgrading to intermediate load use as it ages, and so new equipment is being installed for both the intermediate and peak load regimes. This new equipment is predominantly oil fired turbines which can be very easily switched on and off to match the load curve. However, if the present trends continue, there simply will not be enough oil to supply all the electricity generation requirements and the propulsion needs (automobiles, aircraft, etc.) of this country. These heavy oil demands and the resultant shortages are expected to be felt in any event in the very near future.

An alternative to burning oil must therefore be found to provide this peaking power. An answer is to remove the peaks in the generation scheme by filling in the dips with the peaks, that is to say, by energy leveling. In this the excess energy generated at night is stored for use at the peak midday period, so that the base load capacity can provide a far larger percentage of the total load. There are a number of storage systems that can be used, and the point at which they input into the transmission system is indicated in Fig. 2. Pumped water storage requires both a series of large reservoirs, which are unsightly and therefore open to attack by environmentalists, and distant placement from large metropolitan areas, thus leading to high trans-

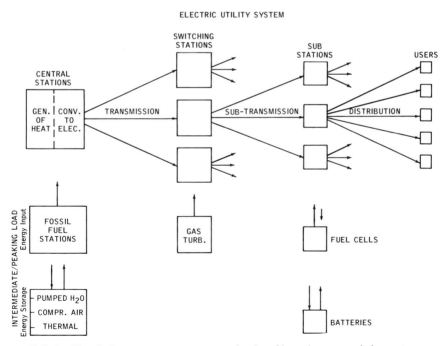

FIG. 2.   Electrical energy storage systems and point of input into transmission system.

mission costs. Water storage using deep wells might reduce these problems. Compressed air and thermal storage are also envisioned on a large scale and to be sited at the central generating station. The thermal off-peak residential storage systems as used in Great Britain have not proved to date to be economically viable. Gas turbines can be situated at switching stations, and fuel cells and batteries right at the substation level. The advantages of a battery system are

(1) its predetermined cost;
(2) its lack of dependence on a fuel supply—oil—of variable cost and availability; and
(3) its cleanness and quietness; thus it can be located very close to the load, as for example, in the basements of large buildings.

Fuel cells have the same advantages as batteries if they can be operated reversibly (for example, storage of power might involve the electrolysis of water, and the generation of power, the reaction of hydrogen and oxygen); they are however fractionally noisier, having some moving parts, and might therefore be located not quite so close to the load. Fuel cells, operated in this reversible manner, will in this paper be considered as a battery subset, having many of the same characteristics.

The investment credit for fuel cells or batteries versus gas turbines due to the reduction in transmission costs is closely related to the land use density, being a factor of 2–4 higher for an inner city, like New York, than for a rural area (in 1973, $150 or $50 per kW). However, at the present time the economics cannot be realistically assessed as other components of the system, such as the dc–ac converters, are not available at viable prices; figures of around $25/kWh for the battery itself are frequently quoted [2].

Other potentially large scale uses of batteries are for automotive propulsion, rail traction, and on a smaller scale for the storage of electricity generated locally by solar cells. These uses are probably the more realistic in the short term as the individual systems are much smaller and may therefore be a stepping stone to the load leveling goal. The basic properties required of batteries for these purposes are essentially the same. They should have the following characteristics:

(1) Full depth discharge and charge;
(2) 24 hr complete cycle;
(3) extremely long life, 5–10 yr minimum (2000–3000 cycles);
(4) essentially zero maintenance;
(5) small size and weight.

In addition, the battery should be capable of operating at elevated temperatures, as on charge and discharge self-heating will occur. The above characteristics mean that the system must have

(1) high power density (both gravimetric and volumetric);
(2) high energy density (both gravimetric and volumetric);
(3) high reliability.

The last will be determined to a major extent by the choice of electrochemically active as well as inactive materials, and this choice is fixed by the former requirements. As will be seen in the following sections, these requirements are not readily compatible.

II. COMPONENT REQUIREMENTS

A battery consists of three active components as shown in Fig. 3. The anode is the electropositive half of the cell, typically a metal, and the cathode the electronegative half, normally a halogen or chalcogen. The electrolyte

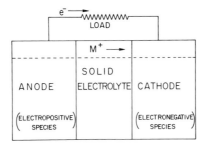

FIG. 3.   Active components of a battery.

must be able to transport ions rapidly from one electrode to the other, preferably only a single ionic species, and at the same time be an electronic insulator so that self-discharge is minimized.

A. Anode

The requirement of a high energy density demands a high voltage, that is, a high free energy change during the anode–cathode reaction. Thus the materials required as the anode are the most electropositive elements, as for example, the alkali metals, and as the cathode the most electronegative, as for example, the halogens, or oxidizing compounds, such as CrO_3. The energy densities of batteries using a variety of pure metal anodes are given in Table 1. The theoretical energy density is given by the expression

$$E.D. = E \times 1000/W \quad W \ hr/kg$$

where W is the equivalent weight of anode and cathode and E the cell emf which is related to the free energy, ΔG, of the cell reaction by

$$\Delta G = -nEF$$

where n is the number of electrons passed per mole of reactant and F is Faraday's constant. In Table 1 it has been assumed that (1) the cathode has a weight of 50 gm/equivalent, (2) the dead weight (electrolyte, casing, etc.) is 100 gm/equivalent (probably optimistic), and (3) the operating cell voltage is that corresponding to formation of the simple sulfide. This last assumption does not effect the trends given in Table 1 and so is representative of almost any cathode material.

TABLE 1

Energy Densities of Some Metal Sulfide Cells

Metal	Emf (V)	Energy density (W hr/kg)	Melting point (°C)
H	0.11	20	—
Li	2.48	439	179
Na	1.85	297	98
K	2.07	304	64
Rb	1.74	206	39
Mg	1.80	309	651
Ca	2.44	399	845
Ag	0.20	22	960
Zn	0.99	151	420
Al	1.35	236	660

The desired practical energy density for automobile propulsion systems is 200 W hr/kg (100 W hr/lb), although for commuter vehicles a value of 100 W hr/kg may be sufficient. The anode materials that clearly in theory satisfy this are the three alkali metals, Li, Na and K, and Mg, Ca and Al. From a cost standpoint sodium metal is to be preferred. Magnesium, calcium, and aluminum, although being polyvalent they carry more charge, have high melting points, diffuse much more slowly through solid lattices (electrolyte or cathode) due to their high charge, and form very stable oxide films which could be a problem. Thus over-all lithium and sodium are the preferred anodes with potassium a runner-up, potassium having a melting point advantage but weight and cost disadvantage.

B. Electrolyte

Electrolytes for use in batteries must exhibit a number of properties, among which are

(1) high ionic conductivity (ideally of only one species), σ_i;
(2) low electronic conductivity, σ_e;
(3) extreme chemical inertness;
(4) the exhibiting of the above properties over a wide range of thermodynamic activity (e.g., in the reducing environment of sodium and in the oxidizing environment of chlorine).

The exact requirements of (1) and (2) above depend on the use of the battery. Thus a battery to be used as a heart pacemaker, where long life and very low current densities are required, must not exhibit any electronic leakage current. Hence, the electronic transference number t_e, which is given by $\sigma_e/(\sigma_e + \sigma_i)$, should be less than about 10^{-6}. On the other hand, for load leveling purposes where the system is cycled daily, t_e could be as high as a few per cent if other advantages such as enhanced ionic conductivity can be obtained. As very much smaller currents are required in the former case, the electrolyte resistance can be higher. In the high power situation, if the electrolyte has a resistance of 1 Ω per unit area of the electrolyte–electrode interface, the cell has a potential of 2 V and a current of 1/2 A/cm^2 flows, then there will be an IR loss of 1/2 V, which is 25% of the system's capacity. This translates to an almost 45% loss on a single charge–recharge cycle; thus an electrolyte resistance of no more than 1/4 Ω/cm^2 is preferred for the highest current density applications. This translates to a resistivity of 2.5 Ω-cm if the electrolyte can be used in a 1 mm thick form. Hence, a small increase in t_e can be tolerated if σ_i is thereby increased.

Although from an energy density standpoint, one would prefer the anode–cathode couple which generates the highest emf, in practice the emf must be restricted. This emf must be related to the decomposition potential of the electrolyte system. To allow for recharge, the voltage of the couple should be around 0.5 V less than this decomposition potential. For presently known electrolytes, this restricts the cell voltage to around 3.0 V. Figure 4 shows the free energy of formation (which is equivalent to decomposition potential) at 350°C of the halides of the group IA and IIA metals, [3] which are the most common molten salts considered for electrolytic use. Clearly here the cell emf should not exceed 3 V. And indeed, mixtures of these molten salts are now being used in the lithium sulfur high temperature cells, where the cell voltage is less than 2.5 V. Even sodium beta alumina has a decomposition potential of around 3.7 V [4]. Thus decomposition of the

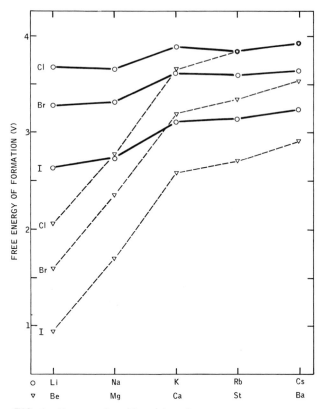

FIG. 4.   Free energies of formation of group IA and IIA halides.

electrolyte will restrict the maximum energy density obtainable in a reversible cell unless the electrolyte is to actively participate in the reaction. There
are some exceptions, such as the acid in the lead acid battery and some
organic solvent systems, which operate outside their thermodynamic stability fields.

C. Cathode

The requirements of a cathode material for high energy density batteries
are

(1) high oxidizing power;
(2) electronic conductivity;
(3) low equivalent weight;
(4) ability to form product reversibly;
(5) no reactivity with electrolyte;

(6) no solubility in electrolyte; and

(7) low corrosive properties.

There is low compatability among a number of these requirements; thus materials with a high oxidizing power, such as the halogens and the chalcogens, tend to be both corrosive and electronic insulators. Then an electrochemically inactive conducting matrix must be added to the active material, which will markedly reduce the energy density; graphite is the most commonly used conductive diluent. A major problem with highly oxidizing cathodes is their containment. This is at its worst for molten systems, such as sulfur and the halogens, and is aggravated by elevated temperature operation. Thus, Argonne National Laboratory [5] in their 450°C Li–S molten salt battery have been forced to switch to solid cathodes, such as FeS_2, NiS, and the like, despite the significant loss of theoretical energy density.

The cathode must be capable of cycling daily with high efficiency for a minimum period of several years. This dictates that the cathodic reaction must be readily reversible, which in turn demands that the cathode structure must maintain its structural integrity in low temperature systems. Although structural integrity is bypassed if a liquid cathode such as sulfur is used, such systems usually require higher temperatures and conductive matrices with all their associated problems. It would be preferable to operate at around 25°C, but it will almost certainly be necessary to work at higher temperatures to jointly optimize the power and energy densities and at the same time minimize polarization losses. In addition, under high current drain or charge the system will self-heat due to IR losses so that all components must be capable of operating above ambient temperatures. On the other hand, the economics of production make it preferable to use plastic components wherever possible, which places an upper temperature limit of no more than 200°C. The ideal is probably around 100°C with the capability of operating, even if under reduced power, at ambient temperature, as in the case of automotive propulsion, in order to allow for "cold starts" (thus cutting out the power wastage that may plague the high temperature Na–S–beta alumina–350°C and Li–S–molten salt–450°C battery systems which must be kept hot at all times).

Operating at low temperatures, it is unlikely that the cathode integrity will be maintained if the crystal structure of the cathode material is destroyed and rebuilt on each cycle (e.g., $4Li + FeS_2 \rightleftarrows 2Li_2S + Fe$). It is therefore likely to be preferable if the structure is essentially maintained during reaction; likewise the volume change of the cathode structure during reaction should be minimized.

The cathode, besides being the oxidizing component of the cell, is also the current collector and so must be an electronic conductor. Most active cathode materials are electronic insulators and so must be mixed with a nonelectrochemically active conductive diluent, such as graphite. These are typically present to the extent of 20–30% of the total cathode weight resulting in a considerable energy density penalty. In addition, good contact must be maintained between the active material and the conductive diluent in order to utilize all the cathode. These deficiencies are eliminated if the active material is also an electronic conductor. The preferred characteristics for the cathode material are

(1) anode–cathode couple of 3.0 V;
(2) solid cathode;
(3) minimum structural charge during reaction;
(4) operating temperature 100°C ± 100°C; and
(5) electronic conductor.

III. THE ANODE

There is little materials research related directly to the anode. That is not to say that there are no problems to be solved here. These problems are mostly related to the electrodeposition of the metal ions on the solid anode during system recharge. Where the electrolyte is a solid, the anode may be a liquid, thus removing even electrodeposition problems; such is the case in the sodium–sulfur battery using beta alumina as the electrolyte.

However, for lithium there is no suitable solid electrolyte and so liquid electrolytes—molten alkali halides at high temperatures and organic liquids at ambient temperatures—must be used. Most of the published work on the lithium–sulfur molten salt system has been performed at the Argonne National Laboratory in the United States [2, 5, 6]. Their initial studies used lithium, which is molten at the temperature of operation, 400°C; to prevent mixing with the electrolyte, LiCl–KCl eutectic, and subsequent shorting the metal was contained. This was done by wicking the molten metal into Feltmetals, which are proprietary metal sponges. However, considerable loss of lithium occurred and the Feltmetal was too expensive. The latter problem could be solved by using cobalt plated steel wool which is preferentially wetted by the lithium rather than by the salt. Here again, however, the lithium loss was unacceptably high and corrosion of the cell components by the lithium metal was marked.

Most of these problems were reduced by using lithium aluminum alloys, which are solid at 400°C and which lower the lithium activity by about 0.3 V [7]. The phase diagram [8] for the lithium aluminum system is shown in Fig. 5. The maximum useful lithium content of the alloy system is around

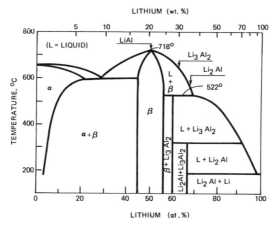

FIG. 5. Phase diagram of the lithium–aluminum system (from Settle *et al.* [8]).

60 at.%; at higher values a liquid phase would be present. The reduction of lithium activity cuts the lithium corrosion of the cell but also causes a loss in the theoretical energy density on two counts. First, over half the weight of the anode, the aluminum, is now electrochemically inactive, and second, there is the 0.3 V loss in cell emf. However, the improved retention of the lithium in the solid actually causes an increase in the energy density of practical cells. Thus the reported cell self-discharge rates are 0.69 mA/cm^2 and 3.0 mA/cm^2 for LiAl and Li, respectively; this is mostly due to the solubility of the lithium in the electrolyte. A further cause of lithium loss is caused by the reaction

$$Li + KCl \rightarrow K + LiCl$$

which is driven by the high equilibrium vapor pressure of potassium, ~ 0.3 mm at 400°C. This presumably also causes self-discharge of the system. The reduction in the lithium activity on alloying will also cut the potassium activity and hence the potassium and lithium loss.

The anode problems in ambient temperature organic solvent systems are somewhat different but can also be solved the same way. When lithium metal is redeposited on the lithium anode on charge it frequently grows as dendrites or in spongy form, and a proportion is thereby permanently lost on each cycle. More importantly, however, the dendrites can penetrate the separator material, thus shorting the cell. By reducing the lithium thermodynamic activity below that of pure lithium by using an alloy, dendritic formations are not possible. In addition, most, if not all, polar organic solvents are thermodynamically unstable in the presence of pure lithium. Their operation is due to the formation of a thin protective film on the surface of the lithium, and its effectiveness varies from solvent to solvent. This breakdown of the electrolyte can be reduced by a lowering of the lithium activity, again by alloying. Thus Kegelman [9] has claimed that for an electrolyte of $LiPF_6$ in dimethyl formamide, an alloy containing 5–16 wt. % of lithium performs very well in a reversible manner for sulfur containing cathodes. This alloy composition covers the broad two phase region, $\alpha + \beta$, of Fig. 5, so that the cell voltage is independent of the lithium content of the anode and is about 0.3 V less than that of lithium itself.

Dey [10] studied the electrochemical alloying of lithium with a number of metals (Sn, Pb, Al, Au, Pt, Zn, Cd, Hg and Mg) from a propylene carbonate electrolyte, and found that it could be rather readily accomplished. The LiAl formed this way gave the same x-ray pattern as the thermally formed, 500–700°C, intermetallic compound. Although these compounds may have some advantages over pure lithium, particularly at elevated temperatures, they do not allow current densities as high as pure lithium does. Therefore, the power densities of cells utilizing them are likely to be slightly depressed. On the other hand, this lower current density is an added safety factor in case of cell failure and thermal runaway; in addition, the alloys do not become molten until much higher temperatures. In the case of sodium, the electrolytes are much more susceptible to decomposition, and in fact no system has been seriously proposed utilizing the combination of sodium and an organic solvent.

IV. SOLID ELECTROLYTES AND SEPARATORS

A. Present Status

Since 1967, when the properties of beta alumina were publicized, a lot of effort has been expended on searching for other electrolytes. This search has been particularly successful in the case of silver. However, silver-based systems are completely impractical for any but very specialized, cost independent, uses. They have been reviewed elsewhere [11] and will not be further discussed here. Some copper electrolytes have also been found [12], but they suffer from several disadvantages, amongst which are rather low decomposition potentials. This restricts the cell potential and hence the energy density. The reason for this is probably associated with the ready disproportionation of the cuprous ion:

$$2Cu^+ \rightleftharpoons Cu^{2+} + Cu$$

Figure 6 compares the ionic conductivity of several electrolytes and although many of these are silver or copper based, a few are not. Two of

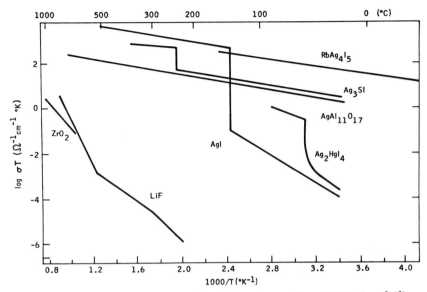

FIG. 6.   Ionic conductivity of several fast ionic conductors (from Whittingham [13]).

these, zirconium oxide, which is an oxygen ion conductor, and beta alumina will be discussed in the next two sections, the latter in detail.

One objective which has not been successful is the search for a lithium conductor to be used as the electrolyte in the lithium–sulfur cell. Figure 7 compares the conductivities of lithium beta alumina [14], lithium iodide

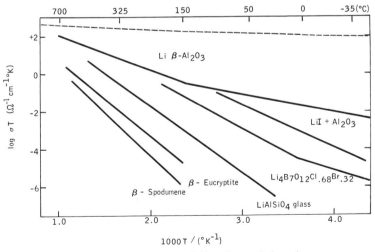

FIG. 7.   Lithium ion conductivities of several electrolytes.

dispersed on alumina [15], and three aluminosilicates [16]; the dashed line indicates the critical value, 2.5 Ω-cm, as discussed in Section II,B. Clearly, none of these approach the desired conductivity at 300°C, the probable operating temperature of such a lithium–sulfur cell. Although lithium beta alumina comes the closest, the data quoted here is for single crystals, so that in ceramic form a loss of conductivity will occur, as described later for sodium beta alumina. The results on the lithium iodide are particularly interesting as the intrinsic lithium ion conductivity of the iodide was found to be enhanced by simply mixing it with alumina. The reason for this is not clear but may well be associated with enhancement of grain boundary diffusion. Johnson *et al.* [16] found that for the eucryptites, the ionic conductivity was higher in the glassy than in the crystalline state; this is important because it should be much easier to fabricate large sheets of electrolyte by glass technology than by the usual ceramic techniques.

As there is at present no suitable solid electrolyte for the lithium sulfur cell, a molten salt eutectic is used as the electrolyte, most commonly a LiCl–KCl eutectic. The temperature of operation may be reduced by the addition of a third salt, such as LiBr. However, a separator has to be used to separate the anode and cathode compartments. This must fulfill all the requirements of a solid electrolyte except that the molten salt will provide the ionic conductivity. A major problem is that lithium reduces almost every known insulator, even zirconia, converting them into conductors. To date only high purity oxide-free boron nitride is stable in this environment [17]. This may be woven into a cloth form for use as the separator, but is presently more than an order of magnitude too expensive. Hence, the search for a good lithium ion conductor is still on.

B. Zirconia

The simplest battery of all to visualize is one in which hydrogen and oxygen react to form water; on recharge the water is simply electrolyzed. This however requires a suitable electrolyte, which must be either an oxygen or a hydrogen ion conductor. Various mixed oxides—those of zirconium, calcium and yttrium—were found as early as 1900 by Nernst [18, 19] to be oxygen ion conductors, and Schottky in 1935 [20] proposed their use in fuel cells. However, it was primarily the work of Kiukkola and Wagner [21, 22] in 1957 that created interest in solid electrolytes.

Although many oxide conductors have been studied, predominantly in the 1960s, basically no improvement of the original calcia- or yttria-doped zirconia was found. Zirconia itself has the fluorite structure, in which the anions reside in the tetrahedral interstices of a face centered cubic arrangement of the cations. Ionic conduction then occurs by movement of the anions amongst these interstitial sites. The conductivity may be enhanced by creating vacancies on the oxygen sites by substituting lower valence cations, such as calcium and yttrium, for the zirconium; the most commonly used compositions are those containing 15 mole % calcia or 8 mole % yttria, which correspond to the maximum conductivities for these dopants. As the resulting vacancy concentration is so high, $\sim 8\%$, these defects interact, with the result that the diffusion mechanism is not a simple vacancy jump but is complicated by an apparent clustering of these vacancies [13]. This high defect concentration may also be the cause of the reported reduction in the conductivity with time at high temperatures, $\sim 1000°C$ [23], as either they or the metal dopant might aggregate into a second phase.

The conductivity of calcia-doped zirconia is shown in Fig. 6. As can be seen, it exhibits high ionic conductivity only at elevated temperatures so that fuel cells or batteries must themselves be operated in this temperature range. This causes considerable materials corrosion problems, and unless the temperature of operation can be markedly reduced, it seems unlikely that such cells will ever be produced. At the present time, all development work on these high temperature fuel cells has ceased. The one advantage of these temperatures is that the overpotential is due only to the ohmic resistance of the electrolyte itself. As the cell configurations are amply discussed in the chapter on fuel cells in this volume and elsewhere [24], they will not be discussed here.

The major use, therefore, of oxygen-ion-conducting electrolytes is technologically and scientifically as oxygen sensors. Other anion conductors include CaF_2, first studied by Ure [25], and LaF_3, which is one of a family of fluorides with the tysonite structure that exhibit appreciable fluorine mobility. O'Keefe [26] has been active in this area recently, but it seems even less likely that any battery system based on fluorine will be viable because of its extreme corrosive power.

C. Beta Alumina

Beta alumina is an aluminum oxide originally so named because it was though not to contain any sodium. However, the structural studies of Bragg *et al.* [27] and Beevers and Ross [28], amongst others in the 1930s, indicated that the alkali metal was a critical structural member. It was not until the work of Yao *et al.* [29, 30] was announced in 1967 that interest was revived. They found that beta alumina could be used as the electrolyte in a sodium–sulfur battery. These electrolytic properties are due to the special characteristics of the crystal structure.

1. Structural Aspects

There are two major types of structure found for the beta aluminas. Beta alumina itself, which has the nominal formula $MAl_{11}O_{17}$ (where M is a monovalent cation), but is more correctly represented by $M_{1+x}Al_{11-x/3}O_{17}$ [31] with x in the range $0.15 \leq x \leq 0.30$, is one of these. The other is beta" alumina with the nominal formula MAl_5O_8; in this case the content of M is less than that expected for the above formula. Both of these structures comprise close packed blocks of oxygen and aluminum, four layers thick, with the $\gamma-Al_2O_3$ spinel structure. These blocks are held together by Al–O–Al bridging groups, and the M ions are then located in the bridging layers between these groups. The packing in this layer is very loose, there being sufficient room for three times the concentration of M ions found in the formula $MAl_{11}O_{17}$; there is one M and one O per layer. This structure is shown schematically in Fig. 8. The close packed version of this structure is

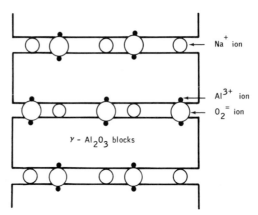

FIG. 8.   Schematic representation of the structure of beta alumina.

magnetoplumbite, $PbFe_{12}O_{19}$, where one Pb and three O are found in the bridging layer. The principal difference between the β and β'' structures, besides the composition, is in the stacking sequence of the $\gamma-Al_2O_3$ blocks. In β these are arranged so that the bridging layers are mirror planes, giving a structure two blocks thick. This arrangement, as will be discussed below, causes the excess M ions to reside on sites different to the stoichiometric M ions; this is very important to the ionic conductivity. In β'' there are no mirror planes; the structure repeats itself every three blocks, giving it rhombohedral symmetry [32–34], and all the M ions can reside on identical sites. Additional structures have been reported [35] which apparently differ in having more layers—six—in the spinel blocks and resemble the well-known ferrites. [36]

The arrangement of the ions in the conducting plane of β alumina is shown in Fig. 9; one quarter of the sites are taken up by the bridging oxygens in a hexagonal net. At the centers of triangles of these oxygens are trigonal antiprismatic (extended octahedral) sites which are filled by the M ions in the stoichiometric structure. The excess M ions can then reside in two other sites. One of these is at the center of the remaining oxygen triangles; these sites, centered between two oxygen ions, were those originally suggested by Beevers and Ross [28]. Peters *et al.* [37] have, however, recently suggested that in the case of sodium beta alumina, the excess sodium ions are located between pairs of oxygen ions, placing them in an eight coordinate site. For each ion placed on one of these sites, an ion must also be displaced from the regularly occupied site so that for 20% excess sodium, 40% ($x = 0.4$) sodium is found on these sites.

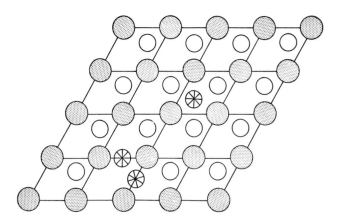

FIG. 9.   Arrangement of ions in the conducting plane of beta alumina.

2. Conductivity

The critical parameter of any solid electrolyte is its ionic conductivity; as discussed earlier this should be no more than a few ohm-centimeters at the temperature of operation. A number of investigators have investigated β alumina both in single crystal and ceramic form, pure and doped, and as a function of phase. Figure 10 shows the values for sodium β'' in single crystal and ceramic form; the techniques used for making these measurements will be described later. Two features are immediately apparent. β'' samples have higher conductivities as also do single crystals. The latter is associated with a higher impedance at the grain boundaries and the tortuosity of the diffusion path due to the anisotropic nature of the conductivity. In a perfectly randomly oriented sample, the diffusion length is about double that in an

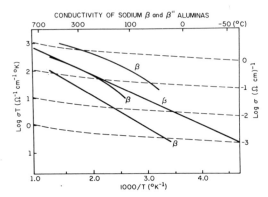

FIG. 10. Conductivity in single crystal and ceramic β and β'' alumina. The upper line of each pair is for the single crystal specimens.

aligned single crystal. The higher conductivity of β'' is related to the greater number of sodium ions available for the conduction process; in this case all the ions are equivalent and there is an excess of crystallographic sites available to them, so that all the ions are able to move. The number involved in any one jump process will be a function of the concentration of ions, or more accurately, the number of vacancies, and the temperature, much as has been proposed for the silver halides [38]. It is this temperature dependence that causes the activation energy to be a function of temperature. In contrast, in β alumina itself the ions reside on two different sites, and it appears that the conductivity is proportional to the number of excess ions over the stoichiometric number, $NaAl_{11}O_{17}$ [39]. The diffusion mechanism here simply involves jumps of these excess ions to the regularly occupied sites, ejecting the incumbent ion to another interstitial site [31]; this is the well-known interstitialcy mechanism. More complex arrangements of these excess ions are possible and have been described by the author elsewhere [13] and will not be discussed here.

Figure 10 shows that pure beta alumina has too high a resistivity at 300°C to be useful in the sodium sulfur battery, whereas β'' is suitable. The conductivity of beta alumina may however be improved by doping the aluminum with lower valent ions such as magnesium when one would expect the addition of one sodium for each magnesium to change neutralization [31]. And indeed, some preliminary experiments at 25°C [40] indicate that this is the case. However, magnesium might also have an effect on the sinterability of powders and thereby indirectly affect the conductivity.

β'' alumina, although it has the required conductivity, cannot be sintered at high temperatures because of decomposition to β alumina. It can, however, be stabilized by the addition of magnesia or lithia and indeed most developmental studies on the ceramic now use the β'' form with various proprietary amounts of magnesia and/or lithia. In some cases a mixture of the two phases are preferred and a linear change in the resistance is found from ~ 16 Ω-cm for all β to ~ 3 Ω-cm for 100% β'' at 350°C [41]. The stabilizing effect of magnesium may be related to its occupying the tetrahedral site at the center of the spinel block, which in γ-Al_2O_3 is vacant but in $MgAl_2O_4$ is occupied by magnesium. At the dopant levels used, 1–4 wt. % MgO, all the magnesium could be fitted into this site which, if full, would contain ~ 7 wt. % MgO.

A second important parameter of any solid electrolyte is its electronic conductivity. This has been determined for silver beta alumina [31] using the Wagner asymmetric polarization technique [42] in the range 550–790°C and over a wide range of oxygen partial pressure. This was only 3×10^{-5} that of the ionic conductivity even at the highest temperature, and was essentially independent of oxygen pressure. Thus no self-discharge of cells is to be expected due to intrinsic electronic leakage through the electrolyte. However, penetration of sodium metal through pores in the ceramic can cause shorting and is though to be one of the major degradation mechanisms.

Beta alumina is not only a conductor for sodium ions but also for almost any other monovalent cation. The conductivity is however a strong function of the ionic radius of the diffusing specie as shown in Fig. 11 [14]. This change in conductivity is also reflected in the lattice parameter of the struc-

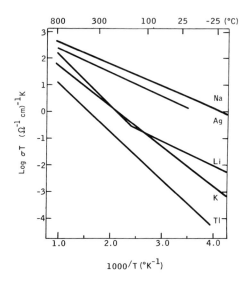

FIG. 11. Variation in conductivity of beta alumina with ionic radius of the mobile cation.

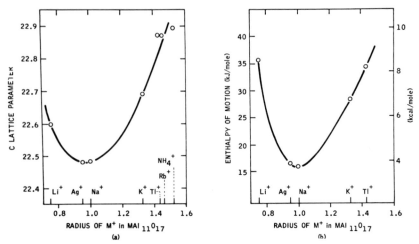

FIG. 12.   Variation in lattice parameter and activation energy for diffusion in beta alumina withe ionic radius of the mobile specie (from Whittingham [13]).

ture perpendicular to the diffusing planes and in the activation energy for the diffusion process (Fig. 12 [13]). These figures show the criticality of matching the size of the diffusion path in the host matrix to the radius of the diffusing ion. It is not surprising that the larger ions are impeded, because they are too large to diffuse through the structure which is strongly pinned together by the Al–O–Al bridging groups. That lithium is not very mobile is also associated with this size problem, but here the bridging groups prevent the structure from collapsing to accommodate the smaller ion. Hence, the lithium resides in an off-center plane position in a potential well, reducing the electrostatic bonding holding the Al_2O_3 blocks together and causing the structure to expand.

This behavior with different cations has serious consequences for the lifetime of ceramics. Any cationic impurities in the sodium, or elsewhere in the cell, can be exchanged for sodium ions in the electrolyte with a resulting change in the lattice dimensions and conductivity. The former may lead to actual mechanical breakdown of the ceramic at even low impurity levels, and can be guaranteed to crack the ceramic if total ion exchange is attempted, as for example, by substituting potassium or lithium for sodium. This therefore dictates stringent purity standards for all cell components, and is likely to be a problem in any solid electrolyte system. Such restrictions are not normally necessary for liquid electrolytes.

3. The Sodium–Sulfur Battery

For use in the sodium–sulfur battery the beta alumina must be fabricated into suitable shapes. The two shapes used are cylindrical tubes, normally closed at one end, and plates. The former is under most intense development at the present as it minimizes the area needing sealing to other cell components and is mechanically stronger. The tubes can be made by a variety of the standard ceramic techniques. Hydrostatic pressing around a mandrel to form the green shape followed by a high temperature sinter has been successfully used by the Electricity Research Council in Great Britain, amongst others. They pioneered a rapid zone-sintering technique in which the green tube is passed rapidly through the heating zone [43]; this allows large production runs and economies and apparently gives good ceramic. In addition, the short time at temperature minimizes sodium loss by evaporation. Other methods of forming the green shape include electrophoretic deposition [44], plasma spraying, extrusion and uniaxial pressing [45]. Hot pressing, which in contrast with the foregoing is a one-step process, can be used for making discs and rods, but the author has found [46] that in this case preferred crystallite orientation occurs, as might be expected.

In the currently used cell configuration the beta alumina tubes contain the sodium and are immersed in a graphite felt that acts as the current collector for the electronically insulating sulfur–polysulfide cathode. This is believed to be the most intensive energy density arrangement geometrically, although the lower temperature, $\sim 200°C$, molten salt scheme of ESB [47] places the cathode reactants in the tube:

$$Na–beta \; alumina–NaAlCl_4 + SbCl_5$$

(In this cell, sodium chloride is formed on discharge.)

D. Other Solids

Since the discovery of the electrolytic properties of sodium beta alumina in 1967, much effort has been expended in searches for other good ionic conductors, particularly for the two alkali metals, lithium and sodium, and to a lesser extent for potassium. The major problems of the lithium–sulfur cell might well be minimized if the molten salt could be replaced with a solid. The problems associated with the anisotropic structure of beta alumina have led to a desire for a three-dimensional sodium diffuser. Potassium metal has a particularly low melting point, 64°C, and would therefore make an interesting high power battery if a suitable electrolyte were available.

The state of knowledge as of 1974 for lithium has been given in Fig. 7; clearly none of these materials are suitable. The major problem with lithium is associated with its small size and high polarizing power so that it will tend to exist in off-center positions in open lattices. Recently Bither [48] found ionic mobility in lithium haloboracites, which is an improvement over the aluminosilicates. Even in the oxide bronzes [49] the activation energy of lithium motion is higher than for sodium species, and similarly in the layered sulfides where the structure is able to relax to fit the lithium ion [50]. Both these are electronic conductors and would therefore not be of interest as electrolytes, but they might suggest that the chances of finding a lithium conductor at ambient temperature is relatively small. On the other hand, the higher activation energy means that at sufficiently high temperatures the required magnitude will be found. However as will be discussed in the next section, few materials are stable in the presence of lithium at elevated temperatures.

The search for sodium conductors has been somewhat more successful. Hong *et al.* [51] found that the antimonates, such as $NaSbO_3$, have a cubic structure with intersecting channels through which the cations can diffuse. Although these particular compounds were not suitable for electrolyte use due to their ready reduction by sodium metal, it was found that the zirconium phosphates of general formula

$$Na_{1+x}Zr_2Si_xP_{3-x}O_{12}$$

were as good ionic conductors as beta alumina at 300°C [52]. Moreover they have a cubic structure and can be made into ceramic form at lower temperatures than β-Al_2O_3. Time will tell whether they can supplant it, but this finding indicates that new electrolytes can be found and that well-planned programs will probably succeed.

No success has been had in searches for a potassium conductor, and potassium beta alumina is still the only candidate [14].

E. Separators

In any battery system employing a liquid electrolyte, a separator must be placed between the electrodes; solid electolytes, naturally, are also the separator. For low temperature aqueous or organic electrolytes, the separators are frequently organic polymers, and amongst these polypropylene is the most common. The separator has to have all the characteristics as described for a solid electrolyte, except that of ionic conductivity. However, the separator should not impede the flow of ions in the electrolytic medium.

The major problem to be solved in this area is to find a material that can be used in high temperature lithium cells. Figure 13 gives the free energy of formation of some of the most stable binary oxides and it can readily be seen that lithia is one of these, so that only a few compounds are thermodynamically stable to reduction in the presence of lithium. The cheapest and most readily available compounds, such as silica and alumina, are not stable. Magnesia and calcia are stable, together with yttria and some closely related oxides like scandia and ceria (also thoria). Beryllia should be resistant to reduction but is ruled out because of its extreme carcinogenic nature. Yttria has been successfully used in the lithium–sulfur cell by Argonne National

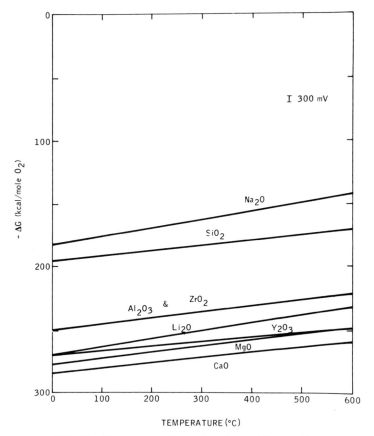

FIG. 13.   Free energy of formation of some binary oxides.

Laboratory, but the most promising compounds are found among the nitrides. Argonne [17,53] found that both boron nitride and aluminum nitride, if rigorously freed of oxide by heating at high temperatures in nitrogen behave excellently. These are used as woven cloths in laboratory cells, but different forms will have to be found before commercialization because of the high cost of the weaving process; papers have been suggested but their strength still has to be proven. These ceramic separator materials have a tendency to wick molten lithium so that they must still be used with the solid lithium–aluminum alloys.

Inert insulating materials are also required for feed-throughs, sealants and protective coatings; in some cases the coatings may be preferred to be conductive, as for example, where the cell container needs protecting from corrosion but is also the electron collector. Many of those compounds described for the separator can also fulfill these functions. The conditions here are somewhat less arduous as many of the components will only have to withstand either the oxidant or reductant, not both. Searches for new, cheaper materials are necessary and it is likely that many are to be found, particularly amongst ternary compounds, which have for the most part been overlooked to date. Those that come to mind are aluminas or aluminosilicates that contain lithium, as for example, lithium beta alumina, and are therefore less susceptible to reduction. Materials such as these that are potentially lithium ion conductors, but not of a high enough value to be used in their own right, could be prepared in a porous form so that the molten salt carries the major part of the ionic current.

F. New Materials Search

1. Techniques

In any search for new solid electrolytes one requires both a rapid initial survey tool and a subsequent accurate technique to determine the diffusivities of the mobile species. Two methods that are particularly amenable to a rapid diagnostic survey are ion exchange and nuclear magnetic resonance (NMR) spectroscopy. For the final accurate analysis a direct measurement of the ionic conductivity must be made.

In ion exchange, as used on beta alumina by Yao and Kummer [29], the solid is simply immersed in, for example, a molten salt of an ion different from that in the solid and the solid is then analyzed for any ion exchange. This is very simple to perform, particularly for single crystals, but care must be taken that a negative result is not due to the nonexistence of the compound that is supposedly being formed. This experiment can also be electrically driven, as was used by Whittingham *et al.* [34] to form copper beta alumina. Such a technique was successfully used in the determination of ionic mobility in the antimonates [54].

The simplest of all the NMR methods is that of determining the linewidth of the spectrum as a function of temperature. When the ions are moving rapidly, the linewidth is narrowed to the inhomogeneity limit of the magnet, ≤ 0.1 G; but as the motional processes drop below an occurrence frequency of $\sim 10^4$ sec^{-1}, an abrupt increase in the linewidth is observed [55]. By fitting this motional narrowing part of the temperature dependence of the linewidth to the equation

$$v_i = [\gamma/\alpha(\Delta H - \Delta H_r)]/2\pi \tan[(\pi/2)/((\Delta H - \Delta H_r)/(\Delta H_1 - \Delta H_r))^2]$$

jump frequencies, v_j, can be calculated [55]. In this equation ΔH is the measured linewidth, ΔH_r the high-temperature-rendered linewidth, ΔH_1 the low temperature linewidth, γ the gyromagnetic ratio of the nucleus, and α a constant with a value of 1–10. Values of v_j can then be fitted to the Arrhenius equation

$$\ln v_j = \ln v_0 - Q/RT$$

from which the activation energy of motion can be computed. Values of ΔH_r and ΔH_l can be optimized by a least squares analysis.

As an example, the proton line narrowing data for ammonium ferrocyanide, $(NH_4)_4Fe(CN)_6 \cdot H_2O$, are shown in Fig. 14. The circles are the experimental data points and the curve was drawn using the above equation with

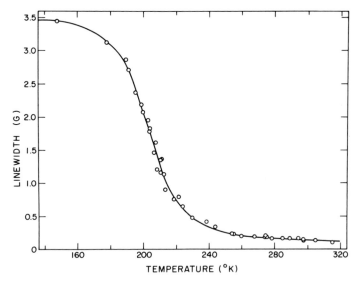

FIG. 14. Proton line narrowing data for ammonium ferrocyanide, $(NH_4)_4Fe(CN)_6 \cdot H_2O$ [56].

$Q = 19.0$ kJ/mole, $v_0/\alpha = 1.52 \times 10^9$ sec^{-1}, $\Delta H_r = 0.085$ G and $\Delta H_1 = 3.45$ G [56].

In the preliminary survey it is unnecessary to determine the entire narrowing curve, as the linewidth at room temperature can be a good indicator of ionic motion. Thus, if we take 2 G and 2.6 as the values of $\Delta H_1 - \Delta H_r$ and α, respectively, the above equation indicates that the half-narrowed position of the line occurs for

$$v_j = \gamma \approx 10^4 \text{ sec}^{-1}$$

for H, Li and Na. For a good solid electrolyte, a typical minimum value of v_j should be $\sim 10^6$ sec^{-1}, so that the line should be essentially fully narrowed at the intended temperature of operation.

This technique is most easily carried out for protons where there are no complications due to quadrupole coupling. When these are found to be a problem for lithium or sodium, then a variety of pulsed NMR techniques can be used. A further complication with NMR techniques and other relaxation methods, such as dielectric loss, is that one can never be certain that true long range motion is being measured; it is possible for relaxation to occur when the ions simply jump a few lattice sites. Confirmation of motion is therefore always necessary.

Ionic conductivity can be determined in three main ways, by tracer diffusion, by potentiostatic techniques, and by using direct conductivity studies. The second is used where the material's composition can be varied and it is an electronic conductor; thus it is used for nonstoichiometric electrode materials rather than for electrolytes. The technique simply involves pulsing the voltage of a suitable cell and measuring the decay in the current flow. Tracer diffusion can be used for almost any material but tends to be very time consuming.

Although conductivity studies can be made on poor ionic conductors such as NaCl using ac or dc techniques and ionically blocking electrodes, such as platinum, this is not feasible when the ionic conductivity is substantial. The electrodes applied to the solid must be reversible to the ions of interest. This may be accomplished in a number of ways, including simply using electrodes of the pure element, such as sodium for beta alumina. The experimental problems of handling the alkali metals are however substantial. A second method that allows the determination of σ_i over a rather restricted temperature range is to use molten salts, such as sodium nitrate, to contact the beta alumina. This is particularly convenient for routine testing of production samples which are ultimately to be used at essentially a single temperature. A method that is operative over wide temperature ranges was developed by the author [55] and uses a highly nonstoichiometric solid, such as Na_xWO_3, which is a mixed ionic and electronic conductor. Being nonstoichiometric, the electrode is able to act as a source and sink of ions for the electrolyte under test. This technique has been used on a number of the beta aluminas by Whittingham and Huggins [14, 57] and the results obtained on the sodium compound are compared in Fig. 15 with tracer [29] and dielectric loss [58] measurements. The agreement in this case is remarkably good.

Two further tests of any potential solid electrolyte are necessary. It must be checked for compatibility with the intended anode and cathode, and the electronic conductivity must be ascertained. The latter can be determined to a first approximation by simply constructing a battery cell and measuring the emf generated and comparing that with the calculated value. Then, the ionic transference number, t_{ion}, is given by

$$t_{ion} = \sigma_{ion}/(\sigma_{ion} + \sigma_{elec}) = E_{meas}/E_{calc}$$

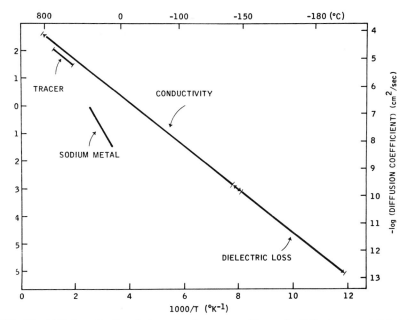

FIG. 15.   Diffusion of sodium in beta alumina measured by conductivity, tracer and dielectric loss techniques (from Whittingham [13]).

Accurate values of the electronic conductivity can be determined—as they need to be to obtain knowledge of long term shelf life—using the Wagner asymmetric polarization technique [42]. In this a dc potential is applied to a cell such as the following [31]

$$- \text{Ag–Ag beta alumina–Pt} +$$

so that at equilibrium there is no ionic flow. Determination of the current flow as a function of cell voltage, which must not exceed the decomposition potential of the electrolyte, allows calculations of both the electron and hole conductivities.

2. *Areas and Materials*

There are a number of general criteria that must be met in a material that is hoped to have a high ionic conductivity. There should be substantially more available sites, of similar energy, than ions of the mobile specie and the energy required to jump between these sites must be small, typically around 4 kcal/mole (20 kJ/mole). The former does not necessarily imply the latter. And third, none of the ions in the compound should be easily reduced or oxidized. This last requirement alone tends to rule out the majority of the transition metals, which readily exhibit multivalency, with the possible exception of zirconium and tantalum. As discussed earlier, the mobile ion must be closely matched to the size of the lattice through which it is to diffuse. If it is either too small or too large, it will tend not to diffuse rapidly; this is the reason that the zeolites exhibit very low ionic mobilities in the anhydrous site. Although much effort has been expended on tunnel structures such as bronzes [49] and hollandites [59], it now appears that there is very little long range movement in such materials. This is not at all surprising considering that a *single* defect in a tunnel could block all diffusion and that there are $\sim 10^7$ ion/cm [46]. Effort has therefore shifted to channel structures or those containing intersecting tunnels in two or three dimensions such as the zirconium phosphates discussed earlier. Classifications of fast ionic conductors have been discussed in detail elsewhere [13, 60] and the reader is referred to them for further information. This area of materials science offers immense opportunities for pioneering work of both a basic scientific and applied nature.

V. THE CATHODE

A. Present Status

The requirements of a cathode were enumerated on page 000. As noted there, none of the existing systems fulfill all the preferred criteria. In both the present high temperature systems, the cathode is molten sulfur or polysulfide. These are extremely corrosive, electronic insulators (and ionic in the case of sulfur), and tend to be lost by evaporation. In the case of the sulfides now in use in the Li–S cells, a considerable reduction in theoretical energy density and potential current density has had to be taken to overcome the sulfur loss. Even then FeS_2 is so corrosive that only molybdenum or carbon current collectors can be used [17]; cheaper current collectors such as iron can be used only for FeS with a resultant further loss in energy density. In the case of the sodium–sulfur cell the polysulfides are so corrosive that great difficulty is being had in finding a suitable container material. This raises the question, are the final energy densities obtained going to warrant all the problems associated with high temperature operation? Unless or until a reversible cathode system can be found that will operate at much lower temperatures and could then present a viable alternative, the question is however academic. The next few pages will therefore review past and present cathodes, their mechanisms of operation, and then possible new areas for research.

B. Materials

1. Displacement Reactions

A displacement reaction will be defined here as one such as is found in the reaction

$$CuF_2 + 2Li \rightarrow Cu + 2LiF$$

where the crystalline structure of the cathode reactant is completely broken down. In this mechanism, which is that operative in all the present high temperature systems, the crystal structures of the reactants and products bear no relation to one another. Thus the crystal lattice must be completely rebuilt on each charge or discharge cycle of the cell. This is not likely to occur at any reasonable rate in the solid state, except at very high temperatures, unless the reaction occurs by a dissolution–recrystallization process involving a liquid electrolyte. The partial solubility that this mechanism requires is, however, likely to cause self-discharge on standing, which may or may not be serious depending on the system and the application in mind. It should be noted though that a number of systems utilize soluble cathodes; thus the zinc halogen cells in which the chlorine or bromine is partially immobilized by hydrate and quarternary ammonium salt formation, respectively, are being seriously pursued for large scale use despite some self-discharge problems.

If a solid electrolyte is used in the cell, then the self-discharge problem does not arise, but then a liquid must be added to a solid cathode to enable high currents to be passed through the electrolyte–cathode interface. If only the discharge product and not the cathode active material is partially soluble in the liquid phase, rechargeability will be enhanced without causing severe corrosion problems. The need for some solubility is however removed if the oxidant is reduced with essential retention of crystalline structure.

2. Single Phase Reactions

When the cathode reacts by incorporation of the anode reactant into its crystalline structure, there is a reasonable chance that the reaction may be reversed. The actual degree of reversibility is related to the amount of disruption of the crystal lattice. The simplest case to consider is that of the reaction of lithium with the early group transition metal sulfides [61, 62], such as titanium disulfide, in which essentially no structural changes occur except for an expansion perpendicular to the basal planes as shown in Fig. 16. This is an intercalation reaction, and may be readily reversed many times even at ambient temperatures. In this particular example, the cathode system represents a single phase from the completely charged to the com-

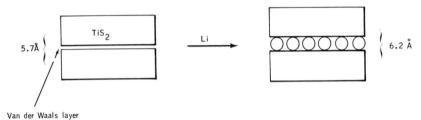

FIG. 16.   Reaction of titanium disulfide with lithium.

pletely discharged state; that is,

$$x\mathrm{Li} + \mathrm{TiS}_2 \rightleftharpoons \mathrm{Li}_x\mathrm{TiS}_2$$

for $0 \leq x \leq 1$. Thus the cell emf as well as the lattice expansion change in a continuous manner with state of discharge as indicated in Fig. 17 [61].

In the above example, only weak Van der Waals bonds are broken during the intercalation reaction. No bonds within the TiS_2 layers are broken or rearranged. On the other hand, for the trisulfide, which also reacts

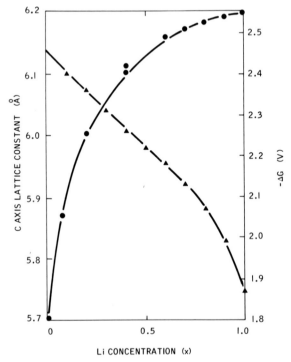

FIG. 17.   Emf of Li–TiS_2 cell and lattice parameter of $\mathrm{Li}_x\mathrm{TiS}_2$ variation as a function of state of discharge, $0 < x < 1$.

by incorporation of lithium [63] and the structure of which bears a superficial resemblance to the disulfide, some chemical bonds are broken and only marginal rechargeability is found [61]. The trisulfide has the formula TiS(S–S); that is, it contains one sulfide sulfur and one polysulfide group, and reacts with three lithiums in the following manner:

$$2Li + TiS_3 \rightarrow Li_2TiS_3$$

$$xLi + Li_2TiS_3 \rightarrow Li_{(2+x)}TiS_3, \qquad 0 \le x \le 1.$$

The first of these reactions involves the breakage of the polysulfide group and probable rearrangement of the sulfur atoms around the titanium. This reorganization is apparently not reversible. The second reaction, which involves the reduction of Ti(IV) to Ti(III), is reversible but only at very low rates, < 1 mA/cm^2. In contrast, for NbSe$_3$, in which there is no normal Se–Se bond and where the niobium already has its preferred atomic environment, all three lithiums are reversible [61, 64]. But because of high equivalent weight, cost and toxicity there is no future for this complex; thus the Li–TiS$_2$ cell has a higher energy density.

Many other oxides and sulfides of the group IVB, VB and VIB transition metals will also react in this same manner [61] with the alkali metals, whereas those of metals further to the right in the periodic table tend to be reduced directly to the metal. Amongst those of the former type are the other layered dichalcogenides with the same structure as TiS$_2$, and MoO$_3$, V$_2$O$_5$, and so on; amongst the latter are NiS, CuS, CuO, FeS$_2$, and so on. None of these are presently used in commercially available high energy density secondary batteries but this is likely to change in the future when suitable electrolytes are found.

3. Future Trends

To date, essentially all the published work on advanced batteries using solid cathodes has evolved around the molten salt–lithium cell in which the iron sulfides have been extensively investigated. Although these compounds meet the requirements of low cost, the remainder of the cell does not. Recently the ternary phase compounds, such as Li$_x$TiS$_2$, have been receiving some interest and it is likely that it is in this direction that future work on solid cathodes will go. These compounds not only meet the reversibility requirements at the preferred low temperatures, but also are frequently good electrical conductors so that there is no need for the addition of an electrochemically inert conductive diluent. As the group VIA anions form these ternary phases more often than the halides, it is amongst the former that reversible ambient temperature cathodes are likely to be found.

VI. BATTERY SYSTEMS

A. Present Status

There are essentially only three secondary batteries commercially available today. They are the lead–acid, the nickel–iron (the Edison Cell) and the nickel–cadmium. The first is available in a variety of configurations ranging from the well-known automobile starter battery, which is the cheapest of the line, to the heavy duty electromotive cells used in milk floats and forklift trucks amongst others. These last, which can withstand many hundreds of cycles, are at least double the cost of the former. Although it is thought that the energy density of the lead–acid cell can be substantially improved (to 20–30 kWh/lb) by removing the inactive lead support structure, this will probably involve an increased cost and may cause problems with cycle life. The Edison cell, which uses an alkaline electrolyte, is much too expensive to be used on a large scale. The same holds true for the nickel–cadmium system, which has the highest power density of the three, and is used in many consumer applications as well as in special high power situations such as starter power sources for aero engines.

Aqueous systems are particularly attractive because there is no need, in most cases, to hermetically seal them. Thus much effort has been expended looking at various cells based on iron and zinc coupled with oxygen or the halogens. Although the air cathodes are at first sight the cheapest, they require expensive electrocatalysts at the air electrode (as in fuel cells) and the electrodeposition of the metal on recharge has not been successfully accomplished. The zinc–halogen cells, although known since the turn of the century, have received renewed attention in the last few years, particularly from the aspect of complexing the halogen; most of the data here is, however, only available in the patent literature. These aqueous based cells hold promise of doubling the energy density of the lead–acid battery and at lower cost. If their problems of self-discharge and metal electrodeposition can be solved, they will undoubtedly capture some of the lead–acid market although they do not have the ultimate capability for long range automobile propulsion.

The two alkali metal–sulfur systems, despite all signs of technical feasibility, are unlikely to meet the economic criteria required ($< \$50/kWh$ in 1975) because of the need for expensive corrosion resistant construction materials. The proponents of these cells are however optimistic of attaining the required goal, and over 5000 discharge–charge cycles have been achieved on single Na–S cells in addition to a demonstration truck run by the Electricity Research Council.

B. Trends and Research Opportunities

In the short term it is likely that improved lead–acid cells will be used to power the second generation road vehicles (milk floats are considered first generation) that might reasonably be used by postal, telephone repair and metropolitan delivery fleets. In the long term, advanced batteries will probably utilize the alkali metals as anodes and will operate at or around the ambient temperature. Before this can become a reality, however, much research will be needed to find suitable electrolytes and cathodes. Present trends would seem to suggest that a liquid electrolyte will be used in conjunction with a solid cathode.

A liquid electrolyte does, however, still require the use of a solid separator. Very little is known about suitable separator materials for alkali systems, but organic polymers are clearly preferred because of both their inherent ease of handling and potentially low cost. An added advantage of this arrangement is that the electrode volume changes during cycling can be readily accommodated by movement of the liquid and separator. This may be a major problem in solid electrolyte cells. Organic polymers are also preferred for cell encasement, insulators, and so on. Fundamental research on polymers resistant to highly oxidizing and reducing conditions is urgently needed, and there is still hope that a polymeric electrolyte might be found. In the meantime, a search must be made for organic based electrolytes and low temperature—< 200°C—molten salts.

In the cathode area, much solid state chemical research is needed to ascertain the nature of reactions between the alkali metals and inorganic solids that form solid solutions. We are only now beginning to understand

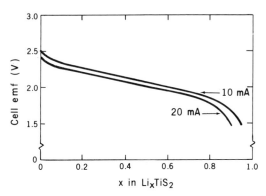

FIG. 18.   Discharge behavior of a Li–TiS$_2$ cell. The cathode was 2 cm^2 in area and the electrolyte a solution of lithium aluminum chloride in methylchloroformate (from Whittingham [66]).

the phenomena of intercalation and its role in the reversibility of chemical reactions. Similarly, emphasis should also be placed on means of forming chemical sinks for the cathodically active material, such as chlorine in the zinc–chlorine cell. Thus, just as TiS_2 acts as an electrochemically active cathode sink for lithium, materials might be found that will hold the chlorine in a nonreactive, that is, noncorrosive, form until required for reaction with the zinc. Intercalation compounds such as C_8Br exemplify where graphite would be the inactive sink, but in this case the weight penalty of the carbon is too high. For alkali cells Armand has proposed that graphite might act in this manner for a number of oxides and halides, in particular for chromium trioxide [65]. The high rates of discharge found for TiS_2 are shown in Fig. 18 [66]. These, combined with the remarkable reversibility of this system, will lead to a greater emphasis on low temperature batteries utilizing solid electrodes.*

REFERENCES

[1]   Federal Power Commission, National Power Survey Report (1970).
[2]   Argonne National Laboratory, Rep. No. ANL-8058.
[3]   J. G. Gibson and J. L. Sudworth, "Specific Energies of Galvanic Reactions and Thermodynamic Data." Chapman and Hall, London, 1973.
[4]   R. Galli, in "Fast Ion Transport in Solids" (W. van Gool, ed.), p. 553. North-Holland Publ., Amsterdam, 1973.
[5]   D. R. Vissers, Z. Tomczuk, and R. K. Steunenberg, J. Electrochem. Soc. 121, 665 (1974).
[6]   Argonne National Laboratory, Rep. No. ANL-7958.
[7]   N. P. Yao, L. A. Heredy, and R. C. Saunders, J. Electrochem. Soc. 118, 1039 (1971).
[8]   J. L. Settle, K. M. Myles, and J. E. Battles, U.S. Patent Appl. 481,285 (1974).
[9]   M. R. Kegelman, U.S. Patent 3,639,174.
[10] A. N. Dey, J. Electrochem. Soc. 118, 1547 (1971).
[11] B. B. Owens, Adv. Electrochem. Electrochem. Eng. 8, 1 (1971).
[12] M. Lazzari, R. C. Pace, and B. Scrosati, Electrochim. Acta 20, 331 (1975).
[13] M. S. Whittingham, Electrochim. Acta 20, 575 (1975).
[14] M. S. Whittingham and R. A. Huggins, Nat. Bur. Std. Spec. Publ. 364, 139 (1972).
[15] C. C. Liang, J. Electrochem. Soc. 120, 1289 (1973).
[16] R. T. Johnson, R. M. Biefield, M. L. Knotek, and B. Morosin, J. Electrochem. Soc. 123, 680 (1976).
[17] E. J. Cairns and R. A. Murie, and J. E. Battles, F. C. Mrazek, W. D. Tuohig, and K. M. Myles, in "Corrosion Problems in Energy Conversion and Generation" (C. S. Tedmon, ed.). The Electrochemical Society, New York, 1974.
[18] W. Nernst, Z. Elektrochem. 6, 41 (1899).
[19] W. Nernst and W. Wald, Z. Elektrochem. 7, 373 (1900).
[20] W. Schottky, Wiss. Veröeff. Siemens-Werken 14, 1 (1935).
[21] K. Kiukkola and C. Wagner, J. Electrochem. Soc. 104, 308 (1957).

* The intercalation chemistry of the dichalcogenides and their use in batteries has recently been reviewed in detail by the author [67]. For further details on solid electrolytes the reader is referred to the book edited by Hagenmuller and van Gool [68].

[22] K. Kiukkola and C. Wagner, *J. Electrochem. Soc.* **104**, 379 (1957).
[23] W. Baukal *in* "From Electrocatalysis to Fuel Cells" (G. Sandstede, ed.). Univ. of Washington Press, Seattle, Washington, 1972.
[24] W. Vielstich, "Fuel Cells." Wiley (Interscience), New York, 1970.
[25] R. W. Ure, *J. Chem. Phys.* **26**, 1363 (1957).
[26] M. O'Keefe, *Science* **180**, 1276 (1973).
[27] W. L. Bragg, C. Gottfried, and J. West, *Z. Kristallogr.* **77**, 255 (1931).
[28] C. A. Beevers and M. A. S. Ross, *Z. Kristallogr.* **97**, 59 (1937).
[29] Y. F. Yao and J. T. Kummer, *J. Inorg. Nucl. Chem.* **29**, 2453 (1967).
[30] N. Weber and J. T. Kummer, *Proc. Ann. Power Sources Conf.* **21**, 37 (1967).
[31] M. S. Whittingham and R. A. Huggins, *J. Electrochem. Soc.* **118**, 1 (1971).
[32] G. Yamaguchi and K. Suzuki, *Bull. Chem. Soc. Jpn.* **41**, 93 (1968).
[33] M. Bettman and C. R. Peters, *J. Phys. Chem.* **73**, 1774 (1969).
[34] M. S. Whittingham, R. W. Helliwell, and R. A. Huggins, U.S. Govt. Res. Develop. Rep. 69, p. 158 (1969).
[35] M. Bettman and L. Terner, *Inorg. Chem.* **10**, 1442 (1971).
[36] P. B. Braun, *Philips Res. Rep.* **12**, 491 (1957).
[37] C. Peters, M. Bettman, J. Moore, and M. Glick, *Acta Crystallogr.* **B27**, 1826 (1971).
[38] R. J. Friauf, *Phys. Rev.* **105**, 843 (1957).
[39] M. S. Whittingham, quoted by W. Van Gool in *Ann. Rev. Mater. Sci.* **4**, 324 (1974).
[40] J. H. Kennedy and A. F. Sammells, *in* "Fast Ion Transport in Solids" (W. van Gool, ed.), North-Holland Publ., Amsterdam, 1973.
[41] J. Sudworth, A. R. Tilley, and K. D. South, *in* "Fast Ion Transport in Solids" (W. van Gool, ed.). North-Holland Publ., Amsterdam, 1973.
[42] C. Wagner, *Z. Elektrochem.* **60**, 4 (1956).
[43] I. Wynn Jones and L. J. Miles, *Proc. Brit. Ceram. Soc.* **19**, 161 (1969).
[44] R. W. Powers, *J. Electrochem. Soc.* **122**, 490 (1975).
[45] S. P. Mitoff, *in* "Fast Ion Transport in Solids" (W. van Gool, ed.). North-Holland Publ., Amsterdam, 1973.
[46] M. S. Whittingham, *in* "Fast Ion Transport in Solids" (W. Van Gool, ed.), p. 427. North-Holland Publ., Amsterdam, 1973.
[47] J. Werth, I. Klein, and R. Wylie, *J. Electrochem. Soc.* **122**, 265C (1975).
[48] T. A. Bither and K. Jeitschko, U.S. Patent 3,911,085.
[49] M. S. Whittingham and R. A. Huggins, *in* "Fast Ion Transport in Solids" (W. van Gool, ed.). North-Holland Publ., Amsterdam, 1973.
[50] M. S. Whittingham and B. G. Silbernagel, unpublished work.
[51] H. Y-P. Hong, J. A. Kafalas, and J. B. Goodenough, *J. Solid State Chem.* **9**, 345 (1974).
[52] H. Y-P. Hong, paper presented at the American Chemical Society Spring Meeting (April 1976).
[53] J. E. Battles and F. C. Mrazek, U.S. Patent 3,915,742.
[54] H. Y-P. Hong, *Acta Crystallogr.* **B30**, 945 (1974).
[55] M. S. Whittingham, *in* "Fast Ion Transport in Solids" (W. van Gool, ed.). North-Holland Publ., Amsterdam, 1973.
[56] M. S. Whittingham, P. S. Connell, and R. A. Huggins, *J. Solid State Chem.* **5**, 321 (1972).
[57] M. S. Whittingham and R. A. Huggins, *J. Chem. Phys.* **54**, 414 (1971).
[58] R. H. Radzilowski, Y. F. Yao, and J. T. Kummer, *J. Appl. Phys.* **40**, 4716 (1969).
[59] J. Singer, H. E. Kautz, W. L. Fielder, and J. S. Fordyce, *in* "Fast Ion Transport in Solids" (W. van Gool, ed.). North-Holland Publ., Amsterdam, 1973.
[60] W. Van Gool, *Annu. Rev. Mater. Sci.* **4**, 311 (1974).
[61] M. S. Whittingham, *J. Electrochem. Soc.* **123**, 315 (1976).

[62] M. S. Whittingham and F. R. Gamble, *Mater. Res. Bull.* **10**, 363 (1975).

[63] R. R. Chianelli and M. B. Dines, *Inorg. Chem.* **14**, 2417 (1975).

[64] J. Broadhead, F. J. Disalvo, and F. A. Trumbore, U.S. Patent 3,864,167.

[65] M. B. Armand, *in* "Fast Ion Transport in Solids" (W. van Gool, ed.). North-Holland Publ., Amsterdam, 1973 and French Patent Appl. 7,229,734 (1972).

[66] M. S. Whittingham, *Science* **192**, 1126 (1976).

[67] M. S. Whittingham, *Progr. Sol. State Chem.* **12**, 41 (1978).

[68] P. Hagenmuller and W. van Gool, eds., "Solid Electrolytes." Academic Press, New York, 1978.

Chapter 10

Superconducting Materials for Energy-Related Applications*

T. H. GEBALLE

DEPARTMENTS OF APPLIED PHYSICS AND MATERIALS SCIENCE
STANFORD UNIVERSITY
STANFORD, CALIFORNIA
AND
BELL LABORATORIES
MURRAY HILL, NEW JERSEY

M. R. BEASLEY

DEPARTMENTS OF APPLIED PHYSICS AND ELECTRICAL ENGINEERING
STANFORD UNIVERSITY
STANFORD, CALIFORNIA

* Written under the support of the National Science Foundation, the U.S. Energy Research and Development Administration, and the Institute for Energy Studies at Stanford University.

491

I. INTRODUCTION

Superconductors embrace a remarkable set of electric and magnetic properties, the most startling being a total lack of dc resistance. Transitions into
the superconducting state occur at a temperature T_c, which may be from less
than $0.01°K$ to a present-day high of $23°K$. The potential of superconductivity for important technological applications has tantalized scientists since
the discovery of superconductivity in frozen mercury by Kamerlingh Onnes
in 1911. However, not until relatively recently have superconducting materials suitable for serious large scale applications of superconductivity become available. At present, a wide variety of such applications are being
vigorously pursued both in the United States and abroad.

Early workers in superconductivity focused their attention on the challenge of understanding superconductivity. The first insights came from
London [1], who correctly anticipated that superconductivity was a "macroscopic quantum phenomenon." Later, this macroscopic quantum mechanical point of view was developed further by Ginzburg and Landau [2] in their
famous phenomenological theory of superconductivity. The high point came
in 1957 when Bardeen et al. (BCS) [3] put forth their successful microscopic
theory of superconductivity based on electron–electron pairing via an
attractive phonon-mediated interaction first suggested by Fröhlich. A few
years later, with the theoretical discovery of the Josephson effect [4], the
macroscopic quantum nature of superconductivity was forcefully reem-
phasized, and our understanding of superconductivity took on its modern
form.

Concurrent with the development of our fundamental understanding,
superconductivity was found to exist in a wide variety of elements, compounds and solid solutions. The materials of interest in this review were,
interestingly enough, discovered during the period of the chief theoretical
advances, that is, 1952–1962, although the two fields developed quite independently. Great progress was made in purifying the refractory metals and
lanthanides so that new areas of the Periodic Table were opened for investi-

gation. A pattern of the occurrence of superconductivity in the periodic system (Fig. 1) was noted even earlier by Schoenberg [6]. Many new superconductors were predicted and discovered by Matthias and by Hulm [5]. Technologically and scientifically a major turning point came in 1961 when Kunzler *et al.* [7] found that Nb_3Sn could support current densities as high as 10^5 A/cm^2 in a magnetic field of 88 kOe and still remain superconducting. This discovery stimulated a healthy interplay between theory and experiment, resulting in a rapid development of both the understanding of high field, high current superconductors and in methods of fabricating these superconductors into useful devices.

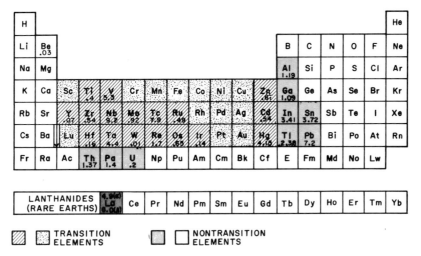

FIG. 1. Occurrence of superconductivity in the Periodic Table. The transition temperatures of elements when cooled from their standard states at atmospheric pressure are given in °K. Other elements of the first, second, fourth, fifth and sixth periods can be superconducting when subjected to high pressures, or as low temperature (amorphous) films. For further details, see Roberts' compilation [5].

Practical superconducting composite wires are presently commercially available and have been successfully used to construct large magnets, as for example, for use in bubble chambers. Not only are these large superconducting magnets much more efficient than conventional electromagnets, but they generate substantially higher fields over much larger volumes than would otherwise be possible by any means. As a result, superconductivity offers possible solutions to a number of difficult technological problems, many of them related to energy generation, storage and transmission [8a,b].

In any full scale thermonuclear fusion reactor (see Chapter 6), superconducting magnets will be necessary to confine the plasma. Such confinement magnets are perhaps already within the state of the art and prototypes are being developed. Large MHD power generating stations also would clearly require superconducting magnets. Huge superconducting magnets are also being considered for energy storage in our electrical power system. The transmission of underground power, both ac and dc, by superconducting cables is being actively pursued. Here, as elsewhere, the superconducting route becomes more attractive vis-a-vis more conventional methods such as oil-, water- and gas-filled lines as the magnitude of the power transmitted goes up.

There are other uses for superconductivity for which more immediate application can be anticipated because the systems are more developed and self-contained. Included in this category are ac and dc power generators with superconducting field windings. These can be made more compact and hence of larger capacity and more efficient than conventional generators. Prototype generators already have been built successfully [9]. In the transportation field, superconducting generators and motors are being developed for ship propulsion that will permit great flexibility in ship design and control. High speed ground transportation using superconducting magnets to provide levitation also has attracted serious attention. A critical listing of possible large scale applications of superconductivity, as summarized by Powell [10], are given in Table 1.

The various energy-related applications of superconductors have different operating conditions and hence call for different optimization of the superconducting material characteristics. Development is proceeding rapidly and precludes the likelihood of having any review such as this remain up to date in an engineering sense for very long. We have attempted therefore to emphasize as much as possible the principles underlying the synthesis and fabrication of practical superconducting materials and composites, to identify the important material characteristics, and to indicate how they can be favorably altered by chemical or physical means. Whenever possible, we have also attempted to identify areas where additional research or development is especially needed or particularly likely in our view to lead to significant advances in the state of the art.

In Section II we present a sufficiently detailed discussion of the thermodynamics and the electrical and magnetic properties of superconductors within the framework of the phenomenological theories of superconductivity to identify how the properties of principal importance (e.g., critical fields, critical currents, stability, losses) are expected to depend upon materials parameters.

In Section III we discuss the superconductors that are of particular or potential importance in energy-related areas. Their occurrence and properties are covered in enough detail to perhaps be suggestive of new materials to the reader.

TABLE 1

Assessment of Large Scale Superconductor Applications[a,b]

Application	Capability of current superconductors and cryogenic system to meet desired application			Projected cost of superconductor and crogenic system	Probability of significant application
	Field	Losses	Stability		
Fusion (dependent on confinement method)	G–P	G–P	G–P	G–P	VG
MHD	G	G	G	VG	F–G
Magnetic storage	G	G	G	F–G	F–G
Ac gen. and motors	G	G	G	G	G
Dc gen. and motors	G	G	G	G	G
Transformers	G	P	F	P	P
Dc transmission	G	G	G	F	F–G
Ac transmission	G	F	G	F	F–G
High speed train	G	F	G	G	G
Ore separation	G	G	G	G	G

[a] After Powell [10].
[b] VG—Very Good; G—Good; F—Fair; P—Poor.

In Section IV we discuss the methods of preparing superconducting materials and fabricating them into suitable engineering composite wires and cables. The important characteristics of the composite that are necessary to secure reliable performance are illustrated. We briefly present representative examples of the state of the art as we know it at present, realizing full well that this is the section that (hopefully) will be outdated soon.

II. THE ELECTRICAL AND MAGNETIC PROPERTIES OF SUPERCONDUCTORS

Within two years of his discovery of superconductivity, Onnes first investigated the possibility of high field superconducting magnets with great expectations. The results were quite discouraging, as reported in Onnes' own words [11]:

The experiments on the question if the magnetic field develops resistance in superconductors, to which we alluded above have been made in the mean time. They have given a startling result. In fields below a threshold value (say 600 gauss for lead at the boiling point of helium) there is developed no resistance at all by the field. In fields above that threshold value a relatively considerable resistance is developed, which increases with the field.

Of course the now acquired knowledge of this property completely changes our view on the problem of obtaining intense magnetic fields with coils of superconductive material.

The text was written in the idea founded on analogies that the magnetic resistance would increase in a continuous way with the field. Starting from what was observed with our small coil and even accepting an increase with the square of the field it seemed probable that the magnetic resistance would not yet come seriously into account with a field of 100,000 gauss.

An unforeseen difficulty is now found in our way, but this is well counterbalanced by the discovery of the curious property which is the cause of it.

Today, superconducting magnets which produce fields up to ~ 175 kOe are commercially available, and superconductors with critical fields as high as 600 kOe are known. Clearly, our understanding of high field superconductivity—or put more generally, the electrical and magnetic properties of superconductors—has come a long way. In this section we review this understanding, stressing in particular those aspects which are of critical importance in energy-related applications.

A. Type-I and Type-II Superconductors

1. Type-I Superconductors and the Magnetic Penetration Depth

Today, we understand Onnes' difficulties as arising because of the low critical magnetic field of superconducting lead, which for the most part is typical of so-called type-I superconductors. Most elemental superconductors (e.g., Sn, In, Pb, but not Nb) are in this class. To understand the physical origins of this critical field, we recall that in addition to having zero resistance, superconductors also exhibit perfect diamagnetism $(B = 0)$, more commonly referred to as the Meissner effect. This spontaneous expulsion of flux in the presence of a magnetic field, H, raises the thermodynamic free energy of the superconductor, and so eventually as H increases it becomes energetically favorable to revert to the normal state $(B \cong H, R \neq 0)$ and let flux enter the material. This behavior is illustrated in Fig. 2(a). The B–H curve of a type-I superconductor is shown in Fig. 2(b). On the figure the critical field is denoted by $H_c(T)$.

Of course, nothing is truly discontinuous in nature and near the surface of a superconductor the magnetic induction B is not zero but rather falls to zero exponentially in a characteristic length λ. This magnetic penetration depth λ is temperature dependent and becomes infinite as the temperature T approaches the transition temperature T_c. For most purposes the temperature dependence of λ can be well approximated by

$$\lambda = \lambda(0)/(1 - t^4)^{1/2} \tag{1}$$

where $t = T/T_c$. The magnitude of the zero temperature penetration depth

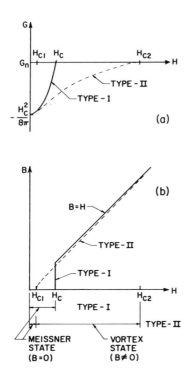

FIG. 2.   (a) Field dependence of the thermodynamic free energy of type-I and type-II superconductors. (b) *B–H* curves for superconductors indicating various regimes of behavior. The critical fields are defined in the text.

$\lambda(0)$ depends on the particular material and ranges from a few hundred angstroms in elemental superconductors up to perhaps nearly 1 μm in some alloys. Intermetallic compounds also tend to have relatively large penetration depths, perhaps a few thousand angstroms, although for these materials they are rarely known with precision.

The critical field $H_c(T)$ is known as the thermodynamic critical field because it corresponds to the field at which the magnetic energy due to complete flux expulsion, $H^2/8\pi$ per unit volume, equals the condensation energy of the superconductor (for $H = 0$) at the given value of T, that is

$$H_c^2(T)/8\pi \equiv \text{condensation energy} \tag{2}$$

More specifically, from electromagnetic theory we know that the increase in energy per unit volume equals $-\int M\, dH$, where M is the magnetization. Since $M = -H/4\pi$ for a type-I superconductor, the above relations are obtained.

Experimentally, it is found that $H_c = H_c(0)[1 - (T/T_c)^2]$ to an excellent approximation, and from the BCS microscopic theory we have

$$H_c^2(0) = 5.95\gamma T_c^2 \tag{3}$$

where γ is the coefficient in the expression $C = \gamma T$ for the electronic specific heat of the normal state, and where we have neglected strong coupling effects in the electron–phonon interaction. Equation (3) shows that, roughly speaking, $H_c(0)$ scales with T_c. It is found that H_c does not exceed a few thousand oersted even for superconductors with critical temperatures as high as 20°K.

Thus we see that type-I superconductors are inherently low field superconductors. In order to preserve superconductivity in high fields, it is necessary that somehow flux be allowed to penetrate into the bulk of the material. Clearly, in the case of a field parallel to a thin film of a type-I superconductor with thickness $d < \lambda$, the flux does effectively penetrate, and correspondingly the critical field is found to exceed the bulk thermodynamic critical field H_c. Specifically,

$$H_c(\text{thin film}) = 2\sqrt{6}\,H_c(\lambda/d) \tag{4}$$

This means of attaining high field superconductivity has not been used in any technologically important application, although (porous) vycor glass impregnated with mercury does have a very high critical field. The practical road to high field superconductivity is described in the next section.

2. Type-II Superconductors and Quantized Flux Lines

One might fairly ask the question, if it is energetically so favorable to let flux into a superconductor, why do type-I superconductors not spontaneously break up into superconducting and normal regions with the field passing through the normal regions? The reason they do not is that (indeed by definition) type-I superconductors have a positive surface energy at such superconducting–normal (s–n) interfaces. In the late 1950s and early 1960s it became apparent that superconductors with negative s–n surface energies existed. These are called type-II superconductors. They do allow flux to penetrate, and they can have very high critical fields. Typically, superconducting alloys and intermetallic compounds are type-II superconductors. Niobium is a borderline case, but is type-II. Also, virtually all high T_c superconductors known are type-II superconductors, although there is no known fundamental reason why this should be so.

The free energy and B–H curves typical of type-II superconductors are shown in Fig. 2, along with those of a type-I superconductor discussed above. At low fields, flux is excluded as in a type-I. However, at a field H_{c1}, the so-called lower critical field, flux penetrates. The fully normal state is not reached, however, until a usually very much higher field H_{c2}. In the mixed or vortex state between H_{c1} and H_{c2}, it is now known that the flux penetrates in the form of discrete current vortices with a total associated magnetic flux that is quantized. This is schematically illustrated in Fig. 3(a). The structure of an isolated vortex is shown in Fig. 3(b) and discussed in greater detail below. The quantum of flux is $\Phi_0 = hc/2e = 2.07 \times 10^{-7}$ G cm². In terms of Φ_0, the magnetic induction B of the material can be expressed $B = n\Phi_0$, where n is the local density (number per unit area) of vortices.

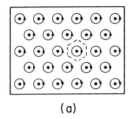

(a)

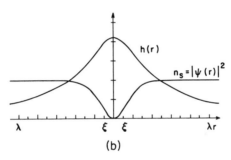

(b)

FIG. 3.   (a) Quantized vortex lattice of a type-II superconductor. Figure shows normal cores surrounded by circulating currents. Dashed line indicates unit cell of lattice. The area of this unit cell contains one quantum Φ_0 of magnetic flux. (b) Structure of isolated vortex. The normal core and the decay of the magnetic field away from center are clearly evident. Curves shown are for $\kappa = 10$.

While superconductivity persists to high fields in type-II superconductors, a heavy price has been paid. Namely, above H_{c1} when B no longer equals zero and vortices are present, type-II superconductors exhibit dc resistance due to the motion of these vortices (a dissipative process) under the influence of an applied current. In order to prevent this dissipation, it is necessary to pin the vortices on physical or metallurgical inhomogeneities, in analogy with dislocation pinning and Bloch wall pinning in materials with high mechanical strength and in permanent ferromagnets, respectively. For suitable engineering materials, even this is not enough, however. It is also necessary to form the material properly (usually in small filaments) and make it part of composite material with a normal metal in order to render the material stable against premature transitions into the normal state due to thermal instabilities. Additional procedures are required to insure low ac losses. These matters are discussed in greater detail below. First, we review the important intrinsic properties of type-II supesconductors in greater detail. More complete discussions of the properties of type-II superconductors are available in several standard references [12a–d].

B. The Ginzburg–Landau Theory and the Characteristic Lengths of Superconductivity

The quantized nature of the current vortices in a type-II superconductor illustrated schematically in Fig. 3(a) is a direct manifestation of the macroscopic quantum nature of superconductivity. As shown in the BCS theory, superconductivity arises from an ordering of the electrons into a macroscopic quantum state of paired electrons—the famous Cooper pairs. The range ξ_0 over which any given pair of electrons remains correlated—a kind of microscopic dimension of a Cooper pair—is known as the BCS coherence length and is one of the important characteristic lengths in superconductivity. As shown by BCS,

$$\xi_0 = 0.18 h v_F / k T_c \tag{5}$$

where v_F is the Fermi velocity of the electrons and T_c the zero-field transition temperature of the superconductor.

In practical superconducting materials, it is rarely necessary to have a detailed microscopic picture of the electronic behavior. It is sufficient and more illuminating in many ways to describe the superconductor in terms of a macroscopic wave function ψ of the center of mass motion of the pairs. The density of superconducting pairs then becomes, for example, $n_s^* = |\psi|^2$, where the asterisk denotes that we refer to pairs of electrons.

The simplest theory of superconductivity which is developed in these terms is the Ginzburg–Landau (GL) theory [2]. Originally introduced totally phenomenologically, we now know from the work of Gorkov [13] that the GL theory follows from the BCS theory under certain conditions. We should also mention that using the GL theory Abrikosov [14] predicted nearly all the essential features of type-II superconductors even before their existence was clearly established experimentally.

The customary starting point of the GL theory is an expansion of the free energy functional of a superconductor in a kind of Taylor series in terms of the order parameter or pair wave function ψ,

$$\Delta F(\psi) = F_s(\psi) - F_n$$

$$= \int d^3r \left\{ \alpha(T)|\psi|^2 + \frac{\beta}{2}|\psi|^4 + \frac{\hbar^2}{2m^*}\left|\left(\frac{\nabla}{i} - \frac{e^*}{\hbar c}\mathbf{A}\right)\psi\right|^2 + \frac{H^2}{8\pi} \right\} \tag{6}$$

Here $\alpha = \alpha_0(T - T_c)/T_c$ and $\beta > 0$ are material-dependent parameters, $e^* = 2e$, by the usual convention $m^* = 2m$, and $\mathbf{A}$ is the vector potential $(\mathbf{H} = \nabla \times \mathbf{A})$. The first two terms in $\Delta F(\psi)$ give the gain in free energy due to the superconducting order. The gradient term gives the increase in that energy due to spatial variations in n_s^* (i.e., $\nabla|\psi|$) and that due to the kinetic energy of the supercurrents. Minimization of the free energy ΔF when no gradients, currents or fields are present leads to the result that $n_s^* = -\alpha/\beta$, and the gain in free energy in this case, which then by definition equals $H_c^2/8\pi$, is given by $\Delta F = \alpha^2/2\beta$.

Variation of the functional $\Delta F(\psi)$ with respect to arbitrary spatial variations of ψ^* and $\mathbf{A}$ yields the famous GL equations

$$\frac{\hbar^2}{2m^*}\left(\frac{\nabla}{i} - \frac{e^*}{\hbar c}\mathbf{A}\right)^2\psi + \beta|\psi|^2 = -\alpha(T)\psi \tag{7a}$$

and

$$\frac{c}{4\pi}(\nabla \times \mathbf{H}) = J_s = \frac{e^*\hbar}{2m^*i}(\psi\nabla\psi^* - \psi^*\nabla\psi) - \frac{e^{*2}}{m^*c}|\psi|^2 A \tag{7b}$$

which together with Eq. (6) provide the complete GL theory of superconductivity. Here J_s is the supercurrent flow due to the superconducting pairs. Note that Eq. (7b) is the familiar quantum mechanical expression for the current density due to a particle of charge e^* and mass m^* described by a wave function ψ.

These equations provide the principal framework in which the analysis of the static properties of type-II superconductors proceeds. More generally, they are useful in situations in which ψ varies spatially or when $|\psi|$ is reduced from its usual value $-\alpha/\beta$ (commonly referred to as depairing) due, say, to the presence of strong fields and/or currents. They are valid only when the expansion is valid, namely, when $|\psi|$ is small and when the gradients of ψ vary slowly in space. In practice, this means when T is near T_c, and in small fields, although qualitatively the GL theory has a much wider range of usefulness. When these conditions are not met and a quantitative theory is required, it is necessary to include higher order terms or to resort to the complete Gorkov theory, which is equivalent to BCS. The Gorkov theory also provides the connection between the material-dependent parameters (e.g., α and β) of the GL theory and the microscopic normal electronic properties of the material.

Examination of Eq. (7a) indicates that the characteristic length scale or healing length for variations in ψ, the so-called GL coherence length, is

$$\xi(T) \equiv \frac{\hbar^2}{2m|\alpha|} \equiv \xi(0)\left(\frac{1}{1-t}\right)^{1/2} \tag{8}$$

where again we have used the notation $t = T/T_c$. This coherence length for ψ is strongly temperature dependent, diverging as $T \to T_c$. It results from an energy balance between the condensation energy gained by the superconductor (which goes to zero as $T \to T_c$) and the kinetic energy associated with gradients of ψ. This is in contrast to the "dimensions of the pair" ξ_0, given by Eq. (5), which is temperature independent. The length $\xi(0)$ is known as the zero-temperature GL coherence length and is a useful material parameter for the superconductor.

Taking the curl of both sides of Eq. (7b), we obtain the result

$$\nabla^2 \mathbf{H} = -\nabla \times \nabla \times \mathbf{H} = \lambda^{-2}\mathbf{H} \tag{9}$$

where $\lambda^{-2} = 4\pi e^{*2} n_s^*/m^* c^2$. Solutions of Eq. (9) are of the form $e^{-x/\lambda}$ and thus the identification of λ as the penetration depth introduced above is evident.

The lengths ξ_0, $\xi(T)$ and $\lambda(T)$ are the important length scales in superconductivity of interest in this review. The ratio $\kappa \equiv \lambda/\xi$ also plays a particularly important role since it determines whether a superconductor is of type-I or type-II. For $\kappa < 1/\sqrt{2}$, the surface energy (as calculated using the GL theory) is positive, and the superconductivity is type-I. For $\kappa > 1/\sqrt{2}$, the surface energy is negative, and type-II superconductivity pertains.

The relationships between these lengths and the microscopic material parameters as derived from the Gorkov theory for T near T_c (or $T/T_c = t \lesssim 1$) are:

$$\xi(T) = 0.74\xi_0/(1 - t)^{1/2} \qquad \text{clean limit } (l \gg \xi_0) \tag{10a}$$

$$\xi(T) = 0.86(\xi_0 l)^{1/2}/(1 - t)^{1/2} \qquad \text{dirty limit } (l \ll \xi_0) \tag{10b}$$

and

$$\lambda(T) = \lambda_L(0)/[2(1 - t)]^{1/2}$$
$$\qquad \text{clean limit (and local electrodynamics [15])} \tag{11a}$$

$$\lambda(T) = \lambda_L(0)(\xi_0/1.33l)^{1/2}/[2(1 - t)]^{1/2} \qquad \text{dirty limit} \tag{11b}$$

Here l is the electron mean free path and $\lambda_L(0) = (mc^2/4\pi ne^2)^{1/2}$ is the so-called London penetration depth, where n is the total number density of single electrons in the material.

As seen from these results, the London penetration depth $\lambda_L(0)$ sets the basic length scale for λ in superconductors. The temperature dependence arises from the temperature dependence of the pair density n_s^* (see the GL expression for λ above). Actual penetration depths are frequently larger than $\lambda_L(0)$, however, as seen for alloys in Eq. (11b), due to nonlocal electrodynamics (as discussed in Tinkham [12b]), and also if there is depairing so as to reduce n_s^* from its usual value.

From these relations we see immediately that if $l \rightarrow 0$, then $\xi \rightarrow 0$, $\lambda \rightarrow \infty$, and hence $\kappa \rightarrow \infty$. Thus, it is evident why most alloys are type-II superconductors. Most intermetallic compounds are also type-II superconductors, but the reason is because ξ_0 is intrinsically small, probably due to a low Fermi velocity [see Eq. (5)] in these metals.

For alloys well in the dirty limit, Eqs. (10b) and (11b) for ξ and λ can be rewritten in terms of more readily measured quantities. Specifically,

$$\xi(t) \cong 0.86 \times 10^{-6}(\rho_n \gamma T_c)^{-1/2}/(1-t)^{1/2} \tag{12}$$

$$\lambda(t) \cong 0.91 \times 10^{-2}(\rho_n/T_c)^{1/2}/[2(1-t)]^{1/2} \tag{13}$$

where ξ and λ are given in centimeters and ρ_n (Ω-cm) is the normal state resistivity and γ(erg cm^{-3} °K^{-2}) is the normal state electronic specific heat coefficient.

C. The Structure of the Vortices and the Critical Fields of Type-II Superconductors

Certainly, the most characteristic feature of type-II superconductors is the presence of current vortices, each containing a single quantum Φ_0 of magnetic flux. The structure of an isolated vortex as calculated from the GL theory is shown in Fig. 3(b). There is a "normal" core ($|\psi|^2 \cong 0$) of radius $\cong \xi(T)$, and outside this core the magnetic field associated with the vortex falls off exponentially with a range determined by the magnetic penetration depth λ. At the center of an isolated vortex the field is $\sim \Phi_0/2\pi\lambda^2$, as follows from the requirement that the total flux associated with the vortex equal Φ_0. As seen just below, this field is $\sim 2H_{c1}$.

Typical critical field behavior for a type-II superconductor is shown in Fig. 4. (The region labeled surface superconductivity will be discussed below.) Near T_c and for $\kappa \gg 1$, H_{c2} and H_{c1} are found from the GL theory to be given by

$$H_{c2} = \sqrt{2}\,\kappa H_c = \Phi_0/2\pi\xi^2 \tag{14}$$

$$H_{c1} = (H_c/\sqrt{2}\,\kappa)\ln\kappa = (\Phi_0/4\pi\lambda^2)\ln\kappa \tag{15}$$

As seen from the figure, the first equality for both H_{c2} and H_{c1} is qualitatively correct at all temperatures (i.e., H_{c2} and H_{c1} have roughly a parabolic

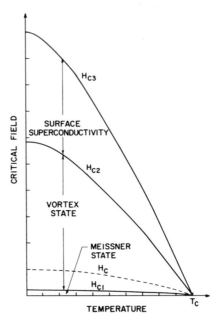

FIG. 4.   Temperature dependence of the critical fields of a type-II superconductor. For a type-II superconductor the bulk thermodynamic critical field H_c retains its definition $H_c^2/8\pi \equiv$ condensation energy, but has no further physical significance.

temperature dependence). The more precise results obtained from the Gorkov theory require a slightly temperature-dependent κ defined separately for H_{c1} and H_{c2}. Further refinements are necessary for clean materials. The resulting κ's reduce to a common value as $T \to T_c$, however, as they must be consistent with the GL results quoted above. The GL result for the upper critical field H_{c2} has a simple physical interpretation. It corresponds to the field at which the normal cores of the vortices begin to overlap and hence the material becomes essentially full of normal electrons. In fact, as $H \to H_{c2}$, n_s^* goes continuously to zero and the transition to the normal state is second order in the thermodynamic sense. If this were not the case, the simple GL theory would clearly not apply.

In the case of very high-κ superconductors for which H_{c2} becomes quite high, one must take into consideration the Pauli paramagnetism of the metal [16]. In the simple BCS picture the electrons in a Cooper pair are paired with opposite spins. Thus, as H increases, eventually the Zeeman energy μH of the electron with its spin aligned against the field exceeds the binding energy of the pair and the precise pairing required for superconductivity becomes energetically unfavorable. This leads to a Pauli paramagnetic limiting field H_p such that $\mu H_p \cong \Delta$, where μ is the magnetic moment of the electrons and

Δ is the superconducting energy gap or half the binding energy per pair of electrons. Detailed calculations for $T = 0$ give

$$H_p(0) = \sqrt{2}\Delta(0)/\mu = 18.4 \text{ kOe} \times T_c \tag{16}$$

where $\Delta(0) = 1.76k_B T_c$ as shown by BCS.

An important modification of this result is usually required in practice, however, due to spin–orbit scattering from impurities. When such spin–orbit scattering is present, spin is no longer a good quantum number and Cooper pairing no longer involves strictly spin-up and spin-down pairs. When the spin–orbit scattering is strong, it is found that

$$H_p^{(so)}(0) = (\hbar/6\tau_{so}\Delta(0))^{1/2}H_p(0) \tag{17}$$

which can be much larger than $H_p(0)$. Here τ_{so} is the spin–orbit scattering relaxation time. While Pauli paramagnetic limiting has been observed to reduce the observed critical fields in some special cases, it has not been so far a serious impediment to high field superconductivity in compounds or alloys containing niobium or high atomic number elements for which one expects strong spin–orbit scattering.

D. Surface Effects

When a magnetic field is applied parallel to the surface of a superconductor, several important effects take place that can significantly influence the overall properties of the superconductor. First, under these conditions the presence of a surface tends to stabilize the superconducting state. A novel manifestation of this fact is that superconductivity can persist near the surface [to a depth $\sim \xi(T)$] even for fields above H_{c2}. (When the field is perpendicular to the surface, this surface effect does not exist.) As can be shown using the GL theory, surface superconductivity is not destroyed until the applied field reaches a field $H_{c3} = 1.69H_{c2}$ (see Fig. 4). In addition, this surface region or surface sheath of superconductivity can carry a finite lossless supercurrent. In fact, the surface sheath retains its ability to carry a lossless surface current even below H_{c2} where vortices are present in the interior of the superconductor.

At low fields in the Meissner state below H_{c1} surface effects are also important. Here the surface exhibits a barrier for vortex entry that can in principle delay the entry of the flux into the bulk of the superconductor up to fields very much in excess of H_{c1}. After flux has entered, this surface barrier then impedes the exit of vortices and consequently can lead to hysteresis in the magnetic behavior. The current-carrying capacity of the surface sheath at high fields also can lead to hysteresis, and at some basic level these two surface effects must be related, although the relationship has never been developed in detail theoretically. For a high-κ superconductor this surface barrier can theoretically delay the entry of flux until $H = 0.75H_c$. Even larger barrier fields exist for low-κ materials. In practice, such large barrier fields are rarely observed, however.

E. Vortex Pinning and the Critical State

As discussed before, superconductors only exhibit truly zero resistance when in the Meissner state $(B = 0)$. In the vortex state the presence of a net current (either a transport current or a magnetization current) produces a Lorentz force $\mathbf{J} \times \mathbf{\Phi}_0 /c$ (per unit length) on the vortices (analogous to the macroscopic Lorentz force density $\mathbf{J} \times \mathbf{B}/c$), where $\mathbf{J}$ is the current density evaluated at the center of the vortex. The subsequent motion of the vortices, usually referred to as flux flow, is a dissipative process and results in a net electrical resistance in the superconductor. A complete understanding of this dissipation is not yet available, and the precise origins of flux flow resistance remain one of the outstanding incompletely solved problems of basic super-conductivity, although clearly a good part of it arises from Joule heating of the normal electrons in the cores of the vortices. By pinning the vortices on physical or metallurgical defects, this vortex motion can be prevented up to some critical current density J_c which depends on the strength of the pinning. The electrical resistance is zero only with dc, however. For ac, hysteretic losses arise which can only be minimized with considerable difficulty, as will be discussed later.

It is not difficult to see how flux pinning can arise. For example, local variations of T_c (or equivalently the condensation energy $H_c^2/8\pi$) through-out the superconductor will lead to specific free energy minima for the vortices, since clearly it is energetically favorable to have the "normal" cores of the vortices located where the superconducting condensation energy is minimal. Even local changes in electron mean free path can produce pinning through the resulting changes in the GL coherence length ξ and con-sequently in the diameter of the vortex core. Two important examples of pinning barriers are dislocation cell walls (important in NbTi alloys) and grain boundaries (important in Nb_3Sn).

While it is easy to imagine how pinning can arise, and in fact to estimate the pinning strength of a particular defect, it is a much more subtle proposi-tion to calculate the resulting critical current. The difficulty arises because when the vortices overlap, as is the case in high fields, they tend to act collectively and move in so-called flux bundles rather than individually. As a result, the net pinning force is not a simple sum of the individual pinning interactions. For a knowledgeable and thorough account of flux pinning in superconductors, see Campbell and Evetts [17] and the references cited therein.

Fortunately, these complicated details can be avoided in practice by using the critical state model first introduced by Bean [18]. In this model, it is assumed that beginning at the surface, flux penetrates (or leaves) the superconductor in such a way that at every point inside the material to which the flux has penetrated, the gradient of the field (or the magnitude of the current, since $J \propto dB/dx$) is at its critical value. This picture follows immediately from the observation that if the gradient were less than the critical value, the flux would be pinned firmly and therefore would not move at all. If the gradient were larger, flux would quickly flow until the gradient relaxed to that value which could just be sustained by the pinning barriers. For simplicity, Bean assumed J_c to be a constant.

Some typical contours of B as a function of position for an increasing (solid lines) and then decreasing (dashed lines) applied field are shown in Fig. 5 for the case of the Bean model and the field applied to a slab of thickness d. The field $H_s = 2\pi J_c d/c$ is of particular importance since it represents the applied field at which in the Bean model flux first completely penetrates the superconductor. For simplicity H_{c1} and any possible surface barrier have been neglected in plotting Fig. 5.

Subsequent workers have generalized the critical state model to cases where J_c is a function of the local induction B. The most prominent of these is the Kim–Anderson model [19] in which J_c has the form $J_c = \alpha/(B + B_0)$, which correctly reflects the property that J_c is a decreasing function of B.

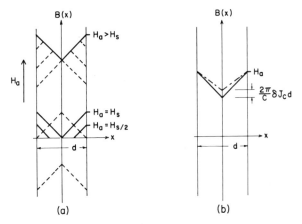

FIG. 5.   (a) Field contours $B(x)$ in a flat superconducting slab for the Bean model. For increasing applied fields flux penetrates from the surface (solid lines) eventually reaching the center of the slab at the field $H_s = 2\pi J_c d/c$. When the applied field is reduced, flux first leaves near the surface (dashed lines). Note that when applied field returns to zero, flux is trapped in the interior of the superconductor. (b) Change in field contour (solid-to-dashed line) after small decrease in J_c associated with a flux jump (see Section IIF).

Neither model properly shows that $J_c \to 0$ as $B \to H_{c2}$, however. For precise quantitative work experimental values of $J_c(B)$ can be used.

In the critical state the magnetic behavior of the superconductor becomes hysteretic. The $\bar{B}$–H curve resulting from the Bean model is shown in Fig. 6. Here $\bar{B} = (1/d) \int B(x)\, dx$ is the volume average of the induction for the slab geometry shown in Fig. 4. Also, we continue to neglect H_{c1} and any surface barrier. The more realistic relations for $J_c(B)$ and the inclusion of H_{c1} and/or a surface barrier lead to similar $\bar{B}$–H curves. We shall continue to use the Bean model, however, because of its computational simplicity.

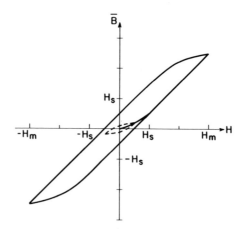

FIG. 6. Hysteresis loop calculated from Bean model for a type-II superconductor with flux pinning. $\bar{B}$ is the volume average of the magnetic induction.

Note that the existence of a finite critical current in the vortex state and magnetic hysteresis go hand in hand in a fundamental way. If one wants large J_c, one must live with the negative effects of hysteresis, as in ac losses discussed below.

F. Instabilities and ac Losses

Flux pinning does lead to superconductors with high critical currents, frequently even in very high magnetic fields. Strong pinning by itself is not enough to produce a practical engineering material, however. There are two other important considerations: stability and losses in ac or swept dc fields. We shall first consider the problem of instabilities.

The critical state is only metastable: in equilibrium $B(x)$ would be uniform. The question thus arises how stable is the critical state? At one extreme is the phenomenon of flux creep in which the critical state very slowly relaxes due to the thermal activation of the vortices over the pinning barriers. This process is analogous to creep in mechanically and ferromagnetically strong materials and is quite negligible in practice. Very important in practice, on the other hand, is the catastrophic collapse of the critical state and concomitant loss of zero resistance due to a thermal instability initiated by a small flux jump.

Imagine that locally in the superconductor there is a small temperature rise. Since, except for some very special cases, $J_c(T)$ is a decreasing function of temperature, J_c will be reduced by an amount $\delta J_c = (-dJ_c/dT)\,\delta T$, and the critical state will relax slightly (see Fig. 5b) generating heat in the superconductor. If this heat generated produces a temperature rise $\delta T' > \delta T$, thermal runaway occurs and the critical state collapses. Mechanical shock or small motion of the superconductor in the field can similarly lead to thermal runaway.

Using the Bean model it is straightforward to estimate the heat generated by a small flux jump. The heat generated per unit volume is given by $\int J_c \cdot E\, dt$, where E is the electric field produced as the flux moves into the material. The additional temperature rise due the initial change δT is

$$\delta T' = (\pi/3c^2)(d^2 J_c(-dJ_c/dT)\,\delta T/C) \tag{18}$$

where C is the heat capacity per unit volume of the superconductor.

The only convenient design variable in Eq. (18) is the thickness d of the superconductor. Thus, stability $(\delta T' < \delta T)$ requires

$$d^2 < (3c^2 C/\pi J_c)(-dJ_c/dT)^{-1} \tag{19}$$

When d satisfies this condition the superconductor is said to be intrinsically or adiabatically stable since no heat transfer away from the superconductor has been assumed. Taking NbTi alloys as an example, Eq. (19) leads to the requirement $d \lesssim 0.01$ cm for stability. As a rule stability is better at high temperatures (less than T_c, of course) due to the rapid increase in the lattice heat capacity $C \propto T^3$ with temperature.

In practice it is customary to imbed these filaments in a matrix of high conductivity normal metal, usually copper. This normal metal provides additional stability in a variety of ways. First, it provides a low resistance path for the applied current during a flux jump. Second, it provides good thermal conductivity between the filaments and the liquid helium coolant and at the same time slows down the flux motion (and the resultant rate of heat generation) by means of eddy current damping. Put in more quantitative terms, it is desired that heat diffuse more rapidly than magnetic flux. Since superconductors generally have both low thermal and electrical conductivity (in the normal state), the desired relationship of the net thermal and magnetic diffusivity can only be accomplished in a composite material incorporating a good normal metal conductor.

In some cases enough normal metal is used so that it can carry the entire current in the event of a failure of the superconductor, with sufficiently small joule heating that the temperature rise does not exceed the T_c of the super-conductor. This is the ultimate in conservative design and is known as cryostatic stabilization. It is normally quite undesirable because of the large amount of normal metal required and the resultant degradation of the net critical current density of the composite conductor. It has been used in the big bubble chamber magnets, however, where large volumes rather than high fields are of prime importance.

Additional requirements are placed on a superconductor with strong flux pinning if it is to experience substantial ac or rapidly swept dc fields. From Fig. 5 it is evident that in the critical state, superconductors are subject to hysteretic losses. In the case of the slab geometry for which that figure applies, the loss per cycle averaged over the volume of the superconductor is $Q = (1/4\pi d) \oint H \cdot dB \, dx = (1/4\pi) \oint H \cdot d\bar{B}$. Using the Bean model

$$Q = (1/2\pi)H_m H_s, \qquad H_m \gg H_s, \tag{20a}$$

$$Q = (1/6\pi)H_m{}^3/H_s, \qquad H_m < H_s, \tag{20b}$$

where H_m is the peak amplitude of the ac field. Note that Eq. (20b) applies even if there is a large dc bias field. For swept fields Eq. (20a) becomes $P = H_s \dot{H}/8\pi$, where P is the volume average of the power generated.

Equation (20a) shows that when the ac field fully penetrates, low losses can only be obtained by making d small (since $H_s \propto d$). This is in fact a considerably more demanding requirement on filament size than the stability requirement discussed above. Equation (20b) shows that when the ac field only partially penetrates, the loss per unit area $Qd/2 \propto J_c{}^{-1}$ and thus can only be reduced by increasing J_c. The first case applies to swept field magnets and the second, for example, to surface losses in an ac supercon-ducting power line or to ac ripple losses in a magnet.

In multifilamentary magnet conductors the presence of the normal metal matrix aggravates the losses unless additional precautions are taken. The problem arises because of the tendency of the normal matrix to allow circu-lating currents to flow between parallel filaments as a reaction to a time-varying applied field. This tendency is simply the composite's attempt to satisfy Lenz's law and results in greatly enhanced losses unless the filaments are decoupled by twisting or better yet fully transposing the filaments as they proceed along the conductor. Approximate calculations show that the filaments will be decoupled only for $\dot{H} < \dot{H}_1 \cong 2c\rho J_c d/l^2$, where l is the period of the transposition and ρ is the resistivity of the normal matrix. For $\dot{H} > \dot{H}_1$ the filaments are said to be coupled, and the losses are still given by Eq. (20a) but with the filament diameter d replaced by the diameter d' of the entire composite conductor. Thus, when the filaments become coupled, the losses increase by a factor d'/d. For a typical multifilamentary conductor ($d \cong 10 \ \mu m$, $d' \cong 0.1$ cm, and $l \sim 1$ cm) with a copper matrix, $\dot{H}_1 \cong 200\text{--}2000$ Oe/sec, and the losses increase by a factor of ~ 100 when the filaments become coupled. At very high sweep rates (typically greater than $\cong 10^6$ Oe/sec) the overall losses of the conductor became dominated by eddy current losses in the normal matrix. At high sweep rates, it is clear that eddy current losses must ultimately dominate since they increase as ω^2 (or $\dot{H}^2$), whereas the hysteresis-type losses associated with the superconductor increase only as ω (or $\dot{H}$).

The above relations are only approximate and not generally suitable for quantitative calculation of the losses. They do clearly indicate, however, the critical parameters involved in the design of practical, stable, low loss magnet conductors. Clearly one wants very small filaments with as short a transposition period as possible. A high matrix resistivity also helps but can be introduced only at the expense of stability. A small composite conductor diameter d' is also desirable.

When Eq. (20b) is applied to superconducting power lines, it must be modified to include the effects of H_{c1} and surface barriers. When these are included, the power generated per unit area on the surface

$$W = fQd/2 = 0, \qquad\qquad\qquad\qquad H_m < H_{c1} + \Delta H,$$

$$= \frac{fcH_m^3}{24\pi^2 J_c}\left(1 - \frac{H_{c1} + \Delta H}{H_m}\right)^2\left(1 + \frac{H_{c1} + 4\Delta H}{2H_m}\right), \qquad H_m > H_{c1} + \Delta H,$$

$$(21)$$

where f is the frequency and ΔH is the field change required to overcome the surface barrier. The first equation is obvious since there can be no losses until flux penetrates. The second equation shows that in the limit of large H_m the losses go to the Bean model result. As discussed later in this review, sufficiently low losses for superconducting power line application can only be obtained when there is a sufficiently large H_{c1} or surface barrier to reduce the losses substantially below the Bean level. Also, it is important to note that losses can arise even for $H_m < H_{c1} + \Delta H$ if surface irregularities produce local field enhancements due to demagnetizing effects.

III. OCCURRENCE AND PROPERTIES OF IMPORTANT AND POTENTIALLY IMPORTANT SUPERCONDUCTORS

Superconductivity is found to occur among the metallic elements in a systematic way throughout the Periodic Table (Fig. 1). Niobium is the element with the highest T_c, 9.4°K. It is intrinsically type-II with a high H_{c1}.

Figure 7 shows how heat capacity (alternatively, magnetization) curves can be used to produce a phase diagram for Nb. Below 6°K and a field of 1000 Oe niobium remains in the Meissner state. It is essentially lossless for dc and all ac applications involving photon energies well below the energy gap. For applications of superconducting materials at higher fields and temperatures than offered by niobium, one must turn to alloys and compounds.

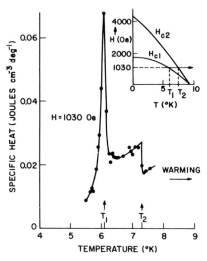

FIG. 7. Specific heat of niobium at 1030 Oe [after T. McConville and B. Serin, *Phys. Rev.* **140**, A1169 (1965)]. The singularity at 7.2°K is the transition from the normal to the vortex state, and that at 6.1°K is from the vortex to the Meissner state; $H_{c1}(6.1°K) = H_{c2}(7.2°K) = 1030$. The phase diagrams (inset; see also Fig. 4) can be constructed from a series of constant field heat capacities or from reversible magnetization isotherms, as illustrated in Fig. 2.

Those presently being used for the large scale applications of superconductivity contain niobium as a major constituent (except for V_3Ga which is useful for applications above 150 kOe). Other elements, namely, the soft nontransition metals such as lead, tin and indium, are used to produce quantum phase detectors (Josephson junction devices) which can provide very low level signal detection and fast signal processing.

The present section is confined to materials for large scale, energy-related applications and is divided into three parts corresponding to different classes of superconductors. First is the solid solution range of the transition metals. They are ductile, easily prepared as useful wire and have been available for some time, as can be seen in Fig. 8 (below the dotted line). Secondly comes compounds of niobium and vanadium in the A15 structure wherein the best superconducting properties are found [above the dashed line (Fig. 8)]. They are all brittle and present challenges to their synthesis and processing.

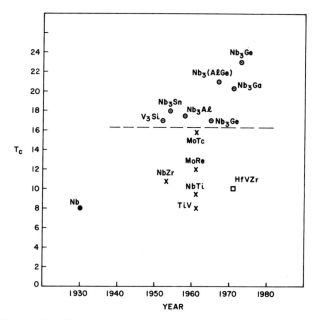

FIG. 8.   Chronological discovery and development of practical and potentially important superconductors.

Finally, we mention briefly some miscellaneous superconductors which are not at present serious candidates for commercial use but which have some unusual characteristics which give incentive for future development. We will also use the occasion to classify the materials according to microscopic parameters.

The basic superconducting parameters T_c, H_{c2} and J_c discussed in Section II set the operational limits of any technology. The three parameters can be used to define a critical surface below which superconducting technology can be utilized, as illustrated in Fig. 9.

The space defined is not to be considered the true phase diagram because of the materials-dependent nonintrinsic nature of H_{c2} and J_c. H_{c2}, it should be recalled, scales as $(l)^{-1}$ for dirty (real) superconductors [Eqs. (10b) and (14)], while J_c depends upon the size and distribution of pinning centers which can be distributed as second phase, nonsuperconducting precipitates, or grain boundaries. It can be seen (Fig. 9) that the technology based upon ductile Nb alloy wires must operate at temperatures near the boiling point of He (4.2°K). A15 compounds offer promise of extending the temperature of operation to $\gtrsim 14°$K (liquid hydrogen) or superheated helium, if desired. At present, however, only operations in liquid or pressurized helium gas near 4°K are being seriously considered.

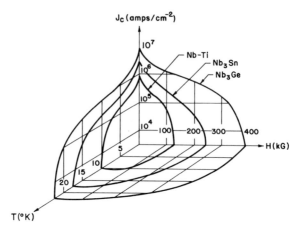

FIG. 9.   Parameter space for the practical superconductors Nb–Ti and Nb$_3$Sn and the potentially important Nb$_3$Ge (after Gavaler *et al.* [20]).

Upper critical fields somewhat in excess of 100 kOe are attainable in Nb-based ductile alloys and in excess of 250 kOe in the A15 structures. In practical magnet technology where transport currents of the order of 10^5 A/cm^2 are required, the ductile alloys are limited to uses below about 80 kOe or up to perhaps 100 kOe at lower temperatures (say $\sim$ 2.3°K) while the A15s are potentially useful to 150 kOe (Nb$_3$Sn) or 180 kOe (V$_3$Ga).

A.   Ductile Alloys

The occurrence of superconductivity in the alloys of the transition metals can be predicted according to Matthias [6] from the Periodic Table, that is, from the average number of valence electrons per atom. This rule has led to the discovery of numerous new superconductors. The first comprehensive study of solid solution alloys by Hulm and Blaugher [21] provided the base from which the technology of ductile superconducting wire and cable has evolved. The body-centered solid solutions were investigated from groups IV to VII of the Periodic Table and two distinct maxima in T_c were found. The first is for a less-than-half-filled d-shell with an average valence number of 4.7. An updated version of the data for the elements and near neighbor alloys is shown in Fig. 10. The Matthias rule can be used as a guide to predict, for example, that T_c will be increased if tantalum is added in small quantities to Nb–Ti solutions with > 50% Ti, an addition which might be undertaken to benefit from the enhanced spin–orbit coupling which in turn increases H_{c2} (see Section II).

The maximum T_c reached ($\sim$ 14°K) in the binary ductile solutions can be seen to occur in the Mo–Re system right at the bcc phase boundary.

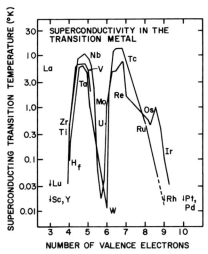

FIG. 10. Variation of T_c among the elements and near neighbor alloys of the transition metals [after Gladstone *et al.*, see Parks [12d], where references are given].

Mo–Re ductile wire was used to make the first high field prototype magnet out of ductile alloy wire and to illustrate that its properties could be predicted from measurements of $J_c(H)$ on short test samples [7]. Typically, J_c versus H measurements show the knee evident in Fig. 11 beyond which

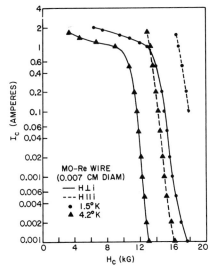

FIG. 11. Critical current–critical field isotherms for Mo–Re alloy displaying the typical knee which defines the maximum useful parameter limits more realistically than Fig. 9 (after Kunzler [7]).

the critical current drops so rapidly that a practical operating limit is reached. Kunzler was able to use the data from Fig. 11 to design a magnet that produced a predicted 15 T. Even though Mo–Re system reaches a higher T_c's than the Nb–Ti and Nb–Zr systems (9.8°K and 10.7°K), it has a much lower H_{c2}. The latter would be substantially raised if a suitable ternary component could be found which increased the residual resistance and hence H_{c2} according to Eqs. (12) and (14). Iron would not be suitable, however, because it forms magnetic state in Mo–Re which makes even traces of it very detrimental to the superconductivity.[1] In contrast, niobium-based alloys can be handled with less precaution because in them iron is nonmagnetic and has little effect. Thus, even aside from cost, which ultimately would preclude the large scale commercial use of rhenium, it can be seen that T_c alone is not the determining property in deciding which of the many known superconductors have commercial importance.

The first generation magnet wire was produced from Nb–Zr alloys. Figure 12 shows the phase diagram from which processing procedures can

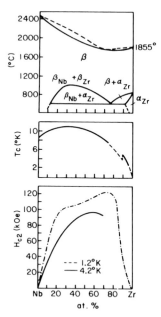

FIG. 12. Phase diagram, superconducting transition temperature, and critical magnetic field of Nb–Zr alloys (after Savitskii *et al.* [22], and the references given therein).

---

[1] Impurities with well-defined magnetic moments reduce T_c and other superconducting properties drastically because scattering off them does not obey time-reversal symmetry and thus destroys the Cooper pairing. See, for example, Tinkham [12c] or Parks [12d, Chapter 18].

be understood. Annealing below 1000°C can be used to precipitate niobium rich β (bcc) and zirconium rich α (hcp) phases from completely homogeneous solid solutions. The zirconium rich phase becomes normal at much lower values of T_c, H_{c2} and J_c and thus, if advantageously dispersed, can serve as pinning centers. Nb–Zr might find application in low fields, such as for handling fault (overload) currents in power line applications, since it has a high J_c near zero field. Otherwise, it has been superseded by Nb–Ti.

Niobium–titanium has the best properties for making ductile multifilamentary cable and is the most prevalent superconductor today. Figure 13 shows a complete bcc solid solution range above 885°C. Upon

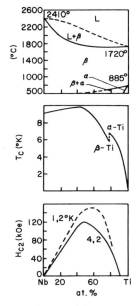

FIG. 13. Phase diagram, superconducting transition temperature, and critical magnetic field of Nb–Ti alloys (after Savitskii *et al.* [22]).

slow cooling, as can be seen from the phase diagram, a bcc solution, say 90 at.% with $T_c = 4.5°$, will decompose into an α (hcp) titanium solution, $T_c < 1°K$ and a β bcc solution $\sim 70\%$ titanium with a $T_c \sim 6°K$. Figure 14 shows the effect of a slow cool on the measured T_c. Superconducting shielding currents, perhaps flowing through a minor amount of higher T_c phase, can cause the whole sample to appear superconducting when the measurement is made by electrical means. Heat capacity measurements, on the other hand, are sensitive to the atom fraction condensing into the superconducting state and thus would give a broad or perhaps resolved double transition on

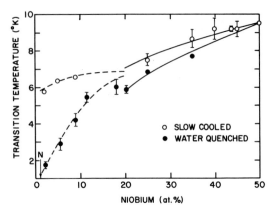

FIG. 14. Transition temperature versus composition for titanium–niobium alloys prepared by different types of heat treatment (after Hulm and Blaugher [21]).

slowly cooled samples. It should be appreciated that small precipitates of normal material make good pinning centers if of the proper size and distribution, and thus can increase $J_c(H)$.

One can always hope to find an optimum ternary system, for instance one which combines the higher T_c and J_c of the Nb–Zr system with the higher H_{c2} (and better technological properties) of Nb–Ti. It can be seen from Figs. 15 and 16 that a slight decrease in T_c is accompanied by a measurable increase in H_{c2} when zirconium is first added to $Nb_{0.4}Ti_{0.6}$. A number of other bcc ternaries have been investigated, but none seems yet to have demonstrated sufficient advantage to overcome the inherently more complex processing of the ternary system.

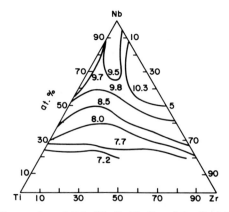

FIG. 15. Curves of equal T_c in Nb–Zr–Ti alloys (after Savitskii et al. [22]).

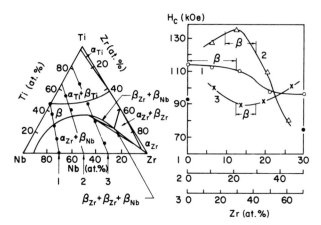

FIG. 16.   Isothermal cross section of the Nb–Ti–Zr system at 550°C and values of the critical magnetic fields for alloys lying along sections 1, 2 and 3 annealed at 550°C for 1–3 hr (after Savitskii *et al.* [22]).

Nb–Ti ∼ 63 at.% is presently the most widely used alloy [23]. Both $H_{c2}(0)(\sim 120 \text{ kOe})$ and T_c are slowly varying functions of composition and are unaffected by impurities introduced in processing. The chief metallurgical challenge (before considering stability of composites discussed in the next section) is to increase J_c. This is achieved by a combination of cold working in which the cross-sectional area is reduced by more than 99% and low temperature ($\lesssim 400°C$) heat treatments. There are conflicting reports [24] as to the precise microstructural state, but present evidence seems to indicate that the 63 at.%Ti alloy is single phase bcc. (Unworked alloys with < 60 at.%Ti undergo a transition to a metastable ω phase with heat treatments below 500°C and also may have some hcp precipitate.) The cold work leads to a well-defined structure with pencil-shaped cells of roughly 450 A in diameter. During the low temperature anneal, the initial stages of recovery occur and the dislocations move from the cell interiors to the cell walls [25].

The sensitivity of the J_c of $Nb_{0.4}Ti_{0.6}$ to a low temperature anneal is illustrated in Fig. 17. The curves are for drawn wire with an area reduction of 5×10^4. It seems likely that the cell structure is responsible, even though there is no obvious change in the cell morphology with anneal accompanying the marked change in pinning. Hampshire and Taylor [25] show that variations in κ (the Ginzburg–Landau parameter of Section IIB) of the order of 0.1 (κ itself is ∼ 40) between cell and cell wall can account for the pinning of the optimum (385°C anneal) material and suggest that rearrangement of dislocations within the cell walls can account for the sharp decrease in pinning observed after the 500°C anneal. Zubeck [26] has found by direct

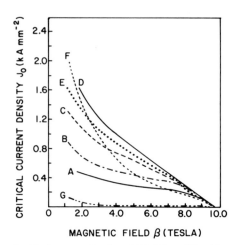

FIG. 17. Variation of critical current density of Nb–Ti at 5°K with magnetic field and heat treatment. A, the as-worked specimen; B, C, D, E, F and G, annealed for 1 hr at 250°C, 300°C, 385°C, 450°C, 500°C and 600°C, respectively. Note the precipitous fall-off between the 385°C anneal (Curve D) and the 500°C anneal (Curve G) (after Hampshire and Taylor [25]).

heat capacity measurements that κ in heavily cold-worked pure niobium varies by about 0.02.

A composition richer in Ti, Nb–Ti 70 at. %, is used for $H < 40$ kOe applications because it has a higher J_c at low fields. In this composition precipitates (?) of the α-phase are found and presumably are responsible for the increased pinning at low fields. The various models [24, 25] which have been proposed to account for the pinning might have some predictive value but are still in the process of being tested. The difficulty arises as we have already noted in describing the vortex–vortex interactions.

B. A15 Compounds

Ordered intermetallic A_3B compounds of the A15 or beta–tungsten structure have the highest T_cs of any known superconductors and also they can have very high H_{c2}s and J_cs. They will undoubtedly play a major role in applications of superconductivity for energy related uses. Of the 70-odd known compounds with the A15 structure [27], those with $T_c > 15°K$ are listed in Table 2 along with some other useful properties.

In addition to unique superconducting properties, A15 structures frequently possess anomalies in their electrical, elastic, and structural properties. A great deal of experimental and theoretical effort has gone into investigating these anomalies, partly because of their bearing upon metal physics in general and partly because of the belief that they must somehow be related

TABLE 2

High Field Superconducting Materials Having Potential for Technological Application

Material	Crystal structure	T_c (°K)	H_{c2} at 4.2°K (kOe)	J_c at 4.2°K (A/cm^2)	Reference[a]
V$_3$Ga	A15	16.5	22	3×10^5 (at 6.5 kOe)	b
V$_3$Si	A15	17.0	24		
Nb$_3$Sn	A15	18.3	23	$> \begin{cases} 5 \times 10^6 \text{ (in 0 field)} \\ 3 \times 10^5 \text{ (at 10 kOe)} \end{cases}$	c
Nb$_3$Al	A15	18.9	33	$> 2 \times 10^5$ (at 14 kOe)	d
Nb$_3$Ga	A15	20.3	34	5×10^5 (at 7 kOe)	e
Nb$_3$(Al$_{0.8}$Ge$_{0.2}$)[f]	A15	20.5	41	1×10^4 (at 12 kOe)	
Nb$_3$Ge	A15	23.2	38	$> 1 \times 10^4$ (at 20 kOe)	g

[a] Data taken from review articles such as those listed below.

[b] K. Tachikawa, K. Itoh, and Y. Tanaka, *IEEE Trans. Magn.* **MAG-11**, 240 (1975).

[c] Stanford University Superconducting Power Transmission Line Project (to be published).

[d] T. W. Eagar and R. M. Rose, *IEEE Trans. Magn.* **MAG-11**, 214 (1975).

[e] Webb and Englehardt [28].

[f] Other ternaries could also be listed here with high T_cs, but none other than Nb$_3$(Al,Ge) has been extensively studied insofar as J_c and H_{c2} are concerned.

[g] Gavaler *et al.* [29].

to the high T_c. An understanding of the relationship might suggest ways of obtaining even more favorable superconducting properties.

The structure itself has some special features [30, 31]. The A sites (Fig. 18) run throughout the crystal forming three sets of orthogonal nonintersecting chains. The near neighbor distances along the chains are $\sim 10\%$ shorter

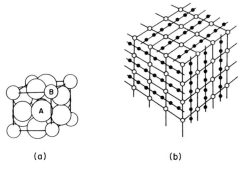

(a)   (b)

FIG. 18.   (a) Unit cell of the cubic A15 (beta–tungsten or Cr$_3$Si) structure. (b) Extended arrangement of atoms in the stoichiometric A15 structure emphasizing the three orthogonal, nonintersection linear chain of A atoms.

than expected from the elemental metallic radius of the A atom, which, in the high T_c materials, is either niobium or vanadium. All the high T_c A15 structures have nontransition elements on the B sites, although many A15 compounds also exist where the B atoms are transition metals. It appears that once a transition metal with available d orbitals occupies the B sites, the special properties of the A15 structure are lost. Geller has derived a set of radii for the A15 phases based upon the premise that the A–B distance determines the size of the unit cell, and thus, that the lattice parameters of all the compounds can be specified by assigning an arbitrary, but self-consistent, value for each A and B atom. These radii reproduce the observed unit cells to within 3% and are quite useful in identifying new A15 phases and in studies of stoichiometry. An enlarged set of Geller radii has recently been published [32]. For example, it was possible to devise means of raising the T_c of Nb_3Ge from 6°K to 17°K (and subsequently to 23°K, see below) [33] by recognizing that the measured lattice constant as prepared from the melt was larger than predicted from the Geller radii (5.17 Å and 5.14 Å). The method of rapid quenching was used to increase the Ge concentration which, in turn, brought a distribution in cell size that included its predicted value. However, a great deal of disorder was also quenched in (i.e., A atoms on B atom sites and vice versa), and a very broad transition resulted. The broadening of the transition with disorder in Nb_3Sn had been previously documented by Hanak *et al.* [34] using material prepared by chemical vapor deposition (CVD).

Studies have shown [35] that if the B atom is a nontransition element T_c increases rapidly with order, presumably reaching a maximum value when the A atoms are all on the A sites and the B atoms are on the B sites. Irradiation by neutrons and heavier particles lowers T_c dramatically [36]. Doses in excess of 10^{19} neutrons/cm^2 reduce the transition of Nb_3X compounds 10°K or more without broadening it. When T_c for Nb_3Sn is reduced from 18°K to ~ 3°K, the long range order parameter[2] is estimated to drop from ~ 1 to 0.6. Annealing at moderate (700°C) temperatures restores the original transition. The results of Sweedler and Cox [37] for Nb_3Al are shown in Fig. 19.

Gavaler [38] and subsequently Testardi *et al.* [39] have shown that Nb_3Ge can be prepared by sputtering, resulting in what is at present the highest known $T_c \cong 23°K$. The resulting thin films are not necessarily single phase nor has it been possible to characterize the high T_c material as to order and stoichiometry. More recent developments have shown that CVD [40]

---

[2] Long range order parameter is defined in the usual Bragg–Williams way, i.e., $S = (r_A - F_A)/(1 - F_A)$, where r_A is the fraction of A sites occupied by A atoms and F_A is the fraction of sites that are A.

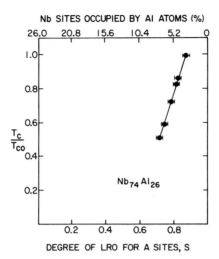

FIG. 19. The fractional reduction in critical temperature for irradiated Nb$_3$Al versus degree of long range order (after Sweedler and Cox [37]).

and electron beam coevaporation [41] can also be used to produce Nb$_3$Ge with $T_c > 21°K$. All these vapor phase methods have the advantage over the more traditional melting and diffusion methods in that the rates of arrival at the substrate of the A and B atoms separately (i.e., stoichiometry) and the substrate temperature (i.e., order) can be independently controlled. Since the highest T_c Nb$_3$Ge produced so far has not yet been characterized as to order and stoichiometry, it may be that 23°K is indeed not the upper limit.

There have been many correlations and predictions of higher T_c materials [42] based upon extrapolations and correlations of various kinds. The reader is invited to invent his own! Figure 20 illustrates how the presently measured T_c correlates with the position in the Periodic Table. Nb$_3$Si appears to be strikingly low. It has been synthesized only by vapor deposition [41]. Comparison of the measured lattice constant ($a_0 = 5.17$) with that prediced ($= 5.09$) from the Geller radii shows that the stoichiometry and/or order are poor. An empirical condition for the formation of the A15 structure is the near equality of the A and B atomic radii [27], and it may be that Si is too small for a well-ordered compound to exist without ternary additions.

It is naturally more difficult to achieve order in a ternary than in a binary compound. Nevertheless, substitution on the B atom site, as can be seen in Fig. 20, has proved to be advantageous upon occasion (NbAlGe) [33]. So far as is known, substitution on the A atom sites always results in lower T_c material than expected by linear interpolation between the end members. It is quite likely that further improvements in T_c will be realized as the B atom substitutions are more carefully investigated. Sequential annealing was found by Webb [28] to be important in raising the T_c of Nb$_3$Ga to above 20°K and is likely to be necessary in ternaries as well.

The effect of B atom substitution is difficult to calculate or predict in advance of actual experiment. The effect upon the free energy of the phases adjacent to the intermediate A15 phase is significant in determining the range of stoichiometry [42]. Evidently the bcc niobium rich solid solution phase has more entropy than the adjacent A15 phase and tends to be favored

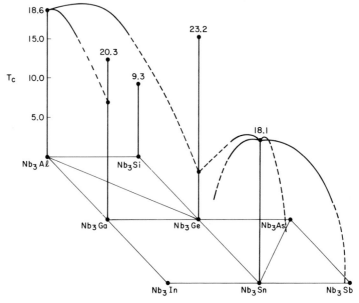

FIG. 20. Transition temperature of binary and pseudobinary A15 superconductors. The dotted curves approaching Nb$_3$Ge are current values that are obviously subject to improvement; after [48].

at high temperatures; the A15 phase forms by a peritectic or peritectoid reaction. The stoichiometric composition may be difficult to maintain due either to retrograde solid solubility (Nb$_3$Al, Nb$_3$Ga) or to the stoichiometric composition being thermodynamically stable only at low temperature (perhaps Nb$_3$Ge) where the equilibrium is difficult to reach.

Phase diagrams for the high T_c material have been determined for Nb–Ga [28], Nb–Al [43], V$_3$Si [44], V$_3$Ga [45], and Nb$_3$Sn [46]. A tentative diagram for Nb$_3$Ge, in its metastable vapor-condensed state, is given in Fig. 21.

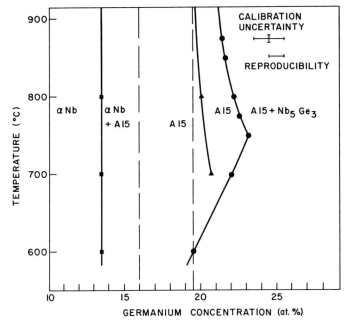

FIG. 21.   A tentative low temperature portion of the phase diagram for Nb₃Ge. The difficulty in obtaining equilibrium well below the melting point and the possibly crucial role of small concentrations of third elements such as oxygen may explain why different methods give different results [A. B. Hallak, R. H. Hammond, and T. H. Geballe, *Appl. Phys. Lett.* **29**, 314 (1976)].

1. Nb_3Sn

Nb_3Sn is presently the most widely used superconductor for high fields and high temperatures. It is fortunately a case where the stoichiometric composition is included in the solid solubility range at all temperatures almost up to the peritectoid temperature (Fig. 22). The T_cs corresponding to various compositions are given by Hammond in Fig. 23 for 3 μm films prepared by codeposition of the elements on a substrate at 720°C. In the solid solution range, which is on the niobium rich side of stoichiometry, a rather broad T_c is observed due to the fact that T_c is a strong function of composition. Slight variations in homogeneity cause correspondingly large changes in T_c. Quenched niobium rich samples may contain phases not shown in the equilibrium diagram [47]. A composition which is nominally tin rich will contain almost stoichiometric Nb_3Sn in equilibrium with the more tin rich phase Nb_6Sn_5. The result is evident in Fig. 23, where it is seen that T_c reaches its maximum at the tin rich phase boundary. Superconductivity is frequently more sensitive to small variations or changes in composi-

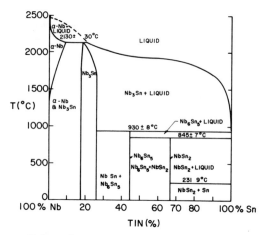

FIG. 22. "Consensus" phase diagram for Nb_3Sn (after Charlesworth *et al.* [46]). The superconducting data in Fig. 23 indicate that the right-hand boundary for Nb_3Sn should be almost at the stoichiometric ratio rather than somewhat greater, as indicated in the figure.

tion than other metallurgical methods of characterization. The slow times necessary to achieve equilibrium in bulk samples and the diffusion barriers at grain boundaries which may arise from phases stabilized by traces of oxygen cause difficulty in reaching equilibrium. Studies of equilibria by codeposition of the elements onto noninteracting substrates avoid some problems connected with reaching equilibrium by long diffusions because the elements are intimately mixed at the start.

Nb_3Sn is commercially available in ribbon form by both chemical vapor deposition and diffusion of tin into niobium. Present efforts to make it a multifilament form (which is important for transient field applications) are meeting with success. An adaptation of the Tachikawa process [48] used for V_3Ga is used to make the multifilament Nb_3Sn. In one process, niobium

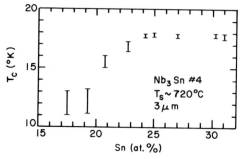

FIG. 23. The T_c versus the at. % Sn along the substrate for codeposited Nb and Sn.

rods are first inserted into bronze billets. After the drawing process is completed, the tin is diffused from the bronze (through the copper) into the niobium forming the Nb_3Sn filamentary copper matrix, which will be discussed in more detail in Section IV (see also Section II,F).

2. Unusual Features

It is worthwhile to discuss the unusual features that have been discovered in other properties of the particularly high T_c A15 compounds. The subject has been reviewed recently by Testardi [30] and by Weger and Goldberg [31]. The unusual features can be ascribed to either electronic or lattice degrees of freedom, although, because of the strong electron–phonon coupling, it is not always clear which is primarily responsible.

a. Magnetic Effects Both the Knight shifts on the A and B atoms and the static magnetic susceptibility are temperature dependent. For V_3X the dependence is stronger for the higher T_c compounds (Fig. 24), but this does not hold for Nb_3Al [49], and perhaps Nb_3Ge [40]. Analysis of the Knight shift as a function of susceptibility (with temperature as the implicit parameter) identifies the almost complete pairing of the d-electrons [50] when the compound becomes superconducting. Knight shift and nuclear relaxation time measurements were used by Ehrenfreund *et al.* [51] to esti-

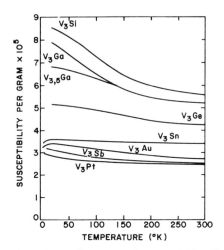

FIG. 24. Temperature dependence of the magnetic susceptibility of some V_3X compounds. The superconducting transition temperatures are: V_3Si, 17.1; V_3Ga, 16; $V_{3.5}Ga$, 13.4; V_3Ge, 6.0; V_3Pt, 2.8; V_3Sn, V_3Au and V_3Sb < 1 (all in °K). Note that the anomalous temperature dependence occurs for the high T_c compounds [after H. J. Williams and R. C. Sherwood, *Bull. Am. Phys. Soc. II* **5**, 430 (1960)].

mate that an upper limit of $< 5\%$ of the total density of states at the Fermi
level could be attributed to the Nb $5s$ band in several Nb_3X compounds.

Anomalous temperature dependences of resistivity have also been observed [52]. The resistivity of Nb_3Sn can be described to an accuracy of
within 1% between T_c and $850°K$ by $\rho = \rho_0 + \rho_1 T + \rho_2 \exp(-T_0/T)$,
where $T_0 = 85°K$ and the exponential term makes an appreciable contribution, as can be seen in Fig. 25.

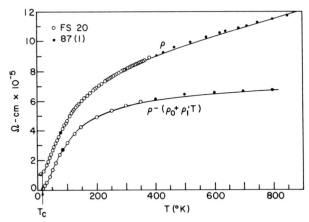

FIG. 25.   The temperature dependence of the electrical resistivity of Nb_3Sn. The lower solid
curve is the $\rho_2 \exp(-T_0/T)$ contribution, as discussed in the test [after D. W. Woodward and
G. D. Cody, *RCA Rev.* **25**, 392 (1964)].

Very large linear heat capacity coefficients for some of the high T_c compounds have been observed. For instance, per transition metal atom, the
coefficient of Nb_3Sn is approximately twice as great as that of niobium, and
for V_3Ga it is nearly three times as great as it is for vanadium [31,53]. On the
other hand, the coefficients of Nb_3Al and Nb_3Ge have values expected for
typical d-band metals such as niobium.

A cubic-to-tetragonal distortion frequently characterized as a Martensitic transition was discovered to occur in single crystal V_3Si between $18°$ and
$25°K$ $(T_c = 17°K)$, by Batterman and Barrett [54]. Subsequently, a similar
effect was found in Nb_3Sn [55]. Not all crystals show the transformation,
which can be inhibited by impurities or perhaps by strain and vacancies.

The lattice parameters of Nb_3Sn as a function of T are shown in Fig. 26
and the heat capacity is shown in Fig. 27. In V_3Si, the distortion is reversed
in that the a axis becomes smaller. An inverse relation has been found [56]
between the pressure dependence of the Martensitic transition, $\partial T_m/\partial p$, and
$\partial T_c/\partial p$ in both V_3Si (where the respective temperatures approach each other

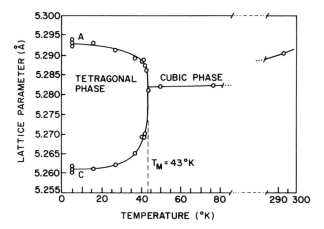

FIG. 26. The c and a lattice parameters of Nb_3Sn versus T showing the cubic-to-tetragonal transformation at $T_m = 43°K$ (after Mailfert *et al.* [55]).

and extrapolate to $\sim 21°K$ at 28 kb) and in Nb_3Sn where the respective coefficients separate. Vieland and Wickland (Fig. 28) have found that about 8% aluminum substituted for tin is just sufficient to suppress the Martensitic transition at which concentration T_c reaches a maximum $= 18.6°$. Testardi summarizes considerable data which, in spite of some claims to the contrary, indicate that the T_c of the tetragonal phase is something less than 1° lower than the related cubic phase.

In both V_3Si and Nb_3Sn any discontinuous volume change at T_m is extremely small. There is no evidence of latent heat. The structural transitions are either weakly first order, or second order. The order is of theoretical importance, because as pointed out by Anderson and Blount [57], a second order transition requires the breaking of some internal symmetry, such as the screw axis associated with the translation $1/2a_0$ along the A site chains.

The acoustic phonon spectrum is also strongly temperature dependent. In Fig. 29, the marked softening of the elastic moduli with temperature (i.e., the decrease in slope of the solid lines) is found to continue far out into the zone although at a reduced level as determined by inelastic neutron scattering.

Other properties related to those described above [30], such as thermal expansion, absorption of sound, dependence of T_c upon strain, and Mössbauer studies of tin recoil, all show features which can be attributed to lattice instability or softening of certain modes as a function of temperature, and/or a temperature-dependent density of states at the Fermi level. Direct evidence for extremely sharp structure in the band density of states near the

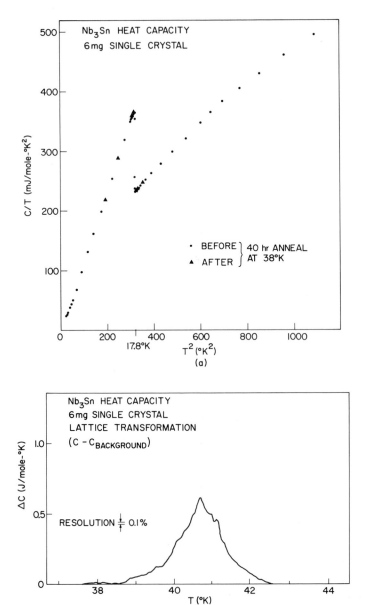

FIG. 27.   Nb$_3$Sn heat capacity measurements through the superconducting transition (above) and through the Martensitic transition (below). No evidence for first-order or time-dependent effects at the Martensitic transition were found [after J. M. E. Harper, Ph.D. Dissertation, Stanford Univ. (1975) (unpublished)].

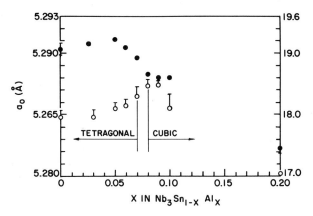

FIG. 28. The Martensitic transition of Nb_3Sn is suppressed by the addition of 8% Al at which concentration a maximum (also apparent in Fig. 20) in T_c is reached. Closed circles, room temperature lattice constant; open circles, transition temperatures. Samples with $x < 7\%$ become tetragonal at low T, as indicated [after L. J. Vieland and A. W. Wicklund, *Phys. Lett.* **A34**, 43 (1971)].

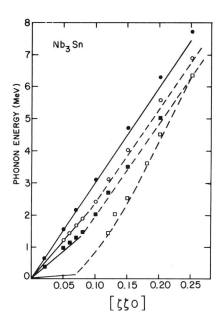

FIG. 29. Acoustic phonon dispersion curves for $q \| [110]$ waves with $[1\bar{1}0]$. Polarization in Nb_3Sn. $q = (\zeta, \zeta, 0)2\pi/a = 1.19$ Å. Note the mode softening on cooling: $295°K$; $80°K$; $46°K$ [after G. Shirane and J. D. Axe, *Phys. Rev. Lett.* **27**, 1803 (1971)].

Fermi level, which is the basis of some of the ad hoc theoretical models to be discussed below, has not been found in optical experiments [58] that have attempted to find it.

3. Theoretical Models

If, below room temperature, the density of states is assumed to have a very sharp structure of the order of kT at an energy near the Fermi level, then the effective density of states will be temperature dependent-thermally smeared—and all the thermodynamic and transport properties dependent upon it will be affected. A simple model invoked by Clogston and Jaccarino [59] with a sharp parabolic peak in the density-of-states at the Fermi level attributed to the d-electrons (and later refined to include the renormalization effects [60] of the electron–phonon interaction) was used to account for magnetic data. Cohen et al. [61] chose a simple step function singularity in $N(E)$ to explain both the electronic and Martensitic transitions. Labbé and Friedel [62] built a one-dimensional model based upon the linear chains of A atom sites, originally emphasized by Weger [63]. The singularities in the density of states are presumed to arise from noninteracting conduction band states which can be described in a tight-binding representation, as shown in Fig. 30. This model was used to account for the anomalous experimental

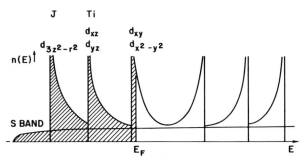

FIG. 30. Density of states for the five d bands and one s band for A_3B compounds with Al5 structure in the Labbé–Friedel model. The singularities result from the one-dimensional nature of the A atom structure in this model. The anomalous temperature dependences and the structural transformation in these materials can be predicted if the Fermi level lies within several millivolts of a singularity (after Labbé and Friedel [62]).

magnetic, electric, and structural behavior outlined above. A word of caution, however, the first principles (APW) calculation of the band structure of V_3X compounds by Matheiss [64] shows no evidence for flat d-band and hence no support for the one-dimensional linear chain model. Because of the attractiveness of the linear chain model in explaining the wealth of data, Weger and Goldberg [31] have preserved the necessary one-dimensional features of the chain model while allowing for some interchain coupling (independent bond model) and hybridization. Gor'kov [65] has recently introduced one-dimensional model without any assumptions of tight binding in which the driving force for the martensitic transition is the splitting of the degeneracy at the X point in the Brillouin zone. The resulting distortion causes a singularity in the density of states which is linked to the enhanced superconductivity. However, more extensive calculations by Mattheiss [66] for V_3Si, V_3Ge, Nb_3Al and Nb_3Sn continue to raise serious doubts about the applicability of any one-dimensional model.

At the present time, after two decades of extensive investigation, the root of the high-T_c aptitude of the A15 structure has not been pinpointed. The linear chain and the sharp density-of-states models do a nice job in correlating all the unusual features observed in Nb_3Sn and V_3Si, both of which compounds are available in single crystal form. The more difficult to prepare compounds, Nb_3Al, Nb_3Ge and Nb_3Ga, which have the highest T_cs, have not yet been prepared as single crystals. However, in polycrystalline form the latter do not show the combination of behavior expected from the above models. For instance, the temperature dependence of the Knight shift and the susceptibility is markedly reduced in Nb_3Al. As already mentioned, the linear heat capacity coefficients are not unusually large. Testardi [30] has shown that a substantial part of the Nb_3Sn and V_3Ga linear heat capacity coefficients can arise from the large anharmonicity observed in the acoustic spectrum. Knapp *et al.* [67] have found the corresponding heat capacities above the Debye temperatures to be lower than expected and attribute the lowering to negative contributions of the anharmonicity.

4. Instabilities

It has been recognized for a long time that high T_c superconductors were close to becoming unstable in one sense or another. Hypothetical compounds believed to have a good chance of having high T_cs cannot be synthesized. Some A15 compounds (such as Nb_3Si) [41] only can be made as disordered vapor-quenched films. It may not be coincidental that Nb_3Ge, Nb_3Ga and Nb_3Al solidify as metal rich A15 compounds with moderate T_cs. They can be prepared with high T_cs only by special means. One can speculate that the overall cubic symmetry combined with a high degree of anisotropy are the essential ingredients which combine to give the A15 structure its favorable superconducting properties. The strong electron–phonon interactions exist over a good fraction of the Brillouin zone and under some special conditions of preparation do not force the lattice to relax into a more stable configuration. The Martensitic transformations at low T have only minor effects on the vibrational spectra and can easily be incorporated within the above viewpoint.

C. Other High Temperature and Potentially Important Superconductors

1. Strong-Coupling Theory

The BCS theory as extended by Eliashberg [68] has been used by McMillan [69] to make contact with real materials. In high temperature superconductors the phonons and electrons are strongly coupled. T_c is an appreciable fraction of the Debye temperature (of the order of 0.1), and the electron–phonon coupling constant λ is appreciable with respect to unity. (Terms are defined below.)

The simple BCS expression for T_c,

$$T_c = 1.14 \langle \omega \rangle \exp(-1/N(0)V) \qquad (22)$$

was derived for a net attractive interaction $N(0)V \ll 1$. The McMillan expression

$$T_c = \frac{\langle \omega \rangle}{1.2} \exp\{-1.04(1 + \lambda)/[\lambda - \mu^*(1 + 0.62\lambda)]\} \qquad (23)$$

is based upon the particular form of the phonon spectrum of niobium, but has applicability to a wide variety of compounds. The various quantities are defined as follows: $\langle \omega \rangle$ is a weighted average of the phonon spectrum roughly given by McMillan as $\theta/1.45$, where θ is the Debye temperature; λ is a weighted average of electron–phonon interaction given by McMillan as $N(0)\langle I^2 \rangle/M\langle \omega^2 \rangle = N(0)V_{ph}$, where M is the ionic mass, $N(0)$ is the band structure density of states, a purely one electron term, and I^2 is the square of the electron–phonon matrix element averaged over the Fermi surface (V_{ph} is referred to as the attractive electron–phonon interaction); μ^* is the screened Coulomb repulsive interaction experimentally found and theoretically [69] expected to be small ≤ 0.15. The brackets indicate the weighting is over the phonon density of states times a coupling constant.

In the cases where the phonon structure can be seen in the dI/dV curves of superconducting tunnel junctions, λ can be evaluated directly from experiment and the McMillan equation is found to be exceedingly accurate, that is, T_c can be calculated to within a few percent [70]. Recent results of Dynes *et al.* [71] indicate that if the coupling constant could be increased without limit (without the lattice becoming unstable) T_c would saturate at some maximum value rather than go over a maximum, as originally suggested by McMillan on the basis of less complete data.

Unfortunately, with the one notable exception [72] tunneling results good enough to evaluate λ for high T_c materials do not yet exist. However, heat capacity measurements through the superconducting transition afford a convenient source of data for classifying all the strong-coupled superconductors by means of the McMillan equation and the relation [69]

$$3\gamma/\pi^2 k_B^2 = N_\gamma = (1 + \lambda)N(0) \tag{24}$$

The left-hand side of Eq. (24) relates the experimental density of states N_γ to the coefficient of the linear dependence of heat capacity, γ, in the usual way. The right-hand side shows how the so-called " bare " or one-electron density of states at the Fermi energy, $N(0)$, is renormalized by λ [68].

McMillan [69] was able to establish that taking $\mu^* \sim 0.13$ introduces little error and permits Eqs. (23) and (24) to be used to obtain approximate values of $N(0)$ and λ from heat capacity data. Figure 31 shows the data so

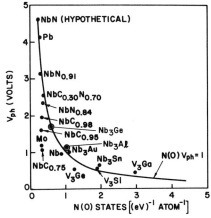

FIG. 31. Attractive phonon interaction V_{ph} versus density of states $N(0)$ (after Geballe *et al.*, [73]).

extracted for a variety of superconductors. It is notable that the hyperbolic boundary $N(0)V_{ph} \sim 1$ (solid curve) acts as a rough limit for both high and low density of states compounds. More recently, efforts have been made to analyze trends of $\lambda = N(0)\langle I^2\rangle/M\langle W^2\rangle$, as defined above, by considering its electronic (numerator) and lattice (denominator) components separately. With good tunneling data, further progress should be obtainable.

The rocksalt structure (so-called interstitial compounds) such as NbN have superconducting properties including T_cs up to 18°K which are of potential interest. It can be seen in Fig. 31 that they have high V_{ph} and low $N(0)$, in contrast to the A15 compounds. Rocksalt compounds not only have large resistivities implying a large $\langle I^2\rangle$, but those with high T_cs show evidence of unstable lattice modes in their vibrational spectra, that is, small $\langle 1/\omega^2\rangle$. In ternary rocksalt compounds, a maximum as a function of composition is found in T_c at intermediate compositions. This includes either anion or cation substitution [74]. Lattice vacancies occur at either the anion or cation sites depending upon stoichiometry. For example, NbN ($T_c \sim 15°K$) has 9% vacancies on the N sites, whereas $Nb(C_{0.3}N_{0.7})$ ($T_c \sim 18°K$) has only 1% vacancies [73].

NbN can be prepared in thin film form by reactive gas sputtering. As such, it has an extremely high $J_c(H = 0)$ of 2×10^7 A/cm² [75], shown in Fig. 32, possibly due to pinning centers resulting from growth morphology. The films are very stable; they can be made as thin as 30 Å and still maintain their properties over long periods of time. Their high resistivity in the normal state has some device application.

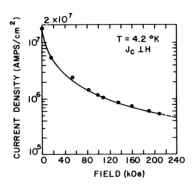

FIG. 32. J_c versus H values for an Nb–N microbridge sample with cross-sectional area of 1.2×10^{-9} cm² (after Gavaler *et al.* [75]).

Flexible fibers of brittle NbN and derivatives have been prepared from fine carbon fibers ($\sim 6 \,\mu$m) [76] by reacting them with $NbCl_5$ in the presence of H_2 and N_2. The hydrogen reduces the $NbCl_5$ to Nb which reacts with the N_2 and the carbon. Boron nitride fibers (6 μm) have been used also. The resulting superconducting NbCN and NbN is available as twisted yarn made from 72 filaments. The resulting superconducting yarn is multiply-connected and thus hysteretic in the presence of changing magnetic fields.

It is possible for superconductivity to persist to very high fields because of dimensional effects alone, as already discussed. In the intercalated transition metal dichalcogenides $TaS_2 \cdot (CH_5N)_{1/2}$, the superconducting layers only 6 Å thick are separated by 6 Å organic layers. When the field is applied parallel to the layers, T_c is suppressed only very slowly, $dH_c/dT > 10^5$ G deg^{-1} [77]. The material has a low T_c ($= 3.5°K$) so it is not of practical importance.

Reduced dimensions have also been produced by impregnating porous glass with indium and with lead–bismuth filaments of the order of 40 Å [78]. These materials have high H_{c2}s but like the NbCN yarn are fabricated in an interconnected way which produces hysteretic losses.

An interesting series of cubic compounds, of which $Pb_xMo_3S_4$ is representative and which are composed of clusters of 6 Mo atoms with metallic binding has recently been synthesized [79] and discovered to have T_cs varying up to 15°K [80] and H_{c2}s up to 600,000 G [81].

The enormous H_{c2}s could be due to properties (high resistivity, small Fermi velocity) inherent in the compounds or perhaps due to some fine scale dimensional effect associated with the Mo clusters which permits field penetration.

IV. THE STATE OF THE ART

In the preceding two sections we have discussed the principles underlying the present understanding of both the physical properties of superconductors and the occurrence of superconductivity in nature. We have also indicated how the important material parameters of a superconductor (e.g., T_c, H_{c2} and J_c) can be altered so as to produce practical superconducting materials. The need for composite materials in practice was also stressed.

In this section we describe the state of the art of practical superconducting materials as we understand it. We include both magnet materials and materials being developed for superconducting power transmission. The two areas are quite distinct in that the former involves the properties of superconductors at very high fields while the latter is concerned with the behavior at low fields. Also, even though pulsed magnets and ac superconducting power lines both involve ac losses, the detailed considerations are different in the two cases. We begin first with magnet materials, treating separately the ductile alloys and the brittle A15 compounds. The former are certainly the most advanced superconducting materials in an engineering sense. The latter have superior superconducting properties, however, and are being developed vigorously. Following this, we turn our attention to power line materials. Since this application has received concerted attention only more recently, the materials are not so well developed. Throughout this discussion, we attempt to indicate where future research and development would be desirable.

A. Composites for High Field Use

In order to construct large solenoids and coils for rotating machinery and other applications such as confinement magnets for nuclear fusion, it is necessary to have available long lengths of superconductor with uniform properties, which must then be incorporated into a composite wire. For a few restricted uses where high effective current densities are not necessary, the composite can be "cryostatically stabilized" (see discussion, p. 510) by incorporating a parallel path containing of the order of 20 times as much normal conductor (Cu or Al) as superconductor. Of more general application is the "adiabatically stabilized" composite where the superconductor has been reduced to micron-sized filaments so that a flux jump in any given filament will not be serious. The filaments are then twisted and interposed in such a way as to reduce magnetic hysteresis losses caused by currents flowing through the matrix between adjacent filaments during magnet excitation or in response to time-varying applied fields. The filaments are imbedded in a matrix of roughly their own volume of copper. Eddy current losses in the (copper) matrix can be reduced by incorporating barriers of high resistivity alloys such as Cu–Ni giving rise to the so-called "mixed matrix" [82]. The evolution of ductile alloy wire, as discussed in detail below, is summarized in Fig. 33. The need to reach higher fields, particularly

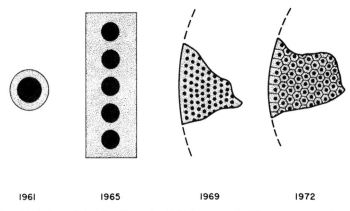

| 1961 | 1965 | 1969 | 1972 |

FIG. 33. Evolution of ductile alloy wire. Initial copper cladding to serve as shunt (1961); cryostatic stabilization with excess Cu (1965); intrinsic stabilization by reduced size of filaments (1969); eddy current stabilization by mixed matrix (1972); after [48].

for nuclear fusion confinement magnets, has led to the development of composites made of brittle A15 compounds using some clever techniques which are discussed below. The rapid progress being made at the time of writing makes it reasonable to assume that commercial supplies of composite for operation at fields up to 15 T will be available (although at an increased cost with respect to Nb–Ti) in the next few years.

1. Ductile Alloys

At present the "workhorse" material for everything other than the highest field superconducting magnets is Nb–Ti. Techniques for producing sophisticated multifilamentary, twisted composites are well established. These composites can have thousands of micron-sized filaments, carefully twisted with a twist period of ~ 1 cm, and embedded in high conductivity copper. Critical current densities (in the filaments) of 10^5 A/cm at 50 kOe and copper-to-superconductor ratios of 2 to 1 are typical. The problems associated with Nb–Ti composites are now largely those of the production of large quantities of uniform material from which huge and special purpose coils can be designed with confidence and built with reliability.

Demand for superconductors has surpassed laboratory production levels. Many thousands of pounds of composite wire are required just for the prototypes to be used before scaling up to the large applications mentioned in Section I. The wire is produced from 2500 lb Nb–Ti ingots [23] which must be homogeneous and ductile since ultimately they will be reduced to ~ 1 μm filaments. The scale of this operation presents certain problems. It is necessary, for example, to avoid excess contamination by carbon, hydrogen, oxygen and nitrogen when the ingots are fashioned from elemental electrodes by consumable arc melting. Typical chemical analysis shows the Ti goes from 45.6 wt. % at the center to 46.7 at the outside of the 16-in.-diam ingots before the final forging and extruding into 1/2-in.-diam rod. At this stage the manufacturing details are proprietary, but the sequence outlined below is probably more or less followed.

Several Nb–Ti rods are inserted into previously drilled holes of copper, cupronickel or aluminum ingots. The composite ingot is reduced and drawn through hexagonal dies. The resulting hexagonal composite is close packed into copper (Al) tubes for final reductions. Heat treatments to insure the proper dislocation cell structure (Section IIIA) are carried out along the way. Up to thousands of micron-sized filaments of Nb–Ti are thus incorporated into a normal metal matrix. The conductor is normally subjected to a twisting operation followed by a final rolling into a flat strip or other suitable shape.

In any given coil design the requirements for stability will vary over the volume of the coil. The composite wire will be subject to regions of varying B, $\dot{B}$, mechanical stress, heat transfer, and so on. Optimization of wire properties by varying parameters such as $J_c(H)$ and the copper-to-superconducting ratio can have important economic consequences. The next several years should see various standard composite wire specifications develop that will permit the advantages of large scale manufacture.

2. A15 Compounds

The brittle A15 compounds cannot be fashioned by the techniques used for the ductile Nb–Ti alloys (although, at least one report shows that the A15 compounds become ductile at high pressures) [22]. Nb_3Sn tape has been available for the construction of laboratory-size magnets for some time, however, using either the chemical vapor deposition process [83] or a straightforward diffusion process [22, p. 394]. In the former method, Nb_3Sn is formed when gaseous $NbCl_5$ and $SnCl_4$ are reduced on a hot metallic ribbon. The metallic ribbon is usually Hastelloy, a high nickel content, nonmagnetic alloy whose coefficient of expansion is slightly greater than Nb_3Sn. This keeps Nb_3Sn in compression with less tendency to crack when it is subjected to thermal contraction or bending.

The formation of Nb_3Sn by diffusion is carried out at about 950°C to avoid formation of Sn rich phases at the interface (see Fig. 22). Either vapor or liquid Sn can be used. Again Hastelloy substrates are employed. Also, in both the CVD and the diffusion process, care must be taken to maintain small grain size since the flux pinning in these materials arises from grain boundaries. The tapes must be clad with copper or aluminum for stability reasons. They are inherently less satisfactory for coils than multifilamentary wires and are being superseded by the newer multifilamentary A15 composites now being produced by diffusion using variations of the Tachikawa method (see Section IIB1, and below) where the diffusion can be carried out at $\gtrsim$ 650°C. Tapes are important, however, for low field applications such as power transmission where they can also be fabricated by high rate vapor deposition such as electron beam evaporation and sputtering.

The Tachikawa process postpones the final reaction, which forms the brittle A15 compound by a diffusion process from an fcc copper-based phase, until after the composite wire is fabricated. The processing is carried out by inserting niobium (vanadium) into billets of copper, or copper alloy, much as is done for the Nb–Ti wire (see above). The final step consists of diffusing tin (gallium) to form the A_3B compound. The tin (gallium) can be either incorporated in the copper or diffused in from the outside after the drawing is completed. The former case suffers from the disadvantage that tin or gallium bronze work hardens during the drawing, but has the advantage that less diffusion time is needed.

Different schemes are used to incorporate some pure copper (or aluminum) in the matrix in parallel with the bronze from which the diffusion takes place. The resulting " mixed matrix " then, by virtue of the pure copper, possesses the necessary rapid thermal time constant for obtaining good dynamic stability (see Section IIF) in rapidly changing fields. One scheme making use of Ta barrier layers to separate the Cu from the Cu–Sn is illustrated Figs. 34 and 35. There are 67,507 filaments in the final conductor (Fig. 34) each $\sim$ 5 μm in diameter. The twist was incorporated before diffusion when the conductor was 4 mm in diameter with a pitch of 12.7 mm. The final reduction is also made to give the conductor a rectangular cross section for achieving a better packing fraction when winding coils and also for better mechanical stability of the coil.

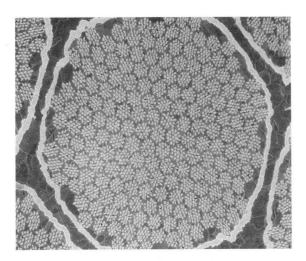

FIG. 34.   Cross section of 67,507 filament Nb₃Sn conductor 3.3 mm (0.130 in.) diameter [after Gregory *et al., IEEE Trans. Magn.* **MAG-11**, 295 (1975)].

A given rectangular composite conductor of outside dimensions 1.75 mm × 5.46 mm carried 5000 A at 6 T and is estimated to carry 1250 A at 12 T. The composite is made of groups of 19 individual 5 μm filaments. One hundred and eighty-seven such groups, encapsulated within a Ta tube barrier are shown in Fig. 34. Nineteen Ta tubes are then drawn in the pure

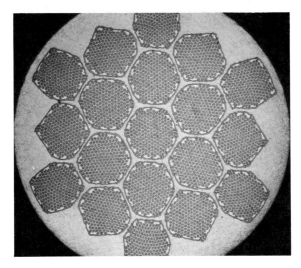

FIG. 35.   Center section of 67,507 filament Nb₃Sn conductor [after Gregory *et al.*, see Fig. 34].

Cu matrix giving the cross section shown in Fig. 35. The niobium comprised 13% of the overall cross section and the diffusion reaction was carried out to produce a 1 μm layer of Nb_3Sn on each Nb filament. The success of the Ta barriers in protecting the copper is measured by the overall resistivity ratio $R_{298°K}/R_{20°K}$ which varies between 100 and 145, a typical value for commercially pure copper. Another scheme for producing a mixed matrix is to use Nb tubes and separate the bronze from the pure copper by the tube itself.

The mechanical properties of the composite still reflect the brittle nature of the Nb_3Sn. Rectangular cross sections with high aspect ratios are being investigated to increase the degree of bending which may be achieved without causing significant deterioration of superconducting properties. At the present writing, commercial composites of overall dimension 6 mm $\times$ 13.75 mm are available which carry 10^4 A/cm^2 at 100 kOe.

In order to achieve the $\lesssim$ 12 mm bend radius needed for some coil configurations, consideration is being given to the original Kunzler scheme [7] of winding the ductile unreacted superconductors with high temperature insulation and carrying out the final reaction as the terminal step.

B. Superconducting Power Lines

Superconducting power lines are very much in the exploratory stage compared with magnets but are progressing rapidly. Both ac and dc lines are under development [84]. As stressed previously, the operating conditions (low H, high J) involve different materials considerations than magnets. The economic considerations are also different. Physically large and/or high field superconducting magnets have no competitors, whereas superconducting power lines do. They must compete with conventional underground lines, and both of these must compete with overhead, high voltage lines, which are basically much cheaper. Higher costs not withstanding, there is a need for underground lines. They are environmentally more acceptable, and where rights of way are restricted, they become essential. The relative virtues of superconducting versus conventional underground lines are continually debated, as one might imagine. Basically, superconducting lines become more competitive as the amount of power to be transmitted increases, since the proportionate cost of refrigeration and vacuum construction becomes less. At present, the crossover point is estimated to be somewhere above 2×10^9 VA which is a factor of two or so larger than the largest existing lines. There are important uses for both ac and dc superconducting lines in the national grid, however, providing they can be made to be reliable and economic. Some specific locations for such lines have even been identified [85].

The materials requirements for a superconducting line are now reasonably clear. For an ac line one requires a superconductor or superconducting composite with sufficiently low losses ($\lesssim 10\ \mu W/cm^2$) at the field levels required (systems considerations dictate a linear current density of ~ 550 A/cm rms, which corresponds to a peak surface field of ~ 1000 Oe) while simultaneously providing adequate fault capacity, that is, the ability to carry three to perhaps as much as ten times the rated current for short periods of time and then to recover successfully. For a dc line the materials problems are somewhat less severe. One wants high current densities in order to carry large amounts of power with a thin conductor. The maximum practical linear current density is determined by the stability properties of the conductor. AC losses enter only as ripple losses due to the 2–3 % current variations at 720 Hz introduced by the ac-to-dc converters. Again, the losses must be kept $\lesssim 10\ \mu W/cm^2$. In contrast to ac lines, fault requirements are relatively mild, 1.1–2 times rated current. For both types of lines it is necessary to fabricate the conductor in a form that can cope with the large thermal contractions present in any cryogenic power line. A comprehensive discussion of the requirements of superconducting power lines and a survey of some of the designs being considered has been given by Bogner [86].

There are only two materials, namely Nb and Nb_3Sn, actively being considered for ac superconducting power transmission, although Nb_3Ge and other A15s may be attractive future prospects. Elemental Nb should have almost no losses below its $H_{c1} \cong 1400$ Oe at 5°K and historically was the first serious candidate. It is under active development as a power line conductor by a number of groups, both in rigid coaxial tubular lines and also for lines using semiflexible, helically wound tapes. The superconducting ac losses of a properly prepared surface of Nb appear to be entirely adequate for operation in pressurized He at temperatures $\lesssim 5°K$. Some representative loss data are shown in Fig. 36. The losses observed below H_{c1} are believed to be due to premature flux penetration at high points on the surface of the conductor where the field is locally enhanced above H_{c1} due to demagnetization effects. The desirability of good surfaces is thus demonstrated.

In these Nb lines it is planned to obtain fault capacity by incorporating a parallel path of high conductivity copper since the Nb itself does not have sufficient total current capacity. This is essentially like cryostatic stabilization although in the present case care must be taken to insure that no ac fields are acting on the copper during normal steady state transmission line conditions. Such stray eddy current losses can easily be prohibitively large. In the coaxial rigid tube designs this is accomplished easily by locating the copper on the inside and the outside of the inner and outer tubes, respectively, where the ac field is zero. Since in these tubular lines thermal contraction is to be dealt with by using invar, the resultant conductor is actually a

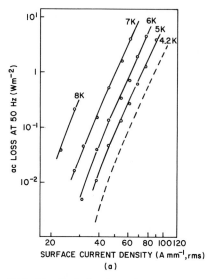

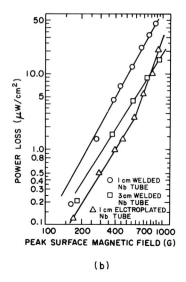

FIG. 36.   Typical ac losses in niobium tubes at 4.2°K [(a) after J. A. Baylis *et al.*, *Cryogenics* **14**, 553 (1974); (b) after R. W. Meyerhoff, *Proc. Appl. Superconduct. Conf. Annapolis, Maryland* IEEE Pub. 72CH0682-TABSC, p. 194 (1972)].

Nb–Cu–Invar concentric tricomposite. For flexible tapes where the contraction problem is dealt with by adjusting the helical pitch [87], it is proposed to shield the copper by completely surrounding it with niobium. This is relatively simple since niobium is ductile. The resulting composite then consists of a niobium tape with a copper core. Both types of composites have been fabricated, but undoubtedly more materials problems will be uncovered when full scale tests of these prototypes are completed.

To operate a line above 5°K, one must go to a high T_c superconductor. With the high temperature superconductors presently known, this necessarily means a type-II superconductor with a relatively low H_{c1}. Thus, hysteretic losses must be taken into account. In this regard it is important to realize that the hysteretic losses expected on the basis of the Bean model when H_{c1} is small and there is no surface barrier ΔH [see Eq. (21)] are too large to be acceptable with the critical current densities that are typically available at low fields (e.g., $\sim 5 \times 10^6$ A/cm^2 for Nb$_3$Sn) [88]. Nevertheless, some early measurements of ac loss in Nb$_3$Sn produced by dual electron beam evaporation codeposition of Nb and Sn demonstrated quite low loss [89]. More recently, these results have been confirmed and extended considerably [90]. Low losses have also been obtained with Nb$_3$Sn produced by the diffusion process [91]. Even commercial Nb$_3$Sn magnet tapes have performed satisfactorily once some excess (lossy) solder was removed [92].

Representative loss data for Nb_3Sn produced using the electron beam and the diffusion methods are shown in Figs. 37 and 38, respectively.

The precise reason why Nb_3Sn exhibits losses well below the Bean model level presently is not well understood. The lower critical field H_{c1} for Nb_3Sn is small, perhaps a few hundred oersted. Surface barriers are a possibility, and some of the data have been interpreted on this basis [20]. Low loss material has been produced, however, where surface barriers are certainly not in any simple way the mechanism leading to low loss (e.g., Fig. 37). The

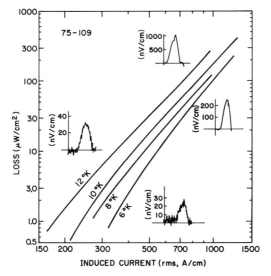

FIG. 37. Typical ac loss data for electron-beam-deposited Nb_3Sn. Loss voltage wave forms at low and high currents for 6°K and 12°K are indicated [after NSF–RANN Semiannual Report, Stanford Univ. M.L. Rep. No. 2498 (December 1975)].

situation illustrates our poor understanding of practical superconducting materials in this low field region. Also, it should be borne in mind that accurate loss measurements below 10 $\mu W/cm^2$, which are essential for establishing the presence of a surface barrier, are not without difficulty [93].

During operation under the normal load, the current flows in the surface region of ≤ 1 μm in depth. The superconducting material beneath that surface region should have a high critical current density in order to handle surges without quenching and without requiring inconveniently large amounts of additional material for cryostatic stabilization. (The even higher fault currents, of course, must be handled by opening the current breakers.) Evaporation techniques for making the Nb_3Sn have an advantage that the grain size and, hence, the critical current of the underlying material can be

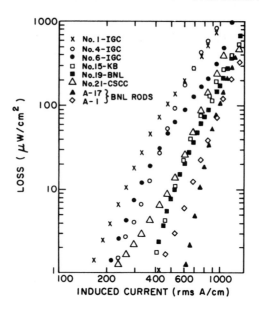

FIG. 38.   Typical ac losses at 4°K for Nb$_3$Sn (tapes and rods) produced by diffusion method (after Bussiere *et al.* [91]).

optimized independently of the film thickness simply by choice of substrate temperature and by the introduction of layers of third component every several hundred angstroms. In the diffusion process the length of heat treatment necessary to achieve the desired thickness of Nb$_3$Sn ($\geq 15\ \mu$m) tends to promote large grain growth and, hence, reduced critical currents. The application of third components such as Zr or ZrO$_2$ has, under some conditions, been used to inhibit grain growth and thus to increase J_c [94], making it likely that the first prototype ac lines using Nb$_3$Sn will utilize presently available commercial technology.

As the present time it appears that providing fault capacity, not achieving low loss, is the principle problem in ac Nb$_3$Sn transmission lines. Unfortunately, a generally acceptable scheme for providing fault capacity in such lines has not yet evolved. A copper backing similar to that used with Nb is of course possible, but the stray loss problem has not been solved. Another possibility is that the high-current-carrying capacity of these materials along with the greater stability inherent in higher temperature operation where the specific heat is larger can be utilized to eliminate all or most of the copper and thereby ameliorate the stray loss problem and give greater flexibility in system design. For dc lines Nb$_3$Sn, Nb$_3$Ge and Nb$_3$(GeAl) conductors have been considered, although multifilamentary Nb$_3$Sn is presently favored. Cryostatic stabilization with copper is being used to insure complete stability, but eddy current losses in the copper due to the ac ripple are presently a problem. Developments are moving rapidly in the superconducting power line materials, and it is expected that the best conductor designs and the materials requirements will be greatly clarified in the next few years.

In closing this review we would like to mention one additional material, helium, which is of obvious importance in large scale applications of superconductivity. Large quantities of liquid helium and supercritical helium gas will be required for refrigeration in the applications considered here. Although it is estimated that the presently stored helium will be adequate for all potential use up to the year 2000 [95], it seems prudent that the United States should continue to recover helium from natural gas whenever possible.

REFERENCES

[1] F. London, "Superfluids." Wiley, New York, 1950.
[2] V. L. Ginzburg and L. D. Landau, *Zh. Eksp. Teor. Fiz.* **20**, 1064 (1950).
[3] J. Bardeen, L. N. Cooper, and J. R. Schrieffer, *Phys. Rev.* **108**, 1125 (1957).
[4] B. Josephson, *Phys. Lett.* **1**, 251 (1962).
[5] See, for example, B. T. Matthias, *Progr. Low-Temp. Phys.* **2**, 138 (1957). For an up-to-date listing of all reported superconductors through mid-1973, see B. W. Roberts, Nat. Bur. Std. Tech. Note No. 724 and 825 (1974); Supplement U.S. Dept. of Commerce/Nat. Bur. of Std.; *J. Phys. Chem. Ref. Data* **5**, 581 (1976).
[6] D. Shoenberg, "Superconductivity." Cambridge Univ. Press, London and New York, 1952.
[7] J. E. Kunzler, *Rev. Mod. Phys.* **33**, 1 (1961).
[8a] For a general discussion of large scale applications of superconductivity, see S. Foner and B. B. Schwartz (eds.), "Superconducting Machines and Devices: Large Systems Applications." Plenum Press, New York, 1974.
[8b] For a discussion of the role of superconductivity in energy-related applications, see Cryogenics (February, 1975) and W. A. Fietz and C. H. Rosner, *IEEE Trans. Magn.* **MAG-10**, 239 (1974).
[9] J. L. Smith, Jr. and T. A. Keim, *in* "Superconducting Machines and Devices: Large Systems Applications" (S. Foner and B. B. Schwartz, eds.), Chapter 5. Plenum Press, New York, 1974.
[10] J. Powell, *in* "Superconducting Machines and Devices: Large Systems Applications" (S. Foner and B. B. Schwartz, eds.), Chapter 1. Plenum Press, New York, 1974.
[11] K. Onnes, *Commun. Kamerlingh Onnes Lab., Univ. Leiden, Suppl. 34b* 65 (1913); reprinted *in* "Superconductivity-Selected Reprints," p. 18. American Institute of Physics, New York, 1964.
[12a] For general references on superconductivity including type-II superconductors, see A. C. Rose-Innes and F. H. Rhoderick, "Introduction to Superconductivity." Pergamon, Oxford, 1969.
[12b] M. Tinkham, "Introduction to Superconductivity." McGraw-Hill, New York, 1975.
[12c] P. G. de Gennes, "Superconductivity in Metals and Alloys." Benjamin, New York, 1966.

[12d] R. D. Parks (ed.), "Superconductivity," Vols. I and II. Dekker, New York, 1969. For specifically type-II superconductivity, see D. Saint-James, G. Sarma, and E. J. Thomas, "Type II Superconductivity." Pergamon, Oxford, 1969. For a materials-oriented review, see G. D. Cody and G. W. Webb, *CRC Crit. Rev. Solid State Sci.* **4**, 27 (1973).

[13]   L. P. Gorkov, *Zh. Eksp. Teor. Fiz.* **36**, 1918 (1959) [*English transl.: Sov. Phys.-JETP* **9**, 1364 (1959)]; *Zh. Eksp. Teor. Fiz.* **37**, 1407 (1959) [*English transl.: Sov. Phys.-JEPT* **10**, 998 (1960)]. See also Refs. 12d.

[14]   A. A. Abrikosov, *Zh. Eksp. Teor. Fiz.* **32**, 1442 (1957) [*English transl.: Sov. Phys.-JETP* **5**, 1174 (1957)].

[15]   For clean superconductors, the electrodynamics can be nonlocal and this effects the value of the penetration depth. Type-II superconductors are almost always dirty and therefore exhibit local electrodynamics. For a discussion of the effects of nonlocality, see Tinkham [12b].

[16]   B. S. Chandrasekhar, *Appl. Phys. Lett.* **1**, 7 (1962); A. M. Clogston, *Phys. Rev. Lett.* **9**, 266 (1962).

[17]   A. M. Campbell and J. E. Evetts, "Critical Currents in Superconductors." Taylor and Francis, London and Barns and Noble Books, New York, 1972.

[18]   C. P. Bean, *Rev. Mod. Phys.* **36**, 31 (1964); *Phys. Rev. Lett.* **8**, 250 (1962).

[19]   P. W. Anderson and Y. B. Kim, *Rev. Mod. Phys.* **36**, 39 (1964).

[20]   J. F. Bussiere, M. Garber, and S. Shen, *Appl. Phys. Lett.* **25**, 756 (1974); see also Schwall *et al.* [90].

[21]   J. K. Hulm and R. D. Blaugher, *Phys. Rev.* **123**, 1569 (1961).

[22]   E. M. Savitskii, V. V. Baron, Yu. V. Efimov, M. I. Bychkove, and L. F. Myzenkova, "Superconducting Materials," pp. 412–413. Plenum Press, New York (1973).

[23]   T. E. Cordier and W. K. McDonald, *IEEE Trans. Magn.* **MAG-11**, 280 (1975).

[24]   J. E. Evetts and P. J. Martin, *Met. Sci. J.* **7**, 179 (1973).

[25]   R. G. Hampshire and M. T. Taylor, *J. Phys. F* **2**, 89 (1972).

[26]   R. B. Zubeck, Ph.D. Dissertation, Stanford Univ. (1973) (unpublished).

[27]   W. B. Pearson, "The Crystal Chemistry and Physics of Metals and Alloys." Wiley (Interscience), New York, 1972.

[28]   G. W. Webb and J. J. Engelhardt, *IEEE Trans. Magn.* **MAG-11**, 208 (1975).

[29]   J. R. Gavaler, M. A. Janocko, A. I. Braginski, and G. W. Rowland, *IEEE Trans. Magn.* **MAG-11**, 192 (1975).

[30]   L. R. Testardi, *Phys. Acoust.* **10**, 193 (1973).

[31]   M. Weger and I. B. Goldberg, *Solid State Phys.* **28**, 1 (1973).

[32]   S. Geller, *Acta Crystallogr.* **9**, 885 (1956); G. R. Johnson and D. H. Douglass, *J. Low-Temp. Phys.* **14**, 565 (1974).

[33]   B. T. Matthias, T. H. Geballe, R. H. Willens, E. Corenzwit, and G. W. Hull, Jr., *Phys. Rev.* **139**, A1501 (1965).

[34]   J. J. Hanak, G. D. Cody, P. R. Aron, and H. C. Hitchcock, *in* "High Magnetic Fields" (H. Kolm, B. Lax, F. Bitter, and R. Mills, eds.), p. 592. Wiley, New York, 1961.

[35]   R. D. Blaugher, R. A. Hein, J. E. Cox, and R. M. Waterstrat, *J. Low-Temp. Phys.* **1**, 531 (1969); T. H. Geballe, *Sci. Am.* **225**, 22 (1971).

[36]   A. R. Sweedler, D. G. Schweitzer, and G. W. Webb, *Phys. Rev. Lett.* **33**, 168 (1974); R. Bett, *Cryogenics* **14**, 361 (1974).

[37]   A. R. Sweedler and D. Cox, *Phys. Rev.* **B12**, 147 (1975).

[38]   J. R. Gavaler, *Appl. Phys. Lett.* **23**, 480 (1973).

[39]   L. R. Testardi, J. H. Wernick, and W. A. Royer, *Solid State Commun.* **15**, 1 (1974).

[40]   L. R. Newkirk, F. A. Valencia, A. L. Giorgi, E. G. Szklarz, and T. C. Wallace, *IEEE Trans. Magn.* **MAG-11**, 221 (1975); J. M. E. Harper, T. H. Geballe, L. R. Newkirk, and F. A. Valencia, *J. Less-Common Met.* **43**, 5 (1975).

[41] R. H. Hammond, *IEEE Trans. Magn.* **MAG-11**, 501 (1975).
[42] D. Dew Hughes, *Cryogenics* **15**, 435 (1975).
[43] C. E. Lundin and A. S. Yamamoto, *Trans. AIME* **236**, 863 (1966).
[44] M. Hansen and K. Anderko, "Constitution in Binary Alloys," 2nd ed., p. 1202. McGraw-Hill, New York, 1958.
[45] J. H. N. Van Vucht, H. A. C. M. Bruning, H. C. Donkersloot, and A. H. Gomes de Mesquita, *Phillips Res. Rep.* **19**, 407 (1964).
[46] J. P. Charlesworth, I. Macphail, and P. E. Madsen, *J. Mater. Sci.* **5**, 580 (1970).
[47] Yu. A. Bashkirov *et al., Phys. Met. Metallogr.* **37**, 60 (1974).
[48] T. H. Geballe and J. K. Hulm, *IEEE Trans. Magn.* **MAG-11**, 119 (1975).
[49] R. H. Willens *et al. Solid State Commun.* **7**, 837 (1969).
[50] A. M. Clogston, A. C. Gossard, V. Jaccarino, and Y. Yafet, *Phys. Rev. Lett.* **9**, 262 (1962).
[51] E. Ehrenfreund, A. C. Gossard, and J. H. Wernick, *Phys. Rev. B* **4**, 2906 (1971).
[52] D. W. Woodard and G. D. Cody, *RCA Rev.* **25**, 392 (1964).
[53] A. Junod, J. L. Staudenmann, J. Muller, and P. Spitzli, *J. Low-Temp. Phys.* **5**, 25 (1971).
[54] B. W. Batterman and C. S. Barrett, *Phys. Rev. Lett.* **13**, 390 (1964); *Phys. Rev.* **149**, 296 (1966).
[55] R. Mailfert, B. W. Batterman, and J. J. Hanak, *Phys. Lett. A* **24**, 315 (1967); *Phys. Status Solidi* **32**, K67 (1969).
[56] C. W. Chu and L. R. Testardi, *Phys. Rev. Lett.* **32**, 764 (1974).
[57] P. W. Anderson and E. I. Blount, *Phys. Rev. Lett.* **14**, 217 (1965).
[58] J. A. Benda, T. H. Geballe, and J. H. Wernick, *Phys. Lett.* **46A**, 389 (1974).
[59] A. M. Clogston and V. Jaccarino, *Phys. Rev.* **121**, 1357 (1961).
[60] A. M. Clogston, *Phys. Rev.* **136**, A8 (1964).
[61] R. W. Cohen, G. D. Cody, and J. J. Halloran, *Phys. Rev. Lett.* **19**, 840 (1967).
[62] J. Labbé and J. Friedel, *J. Phys.* **27**, 153, 303, 708 (1966).
[63] M. Weger, *Rev. Mod. Phys.* **36**, 175 (1964).
[64] L. F. Mattheiss, *Phys. Rev.* **138**, A112 (1965).
[65] L. P. Gor'kov, *Sov. Phys.-JETP* **38**, 830 (1974); L. P. Gor'kov and O. N. Dorokhov, *J. Low-Temp. Phys.* **22**, 1 (1976).
[66] L. F. Mattheiss, *Phys. Rev. B* **12**, 2161 (1975).
[67] G. S. Knapp, S. D. Bader, H. V. Culbert, F. Y. Fradin, and T. E. Klippert, *Phys. Rev. B* **11**, 4331 (1975).
[68] G. M. Eliashberg, *JETP* **11**, 696 (1960); **12**, 1000 (1961).
[69] W. L. McMillan, *Phys. Rev.* **167**, 331 (1968).
[70] R. C. Dynes, *Solid State Commun.* **10**, 615 (1972).
[71] R. C. Dynes and J. M. Rowell, *Phys. Rev. B* **11**, 1884 (1975); P. B. Allen and R. C. Dynes, *ibid* **11**, 1895 (1975).
[72] L. Y. L. Shen, *Phys. Rev. Lett.* **29**, 1082 (1972).
[73] T. H. Geballe *et al., Physics* **2**, 293 (1966).
[74] B. T. Matthias, *Phys. Rev.* **92**, 874 (1953); N. Pessall and J. K. Hulm, *Physics* **2**, 311 (1966); N. Pessall, R. E. Hold, and H. A. Johansen, *J. Phys. Chem. Solids* **29**, 19 (1968); J. K. Hulm, M. S. Walker, and N. Pessall, *Proc. Int. Conf. Sci. Superconduct.* (F. Childton, ed.). North-Holland Publ., Amsterdam, 1971.
[75] J. R. Gavaler, M. A. Janocko, A. Patterson, and C. K. Jones, *J. Appl. Phys.* **42**, 54 (1971).
[76] W. D. Smith, R. Y. Lin, J. A. Coppola, and J. Economy, *IEEE Trans. Magn.* **MAG-11**, 182 (1975).
[77] D. E. Prober, M. R. Beasley, and R. E. Schwall, *Bull. Am. Phys. Soc.* **20**, 342 (1975); S. Foner, E. J. McNiff, Jr., A. H. Thompson, F. R. Gamble, T. H. Geballe, and F. J. Di Salvo, *ibid.* **17**, 289 (1972).
[78] J. H. P. Watson and R. M. Hawk, *Solid State Commun.* **9**, 1993 (1971).

[79] R. Chevrel, M. Sergent, and J. Prigent, *J. Solid State Chem.* **3**, 515 (1971).

[80] B. T. Matthias, M. Marezio, E. Corenzwit, A. S. Cooper, and H. E. Barz, *Science* **175**, 1465 (1972).

[81] R. Odermatt, Ø. Fischer, H. Jones, and G. Bongi, *J. Phys. C* **7**, L13, (1974); S. Foner, E. J. McNiff, Jr., and E. J. Alexander, *IEEE Trans. Magn.* **MAG-11**, 155 (1975).

[82] M. N. Wilson, *Proc. Appl. Superconduct. Conf., Annapolis, Maryland.* IEEE Publ. 72CH0682-TABSC (1972).

[83] J. J. Hanak, K. Strater, and G. W. Cullen, *RCA Rev.* **25**, 342 (1964).

[84] For review of work in U.S., see B. Belanger, *Cryogenics* **15**, 88 (1975); for work in Europe, see G. Bogner, *Ibid.* **15**, 79 (1975).

[85] R. W. Meyerhoff, *Adv. Cryogen. Eng.* **19**, 101 (1974); W. T. Beall, Jr., *IEEE Trans. Magn.* **MAG-11**, 381 (1975).

[86] G. Bogner, *in* " Superconducting Machines and Devices: Large Systems Applications " (S. Foner and B. B. Schwartz, eds.), Chapter 7. Plenum Press, New York, 1974.

[87] The pitch of the helix must be chosen to satisfy a number of other conditions as well. The situation has been analyzed recently and double helics have been suggested as a compromise. See J. Sutton, *Cryogenics* **15**, 541 (1975).

[88] Critical current densities in excess of 10^7 A/cm have recently been achieved in a fine scale layered composite of Nb_3Sn. See Stanford Univ. Microwave Lab. Rep. No. 2487.

[89] D. P. Snowden, C. H. Meyer, and S. A. Sterling, *J. Appl. Phys.* **45**, 2693 (1974).

[90] R. E. Schwall, R. E. Howard, and R. B. Zubeck, *IEEE Trans. Magn.* **MAG-11**, (1975); R. E. Howard *et al.*, *Appl. Supercond. Conf.*, Paper E2, Stanford Univ. (August 1976).

[91] J. F. Bussiere, M. Garber, and M. Suenaga, *IEEE Trans. Magn.* **MAG-11**, 324 (1975); J. F. Bussiere, *Appl. Supercond. Conf.*, Paper E1, Stanford Univ. (August 1976).

[92] J. F. Bussiere, M. Garber, and M. Suenaga, *J. Appl. Phys.* **45**, 4611 (1974).

[93] R. A. Norton, R. E. Howard, and R. E. Schwall, *Adv. Cryogen. Eng.* **22**, 414 (1976).

[94] W. de Sorbo, U.S. Patent Class 75-174, No. 341681 (December 17, 1968).

[95] C. Laverick, *IEEE Trans. Magn.* **MAG-11**, 109 (1975); R. C. Seamans, A Report to the President and the Congress of the United States, ERDA-13 (April 11, 1975).

Index

O

P